National Fire Protection Association

International Association of Fire Chiefs

NVFC
National Volunteer Fire Council
Fire • EMS • Rescue

Fundamentals of Firefighter Skills for Support Personnel

W0113361

JONES & BARTLETT
LEARNING

World Headquarters
Jones & Bartlett Learning
25 Mall Road
Burlington, MA 01803
978-443-5000
info@jblearning.com
www.psglearning.com

National Fire Protection Association
1 Batterymarch Park
Quincy, MA 02169-7471
www.NFPA.org

International Association of Fire Chiefs
4025 Fair Ridge Drive
Fairfax, VA 22033
www.IAFC.org

National Volunteer Fire Council
712 H Street NE, Suite 1478
Washington, DC 20002
https://www.nvfc.org/

Jones & Bartlett Learning books and products are available through most bookstores and online booksellers. To contact the Jones & Bartlett Learning Public Safety Group directly, call 800-832-0034, fax 978-443-8000, or visit our website, www.psglearning.com.

Substantial discounts on bulk quantities of Jones & Bartlett Learning publications are available to corporations, professional associations, and other qualified organizations. For details and specific discount information, contact the special sales department at Jones & Bartlett Learning via the above contact information or send an email to specialsales@jblearning.com.

Copyright © 2025 by Jones & Bartlett Learning, LLC, an Ascend Learning Company and the National Fire Protection Association.

All rights reserved. No part of the material protected by this copyright may be reproduced or utilized in any form, electronic or mechanical, including photocopying, recording, or by any information storage and retrieval system, without written permission from the copyright owner.

The content, statements, views, and opinions herein are the sole expression of the respective authors and not that of Jones & Bartlett Learning, LLC. Reference herein to any specific commercial product, process, or service by trade name, trademark, manufacturer, or otherwise does not constitute or imply its endorsement or recommendation by Jones & Bartlett Learning, LLC and such reference shall not be used for advertising or product endorsement purposes. All trademarks displayed are the trademarks of the parties noted herein. *Fundamentals of Firefighter Skills for Support Personnel* is an independent publication and has not been authorized, sponsored, or otherwise approved by the owners of the trademarks or service marks referenced in this product.

There may be images in this book that feature models; these models do not necessarily endorse, represent, or participate in the activities represented in the images. Any screenshots in this product are for educational and instructive purposes only. Any individuals and scenarios featured in the case studies throughout this product may be real or fictitious but are used for instructional purposes only.

The International Association of Fire Chiefs, the National Fire Protection Association, the National Volunteer Fire Council, and the publisher have made every effort to ensure that contributors to *Fundamentals of Firefighter Skills for Support Personnel* materials are knowledgeable authorities in their fields. Readers are nevertheless advised that the statements and opinions are provided as guidelines and should not be construed as official International Association of Fire Chiefs, National Fire Protection Association, or National Volunteer Fire Council policy. The recommendations in this publication or the accompanying resources do not indicate an exclusive course of action. Variations, taking into account the individual circumstances and local protocols, may be appropriate. The International Association of Fire Chiefs, the National Fire Protection Association, the National Volunteer Fire Council, and the publisher disclaim any liability or responsibility for the consequences of any action taken in reliance on these statements or opinions.

Defined terms from NFPA documents are reproduced with permission of NFPA, Copyright © National Fire Protection Association, Quincy, MA. All rights reserved. For additional information, please go to www.nfpa.org.

30320-9

Production Credits
Director, Product Management: Cathy Esperti
Product Manager: Janet Maker
Manager, Content Strategy: Tiffany Sliter
Content Manager: Jennifer Deforge-Kling
Content Coordinator: Michaela MacQuarrie
Manager, Intellectual Properties and Content Production: Kristen Rogers
Project Manager: Karrie Larsson
Senior Digital Project Specialist: Angela Dooley
Senior Product Marketing Manager: Elaine Riordan
Director, Sales: Brian Hendrickson
Procurement Manager: Wendy Kilborn
Composition: S4Carlisle Publishing Services
Project Management: S4Carlisle Publishing Services
Cover and Text Design: S4Carlisle Publishing Services
Media Development Editor: Faith Brosnan
Rights Specialist: Maria Leon Maimone
Cover Image (Title Page, Part Opener, Chapter Opener): © Eric Scruggs
Printing and Binding: Sheridan Kentucky

Library of Congress Cataloging-in-Publication Data
Library of Congress Cataloging-in-Publication Data unavailable at time of printing.
LCCN: 2024034085

6048

Printed in the United States of America

28 27 26 25 24 10 9 8 7 6 5 4 3 2 1

Brief Contents

Contents

Skill Drills

CHAPTER 8

CHAPTER 9

CHAPTER 10

Preface

A Note on the Relationship between NFPA 1010 and NFPA 1001

Fundamentals of Firefighter Skills for Support Personnel covers the job performance requirements (JPRs) that a candidate must master in order to become certified as a support person. As part of the standard consolidation initiative of the National Fire Protection Association (NFPA), multiple standards that cover firefighter, fire apparatus driver/operator, airport firefighter, and marine firefighting training have been combined into a single document.

NFPA 1010, *Standard on Professional Qualifications for Firefighters, 2024 Edition* was issued by the NFPA Standards Council on December 21, 2023, with an effective date of January 10, 2024. NFPA 1010 represents a consolidation and revision of the following stand-alone standards:

- **NFPA 1001**, *Standard for Fire Fighter Professional Qualifications, 2019 Edition*
- **NFPA 1002**, *Standard for Fire Apparatus Driver/Operator Professional Qualifications, 2017 Edition*
- **NFPA 1003**, *Standard for Airport Fire Fighter Professional Qualifications, 2019 Edition*
- **NFPA 1005,** *Standard for Professional Qualifications for Marine Fire Fighting for Land-Based Fire Fighters, 2019 Edition*

Appendix A includes the knowledge and skills objectives that are correlated to relevant professional qualifications in the following chapter in NFPA 1010:

- Chapter 5: Support Person (NFPA 1001)

A Note on the Relationship between NFPA 1010 and NFPA 470

Fundamentals of Firefighter Skills for Support Personnel covers all JPRs that a candidate must master in order to become certified as a support person. This includes the requirements listed in 5.1 in chapter 5, "Support Person," of NFPA 1010, which requires candidates to meet the JPRs as defined in chapter 5, "Awareness" (NFPA 1072), of NFPA 470.

As part of the NFPA's standard consolidation initiative, multiple standards that cover hazardous materials have been combined in a single document. **NFPA 470**, *Hazardous Materials/Weapons of Mass Destruction (WMD) Standard for Responders, 2022 Edition* was published on August 26, 2021, with an effective date of September 15, 2021. NFPA 470 represents a consolidation and revision of the following stand-alone standards:

- **NFPA 472**, *Standard for Competence of Responders to Hazardous Materials/Weapons of Mass Destruction Incidents, 2018 Edition*
- **NFPA 473**, *Standard for Competencies for EMS Personnel Responding to Hazardous Materials/Weapons of Mass Destruction Incidents, 2018 Edition*
- **NFPA 1072**, *Standard for Hazardous Materials/Weapons of Mass Destruction Emergency Response Personnel Professional Qualifications, 2017 Edition*

Appendix A includes the knowledge and skills objectives from chapters 12 and 13 and is correlated to NFPA 470:

- Chapter 5, "Professional Qualifications for Hazardous Materials/WMD Awareness Level Personnel" (NFPA 1072)

Acknowledgments

Fundamentals of Firefighter Skills for Support Personnel was developed from content included in *Fundamentals of Firefighter Skills and Hazardous Materials Response, Fifth Edition*. The content was adapted to cover the knowledge and skill objectives for the support person included in chapter 5, "Support Person" (NFPA 1001), of NFPA 1010.

The Jones & Bartlett Learning Public Safety Group, the National Fire Protection Association (NFPA), the International Association of Fire Chiefs (IAFC), and the National Volunteer Fire Council (NVFC) would like to thank all of the authors, contributors, advisors, and reviewers of *Fundamentals of Firefighter Skills for Support Personnel*.

Special thanks go out to Greg Otting of the Ohio Fire Academy for his contributions and to the volunteers from the NVFC who reviewed the content included in *Fundamentals of Firefighter Skills for Support Personnel*.

Fundamentals of Firefighter Skills and Hazardous Materials Response, Fifth Edition

Contributors

Toby Ballard
Captain
Missoula Rural Fire District
Missoula, Montana

Bret A. Davidson
Deputy Fire Chief
City of Vista
Vista, California

Marc Davidson
Captain
Fairfax County Fire & Rescue Department
Fairfax, Virginia
Christopher DeSantis
Yonkers, New York Fire Department
Yonkers, New York

Rommie L. Duckworth, MPA, LP, EFO, FO
Ridgefield Fire Department
Ridgefield, Connecticut

Todd Eddy, BS
Alabama Fire College
Tuscaloosa, Alabama

Brandon Hausbeck, BASc, CFI-I
Fire Chief
Saginaw Fire Department
Saginaw, Michigan

Michael Heffner, MS, EMT-P
Eastern Oregon University
La Grande, Oregon

Dustin Housewright, CFO, CEMSO, NRP
Eastman Emergency Services
Kingsport, Tennessee

James P. Kenney
Assistant Chief, Warwick, Rhode Island Fire Department (retired)
Rope Coordinator for the Connecticut State Fire Academy
Rhode Island

Chad Landis
Training Officer
Rapides Parish Fire District 3/Alpine Fire Department
Pineville, Louisiana

Kevin Lewis
Battalion Chief
Cobb County Fire Department
Marietta, Georgia

Aaron Miranda
Poway Fire Department
Poway, California

Guy Peifer, BS, NRP
Captain (retired)
Yonkers Fire Department
Yonkers, New York

Becki Rowan-White
Battalion Chief
Chanhassen Fire Department
Chanhassen, Minnesota

Rob Schnepp
Alameda County Fire Department (retired)
Alameda, California

Michael A. Smith, MPA
Lieutenant
Lynn Fire Department
Lynn, Massachusetts
Adjunct Professor
Anna Maria College
Paxton, Massachusetts
Adjunct Professor
North Shore Community College
Danvers, Massachusetts
Adjunct Professor
Bunker Hill Community College
Boston, Massachusetts

Sean Wilson
Captain
Royal Oak Fire Department
Royal Oak, Michigan
Owner
Rise Above Fire Training
Royal Oak, Michigan

Thank You

Thank you to the Aurora, Colorado, Fire Department and Aims Community College Public Safety Institute for hosting photoshoots and to the following individuals:

Juan Calderoa

Brandon Fryman

Austin Hackbarth

Commander Mark Hays

Captain Rob Hulse

John McDougall

Jamie Meyers

Travis Miller

Susan Moreland

Jason Natzke

Ryan Nelson

Steve Reynoldson

Ronnie Riedel

Jared Scott

Jonathan Timms

Lee Vidal

Andrew Welsh

CHAPTER

1

Support Person

The Fire Service

KNOWLEDGE OBJECTIVES

After studying this chapter, you will be able to:

- Describe the organization of the fire department.
- Describe the mission of the fire service.
- Outline the roles and responsibilities of the support person in the organization.
- Describe the modern fire service.
- Explain the concept of governance, and describe how regulations, standards, policies, and standard operating procedures affect it.
- Explain the importance of physical fitness and a healthy lifestyle to the performance of the duties of a support person.

SKILLS OBJECTIVES

After studying this chapter, you will be able to perform the following skills:

- Locate specific information that identifies current best practices.

ADDITIONAL NFPA STANDARDS AND CODES

- **NFPA 1225**, *Standard for Emergency Services Communications, 2022 Edition*
- **NFPA 1410**, *Standard on Training for Emergency Scene Operations, 2020 Edition*
- **NFPA 1550**, *Standard for Emergency Responder Health and Safety, 2024 Edition*
- **NFPA 1582**, *Standard on Comprehensive Occupational Medical Program for Fire Departments, 2022 Edition*
- **NFPA 1710**, *Standard for the Organization and Deployment of Fire Suppression Operations, Emergency Medical Operations, and Special Operations to the Public by Career Fire Departments, 2020 Edition*
- **NFPA 1900**, *Standard for Aircraft Rescue and Firefighting Vehicles, Automotive Fire Apparatus, Wildland Fire Apparatus, and Automotive Ambulances, 2024 Edition*

CASE STUDY

You Are the Support Person

It is your first day in the station after having completed the requirements and training to join the department. The firehouse crew welcomes you. As you are shown around the station, the doorbell rings, and you are told your first official duty is to answer it. At the door, you find a woman and her young child. Excitedly, she tells you her child has been begging to meet the firefighters because, "Riley wants to grow up to be a firefighter."

You agree to give a quick tour of the station. You first show them the living quarters and then move to the apparatus floor. The child is full of questions and asks, "Do you go to fires all day long? Do you know how to use all those tools? Do you report to the fire chief? Do you work with police officers?" The mother tells her child to give you a chance to explain before asking any more questions.

1. What are the duties of a support person?
2. What tools and equipment do you need specialized training to operate?
3. What are the roles and responsibilities of a support person versus a company officer?
4. With what other agencies will support personnel typically interact?

Introduction

When you join the fire service, you join a profession with a long and noble history of protecting and serving the community (**FIGURE 1-1**). On completion of this course, you will be well prepared with the knowledge and skills to serve your community in the modern fire service while continuing the centuries-old tradition of neighbors coming together to save lives and preserve property.

FIGURE 1-1 The National Fallen Firefighters Memorial is located at the National Fire Academy in Emmitsburg, Maryland. The memorial honors those who have died in the line of duty. The eternal flame at the base of the memorial symbolizes the spirit of all past, present, and future firefighters.

© Bill Ryan/AP Photo

Mission of the Fire Service

Individual departments may have different needs and capabilities, but the core mission of the fire service remains the same. The primary mission of the fire service is to save lives, but the fire service also works to protect property and the environment. The missions of the fire service are accomplished through prevention, education, suppression, rescue, and salvage activities. This is typically presented in a department's mission statement. For example, *We protect the lives and property of residents and visitors through fire prevention; education; and delivery of fire, rescue, and emergency medical services.* A mission statement communicates the purpose and goals of the department both to its firefighters and to the public. This example mission statement makes it clear that the fire department protects lives and property by providing fire prevention and education as well as fire suppression, rescue services, emergency medical services (EMS), and disaster services. The core mission of the fire service remains the same for every firefighter in every department.

Brief History of the Fire Service

Communities have needed to fight fires and prevent them from happening since the first humans walked the earth. As human societies grew into more dense communities, this need became more acute because one person's fire could quickly become an entire community's decimation. Some of the most important lessons have been learned in the aftermath of tragic fires that resulted in a devastating loss of life. Governments became involved in firefighting and fire-prevention efforts early in human history.

TIP

Many resources describing the history of firefighting are available on websites and in books. Learn about how things have changed—and not changed—over the years. For example, buckets made of leather did not hold water for any length of time, so sand was sometimes substituted for water in those buckets. After the sand was thrown on the fire, the bucket would be refilled with water and used in a bucket brigade.

The Firefighter's Cross

The Firefighter's Cross is a symbol of the fire service used across the United States (**FIGURE 1-2A**). You will see this symbol on most fire vehicles, on firefighter uniforms, and on firefighter badges. The origins of the Firefighter's Cross are not clear. It is often referred to as the *Maltese cross*, which was originally a symbol of the Knights of Malta (**FIGURE 1-2B**). The Maltese cross dates back to the 16th century and may have been used to represent the fire service as early as the 12th and 13th centuries. It is an international symbol of the fire service. Each of the eight points on the Maltese cross serves as a reminder of eight key attributes of a good first responder: observant, tactful, resourceful, persevering, dexterous, clear, sympathetic, and always using good judgment. Another cross symbol known as the *Florian cross* (**FIGURE 1-2C**) has also been used to represent the fire service. It is named after a Roman officer in the 4th century who was assigned to organize firefighting brigades in Rome (Martinez-Granata 2017).

The important thing to remember is that the Firefighter's Cross continues to represent the firefighter's life of service, dedication, and sacrifice. When you become a member of the fire department and you wear this cross on your uniform, you must wear it proudly and strive to uphold the best traditions of the fire service.

Fire Protection in England

As early as 1066, William the Conqueror decreed that all home fires in England were to be extinguished and

A.

B.

C.

FIGURE 1-2 A. The Firefighter's Cross is widely recognized as a symbol of the fire service and is displayed on most fire vehicles, on firefighter uniforms, and on firefighter badges. **B.** The Maltese cross has been a symbol of the fire service since about the 12th century. **C.** The Florian cross has represented the fire service since the 4th century and is named after a man charged with organizing firefighting brigades in Rome.

A. © AWesleyFloyd/Shutterstock; **B.** © Zoart Studio/Shutterstock; **C.** © K Barrett York/ Shutterstock covered every evening with a metal lid called a "couvre feu," likely the source of the modern concept of a curfew. In 1500, English cities passed ordinances regulating hazardous trades such as baking and kettle making, as well as governing fire hazards such as wooden chimneys and thatched roofs. Despite early successes in fire prevention, the Great Fire of London struck in 1666, destroying more than 13,000 homes. This disaster led to improvements in fire protection in England for more than 100 years (Coleman 1988, 133).

Early America

The first documented structure fire in North America occurred in Jamestown, Virginia, in 1608. This fire quickly spread in the fort in which the settlers had built their houses and almost burned down the entire settlement. At that time, most structures were built entirely of combustible materials, such as straw and wood. In 1630, the city of Boston, Massachusetts, established the first fire **regulations** in North America when it banned wood chimneys and thatched roofs. In 1648, in the Dutch colony of New Amsterdam (which became New York), Governor Peter Stuyvesant enacted the same ban and required that chimneys be swept out regularly. Stuyvesant also appointed fire wardens to impose fines on homeowners who did not obey these regulations. The money collected was used to pay for firefighting equipment (Merrimack Fire and Rescue n.d.).

The first fire department with paid firefighters in the United States was established in 1678 in Boston. Boston also had the first fire stations and fire engines (Boston Fire Historical Society n.d.). The first volunteer fire department was created in Philadelphia in 1736 under the leadership of Benjamin Franklin (Benjamin Franklin Historical Society n.d.). Franklin recognized the many dangers of fire and continually sought ways to prevent it. For example, citizens in Philadelphia were required to keep buckets filled with water outside their doors, readily available to fight fires. Franklin also developed the lightning rod to help draw lightning strikes (a common cause of fires) away from homes. Fire-prevention and fire-suppression efforts have gone hand in hand since the very inception of the American fire service.

Early Firefighting Equipment

Colonial-era firefighters had only buckets, ladders, and **fire hooks** (long poles with metal hooks at the end used to pull down burning structures) at their disposal. Homeowners were required to keep buckets filled with sand or water and to bring them to the scene of a fire. Some towns also required that ladders be available so that firefighters could access the roof to extinguish small fires. If all else failed, the fire hook was used to pull down a burning building and prevent the fire from spreading to nearby structures. (The hook-and-ladder truck evolved from this early equipment.)

Buckets gave way to hand-powered pumpers around 1721, when Richard Newsham developed the first one in London, England. The pump made it possible to propel a steady stream of water from a safe distance (Noonan 2017). In the early 1800s, more powerful steam-powered pumpers replaced the hand-powered pumpers. Steam-powered pumpers were heavy machines that were pulled to the fire by a trained team of horses. They required constant attention, which limited their use to larger cities that could bear the costs of maintaining the horses and the pumpers.

The advent of the internal combustion engine in the early 1900s greatly changed the fire service and enabled even small towns to have motorized pumpers. Although pumpers powered by an internal combustion engine required regular maintenance, those engines did not require the constant attention that horses or steam pumpers did.

The progress in equipment in the fire service extends beyond trucks. Firefighters without an adequate water supply are severely hampered in their ability to extinguish fires. Romans developed the first municipal water systems, just as they had developed the first fire companies. But it was not until the 1800s that water distribution systems were used for **fire suppression** by fire departments. Frederick Graff Sr., a firefighter in New York City, developed the first modern fire hydrant in 1801. He used a valve to control access to the water in city pipes (FireHydrant.org 2004). This enabled firefighters to tap into the water system whenever a fire occurred. This type of valve, called a **fireplug**, was used with both aboveground and below-ground piping systems.

Early Communications

Good communication is vital for effective firefighting. When a fire is discovered, firefighters must be summoned, and citizens must be alerted to the danger. While they are fighting a fire, officers must be able to communicate with firefighters and summon additional resources. Not surprisingly, improvements in communication systems are tied to improvements in the fire service.

During the colonial period, a **fire warden** or night watchman patrolled neighborhoods and sounded the

FIGURE 1-3 The chief's trumpet was once used to amplify the commander's voice. Today it serves as a symbol of authority in the fire service.

© Jones & Bartlett Learning

FIGURE 1-4 The historic symbol of the officer's trumpet is still used on an officer's badge. This image of crossed trumpets is one of the cherished traditions of the fire service.

Courtesy of Rom Duckworth.

alarm if a fire was discovered. Some towns, including Charleston, South Carolina, built a series of fire towers from which wardens could watch for fires. In many towns, ringing the community fire bell or church bells alerted citizens to a fire.

In the late 1800s, telegraph fire alarm systems were installed in large cities. These systems allowed fires to be reported to the fire departments more quickly. They also made it possible for officers to request additional resources or to let others know when the fire was extinguished. In small towns, community sirens mounted on poles or tall rooftops were used to signal that assistance was needed to fight a fire.

Communication while fighting a fire is also important. One of the earliest methods chief officers used was to shout commands through a speaking trumpet. The **chief's trumpet**, or **chief's bugle**, eventually became a symbol of authority (**FIGURE 1-3**). Although chief officers no longer use trumpets for communicating, the use of trumpets to symbolize ranks in the fire service signifies the officer's need to communicate as well as to lead (**FIGURE 1-4**). The more trumpets, the higher the rank.

Paying for Fire Service

Fire insurance companies were established in England soon after the Great Fire of London in 1666 to help victims cope with the financial loss from fires. Benjamin Franklin, who coined the phrase "an ounce of prevention is worth a pound of cure," organized one of the first insurance companies in the United States, the Philadelphia Contributionship.

Because insurance companies saved money if a fire was put out before much damage was done, they agreed to pay fire companies to extinguish fires at the homes they insured. To identify these homes, early insurance companies marked them with a **fire mark**, which is a plaque that displayed the name or logo of the insurance company (**FIGURE 1-5**).

Sometimes more than one fire company showed up to fight a fire. When that happened, a dispute arose over which company would fight the fire and collect the money. Consequently, municipalities began assuming the role of providing fire protection. Much of today's fire protection is funded by tax dollars. Fundraising and donations are also a significant source of funding for the volunteer fire service.

FIGURE 1-5 Fire marks were originally symbols affixed to the front of a building designating the insurance company responsible for covering that fire.

© Jones & Bartlett Learning. Photographed by Glen E. Ellman.

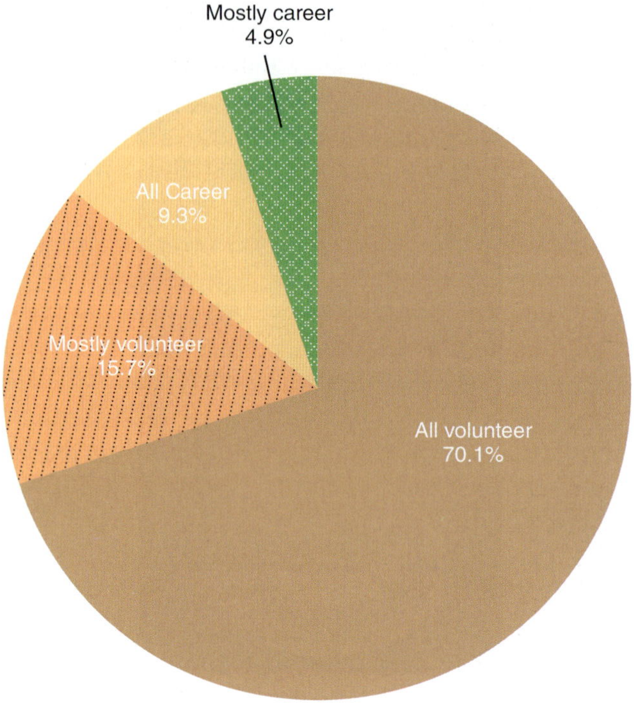

FIGURE 1-6 The majority of fire departments consist of firefighters who are volunteers.

Data from U.S. Fire Administration. 2023. "National Fire Department Registry Quick Facts." Accessed April 15, 2023. https://apps.usfa.fema.gov/registry/summary

Fire Service in the United States Today

Today, the fire service in the United States is the product of an evolution that has occurred over the past 400 years. As a support person, it is helpful to study the past as well as the modern-day U.S. fire service.

According to the **National Fire Protection Association (NFPA)**, there were 1,041,200 firefighters in the United States in 2020. Of this number, approximately 35 percent are full-time, career firefighters, and 65 percent are volunteers. Most career firefighters (71 percent) work in communities with populations of 25,000 or more. Most volunteer firefighters (95 percent) work in fire departments that protect fewer than 25,000 people (Fahy, Evarts, and Stein 2021, 1–2).

The firefighters served in approximately 30,000 fire departments throughout the United States. These fire departments responded to 36,416,000 calls for help (Fahy, Evarts, and Stein 2021, 4–5). Departments are categorized based on whether the firefighters are career, volunteer, or a combination of the two (**FIGURE 1-6**):

- All-career fire departments: A career fire department is one in which all members are paid, full-time firefighters.
- All-volunteer fire departments: A volunteer fire department is one in which all members are volunteer firefighters.
- Combination fire departments: These departments include both paid, full-time firefighters and either on-call firefighters or volunteers:
  - Mostly career fire departments: Between 51 and 99 percent of the members are paid, full-time firefighters.
  - Mostly volunteer fire departments: Between 51 and 99 percent of the members are volunteer firefighters.

In 2021, fire departments in the United States responded to approximately 1.35 million fires. Those fires caused 3,800 civilian fire fatalities, 14,700 civilian fire injuries, and $15.9 billion in property damage. Although only 25 percent of the fires were in residences, that 25 percent caused 75 percent of the civilian fire fatalities and 76 percent of the civilian fire injuries (Hall and Evarts 2022, 1).

> **TIP**
>
> Although career and mostly career fire departments make up only 18 percent of the total number of fire departments in the United States, they serve 69 percent of the U.S. population.

Training and Education

Firefighter training and education have come a long way over the years. The first firefighters simply needed sufficient strength and endurance to pass buckets or operate hand pumps. Today, firefighters require formalized training, attention to safety, and good judgment.

Firefighters operate modern equipment, including million-dollar apparatus, specialized radios, thermal imaging devices, protective gear made of newly developed fabrics, and **self-contained breathing apparatus (SCBA)**, a respirator that provides breathable air to the wearer. These tools, as well as better fire-detection devices, have greatly increased the safety and effectiveness of modern-day firefighters. The most important "machines" at the fire scene, however, remain the intelligent, knowledgeable, well-trained, physically capable firefighters who have the ability and determination to attack the fire. A thermal imaging device may be able to find someone trapped in a burning building, but it takes a smart, able-bodied firefighter to remove the victim safely.

The world and the science of firefighting are constantly changing. This means that firefighters must continually sharpen their skills and increase their knowledge of potential hazards. The need for ongoing education is why training courses, such as this one, are just as important as good physical fitness to today's firefighters.

At the end of your training, you will likely have to pass a test or a series of tests to successfully complete your course of study. Certification is available at all levels of the fire and emergency services, from entry-level firefighters through chief-level officers and everything in between. There are currently two national organizations that accredit or recognize emergency service certification systems: the **National Board on Fire Service Professional Qualifications (Pro Board)** and the **International Fire Service Accreditation Congress (IFSAC)**. These organizations establish the criteria used to ensure that testing does the following:

- Aligns with NFPA professional qualifications standards
- Uses valid and reliable testing methodologies
- Provides a mechanism by which to certify those who successfully complete the testing

Source of Authority

Governments—whether municipal, state, provincial, or national—are charged with protecting the welfare of the public against common threats. Citizens accept certain restrictions on their behavior and pay taxes to protect themselves and the common good. Fire is one such peril. An uncontrolled fire threatens everyone in the community.

The fire service draws its authority from the governing entity responsible for protecting the public from fire, whether it is a town, city, county, township, or special fire district. Federal and state governments also grant authority to fire departments. For example, firefighters can legally enter a locked home without permission to extinguish a fire and protect the public. In some states, private corporations have contracts to provide fire protection to municipalities or government agencies. The head of the fire department (the fire chief) is accountable to the elected leaders of the governing body, such as the city council, the county commission, the mayor, or the city manager. Because of the relationship between fire departments and local government, firefighters are often (but not always) civil servants, working for the citizens of their community.

Culture of the Fire Service

The culture of any organization influences the behavior of its members. Culture is the collection of expectations, values, and practices that guide the actions of the members of an organization. In short, it is the personality of the organization.

Every fire department has its own culture. Often, each group within a department may even have its own subculture, a way of behaving that is somewhat different from the way other groups in the department behave. The culture of the fire service is known for the following characteristics:

- Duty: Firefighters accept the obligation to make performing the requirements of their mission the top priority in all that they do.
- Honor: Firefighters hold a special, trustworthy position in their communities. Firefighters should adhere to the highest standard of moral and ethical conduct to maintain that trust.
- Service: An inherent part of being a firefighter is putting the needs of others above your own.

- Safety: To be a firefighter is to accept an element of danger. However, taking unnecessary risks endangers the firefighting operation, the firefighter, the firefighting family, and the public.
- Excellence: Firefighters must strive to be more than "good enough." Firefighters demand excellence of themselves and those around them. They are prepared to put in the work to achieve that excellence.

Fire Department Governance

Governance is the framework and procedures for managing and operating an organization. The governance of a fire department depends on its policies and standard operating procedures (SOPs). Fire department **policies** outline expectations for firefighter performance and procedures in different circumstances. These policies often require personnel to make judgments to determine the best course of action within the stated policy. Policies governing parts of a fire department's operations may be developed by other government agencies, such as personnel policies that cover all employees of a city or county.

Standard operating procedures (SOPs) provide specific information describing the actions that should be taken to accomplish tasks within the department or at an emergency scene (**FIGURE 1-7**). An example would be your policy and procedure manuals. SOPs are developed within the fire department, are approved by the chief of the department, and ensure that all members of the department perform a given task in the same manner. They provide a uniform way of dealing with emergency situations, enabling firefighters from different stations or companies to work together smoothly, even if they have never encountered one another before. These procedures are vital because they enable everyone in the department to work together and know what is expected for each task. Firefighters must learn and frequently review departmental SOPs.

In some fire departments, **standard operating guidelines (SOGs)** are used. SOGs are similar to SOPs; however, SOGs may vary because of the circumstances surrounding a particular incident. SOGs are not as strict as SOPs because conditions may dictate that the firefighter or officer must use their personal judgment in completing the procedure. This flexibility allows the responder to deviate from a set procedure yet still be held accountable for that action.

Each department member should have access to up-to-date department policies and a physical or digital up-to-date SOP or SOG manual. The SOP or SOG manual should be organized into sections, such as administration, safety, scene operations, apparatus and equipment, station duties, uniforms, and miscellaneous. Many fire departments maintain their SOPs on a computer network, which simplifies the process of providing all employees with up-to-date SOPs. One of the first things you should do as a new recruit in a fire department is ask where the policies and SOP or SOG manual are located.

Fire Apparatus

In the fire service, the term *equipment* generally refers to protective clothing, devices firefighters use to help them breathe, and the tools a firefighter uses to fight fires and rescue victims. The term **fire apparatus**, often shortened to just **apparatus**, refers to the vehicles firefighters drive to an emergency scene or other incident and then use to help fight the fire. (The **emergency scene**, sometimes referred to as simply **on scene**, is the location of an emergency incident and the surrounding area where responders set up and where apparatus are located. When the incident is a fire, this is also called the **fire ground** or **fire scene**.) Modern apparatus carry water; a pumping mechanism; hose; equipment such as the tools a firefighter uses to fight fires and rescue victims, protective clothing, and devices that help them breathe in hazardous environments; as well as personnel. In this sense, a single apparatus has replaced several single-function vehicles from the past. Apparatus continue to evolve as new inventions are adapted to the needs of the fire service.

The two main types of apparatus are pumpers and aerial fire apparatus. A **pumper**, also referred to as an **engine**, is the apparatus that has a pump, carries hose, and maintains a booster tank of water (**FIGURE 1-8**). Pumpers also carry a limited quantity of ladders and hand tools.

An **aerial fire apparatus**, also referred to as a **truck**, **ladder truck**, **aerial apparatus** or just **aerial**, or **tower ladder truck**, is the apparatus that carries several ground ladders, ranging from 8 to 50 feet (2.4 to 15.24 m) in length, as well as an extensive quantity of tools. Trucks are also equipped with an aerial device, such as an aerial ladder, a tower ladder, or an elevating platform (**FIGURE 1-9**). These aerial devices can be raised and positioned above a roof to provide firefighters with a stable, safe work zone.

In addition to pumpers and trucks, most fire departments have additional specialized apparatus, including the following:

- **Quint apparatus**: Often shortened to simply **quint**. *Quint* is short for *quintuple*, meaning "five."

Anytown Fire Department

Standard Operating Procedure
Date: 1/1/24
Section: Administration SOP 01-01
Maintaining Station Logbooks

Page 1 of 1

Purpose
This guideline is provided to ensure that information documented in station logbooks is maintained in a consistent manner, station to station, shift to shift, throughout the department. The station commander shall have discretion in formatting the information.

Scope
This guideline shall be followed by all authorized department personnel entering information into the station logbooks. Any deviations from this guideline will be the responsibility of the individual making the deviation.

Policy
The station log is an important component of the fire department's record-keeping system. All entries must be legible, written in black ink, and the writer identified by name and computer identification number. All members should treat the logbook as a legal document and record all of the station's activities as well as other information deemed important by the station commander.

Procedure
The following guidelines should be used when making logbook notations:

1. The day, date, and shift noted on the top line of the page.
2. A list of each person assigned to that shift to include their computer identification numbers, and the apparatus to which they are assigned for the shift. Personnel on leave should be identified with the type of leave (ie, annual, sick, leave without pay, funeral, military, training) noted. A notation should be made by the name of any member temporarily assigned to another station during that shift.
3. Daily safety talks should be recorded prior to the section dedicated to emergency responses. For more information on safety talks, refer to SOP 31.3
4. The section of the log dedicated to recording chronological activity should have two (2) columns on the side of the book: the far left should note time and the second column the incident number. A column to the far right should be used to note the identification number of the person making the entry.
5. Incident numbers for medic unit responses should be noted in blue ink.
6. Incident numbers for fire suppression responses should be noted in red ink.
7. Units responding, the address to which the response is being made, and the type of situation found should be noted behind the message number (ie, E-5 responded to 304 Albemarle Drive/Gas Leak).
8. The term fill-in should be used in the logbook to denote one station standing by another station to cover a company that has responded to an incident.
9. Entries for medic units should include the hospital to which the patient was transported.
10. Station maintenance/repairs and apparatus maintenance should be documented in the section of the logbook dedicated to that purpose. This documentation may also be maintained in a separate logbook dedicated to those types of entries.

Although the information requirements provided here must be maintained in the station's logbook, the format used for ensuring that documentation is left to the station commander's discretion.

Approved: ______________ Date: 1/1/24
Fire Chief

FIGURE 1-7 A sample SOP.

© Jones & Bartlett Learning

The quint apparatus has five components: a pump, a water tank, a fire hose storage area, an aerial device, and ground ladders.

- **Initial attack apparatus**: The primary purpose of the initial attack apparatus is to initiate a fire suppression attack on structural, vehicular, or vegetation fires. It is sometimes referred to as a **quick attack apparatus**.
- **Mobile water supply apparatus**: The mobile water supply apparatus is designed to transport water to emergency scenes. It may or may not have a pump. This apparatus is also referred to as a **tanker** or **water tender**, or just **tender**.
- **Rescue apparatus**: Rescue apparatus carry an extensive array of regular and specialized tools and equipment that are used to rescue victims. Most of this equipment is too heavy to carry on pumpers.
- **Ambulance** or first-response vehicle: An EMS vehicle carries medical supplies, such as medications, defibrillators, and other equipment that is used to stabilize a critical patient prior to and during transport to a hospital (**FIGURE 1-10**).

FIGURE 1-8 A pumper brings water, a pump, and hose to a fire scene.

Courtesy of Rom Duckworth.

FIGURE 1-9 An aerial fire apparatus carries ground ladders and an aerial device.

© Jones & Bartlett Learning. Photographed by Glen E. Ellman.

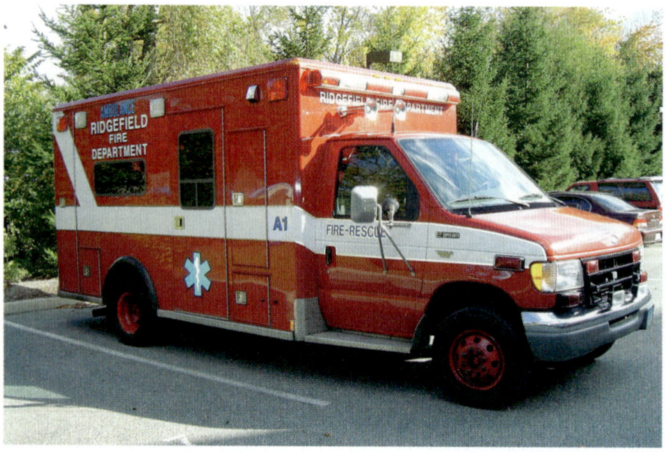

FIGURE 1-10 An EMS company delivers medical services.

Courtesy of Rom Duckworth.

- Community health vehicle: Some departments use a community health vehicle to provide medical care to citizens without bringing them to a hospital.
- **Wildland apparatus**: A four-wheel drive vehicle that is used to transport firefighters closer to wildfires over rough, uneven terrain. It often carries a tank of water and a pump that enables firefighters to pump water while the truck is moving. It also carries special firefighting equipment, such as portable pumps, rakes, shovels, and chainsaws.

Fire Department Organization

A **fire department** is the organization that fights fires and promotes fire prevention in a municipality. A **fire station** or **firehouse** is a building that houses fire apparatus and equipment for a geographic area within a fire department. Departments in larger jurisdictions have multiple stations.

Departments in very large cities are further subdivided. The names of the subdivisions vary from one department to the next. For example, in Boston, Massachusetts, the department is divided into two divisions, and these divisions are divided into districts, each of which includes several fire stations. In New York City, a much larger municipality, the department is divided into boroughs, which are divided into divisions, which are divided into battalions consisting of several fire stations each.

Companies

Fire stations are often made up of smaller units called *companies*, which are groups of people assigned to a specific apparatus who have specific responsibilities at an emergency scene under the command of an officer. A **company** may be composed of various combinations of people and equipment; in smaller departments, one company may fill many roles. The most common types of companies are the following:

- **Engine company**: An engine company rides on the pumper and is responsible for securing a water source, unloading hoses that are held by hand, conducting search and rescue operations, and putting water on the fire.
- **Truck company**: A truck company, also referred to as a **ladder company** or a **tower ladder company**, rides on the aerial fire apparatus and specializes in forcible entry, ventilation, roof operations, search and rescue, and deployment of ground ladders.

- **Rescue company**: A rescue company usually is responsible for rescuing victims from fires, confined spaces, trenches, and high-angle situations.
- **Wildland company**: A wildland company, also called a **brush company**, is dispatched to vegetation fires.
- **Hazardous materials company**: A hazardous materials company responds to and controls scenes involving spilled or leaking hazardous chemicals. These companies have special equipment, personal protective equipment (PPE), and training to handle most emergencies involving chemicals.
- **Emergency medical services (EMS) company**: Sometimes referred to as an **emergency medical services (EMS) squad**, or just **squad**, an EMS company rides in an ambulance or first-response vehicle. These companies treat medical and trauma victims so that they are stable enough to transport to medical facilities for further treatment. Engine or truck companies may be staffed with EMS providers who provide medical care until an ambulance arrives.

The terms **crew** and **unit** are often used in casual conversation as synonyms for *company*. They can also indicate any small collection of firefighters doing a similar job.

Roles in the Fire Department

There are many roles in fire departments. Not every department has someone in every position. No matter what your position is, every member of the fire service will interact with the public. In addition to interacting with citizens at the scene of incidents, firefighters encounter people who visit the fire station requesting a tour or asking questions about specific fire safety issues (**FIGURE 1-11**). Firefighters and support personnel should be prepared to assist these visitors and use this opportunity to provide them with additional fire safety information. Keep in mind that interaction with the public occurs both on and off duty. Use every instance to deliver positive public relations and send educational messages.

Roles in the fire department can be sorted into two categories: general and specialized response. Although every department is different, generally, firefighters seeking specialty roles in a department do not need to do this in a specific order. In other words, a firefighter does not necessarily need to be a driver/operator of an engine before they become a driver/operator of an aerial apparatus or a fire inspector. In fact, a driver/operator of an engine does not necessarily need to be a firefighter.

FIGURE 1-11 Any contact with the public should be used as an educational opportunity.

Courtesy of Rom Duckworth.

General Roles

General roles do not require advanced, specialized training. Common positions include the following:

- **Firefighter**: A firefighter may be assigned any task, including routine cleaning and maintenance duties, placing hose lines to extinguish fires, and assisting with a public fire-prevention program. Generally, firefighters are not responsible for command functions and do not supervise other personnel, except on a temporary basis as a senior firefighter or an acting officer. A firefighter may fulfill many roles during their career through additional training, testing, and promotion.
- **Support person**: A support person is a fire department member who is not trained to the Firefighter I level but who assists members of a fire department by performing duties in environments that are not hazardous. These duties can include assisting with communications, identifying hazardous environments, connecting a pumper to a water supply, opening and closing fire hydrants, operating emergency scene lighting, refilling SCBA air cylinders, and cleaning and checking equipment and tools.
- **Driver/operator**: Often called an **engineer** or a **technician**, the driver/operator is responsible for driving the fire apparatus to the scene safely and then setting up and running the pump on the engine or operating the aerial device on the scene. In some departments, this function is a full-time role; in other departments, it is rotated among members.
- **Company officer**: The company officer is usually a lieutenant or captain who oversees the crew of an apparatus. This person leads the company both

on scene and at the station. The company officer is responsible for the initial firefighting strategy, personnel safety, and the overall activities of the firefighters on their apparatus. Once command is established, the company officer focuses on tactics.

- **Incident safety officer (ISO)**: The ISO observes the overall operation for unsafe practices. The ISO has the authority to halt any firefighting activity, allowing the activity to resume only when it can be done safely and correctly. The senior ranking officer may act as the ISO until the appointed ISO arrives or until another officer is delegated those duties.
- **Training officer**: The training officer is responsible for updating the training of current firefighters and for training new firefighters. They must be aware of the most current firefighting techniques, EMS, and other specialized services provided by the fire department. The training officer coordinates various aspects of the fire department training program and maintains documentation of all training activities.
- **Fire marshal**: The fire marshal delivers, manages, and/or administers fire protection and life safety related **codes** and **standards**, investigations, education, and/or prevention services.
- **Fire inspector**: The fire inspector inspects businesses and enforces public safety laws and **fire codes**.
- **Fire investigator**: The fire investigator responds to fire scenes to help investigate the cause of a fire. The fire investigator may have full police powers to investigate and arrest suspected arsonists and people causing false alarms.
- **Fire and life safety educator (FLSE)**: The FLSE educates the public about fire safety and injury prevention and presents juvenile fire safety programs. The FLSE and the public information officer (PIO) may be part of a larger community risk-reduction program within the department.
- **Telecommunicator**: Also called a **dispatcher**, the telecommunicator takes calls from the public, sends appropriate units to the scene, assists callers with emergency medical information (treatment that can be performed until the medical unit arrives), and assists the incident commander (IC) with obtaining needed resources from the communications center.
- **Emergency vehicle technician (EVT)**: An EVT repairs and services fire and EMS vehicles, keeping them ready to respond to emergencies. These individuals are usually trained by equipment manufacturers to repair vehicle engines, lights, and all parts of the fire pump and aerial ladders.
- **Fire police officer**: Fire police officers are usually firefighters who control traffic and secure the scene from public access. Many fire police officers are also sworn peace officers.
- **Public information officer (PIO)**: The PIO serves as a liaison between the IC and the news media.
- **Fire protection engineer**: The fire protection engineer reviews building plans and works with building owners to ensure that their systems for fire suppression and detection meet the applicable codes and function as needed. Some fire protection engineers design these systems, and most have a degree in fire engineering.

Specialized Response Roles

Many assignments require specialized training. Most large departments have teams of firefighters who respond to specific types of calls. Members of these teams are usually required to be firefighters before they begin additional training. Specialist positions include the following:

- **Emergency medical services (EMS) personnel**: EMS personnel administer care to people who are sick or injured at the scene and during transport to a medical facility. Calls for this type of service account for the majority of responses in many departments today, so EMS personnel are often firefighters with additional training. EMS training levels are normally divided into the following four categories:
  - **Emergency medical responder (EMR)**: EMRs have basic training for providing initial medical assistance. They often perform in an assistant role in the ambulance and have training in bleeding control and cardiopulmonary resuscitation (CPR). Police officers, firefighters, lifeguards, or other rescuers who are often the first to arrive at the scene of an emergency are often trained as EMRs.
  - **Emergency medical technician (EMT)**: Most EMS providers are EMTs. They have training in basic emergency care skills, including bleeding control, CPR, use of automated external defibrillators, oxygen therapy, use of basic airway devices, and assisting patients with certain medications.
  - **Advanced emergency medical technician (AEMT)**: AEMTs have advanced training and can often perform more procedures than EMTs. They

have training in specific aspects of advanced life support, such as intravenous (IV) therapy, interpretation of cardiac rhythms, and advanced airway management.

- **Paramedic**: A paramedic has completed the highest level of training in EMS. They have extensive training in advanced life support, administering drugs, cardiac monitoring, inserting advanced airways, manual defibrillation, and other advanced assessment and treatment skills.

- **Technical rescuer**: Also called a **rescue technician**, these firefighters are trained in special rescue techniques for incidents involving structural collapse, trench rescue, swiftwater rescue, confined-space rescue, high-angle rescue, and other unusual situations. The units they work in are sometimes called *urban search and rescue teams.*
- **Hazardous materials technician**: Hazardous materials technicians have training and certification in chemical identification, leak control, decontamination, and cleanup procedures.
- **Airport firefighter**: Airport firefighters are based at airports. They receive specialized training in extinguishing aircraft fires and extricating victims from an aircraft. They wear special PPE and respond in specialized fire apparatus that protects them from high-temperature fires fueled by substances, such as jet fuel.

Chain of Command

The fire department uses a paramilitary style of leadership. Firefighters operate under a rank system, which establishes the chain of command. The **chain of command** is a rank-based, hierarchical structure that creates an orderly line of authority (**FIGURE 1-12**). Although the precise ranks may vary from department to department, the basic concept remains the same across the fire service. The chain of command creates a structure for managing the department and the fire ground operations. Larger departments have more levels in their chain of command than smaller departments.

Firefighters usually report to a **lieutenant**, who is responsible for a single fire company (such as an engine company) on a shift. The next level in the chain of command is the captain. A **captain** is responsible not only for managing a fire company on their shift but also for coordinating the company's activities with other shifts. A captain may also be in charge of the activities of a fire station. Both captains and lieutenants are company officers.

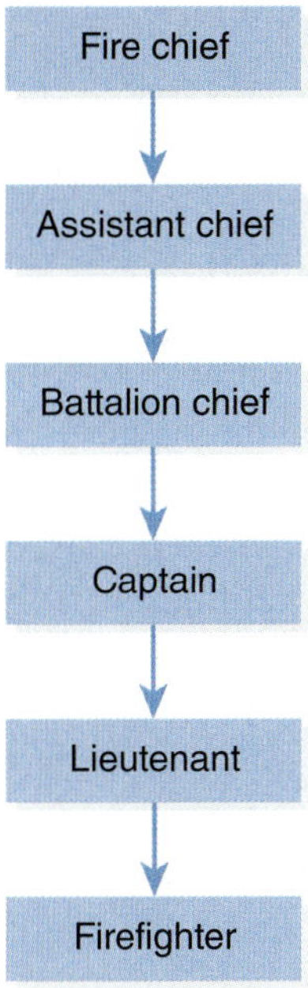

FIGURE 1-12 The chain of command ensures that the department's mission is carried out effectively and efficiently.

© Jones & Bartlett Learning

Captains report to chief officers. A **battalion chief** is the first level of chief officer. They are typically responsible for coordinating the activities of several fire companies in a defined geographic area, such as a district or a division. A battalion chief is usually the officer in charge of a single-alarm working fire.

Above the battalion chief is the **assistant chief**, **deputy chief**, or **division chief**. Chiefs at this level are usually in charge of a functional area, such as training, within the department or in command of a group of battalions or districts. These officers report directly to the chief of the department.

The **fire chief** is the highest-ranking chief officer. The fire chief is in charge of the entire department and has overall responsibility for the administration and operations of the fire department. The chief can delegate responsibilities to other members of the department but is still responsible for ensuring that these activities are carried out properly. The chief usually answers directly to the mayor, town manager, or other designated public official.

Although firefighters often do not need to advance in a specific order to be assigned to a different company or specialty role in a department, to advance in rank, promotion is typically needed. For example, a firefighter must first be promoted to lieutenant before becoming eligible for promotion to captain or battalion chief.

Working with Other Organizations

Modern fire departments do not respond only to fires. They also respond to motor vehicle collisions, elevator emergencies, victims trapped in a collapsed building, accidents in manufacturing plants, and more. Consider the response to a motor vehicle collision (**FIGURE 1-13**). In an area with a centralized 911 call center, the fire department, EMS, law enforcement officials, and tow-truck operators are all notified of this event. If EMS is not part of the fire department, a separate EMS provider is contacted. If the crash affects community infrastructure, the fire department may interact with the local environmental protection agency, utility company, highway department, building department, or other related agencies. If it is a large-scale emergency, the fire department may work alongside local, regional, or state emergency management personnel. If the fire department responds to a school, firefighters will interact with school staff. Firefighters also frequently interact with hospital personnel who assume care for the ill or injured victims who are transported to emergency departments.

When only one agency is involved in an incident, a single **incident commander (IC)** is responsible for all aspects of managing the incident. All personnel, apparatus, and resources work under the authority and direction of the IC. When multiple agencies work together at an incident, a unified command must be established as part of the Incident Command System (ICS). A **unified command** creates a single set of incident goals and objectives and fosters mutual communication and cooperation among agencies. When a unified command is established, each responding agency assigns one or more representatives to the unified command, and these representatives have input on how the incident will be handled instead of having a single IC manage the incident. Even when a formal unified command is not established, emergency and nonemergency personnel from different agencies must work together in a coordinated manner.

FIGURE 1-13 Many agencies work together at the scene of a motor vehicle collision.

© Patrick Kane/The Progress-Index/AP Photo

Nongovernmental organizations may also be involved in incidents. For example, the Salvation Army and the Red Cross may be part of the team responding to a structure fire to assist with care for noninjured victims or to provide fluids, food, or medical care for firefighters. Wildland fires may involve representatives from various government agencies and jurisdictions.

Firefighter Qualifications

Someone who applies to become a firefighter is called a **candidate**. If their application is accepted, they become a **recruit**. Not every candidate will become a recruit. To achieve this accomplishment, a candidate must be in good physical condition and well trained. Candidates must also have a positive attitude toward the community they will serve as well as the teammates with whom they will work. A successful firefighter is a person who has the desire to learn; the will to practice; and the ability to apply the proper knowledge, skills, and attitudes (**FIGURE 1-14**). A firefighter must also constantly learn to stay educated about the expanding body of knowledge about fires and the technology used to fight them.

Each department establishes its own age, background, education, medical, physical fitness, and emergency medical care requirements.

Medical Requirements

Firefighting is both stressful and physically demanding. Before a candidate is accepted as a recruit, they must undergo a medical evaluation. This medical evaluation identifies most medical conditions or physical limitations that could increase the risk of injury or illness to the candidate or other firefighters. Part of the medical requirement is to successfully pass a drug screening test. Medical requirements for firefighters are specified in NFPA 1582, *Standard on Comprehensive Occupational Medical Program for Fire Departments, 2022 Edition.*

Physical Fitness Requirements

Physical fitness requirements ensure that firefighters have the strength, stamina, and flexibility needed

FIGURE 1-14 Firefighters understand the mission of the fire department: to save lives and protect property and the environment through prevention, education, suppression, rescue, and salvage.

Courtesy of Rom Duckworth.

to perform the tasks associated with emergency operations. It is important that firefighters maintain a high level of fitness throughout their careers to meet the physical rigors required of firefighters and to help prevent injuries and illness. NFPA 1010, *Standard on Professional Qualifications for Firefighters, 2024 Edition* allows individual fire departments to choose the fitness-testing method that will be used for firefighter candidates. Many departments require candidates to complete a standardized physical ability test. These tests are designed to measure the strength and endurance required by active firefighters. One of the most widely used tests is the Candidate Physical Ability Test (CPAT). This testing process was developed by the International Association of Firefighters (IAFF) and the International Association of Fire Chiefs (IAFC). The test consists of the following eight scenarios, which are completed while wearing a 50-pound vest to simulate the weight of equipment worn during a fire (IAFC n.d.):

- Stair Climb: Climbing stairs while carrying an additional 25-pound (11-kg) simulated hose pack
- Ladder Raise and Extension: Placing a ground ladder at the fire scene and extending the ladder to the roof or a window
- Hose Drag: Stretching uncharged hose lines and advancing lines
- Equipment Carry: Removing and carrying equipment from fire apparatus to the fire ground
- Forcible Entry: Penetrating a locked door and breaching a wall
- Search: Crawling through dark, unpredictable areas to search for victims
- Rescue Drag: Removing a victim or partner from a fire building
- Ceiling Pull: Locating fire and checking for fire extension

Firefighting candidates have 10 minutes and 20 seconds to complete all eight events. Many fire departments require firefighters to complete this series of events every year to ensure that they maintain the level of physical fitness required for active firefighting.

Emergency Medical Care Requirements

Delivering EMS is an important function of most fire departments. NFPA 1010 allows individual fire departments to specify the level of emergency medical care training required for candidates accepted as recruits. At a minimum, a recruit will need to understand infection control procedures and be able to perform CPR, control bleeding, and manage shock. Many departments require firefighters to become certified as EMS personnel, at least at the EMR level.

Roles and Responsibilities of Support Person, Firefighter I, and Firefighter II

The training and performance qualifications for firefighters are specified by NFPA 1010. This standard defines the parameters of the Support Person, Firefighter I, and Firefighter II courses. As stated earlier, a support person is a fire department member who is not trained to the Firefighter I level but who assists members of a fire department by performing duties in environments that are not hazardous. A **Firefighter I** is a person at the first level of progression in becoming a firefighter as defined in Chapter 6 of NFPA 1010. A Firefighter I has the knowledge and skills to function as a member of a firefighting team under direct supervision.

A **Firefighter II** is a person at the second level of progression in becoming a firefighter as defined in Chapter 7 of NFPA 1010. A Firefighter II has a higher level of training that allows them to more fully assist in mitigating emergency situations. For example, the Firefighter II can coordinate and direct other firefighters and assume a greater level of responsibility for incident management. While the Firefighter I must work under direct supervision, the Firefighter II works under general supervision.

The first step in understanding the organization of the fire service is to learn your roles and responsibilities as a support person. The roles and responsibilities of a support person are a subset of the roles and responsibilities of Firefighter I. As you progress through this text, you will learn what to do and how to do it so that you can take your place confidently among the ranks.

Roles and Responsibilities of a Support Person

- Put on and take off PPE properly.
- Decontaminate PPE in the field and prepare it to be used again.
- Locate information in departmental documents and SOPs.
- Understand and correctly apply appropriate communication protocols.
- Identify hazardous environments that require respiratory protection and avoid those areas.
- Respond on apparatus to an emergency scene.
- Establish and operate safely in emergency work areas.
- Connect a fire department engine to a water supply as a team.
- Extinguish fires in their initial stages using portable fire extinguishers.
- Illuminate an emergency scene.
- Turn off utilities.
- Hoist hand tools using appropriate ropes and knots.
- Refill SCBA cylinders.
- Clean and maintain equipment.
- Operate as part of a team.

CASE STUDY

You Are the Support Person CONCLUSION

It is your first day in the station after having completed the requirements and training to join the department. The firehouse crew welcomes you. As you are shown around the station, the doorbell rings, and you are told your first official duty is to answer it. At the door, you find a woman and her young child. Excitedly, she tells you her child has been begging to meet the firefighters because, "Riley wants to grow up to be a firefighter."

You agree to give a quick tour of the station. You first show them the living quarters and then move to the apparatus floor. The child is full of questions and asks, "Do you go to fires all day long? Do you know how to use all those tools? Do you report to the fire chief? Do you work with police officers?" The mother tells her child to give you a chance to explain before asking any more questions.

1. What are the duties of a support person?

Answer: The specific duties of a support person will depend on the support person's department and their assignment within that department. However, the overall mission of the fire service, and of all emergency responders, is to save lies and protect property and the environment through prevention, education, suppression, rescue, and salvage activities. This is typically reflected in the mission statement of the department.

2. What tools and equipment do you need specialized training to operate?

   Answer: Beyond the qualifications listed in NFPA 1010, some of the equipment that you need additional training to operate includes driving/operating pumping, aerial, and airport rescue apparatus; equipment required for handling hazardous materials; and equipment required to provide medical assistance to people.

3. What are the roles and responsibilities of a support person versus a company officer?

   Answer: The basic responsibilities of a support person are outlined in the job performance requirements of NFPA 1010. Company officers lead the company both on scene and at the station. At the emergency scene, the company officer is responsible for establishing the initial firefighting strategy for their company, monitoring the safety of the personnel, and supervising the overall activities of the personnel on their apparatus.

4. With what other agencies will support personnel typically interact?

   Answer: Support personnel typically interact with other public safety agencies, including law enforcement, EMS, or emergency management. If an emergency involves parts of the community infrastructure, support personnel may interact with highway departments, environmental protection services, utility companies, and other related agencies.

WRAP-UP

SUMMARY

- The core mission of the fire service is to save lives while protecting property and the environment through prevention, education, suppression, rescue, and salvage.
- A support person is a fire department member who is not trained to the Firefighter I level but who assists members of the department by performing duties in environments that are not hazardous.
- Support person duties include, but are not limited to, assisting with communications, identifying hazardous environments, connecting a pumper to a water supply, opening and closing fire hydrants, operating scene lighting, refilling SCBA air cylinders, and cleaning and checking equipment and tools.
- SOPs or SOGs are developed to ensure all members of the fire service perform a given task in the same manner.
- It is important for all fire service personnel to know, understand, and frequently review departmental SOPs and SOGs.
- The fire service is physically demanding, and all personnel must maintain physical fitness to prevent injury and effectively execute duties.

KEY TERMS

advanced emergency medical technician (AEMT) Emergency medical services (EMS) personnel who can do everything an emergency medical technician (EMT) can do and who have advanced training in specific areas of advanced life support, including intravenous (IV) therapy, interpretation of cardiac rhythms, and advanced airway management.

aerial See *aerial fire apparatus.*

aerial apparatus See *aerial fire apparatus.*

aerial fire apparatus A vehicle equipped with an aerial ladder, elevating platform, or water tower that is designed and equipped to support firefighters and rescue operations by positioning personnel, handling materials, providing continuous egress, or discharging water at positions elevated from the ground. (NFPA 1900)

airport firefighter The Firefighter II who has demonstrated the skills and knowledge necessary to function as an integral member of an aircraft rescue and firefighting (ARFF) team. (NFPA 1010)

KEY TERMS CONTINUED

ambulance A vehicle used for out-of-hospital medical care and patient transport that provides a driver's compartment; a patient compartment to accommodate an emergency medical services provider (EMSP) and at least one patient located on the primary cot positioned so that the primary patient can be given emergency care during transit; equipment and supplies at the scene as well as during transport; safety, comfort, and avoidance of aggravation of the patient's injury or illness; two-way radio communication; and audible and visual warning devices. (NFPA 1900)

apparatus See *fire apparatus.*

assistant chief A midlevel chief who often has a functional area of responsibility, such as training, or who is responsible for a group of battalions or districts and who answers directly to the fire chief. Also called *deputy chief* or *division chief.*

battalion chief Usually the first level of chief, the person in charge of running calls and supervising multiple stations or districts within a city.

brush company See *wildland company.*

candidate A person who applies to become a firefighter.

captain The second rank of promotion in the fire service, between the lieutenant and the battalion chief. Captains are responsible for managing a fire company and for coordinating the activities of that company among the other shifts.

chain of command A rank-based, hierarchical structure that creates an orderly line of authority.

chief's bugle See *chief's trumpet.*

chief's trumpet An obsolete amplification device that was a precursor to a bullhorn and that enabled a chief officer to give orders to firefighters during an emergency. Also called *chief's bugle.*

code A standard that is an extensive compilation of provisions covering broad subject matter or that is suitable for adoption into law independent of other codes and standards. (NFPA 1)

company The basic firefighting organizational unit staffed by various grades of firefighters under the supervision of an officer and assigned to one or more specific pieces of apparatus. (NFPA 1410)

company officer The individual responsible for command of a company, a designation not specific to any particular fire department rank (can be a firefighter, lieutenant, captain, or chief officer, if responsible for command of a single company). (NFPA 1026)

crew A collective term used casually to refer to a group of firefighters in a department with similar duties or responsibilities. See also *company* and *unit.*

deputy chief See *assistant chief.*

dispatcher See *telecommunicator.* (NFPA 1225)

division chief See *assistant chief.*

driver/operator A person qualified to operate a fire apparatus. Also called *engineer* or *technician.* (NFPA 1910)

emergency medical responder (EMR) Emergency medical services (EMS) personnel who have basic training for providing initial medical assistance, have training in bleeding control and cardiopulmonary resuscitation (CPR), and often perform in an assistant role within the ambulance.

emergency medical services (EMS) company A company that may include medical units and first-response vehicles and that responds to and assists in the transport of medical and trauma victims to medical facilities for further treatment. Also called *emergency medical services (EMS) squad* or *squad.*

emergency medical services (EMS) personnel Personnel responsible for administering care to people who are sick and injured.

emergency medical services (EMS) squad See *emergency medical services (EMS) company.*

emergency medical technician (EMT) Emergency medical services (EMS) personnel who can do everything an emergency medical responder (EMR) can do and who have training in basic emergency care skills, including oxygen therapy, bleeding control, cardiopulmonary resuscitation (CPR), automated external defibrillation, use of basic airway devices, and assisting patients with certain medications.

emergency scene The area encompassed by the incident and the surrounding area needed by the emergency forces to stage apparatus and mitigate the incident. Also called *on scene.* See also *fire ground* and *fire scene.* (NFPA 901)

emergency vehicle technician (EVT) The individual who repairs and performs service on emergency vehicles.

engine See *pumper.*

engine company A piece of fire apparatus staffed with firefighters that has the primary responsibility to deliver a fire stream or streams to extinguish the fire in coordination with ventilation (truck company) and rescue operations. (NFPA 1700)

engineer See *driver/operator.*

fire and life safety educator (FLSE) An individual who has demonstrated the ability to coordinate, create, administer, prepare, deliver, and evaluate educational programs and information.

fire apparatus A vehicle designed to be used under emergency conditions to transport personnel and equipment or to support the suppression of fires and mitigation of other hazardous situations. Also called *apparatus.* (NFPA 1010)

fire chief The highest-ranking officer in charge of a fire department. (NFPA 1550)

fire code The code that specifies practices and procedures to prevent fires, prevent fires that start from spreading by suppressing them and blocking them, and protect lives in the event of a fire by specifying how occupants will be evacuated.

fire department An organization providing rescue, fire suppression, and related activities, including any public, governmental, private, industrial, or military organization engaging in this type of activity. (NFPA 1010)

firefighter A member of a fire department who is assigned to do routine cleaning and maintenance, place hose line to extinguish fires, and assist with a public fire-prevention program.

Firefighter I A person at the first level of progression, as defined in Chapter 6 of NFPA 1010, who has demonstrated the knowledge and skills to function as an integral member of a firefighting team under direct supervision in hazardous conditions. (NFPA 1010)

Firefighter II A person at the second level of progression, as defined in Chapter 7 of NFPA 1010, who has demonstrated the skills and depth of knowledge to function under general supervision. (NFPA 1010)

fire ground Another name for *emergency scene* when the incident is a fire. Also called *fire scene.*

fire hook A tool used to pull down burning structures.

firehouse See *fire station.*

fire inspector An individual who conducts fire code inspections and applies codes and standards. (NFPA 1030)

fire investigator An individual who has demonstrated the skills and knowledge necessary to conduct, coordinate, and complete an investigation. (NFPA 1030)

fire mark Historically, a plaque displayed on a building with the name or logo of a fire insurance company informing firefighters that the building was insured by that insurance company, which meant that insurance company would pay the firefighters for extinguishing the fire.

fire marshal A person designated to provide delivery, management, and/or administration of fire protection- and life safety-related codes and standards, investigations, education, and/or prevention services for local, county, state, provincial, federal, tribal, or private sector jurisdictions as adopted or determined by that entity. (NFPA 1030)

fireplug Historically speaking, a plug installed to control water accessed from wooden pipes but today is slang for *fire hydrant.*

fire police officer An individual officially deployed who provides scene security, directs traffic, and conducts other duties. (NFPA 1091)

fire protection engineer A member of the fire department or an employee of an architectural firm who is responsible for reviewing building plans and working with building owners to ensure that the design of and systems for fire detection and suppression meet applicable codes and function as needed.

fire scene Another name for *emergency scene* when the incident is a fire. Also called *fireground.*

fire station A building that houses fire apparatus and equipment for a geographic area within a fire department. Also called *fire house.*

fire suppression The activities involved in controlling and extinguishing fires. (NFPA 1500)

fire warden An individual charged with enforcing fire regulations in colonial America.

FLSE See *fire and life safety educator.*

governance The framework and procedures for managing and operating an organization.

hazardous materials company A company that responds to and controls scenes where hazardous materials have spilled or leaked and whose members wear special suits and are trained to deal with most chemicals.

hazardous materials technician A person who responds to hazardous materials/weapons of mass

KEY TERMS CONTINUED

destruction (WMD) incidents using a risk-based response process to analyze a problem involving hazardous materials/WMD, plan a response to the problem, implement the planned response, evaluate progress of the planned response and adjust accordingly, and assist in terminating the incident. (NFPA 470)

incident commander (IC) The individual responsible for all incident activities, including the development of strategies and tactics and the ordering and release of resources. (NFPA 1410)

incident safety officer (ISO) A member of the command staff responsible for monitoring and assessing safety hazards and unsafe situations and for developing measures for ensuring personnel safety. (NFPA 1700)

initial attack apparatus Fire apparatus with a fire pump of at least 250-gpm (946-L/min) capacity, water tank, and hose body, whose primary purpose is to initiate a fire-suppression attack on structural, vehicular, or vegetation fires and to support associated fire department operations. Also called *quick attack apparatus.* (NFPA 1900)

International Fire Service Accreditation Congress (IFSAC) A national organization that accredits or recognizes emergency service certification systems.

ladder company See *truck company.*

ladder truck See *aerial fire apparatus.*

lieutenant The first level of officer and the person who is usually responsible for a single fire company on a shift.

mobile water supply apparatus A vehicle designed primarily for transporting (pickup, transporting, and delivering) water to fire emergency scenes to be applied by other vehicles or pumping equipment. Also called *tanker, tender,* or *water tender.* (NFPA 1900)

National Board on Fire Service Professional Qualifications A national organization that accredits or recognizes emergency service certification systems. Also called *Pro Board.*

National Fire Protection Association (NFPA) A nonprofit association that develops and maintains nationally recognized minimum consensus standards and fire codes for fire safety and handling of hazardous materials.

on scene See *emergency scene.*

paramedic Emergency medical services (EMS) personnel who can do everything an advanced emergency medical technician (AEMT) can do and who have extensive training in advanced life support, including administering drugs, cardiac monitoring, inserting advanced airways, manual defibrillation, and other advanced assessment and treatment skills.

policies Formal statements that outline expectations for performance and procedures in different circumstances but usually require personnel to make judgments to determine the best course of action within the stated study.

Pro Board See *National Board on Fire Service Professional Qualifications.*

public information officer (PIO) A member of the command staff responsible for interfacing with the public and media or with other agencies with incident-related information requirements. (NFPA 1550)

pumper Fire apparatus with a permanently mounted fire pump of at least 750-gpm (1300-L/min) capacity, water tank, and a hose body whose primary purpose is to control structural and associated fires. (NFPA 1900)

quick attack apparatus See *initial attack apparatus.*

quint See *quint apparatus.*

quint apparatus Fire apparatus with a permanently mounted fire pump, a water tank, a hose storage area, an aerial device or elevating platform with a permanently mounted waterway, and a complement of ground ladders. Also called *quint.* (NFPA 1710)

recruit A candidate whose application to become a firefighter is accepted.

regulation A mandate issued and enforced by governmental bodies such as the U.S. Occupational Safety and Health Administration (OSHA) and the U.S. Environmental Protection Agency (EPA).

rescue apparatus Apparatus that carry an extensive array of regular and specialized tools and equipment that are used to rescue victims.

rescue company A piece of fire apparatus staffed with firefighters that is generally used for search and rescue at fire incidents. (NFPA 1700)

rescue technician See *technical rescuer.*

self-contained breathing apparatus (SCBA) An atmosphere-supplying respirator that supplies a respirable air atmosphere to the user from a breathing air source that is independent of the ambient environment and designed to be carried by the user. (NFPA 1970)

squad See *emergency medical services (EMS) company.*

standard Documents, the main text of which contains only mandatory provisions using the word "shall" to indicate requirements, and are in a form generally suitable for mandatory reference by another standard or code or for adoption into law. Nonmandatory provisions are not to be considered a part of the requirements of a standard and shall be located in an appendix or annex, footnote, informational note, or other means as permitted in the NFPA Manuals of Style. (NFPA 1)

standard operating guidelines (SOGs) Written organizational directives that establish or prescribe specific operational or administrative methods to be followed routinely, which can be varied because of operational needs in the performance of designated operations or actions. (NFPA 1550)

standard operating procedures (SOPs) Written organizational directives that establish or prescribe specific operational or administrative methods to be followed routinely for the performance of designated operations or actions. (NFPA 1550)

support person A fire department member who is not a firefighter but who assists members of a fire department by performing duties in environments that are not hazardous.

tanker See *mobile water supply apparatus.*

technical rescuer A person who is trained to perform or direct a technical rescue. Also called *rescue technician.* (NFPA 1006)

technician See *driver/operator.*

telecommunicator An individual whose primary responsibility is to receive, process, or disseminate information of a public safety nature via telecommunication devices. Also called *dispatcher.* (NFPA 1225)

tender See *mobile water supply apparatus.*

tower ladder company See *truck company.*

tower ladder truck See *aerial fire apparatus.*

training officer The person designated by the fire chief with authority for overall management and control of the organization's training program. (NFPA 1401)

truck See *aerial fire apparatus.*

truck company A company of firefighters who are equipped with one or more pieces of aerial fire apparatus. Also called *ladder company* or *tower ladder company.* (NFPA 1700)

unified command A team effort that allows all agencies with jurisdictional responsibility for an incident or planned event, either geographical or functional, to manage the incident or planned event by establishing a common set of incident objectives and strategies. (NFPA 1026)

unit A collective term used casually to refer to a group of firefighters in a department with similar duties or responsibilities. See also *company* and *crew.*

water tender See *mobile water supply apparatus.*

wildland apparatus A four-wheel-drive vehicle used to transport firefighters closer to wildfires over rough, uneven terrain and that carries a tank of water and a pump that enables the firefighters to pump water while the truck is moving and special firefighting equipment, such as portable pumps, rakes, shovels, and chainsaws.

wildland company A company of firefighters who fight vegetation fires where larger pumpers cannot gain access and who are equipped with four-wheel-drive vehicles and special firefighting equipment. Also called *brush company.*

REVIEW QUESTIONS

1. What is the primary mission of the fire service?
2. Where were the first fire regulations in the United States enacted?
3. What precipitated most of the major advances in fire safety codes?
4. What is a combination department?
5. What is a unified command?
6. Which aspect of fire service culture means putting the needs of others before your own needs?
7. What are SOPs?
8. Which company in a fire department specializes in forcible entry, ventilation, roof operations, search and rescue, and deployment of ground ladders?
9. Which fire department member is responsible for setting up and running the fire pump or operating the aerial device?
10. What is the difference between the designations of support person, Firefighter I, and Firefighter II?

DISCUSSION QUESTIONS

1. How should firefighters interact with representatives of other agencies at emergency scenes?
2. Why does a support person need to know and understand departmental SOPs?
3. Why should a support person care about fires that happened decades or even hundreds of years ago?

REFERENCES

Ahrens, Marty, and Ben Evarts. 2021. *Fire Loss in the United States during 2020.* National Fire Protection Association (NFPA), September 2021. Accessed September 6, 2022. www.nfpa.org//-/media/Files/News-and-Research/Fire-statistics-and-reports/US-Fire-Problem/osFireLoss.pdf.

Benjamin Franklin Historical Society. n.d. "Union Fire Company." Accessed March 28, 2023. www.benjamin-franklin-history.org/union-fire-company/.

Boston Fire Historical Society. n.d. "Boston History before 1859." Accessed March 28, 2023. https://bostonfirehistory.org/boston-history-before-1859.

Coleman, Ronny J. 1988. *Managing Fire Services.* Edited by John A. Granito. Washington, DC: International City/County Management Association.

Fahy, Rita, Ben Evarts, and Gary P. Stein. 2021. "U.S. Fire Department Profile 2019." National Fire Protection Association, December 2021. Accessed September 12, 2022. https://web.archive.org/web/20220324020917id_/https://www.nfpa.org/-/media/Files/News-and-Research/Fire-statistics-and-reports/Emergency-responders/osfdprofile.pdf.

Federal Emergency Management Agency (FEMA). 2018. *ICS Organizational Structure and Elements.* March 2018. Accessed February 23, 2023. https://training.fema.gov/emiweb/is/icsresource/assets/ics%20organizational%20structure%20and%20elements.pdf.

Federal Emergency Management Agency (FEMA). n.d. *ICS 100 – Incident Command System.* Accessed March 24, 2023. www.usda.gov/sites/default/files/documents/ICS100.pdf.

FireHydrant.org. 2004. "A Brief History of the Hydrant." Revised January 28, 2003. Accessed March 24, 2023. www.firehydrant.org/pictures/hydrant_history.html.

Hall, Shelby, and Ben Evarts. 2022. "Fire Loss in the United States during 2021." National Fire Protection Association, September 2022. Accessed September 21, 2022. www.nfpa.org//-/media/Files/News-and-Research/Fire-statistics-and-reports/US-Fire-Problem/osFireLoss.pdf.

International Association of Fire Fighters (IAFF). 2007. *Candidate Physical Ability Test,* 2nd ed. Accessed September 14, 2022. www.iaff.org/wp-content/uploads/2019/04/CPAT-2nd-Edition.pdf.

Martinez-Granata. 2017. "The Maltese vs. Florian Cross: Which One Is Correct?" FireRescue1, March 10, 2017. Accessed August 1, 2024. www.firerescue1.com/history/articles/the-maltese-vs-florian-cross-which-one-is-correct-Xr7pnTxy1nwnHe0M/.

Merrimack Fire and Rescue. n.d. "The History of Firefighting." Accessed March 26, 2023. www.merrimacknh.gov/about-fire-rescue/pages/the-history-of-firefighting.

National Fire Protection Association. n.d.-a. "List of NFPA Codes & Standards." Accessed September 23, 2022. www.nfpa.org/Codes-and-Standards/All-Codes-and-Standards/List-of-Codes-and-Standards.

National Fire Protection Association. 2008. "Deadliest Single Building or Complex Fires and Explosions in the U.S." Updated March 2008. Accessed February 20, 2023. www.nfpa.org/Codes-and-Standards/All-Codes-and-Standards/List-of-Codes-and-Standards.

National Fire Protection Association. 2016. *NFPA 1401, Recommended Practice for Fire Service Training Reports and Records.* 2017 Edition. Quincy, MA: National Fire Protection Association.

National Fire Protection Association. 2019. *NFPA 1410, Standard on Training for Emergency Scene Operations.* 2020 Edition. Quincy, MA: National Fire Protection Association.

National Fire Protection Association. 2019. *NFPA 1452, Guide for Training Fire Service Personnel to Conduct Community Risk Reduction for Residential Occupancies.* 2020 Edition. Quincy, MA: National Fire Protection Association.

National Fire Protection Association. 2019. *NFPA 1710, Standard for the Organization and Deployment of Fire Suppression Operations, Emergency Medical Operations, and Special Operations to the Public by Career Fire Departments.* 2020 Edition. Quincy, MA: National Fire Protection Association.

National Fire Protection Association. 2023. *NFPA 1, Fire Code.* 2024 Edition. Quincy, MA: National Fire Protection Association.

National Fire Protection Association. 2020. *NFPA 101, Life Safety Code.* 2021 Edition. Quincy, MA: National Fire Protection Association.

National Fire Protection Association. 2020. *NFPA 901, Standard Classifications for Fire and Emergency Services Incident Reporting.* 2020 Edition. Quincy, MA: National Fire Protection Association.

National Fire Protection Association. 2020. *NFPA 1006, Standard for Technical Rescue Personnel Professional Qualifications.* 2021 Edition. Quincy, MA: National Fire Protection Association.

National Fire Protection Association. 2020. *NFPA 1500, Standard on Fire Department Occupational Safety, Health, and Wellness Program.* 2021 Edition. Quincy, MA: National Fire Protection Association.

National Fire Protection Association. 2020. *NFPA 1700, Guide for Structural Fire Fighting.* 2021 Edition. Quincy, MA: National Fire Protection Association.

National Fire Protection Association. 2021. *NFPA 13, Standard for the Installation of Sprinkler Systems.* 2022 Edition. Quincy, MA: National Fire Protection Association.

National Fire Protection Association. 2021. *NFPA 470, Hazardous Materials/Weapons of Mass Destruction (WMD) Standard for Responders.* 2022 Edition. Quincy, MA: National Fire Protection Association.

National Fire Protection Association. 2021. *NFPA 1225, Standard for Emergency Services Communications.* 2022 Edition. Quincy, MA: National Fire Protection Association.

National Fire Protection Association. 2021. *NFPA 1582, Standard on Comprehensive Occupational Medical Program for Fire Departments.* 2022 Edition. Quincy, MA: National Fire Protection Association.

National Fire Protection Association. 2023. *NFPA 1030, Standard for Professional Qualifications for Fire Prevention Program Positions.* 2024 Edition. Quincy, MA: National Fire Protection Association.

National Fire Protection Association. 2023. *NFPA 1091, Standard for Traffic Incident Management Personnel Professional Qualifications.* 2024 Edition. Quincy, MA: National Fire Protection Association.

National Fire Protection Association. 2023. *NFPA 1550, Standard for Emergency Responder Health and Safety.* 2024 Edition. Quincy, MA: National Fire Protection Association.

National Fire Protection Association. 2023. *NFPA 1900, Standard for Aircraft Rescue and Firefighting Vehicles, Automotive Fire Apparatus, Wildland Fire Apparatus, and Automotive Ambulances.* 2024 Edition. Quincy, MA: National Fire Protection Association.

National Fire Protection Association. 2023. *NFPA 1910, Standard for the Inspection, Maintenance, Refurbishment, Testing, and Retirement of In-Service Emergency Vehicles and Marine Firefighting Vessels.* 2024 Edition. Quincy, MA: National Fire Protection Association.

National Fire Protection Association. 2023. *NFPA 1970, Standard on Protective Ensembles for Structural and Proximity Firefighting, Work Apparel and Open-Circuit Self-Contained Breathing Apparatus (SCBA) for Emergency Services, and Personal Alert Safety Systems (PASS).* 2024 Edition. Quincy, MA: National Fire Protection Association.

National Fire Protection Association (NFPA). 2024. NFPA 1010: *Standard on Professional Qualifications for Firefighters.* 2024 Edition. Quincy, MA: National Fire Protection Association.

National Fire Protection Association (NFPA). 2024. *NFPA 1026: Standard for Incident Management Personnel Professional Qualifications.* 2024 Edition. Quincy, MA: National Fire Protection Association.

Noonan, Travis. 2017. "History of the Modern Fire Truck." DriveZing, November 29, 2017. Accessed March 24, 2023. https://drivezing.com/history-modern-fire-truck.

Chapter Opener: © Eric Scruggs

CHAPTER

2

Support Person

Firefighter Health and Safety

KNOWLEDGE OBJECTIVES

After studying this chapter, you will be able to:

- Describe the purpose of the fire department's employee assistance program.
- Identify the signs and symptoms associated with behavioral and emotional distress.
- Explain the importance of following departmental standard operating procedures (SOPs) for receiving and processing communications.
- Describe how to safely mount an apparatus.
- Describe how to safely ride on a fire apparatus.
- Describe how to safely dismount an apparatus.
- Describe hazards and safety measures associated with riding apparatus.
- List prohibited practices when riding in a fire apparatus.
- Identify types of department-issued personal protective equipment (PPE) and their usage.
- Describe how to manage traffic safely at an emergency scene.
- Explain considerations for hazard and scene control.
- List the common hazards at an emergency scene.
- Describe measures firefighters follow to ensure electrical safety at an emergency scene.
- Describe measures firefighters follow to ensure safe operation at an emergency scene.
- Describe how to safely dismount an apparatus.
- Describe the hazards and safety concerns associated with building utilities.
- Describe when gas service should be shut off.
- Describe when the electrical service should be shut off.

SKILLS OBJECTIVES

After studying this chapter, you will be able to perform the following skills:

- Demonstrate how to mount an apparatus safely.
- Demonstrate how to locate and safely shut off electric utilities.

ADDITIONAL NFPA STANDARDS

- **NFPA 1451**, *Standard for a Fire and Emergency Service Vehicle Operations Training Program, 2018 Edition*
- **NFPA 1550**, *Standard for Emergency Responder Health and Safety, 2024 Edition*
- **NFPA 1581**, *Standard on Fire Department Infection Control Program, 2022 Edition*
- **NFPA 1582**, *Standard on Comprehensive Occupational Medical Program for Fire Departments, 2022 Edition*

CASE STUDY

You Are the Support Person

The firefighters in your crew are at a large fire in an apartment building that started on the third floor, extended into the attic, and is now venting through the roof. Your captain gave the incident commander (IC) the accountability tags, and a crew was assigned to confine the fire on the third floor. The crew members gather their gear, and as they approach the apartment building, you consider the ways firefighters can accomplish their mission while avoiding preventable injury or death.

1. What is the leading cause of firefighter injury and death?
2. What are the potential hazards during this incident?
3. What safety measures should be taken during this incident?

Introduction

This chapter covers the topics of injury and illness prevention; means of reducing firefighter deaths; and safety and health measures needed during all activities performed by firefighters, including training, response, fireground operations, and fire station duties. Firefighting, by its nature, is dangerous. Every firefighter must be aware of the risks inherent in all job responsibilities and activities and must learn safe methods of confronting all the risks.

For firefighters, the goal is never to eliminate all risks. Rather, it is to reduce preventable deaths and injuries by reducing unnecessary risks. For example, potential carcinogens in smoke put firefighters at a high risk of cancer. However, proper decontamination of protective gear, tools, and equipment can greatly reduce that risk without compromising the firefighter's ability to do their job. A commitment to this kind of safety is integral to a firefighter's integrity and professionalism.

Every fire department must do what it can to reduce the hazards and dangers of the job and help prevent firefighter injuries, illness, and deaths. Each fire department must have a strong commitment to firefighter health and safety, with firefighters taking the lead for their own health and safety. When firefighters understand the various risks associated with their job responsibilities and activities and the actual threat that they present, safety measures can become routine, consistent, and fully integrated into every activity, procedure, and job description.

Advances in standards, technology, and equipment require fire departments to review and revise their health and safety policies and procedures regularly. Safety officers are responsible for evaluating the hazards of various situations and recommending appropriate safety measures to the IC. Each accident, injury, or near miss must be thoroughly investigated to learn why it happened and how it can be avoided in the future. After-incident reviews and research by designated health and safety officers can identify new hazards as well as appropriate safety measures. In addition, reports of accidents, fatalities, and near misses from other fire departments can help identify common problems and lead to the development of effective preventive actions.

Because firefighters must be ready to react immediately to an alarm, preparations for response begin long before the alarm is sounded. These preparations include physical and mental readiness, checking personal equipment, ensuring that the fire apparatus is ready, and making sure that all equipment carried on the apparatus is ready for use. Firefighters should also be familiar with their response district, know the buildings under their protection, and understand their department's standard operating procedures (SOPs).

Response actions for the apparatus driver also include considering road and traffic conditions, determining the best route to the incident, identifying nearby hydrant locations or water sources, and selecting the best position for the apparatus at the incident scene.

Causes of Firefighter Deaths and Injuries

The National Fire Protection Association (NFPA) reports that 135 firefighters were killed in the line of duty in 2022 (Campbell and Hall 2023). These deaths resulted from emergency operations such as emergency medical services (EMS), fire suppression, rescue, and hazardous materials calls, as well as nonemergency situations, such as public service calls, training, and responding to or returning from emergency situations (**FIGURE 2-1**).

The largest share of deaths occurred from stress, overexertion, and medical issues (**TABLE 2-1**). Cardiac events accounted for 38 percent of these deaths. The second-leading cause of death was vehicle crashes. In 2022, 15 firefighters died in vehicle accidents, a majority of whom were responding to or returning from incidents (Campbell and Hall 2023).

The NFPA estimates that 64,875 firefighters were injured in the line of duty in 2022, a 7 percent increase from the previous year (Campbell and Hall 2023). **FIGURE 2-2** shows the breakdown of injuries by type of duty.

SAFETY TIP

Prevention of illness and injury is a responsibility shared by each member of the firefighting team. Firefighters must always consider three groups when ensuring safety at the scene:

- Their personal safety
- The safety of their team members
- The safety of everyone at an emergency scene

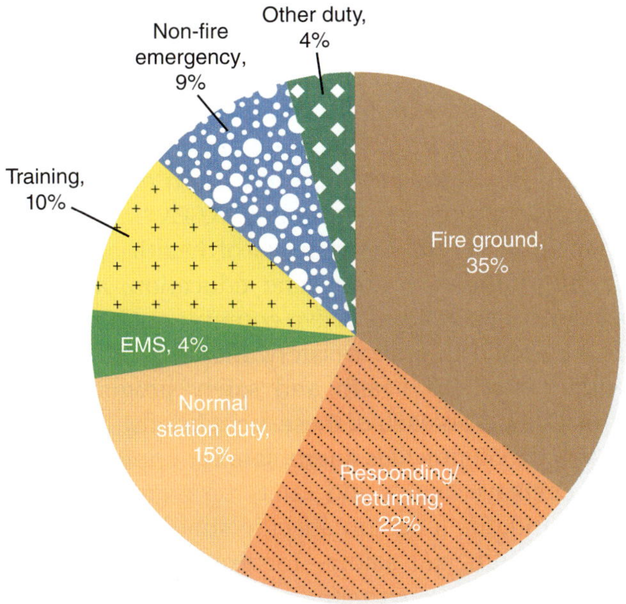

FIGURE 2-1 Firefighter deaths in the United States by type of duty. In 2020, 135 firefighters died in the line of duty.

Reproduced from Campbell, Richard and Jay T. Petrillo. 2023. "Fatal Firefighter Injuries in the United States in 2022." National Fire Protection Association. www.nfpa.org/-/media/files/news-and-research/fire-statistics-and-reports/emergency-responders/osfff.pdf

Fewer than half of the injuries occurred on the fireground. The leading cause of fire-ground injuries was overexertion or strain (**TABLE 2-2**).

TABLE 2-1 Firefighter Deaths by Cause of Injury, 2022

Overexertion, stress, medical	57%
Vehicle accidents	14%
Struck by vehicles	9%
Structural collapse	6%
Rapid fire progress/explosions	4%
Struck by objects	3%
Falls	3%
Exposure to electricity	1%
Fatal assault	1%
Lost inside	1%

Campbell, Richard B., and Shelly Hall. 2023. "United States Firefighter Injuries." National Fire Protection Association. www.nfpa.org/education-and-research/research/nfpa-research/fire-statistical-reports/firefighter-injuries-in-the-united-states.

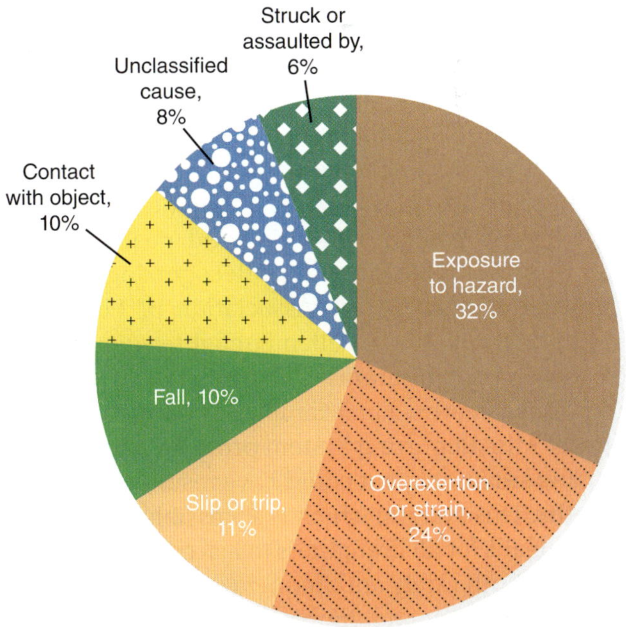

FIGURE 2-2 Firefighter injuries in the United States by type of duty. In 2020, 64,875 firefighters were injured in the line of duty.

Data from Campbell, Richard and Jay T. Petrillo. 2023. "Fatal Firefighter Injuries in the United States in 2022". National Fire Protection Association. www.nfpa.org/-/media/files/news-and-research/fire-statistics-and-reports/emergency-responders/osfff.pdf

TABLE 2-2 Fire-Ground Injuries by Cause

Types of Injury	Percentage of Total Injuries
Overexertion/strain	32%
Fall, jump, slip	22%
Other	15%
Exposure to fire products	10%
Contact with an object	9%
Struck by an object	7%
Exposure to chemicals or radiation	2%
Extreme weather	3%

Campbell, Richard B., and Shelly Hall. 2023. "United States Firefighter Injuries." National Fire Protection Association. www.nfpa.org/education-and-research/research/nfpa-research/fire-statistical-reports/firefighter-injuries-in-the-united-states.

Reducing Firefighter Deaths and Injuries

In 1992, Congress created the National Fallen Firefighters Foundation to lead a nationwide effort to remember the U.S. fallen firefighters. In the years since then, this foundation has expanded its programs to sponsor the annual National Fallen Firefighters Memorial Weekend, offer support programs for family members and other survivors, award scholarships to fire service members' survivors, and work to prevent line-of-duty injuries and deaths.

Although firefighting is an inherently dangerous activity, most firefighter injuries and deaths are the result of preventable situations. Recognizing this fact, organizations such as the Association of Fire Chiefs (IAFC) and the National Fallen Firefighters Foundation have developed programs with the goal of reducing line-of-duty deaths. For example, developed in 2005, the Near Miss Reporting System provides a method for reporting situations that could have resulted in injuries or deaths. This system, which is accessible online, provides a means for all firefighters to learn from situations that occur both rarely and frequently.

In an effort to do more to prevent line-of-duty deaths and injuries, the first National Firefighter Life Safety Summit convened in 2004, uniting fire service leaders and organizations across the United States. The result was the creation of the Everyone Goes Home program and a set of key initiatives, the 16 Firefighter Life Safety Initiatives. The goal of the Everyone Goes Home program is to raise awareness of life safety issues, improve safety practices, and allow everyone to return home at the end of their shift. The 16 Firefighter Life Safety Initiatives describe steps that need to be taken to change the current culture of the fire service to help make it a safe work environment. They are listed in **TABLE 2-3**.

Reducing firefighter injuries and deaths requires the dedicated efforts of every firefighter, of every fire department, and of the entire fire community working together. It also requires a safety program that integrates important components such as regulations, standards, procedures, personnel, training, and equipment. These components are discussed next.

Regulations, Standards, and Procedures

Safety is the highest priority. Ensuring a safe working environment for the members of the fire service is undertaken by several professional organizations and results in various regulations, standards, and procedures.

Standards are issued by nongovernmental entities and are generally consensus based. Standards are discussed in greater detail in Chapter 1, *The Fire Service*. The NFPA develops standards for training courses, apparatus, equipment, and operations. The NFPA's mission is focused on safety: to save lives and reduce loss with information, knowledge, and passion. You will see many NFPA standards referenced throughout this course. Most fire departments have access to these standards.

To reduce the risks of accidents, injuries, occupational illnesses, and fatalities, a successful health and safety program that complies with NFPA 1550, *Standard for Emergency Responder Health and Safety, 2021 Edition*, must be established. NFPA 1550 includes guidance on several key aspects of health and safety, including policies, training and education, apparatus operation, personal protective equipment (PPE), emergency operations, station safety, medical and physical requirements, and health and safety programs. This chapter addresses each of these areas briefly.

NFPA 1550 provides a template for implementing a comprehensive health and safety program. Additional NFPA standards focus on specific subjects directly related to health and safety—for example, NFPA 1581, *Standard on Fire Department Infection Control Program, 2022 Edition*, and NFPA 1582, *Standard on Comprehensive Occupational Medical Program for Fire Departments, 2022 Edition.*

Regulations are issued and enforced by governmental bodies. The federal **Occupational Safety and**

TABLE 2-3 16 Firefighter Life Safety Initiatives

1. **Cultural change:** Define and advocate the need for a cultural change within the fire service relating to safety and incorporating leadership, management, supervision, accountability, and personal responsibility.
2. **Accountability:** Enhance the personal and organizational accountability for health and safety throughout the fire service.
3. **Risk management:** Focus greater attention on the integration of risk management with incident management at all levels, including strategic, tactical, and planning responsibilities.
4. **Empowerment:** All firefighters must be empowered to stop unsafe practices.
5. **Training and certification:** Develop and implement national standards for training, qualifications, and certification (including regular recertification) that are equally applicable to all firefighters based on the duties they are expected to perform.
6. **Medical and physical fitness:** Develop and implement national medical and physical fitness standards that are equally applicable to all firefighters based on the duties they are expected to perform.
7. **Research agenda:** Create a national research agenda and a data-collection system that relates to the 16 Firefighter Life Safety Initiatives.
8. **Technology:** Use available technology whenever it can produce higher levels of health and safety.
9. **Fatality, near-miss investigation:** Thoroughly investigate all firefighter fatalities, injuries, and near misses.
10. **Grant support:** Grant programs should support the implementation of safe practices and procedures and/or mandate safe practices as an eligibility requirement.
11. **Response policies:** National standards for emergency response policies and procedures should be developed and championed.
12. **Violent incident response:** National protocols for response to violent incidents should be developed and championed.
13. **Psychological support:** Firefighters and their families must have access to counseling and psychological support.
14. **Public education:** Public education must receive more resources and be championed as a critical fire and life safety program.
15. **Code enforcement and sprinklers:** Advocacy must be strengthened for the enforcement of codes and the installation of home fire sprinklers.
16. **Apparatus design and safety:** Safety must be a primary consideration in the design of apparatus and equipment.

National Fallen Firefighters Association. Everyone Goes Home. 16 Firefighter Life Safety Initiatives.

Health Administration (OSHA), along with a variety of state and provincial health and safety agencies, develops and enforces government regulations on workplace safety and, in some cases, responder safety. NFPA standards are often incorporated by reference in government regulations.

Each state in the United States has the right to *adopt and/or supersede* workplace health and safety regulations put forth by OSHA. States that have adopted the OSHA regulations are called *state-plan states.* California, for example, is a state-plan state; its regulatory body is called *Cal-OSHA.* About half of the states in the United States are state-plan states. States that have not adopted the OSHA regulations are nonplan states. Nonplan states are called *EPA states* because they follow Title 40 of the **Code of Federal Regulations (CFR)**, *Protection of the Environment, Part 311, Worker Protection.* The CFR is a collection of permanent rules published by the federal government. It includes 50 titles that represent broad areas of interest that are governed by federal regulation.

TIP

It is important to understand the relationship between OSHA regulations and NFPA standards. OSHA regulations are the law; NFPA standards are generally used as guidelines that, under some circumstances, have the force of law.

Every fire department should have a set of SOPs or standard operating guidelines (SOGs) that provide specific information on the actions that should be taken to accomplish a certain task. These procedures are vital because they enable everyone in the department to function properly and know what is expected for each task. Each firefighter is responsible for understanding and following these procedures. This enables firefighters from different stations or companies to work together safely and smoothly.

The fire department chain of command also enforces safety goals and procedures. In particular, the command structure keeps everyone working toward common goals in a safe manner. An Incident Command System (ICS) is a nationally recognized plan to establish command and control of emergency incidents. The ICS is flexible enough to meet the needs of any emergency situation, so it should be implemented at every emergency scene—from a routine auto accident to a major disaster involving responders from numerous agencies.

Many fire departments have a health and safety committee that is responsible for establishing policies and monitoring firefighter health and safety. Members of this committee should include representatives from every area, component, and level within the department, from firefighters to chief officers. The health and safety officer and the fire department physician should also be members of the committee.

Personnel

A health and safety program is only as effective as the individuals who implement it. Safety officers are members of the fire department whose primary responsibility is safety (discussed further in Chapter 1, *The Fire Service*). At the emergency scene, the designated safety officer reports directly to the IC and has the authority to correct or stop any action that is judged to be unsafe. Safety officers observe operations and conditions, evaluate risks, and work with the IC to identify hazards and ensure the safety of all personnel. They also determine when firefighters can work without self-contained breathing apparatus (SCBA) after a fire is extinguished. Safety officers can enhance safety in the workplace, at emergency incidents, and at training exercises. Even so, it is important to remember that each and every member of the fire department shares the responsibility for promoting safety, both as an individual and as a member of the team.

Teamwork is an essential element of safe emergency operations. On the fireground and during any hazardous activity, firefighters must work together to get the job done. Freelancing has no place on the fireground; it poses a danger to the firefighter who acts independently and every other firefighter on the emergency scene. **Freelancing** is acting independently of a superior's orders or the fire department's SOPs. Freelancing is discussed in more detail later in the chapter.

Training

Adequate training is essential for firefighter safety. The initial firefighter training covers the potential hazards of each skill and outlines the steps necessary to avoid injury. Firefighters must avoid sloppy practices or shortcuts that might potentially contribute to injuries. They also must learn how to identify hazards and unsafe conditions.

The knowledge and skills developed during training classes are essential to maintain safety at actual emergency scenes. The initial training course is just the beginning—firefighters must continually seek out additional courses to keep their skills current.

Equipment

A firefighter's equipment ranges from power and hand tools to PPE and electronic instruments. Firefighters must know how to use equipment in the correct manner and then operate it safely at all times. Equipment must also be properly maintained. Poorly maintained equipment can create additional hazards to the user or fail to operate when needed.

Manufacturers usually supply operating instructions and safety procedures for their equipment. These instructions cover the proper use of the equipment, its limitations, and warnings about potential hazards. Firefighters must read and heed these warnings and instructions. In addition, new equipment must meet applicable standards to ensure that it can perform under the difficult and dangerous conditions often encountered on the fireground.

Personal Health and Well-Being

Safety and well-being are directly related to personal health and physical fitness. Although fire departments regularly monitor and evaluate the health of firefighters, each department member is responsible for their own personal health, conditioning, and nutrition. To be an effective firefighter, you must exercise regularly, eat a healthy diet, get an adequate amount of sleep, and take preventive measures to avoid illnesses such as heart disease and cancer.

Physical Fitness

All members of the fire service—whether career or volunteer—should spend at least an hour each day in physical fitness activities of some kind (**FIGURE 2-3**). Each individual should be examined by either a

FIGURE 2-3 Regular exercise will help you to stay healthy and enable you to perform your job effectively.

© Jones & Bartlett Learning. Photographed by Glen E. Ellman.

departmental or personal physician before beginning any new workout routine. An exercise routine that includes weight training, cardiovascular workouts, and stretching, with a concentration on job-related exercises, is ideal. For example, many firefighters use a stair-climbing machine to focus on the muscle groups used for climbing. This type of exercise also builds cardiovascular endurance for the fireground; however, other muscle groups should not be neglected. Physical fitness must be a career-long activity.

Nutrition

You have to give your body the right nutrition, or it will fail when you need it. A healthy diet includes good proportions of fruits, vegetables, healthy fats, whole grains, and lean protein. Pay attention to portion sizes. Unfortunately, most people eat larger portions than their bodies need. Meals can still be satisfying without excess calories.

Hydration

Hydration is an important part of staying healthy (**FIGURE 2-4**). Water is generally the best fluid available because the body absorbs it faster than any other fluid. Avoid fluids that contain high levels of sugar. Sugary fluids can actually slow the rate of fluid absorption by the body and cause abdominal discomfort. A good guideline is to consume 8 to 10 ounces (oz; 0.2 to 0.3 liters [L]) of water for every 5 to 10 minutes of physical exertion. Do not wait until you feel thirsty to start hydrating. The amount of water needed to maintain adequate hydration will depend on the type of work you are doing and the ambient temperature.

One indication of adequate hydration is frequent urination. Infrequent urination or urine that has a deep-yellow color indicates dehydration. Remember, anytime you are working in full PPE, your internal environment will rapidly become hot, and you cannot dissipate body heat normally to the outside environment. Proper hydration enables muscles to work longer and reduces the risk of illness and injuries.

FIGURE 2-4 Consume 8 to 10 oz (0.2 to 0.3 L) of water for every 5 to 10 minutes of physical exertion.

Courtesy of Rom Duckworth.

SAFETY TIP

Maintaining proper hydration is essential to performing at your peak physical level. Be proactive when it comes to hydration, and do not wait until you feel thirsty or too tired to begin drinking water.

Sleep

Good health requires an adequate amount of uninterrupted sleep to maintain alertness, prevent stress, and avoid illnesses and injuries. The nature and intensity of responding to emergencies mean that many members of the fire service find themselves chronically fatigued and experiencing sleep-deprivation issues. The National Sleep Foundation and the American Academy of Sleep Medicine recommend that adults obtain a minimum of 7 to 9 hours of sleep per night. In the long term, sleep deprivation has been shown to lead to hypertension, sleep apnea, respiratory issues, diabetes, depression, and other medical conditions. Sleep deprivation and fatigue issues can increase stress, and in turn, increased stress can contribute to sleep deprivation and fatigue issues. Establish a consistent sleep schedule and sleep routine, such as turning off all electronic devices a half hour before bedtime to allow your mind to wind down and prepare for sleep.

Heart Disease

Heart disease is the leading cause of death in the United States. A healthy lifestyle that includes a balanced diet, effective hydration, physical conditioning, and proper sleep can help reduce many risk factors for heart disease and enable fire department personnel to meet the physical demands of the job. It is also important that you have regular physical examinations to identify heart disease at an early stage.

Cancer

An increase in the use of synthetic products has led to an increase in the toxicity of today's modern fires. Cancer, now considered to be the leading cause of death among firefighters, can be caused by a wide variety of cancer-causing substances (carcinogens) entering the body (International Association of Firefighters 2019). These include exhaust from diesel engines, poisonous gases in smoke, and a wide variety of chemical particles. The dirt and soot that become attached to a firefighter's turnout gear and uniform contain large quantities of substances known to cause cancer. Firefighters' hoods and gloves are thought to contain especially high concentrations of carcinogens. Carcinogens can be ingested through the mouth, injected into the body, absorbed through the respiratory system, or absorbed through the skin. Firefighters are most likely to absorb carcinogens through their skin and through their respiratory systems.

The Firefighter Cancer Support Network (FCSN) estimates that firefighters have a 9 percent higher risk of being diagnosed with cancer than the general U.S. population. Along the same lines, a study conducted by the National Institute for Occupational Safety and Health (NIOSH) concluded that the nearly 30,000 participating firefighters had a greater number of cancer diagnoses and cancer-related deaths than the general U.S. population. Most were digestive, oral, respiratory, and urinary cancers. When comparing firefighters in this study to each other, they found that the chance of lung cancer increased with the amount of time spent at fires, and the chance of death from leukemia increased with the number of fire incidents (Centers for Disease Control and Prevention 2023).

Remember, contaminated objects that are placed in the cab of a fire engine or in the trunk of a firefighter's personal vehicle continue to release cancer-causing substances into the area around them. The longer contaminated objects are present, the more these objects continue to release toxic substances. Firefighters should remove their **structural firefighting protective equipment** (personal protective clothing that provides full-body coverage and provides protection from heat and fire, keeps water away from the body, and helps reduce injuries from cuts or falls) and all other contaminated clothing as soon as possible. All structural firefighting protective equipment should be transported away from the riding compartment in a fire apparatus and thoroughly washed immediately according to the manufacturer's instructions (**FIGURE 2-5**). Additional information on the care and maintenance of PPE can be found in departmental SOPs as well as NFPA 1581, *Standard on Selection, Care, and Maintenance of Protective Ensembles for Structural Firefighting and Proximity Fire Fighting, 2022 Edition.*

The FCSN (2013) suggests actions members of the fire service can take to protect themselves:

1. Use SCBA from initial attack to finish of overhaul. (Not wearing SCBA in both active and postfire environments is the most dangerous voluntary activity in the fire service today.)
2. Do gross field decontamination of PPE to remove as much soot and particulates as possible.
3. Use cleansing wipes to remove as much soot as possible from the head, neck, jaw, throat, underarms, and hands immediately and while still on the scene.
4. Change clothes and wash them immediately after a fire.
5. Shower thoroughly after a fire.
6. Clean PPE, gloves, hood, and helmet immediately after a fire.
7. Do not take contaminated clothes or PPE home or store them in a vehicle.

FIGURE 2-5 Special washing machines are available to launder personal protective clothing.

© Jones & Bartlett Learning. Photographed by Glen E. Ellman.

8. Perform decontamination of fire apparatus interior after fires.
9. Keep structural firefighting protective equipment out of living and sleeping quarters.
10. Stop using tobacco products.
11. Use sunscreen or sunblock.

The importance of annual medical examinations cannot be overstated—early detection and early treatment are essential to increasing survival. These 11 lifesaving actions are excerpted from FCSN's "Taking Action Against Cancer" white paper; it is available free of charge at firefightercancersupport.org.

Some cancers do not present for 20 years or more after exposure to a carcinogen. Therefore, it is important to reduce your exposure to cancer-causing substances starting on your first day in the fire service. More information on performing field reduction of contaminants can be found in Chapter 3, *Personal Protective Equipment*.

Smoking and Tobacco Products, Alcohol, and Illicit Drugs

Many fire departments have adopted policies that prohibit the use of smoking products (including tobacco, marijuana, and vaping devices) by firefighters, both on and off duty. Smoking, especially of tobacco products, is a major risk factor for cardiovascular disease; it reduces the efficiency of the body's respiratory system, and it increases the risk of infections as well as lung and other types of cancer. Firefighters should avoid tobacco products entirely for both health and insurance reasons.

Excessive alcohol use can damage the body and affect performance. In addition, alcohol use increases the risk of mouth, throat, larynx, esophagus, liver, colon, and breast cancers (American Cancer Society 2020).

Both prescription medications and illegal drugs may be abused or misused. Many fire departments have drug-testing programs to ensure that firefighters do not use or abuse legal or illegal drugs. Substance abuse endangers lives.

SAFETY TIP

Everyone is subject to an occasional illness or injury. You should not try to work when ill or injured. Operating safely as a member of a team requires both fitness and concentration. Do not compromise the safety of the team or your personal health by trying to work while you are ill or injured.

Counseling and Critical Incident Stress Management

Many firefighters see more traumatic situations in a short time than most citizens see in their entire lifetime. Firefighting involves not only the stresses directly connected to fighting the fire and rendering emergency medical care but also the added burdens of disrupted sleep patterns, rotating work schedules, unscheduled overtime, and interrupted meals. High levels of stress can produce a variety of symptoms. Some people are not able to sleep well; others tend to gain or lose weight. Many people become irritable when stressed. Overeating, increased consumption of alcoholic beverages, and the use of nonprescribed drugs may be the result of stress. Stress may produce depression or suicidal thoughts in some people.

TIP

Find time for yourself, your family, and friends as a mental buffer from the stressors of the job.

To diminish the effect of these stressors, it is important to get adequate sleep; maintain a healthy, balanced diet; and get adequate exercise. It is also important to balance your work schedule with other activities and monitor your behavioral health. Take vacations to lower your stress levels and improve your physical health so that you will be ready to respond the next time you are needed.

If at any point the stress of work feels overwhelming, say something. Talking with family, friends, and co-workers may be a starting point, and peer counselors or mental health professionals can provide helpful strategies. Remember, seeking help does not make you weak in the eyes of others—it shows that you are in control of your life.

It is essential to identify and use the resources that are available to assist in maintaining behavioral health. Some of these resources are provided by employers, such as employee assistance programs, which will be discussed later. Other resources and assistance are provided by professional organizations. For example, the International Association of Firefighters (IAFF) offers information, available through the IAFF website, as well as the inpatient IAFF Center of Excellence for Behavioral Health Treatment and Recovery. The National Volunteer Fire Council (NVFC) "Share the Load" Program program provides access to critical resources and information to help first responders and their families manage and overcome personal and work-related problems.

Critical Incident Stress Management

Critical incidents challenge the capacity of most individuals to deal with stress. It is important to understand what the stressors in this job are and to learn how to work to diminish their effect. Examples of critical incidents include the following:

- Line-of-duty deaths (police, fire/rescue, EMS)
- Suicide of a colleague
- Serious injury to a colleague
- Situations that involve a high level of personal risk to firefighters
- Events in which the victim is known to the firefighters
- Multiple-casualty incidents/disasters/scenes of violent incidents
- Events involving death or life-threatening injury or illness to a victim, especially a child

This list is not complete, nor is it necessarily a fact that any of these situations will seriously trouble every individual. Normal coping mechanisms help many firefighters to handle many situations. Some individuals deal with stressful situations through exercise, talking to friends and family, or turning to their religious beliefs. These are healthy, nondestructive ways to manage the pressures of being exposed to critical incidents.

Posttraumatic stress disorder (PTSD) may develop after a person has experienced a critical incident. It is characterized by reexperiencing the event and overresponding to stimuli that recall the event. Some of the symptoms of PTSD include depression, startle reactions, flashback phenomena, and dissociative episodes (e.g., amnesia of the event).

Critical incident stress management (CISM) was developed to address acute stress situations and potentially decrease the likelihood that PTSD will develop after such an incident. CISM is used to confront the responses to critical incidents and defuse them, directing crew members toward physical and emotional equilibrium. CISM can occur formally through a debriefing for those who were on the scene. In such situations, trained CISM teams of peers and mental health professionals may facilitate the debriefing.

These symptoms can occur in anyone, even individuals who normally have healthy coping skills. Reactions vary from one individual to the next, both in type and severity. Many times, department personnel do not realize they are affected in a deeply negative manner. A somewhat routine incident may trigger negative reactions from a critical incident that occurred in the past.

A **critical incident stress debriefing (CISD)** is held as soon as possible after a traumatic call. It provides a forum for firefighting and EMS personnel to discuss the anxieties, stress, and emotions triggered by a difficult call. Follow-up sessions can be arranged for individuals who continue to experience stressful or emotional responses after a challenging incident. With peer support teams, specially trained department members help those struggling with day-to-day issues by communicating with them regularly and recommending resources to assist them. Firefighters should understand the resources available to them and know how they can access them. Maintaining good emotional health is a simple but important part of firefighter survival. The aim of counseling, peer support teams, employee assistance

FIGURE 2-6 Group stress defusing sessions are sometimes used to alleviate stress reactions generated by high-stress emergency situations.

© Operation 2022/Alamy Stock Photo

programs, and CISM programs is to prevent these emotional reactions from having a negative impact on the firefighter's work and life over both the short and long term (**FIGURE 2-6**). The major difference between CISM and peer support teams is that CISM teams usually respond immediately after a crisis incident, and peer support teams provide continuous, ongoing support.

SAFETY TIP

Co-workers often notice a change in behavior or attitude before an officer or chief does. This is because close relationships develop between people who work together and share rooms, meals, and social interactions. Being a member of the department means helping a friend. Talk to your partner about changes you may notice in their behavior. If you are the person having trouble dealing with a crisis, remember that you are not alone. Talk with a trusted officer, chief, or member of your crew.

Defusing sessions are the first to occur. These sessions are held during the event or immediately afterward. A group informally discusses events that they experienced together. Defusing sessions are designed to educate the participants about the expectations over the next few days and give guidance on proper techniques to manage the feelings they may be experiencing. One example is to discourage drinking alcohol during this stressful time.

Debriefing sessions are held within 24 to 72 hours of a major incident. These meetings are held by a CISM team consisting of peers and mental health professionals. At the debriefing session, pent-up emotions can be properly expressed. It is more likely you will be ready to express your emotions more freely a few days after the event.

One of the important rules associated with the debriefing session is to not turn it into an operational critique. No one is right. No one is wrong. No one is to blame. Only emotions about the specific event are to be relayed. These debriefing sessions may also need to be repeated at a later time.

TIP

CISM programs are located throughout the United States. You can locate a CISM program in your area via the Internet, or it can be requested through your employer.

TIP

CISM can be a helpful strategy, but it may not be effective for everyone. Some people are not receptive to openly discussing psychologically traumatic memories. When an individual's behavior is noticeably different after a traumatic event and CISM is not an option, private counseling by a mental health professional may be invaluable.

Burnout

Critical incident stress also can be cumulative, building up over time. This condition, which is called *burnout*, cannot be traced to any one incident. Like other negative aspects of stress, burnout affects not only the personal well-being of the crew member, but it also affects their colleagues and the victims they care for because it results in increased errors and decreased performance. Members of the fire service suffering from burnout contribute to decreases in work morale, overall work effort, and effective teamwork, as well as an increase in job turnover.

Compassion Fatigue

Sometimes referred to as "the cost of caring," **compassion fatigue** is common among those who work in health care and disaster and emergency services. Compassion fatigue, also known as *secondary stress disorder*, is a disorder characterized by a gradual lessening of compassion over time. It differs from PTSD in that PTSD is caused by direct exposure to a traumatic incident or a series of traumatic incidents, whereas compassion fatigue is a reaction to caring for others who have experienced trauma themselves. Symptoms of compassion fatigue include the following:

- High absenteeism
- Difficult relationships with colleagues and co-workers
- Inability to work in teams
- Aggressive behavior toward victims
- Strong negative attitudes toward work
- Lack of empathy for patients
- Judgmental attitude toward victims
- Preoccupation with nonwork issues while on duty
- Other symptoms of increased stress

Supporting victims in emergency situations is difficult. All members of the fire service are vulnerable to the stresses that go with the profession. It is critical that everyone recognize the signs of cumulative stress so that it does not interfere with their work or life away from work, including their family life. The signs and symptoms of cumulative stress may not be obvious at first. Rather, they may be subtle and not present all the time, such as the following:

- Irritability toward co-workers, family, and friends
- Inability to concentrate
- Difficulty sleeping, increased sleeping, or nightmares
- Feelings of sadness, anxiety, or guilt
- Indecisiveness
- Loss of appetite (gastrointestinal disturbances)
- Loss of interest in sexual activities
- Isolation
- Loss of interest in work
- Increased use of alcohol
- Recreational drug use
- Physical symptoms such as chronic pain (headache, backache)
- Feelings of hopelessness

Working in the fire service requires close teamwork. It is the responsibility of all crew members to monitor themselves and other members of their team for signs of mental stress. We are all interdependent on the members of our team in order to accomplish our goal. The team needs to function well, not only at emergency scenes but also when performing routine tasks or relaxing between runs. It is important that you respect each member of your team. Bullying or discrimination should never be tolerated. It is everyone's job to be on the lookout for signs of unhealthy behavior or stress. If you think a fellow fire service member is exhibiting signs of stress or experiencing depression, there are several ways in which you may be able to help. Although the ways in which this outcome is accomplished vary, the purpose remains the same.

Sometimes it may be helpful to sit down with a coworker and ask them how they are doing or to let them know that you are concerned about them. You may be able to encourage them to seek help. If you see signs of stress, talk with your officer or someone who can help that person receive assistance.

Suicide Awareness and Prevention

Suicide is the 11th-leading cause of death in the United States, but the rate among emergency responders is much higher. According to the Ruderman White Paper on Mental Health and Suicide of First Responders, firefighters and law enforcement personnel are more likely to die of suicide than from life-of-duty death. In 2016, there were 99 reported firefighter deaths as a result of suicide (Firefighter Behavioral Health Alliance [FBHA] 2017). Estimates reveal that a fire department is three times more likely to experience a suicide in any given year than a line-of-duty death (National Fallen Firefighters Foundation 2022). This fact reminds us that stressful occupations may experience a higher rate of suicides than the general population. Many people are concerned about the incidence of suicide in firefighters and other emergency care providers.

Although firefighters are trained to be the best they can be, many are not prepared for the ill effects or aftermath of stress or a traumatic situation. It is important to understand that by recognizing signs of stress or depression in a team member, you may be able to help them get assistance for their problem before it becomes worse.

Numerous national services are available to firefighters. These include the following:

- Fire/EMS Helpline at 1-888-731-FIRE (3473)
- National Suicide Prevention Lifeline at 988

TIP

Just as 911 is called for help in an emergency, 988 is the number to call for help in a behavioral health emergency. The Suicide and Crisis Lifeline, available by calling 988, is free, confidential, and staffed with trained professionals who are ready to listen.

TIP

According to the FBHA, the top-five warning signs of job-related stress include the following:

- Recklessness/impulsiveness
- Anger
- Isolation
- Loss of confidence in abilities or skills
- Sleep deprivation

If you believe you are suffering from these issues, please seek help from your employee assistance program, chaplain, peer support team, or a qualified counselor in your community.

Employee Assistance Programs

There are many formal programs to support behavioral health within the fire service. These programs, also known as *member assistance programs*, are maintained by fire departments, unions, local governments, and even charitable organizations. Support is provided through special events, projects, peer support, chaplain programs, and education. A traditional and common form of support for firefighters is the employee assistance program. **Employee assistance programs (EAPs)** provide confidential help with a wide range of problems that might affect performance. Many fire departments have established EAPs so that department personnel can get counseling, support, or other assistance in dealing with physical, financial, emotional, or substance use issues. EAPs include a variety of helpful resources, including access to qualified counselors and chaplains who have a working knowledge of the fire service. Some fire departments have qualified counselors available 24 hours a day. The initial counseling may consist of a group session for all firefighters and rescuers; alternatively, it can also be done on a one-on-one basis or in smaller groups. A fire officer may refer a member of the team to an EAP if a problem starts to affect the individual's job performance. Personnel who take advantage of an EAP can do so with complete confidentiality and without fear of retribution.

According to the FBHA, many do not seek help, mainly because of confidentiality, job promotion concerns, or simply because counselors may not know the culture of the fire service. The FBHA recommends inviting counselors to the station and including them on ride time, in training, and during meals. Another suggestion is to have the EAP counselors create a video biography so that department members and their families can see who they are and get to know them.

Safety during Training

During training, fire service personnel learn and practice the skills that they will use later under emergency conditions. Typically, the patterns that develop during training continue during actual emergency incidents. Thus, developing the proper working habits during training courses helps ensure safety later. The use of proper protective gear and good teamwork are as important during training as they are on the fireground.

Instructors and veteran firefighters are virtually always willing to share their experiences and advice. They can explain and demonstrate every skill and point out the safety hazards involved because they have performed these skills hundreds of times and know what to do. But here, too, safety is a shared responsibility. Do not attempt anything you feel is beyond your ability or knowledge. If you see something that you believe is an unsafe practice, bring it to the attention of your instructors or a designated safety officer.

Avoid freelancing. Wait for instructions or orders before beginning any task. Do not assume that something is safe and act independently. Follow instructions and learn to work according to the proper procedures.

Safety during the Emergency Response

Safety during an emergency response must begin before there is an actual response. This process begins by ensuring that your PPE is complete, ready for use, and in good condition. At the beginning of each tour of duty, place your PPE in its designated location, which depends on your assigned riding position on the apparatus (**FIGURE 2-7**).

Every response to an incident has a high potential for accidents, injuries, and death. **Response** actions include receiving the alarm, donning protective clothing and equipment, mounting and dismounting the apparatus, and transporting equipment and personnel to and from the emergency incident quickly and safely.

Alarm Receipt

When an alarm is received, the response should be prompt and efficient. Responding firefighters should walk briskly to the apparatus. There is no need to run; the objective is to respond quickly without injuring anyone or causing any damage. Follow established

FIGURE 2-7 Protective clothing should be properly positioned so that you can quickly don it.

© Jones & Bartlett Learning

procedures to ensure that stoves, faucets, and other appliances at the station are shut off. Wait until the apparatus bay doors are fully open before leaving the station.

SAFETY TIP

Volunteer firefighters who are not assigned to specific tours of duty or riding positions should check their PPE, SCBA, and associated tools and equipment on a regular basis to ensure that these items are ready for use whenever an emergency response becomes necessary. After each use, all PPE should be carefully cleaned and checked before it is put away.

Riding the Apparatus

Expectations for donning PPE are frequently established by the company officer. A common practice is for firefighters to don PPE prior to mounting the apparatus. While a fire apparatus is in motion, all crew members

FIGURE 2-8 Check the tools assigned to your riding position.

© Jones & Bartlett Learning. Photographed by Glen E. Ellman.

should be wearing seat belts properly. Do not attempt to don PPE while the apparatus is on the road. Wait until you dismount at the incident scene to don any protective clothing that was not donned prior to mounting the apparatus.

All equipment should be properly mounted, stowed, or secured on the fire apparatus (**FIGURE 2-8**). Unsecured equipment in the crew compartment can prove dangerous if the apparatus must stop or turn quickly because a flying tool, map book, or PPE can seriously injure a firefighter.

Be careful when mounting the fire apparatus because the steps are often high and can be slippery. Use handrails when mounting or dismounting the apparatus. Follow the steps in **SKILL DRILL 2-1** to mount an apparatus properly.

All crew members must be seated in their assigned riding positions with seat belts and/or harnesses fastened before the apparatus begins to move. NFPA 1550, *Standard for Emergency Responder Health and Safety, 2021 Edition*, and NFPA 1010, *Standard on Firefighter Professional Qualifications, 2024 Edition*, require all members to be in their seats, with seat belts secured, whenever the vehicle is in motion. Do not unbuckle your seat belt to don any clothing or equipment while the apparatus is en route to an incident. Vehicle occupants who are not wearing seat belts are much more likely to suffer serious injuries or death than occupants who have their seat belts properly fastened. Air bags are most effective when seat belts are properly fastened; they are not effective without properly applied seat belts. Seat belts also greatly reduce the possibility that vehicle occupants will be ejected from the vehicle.

SKILL DRILL 2-1

Mounting Apparatus Support Person, NFPA 1010: 5.3.2

1. When mounting (climbing aboard) the fire apparatus, always have at least one hand firmly grasping a handhold and at least one foot firmly placed on a foot surface. Maintain the one-hand and one-foot placement until you are seated.

2. Fasten your seat belt and leave it fastened until the apparatus is stopped at its destination. Don any other required safety equipment for the response, such as hearing protection and intercom systems.

© Jones & Bartlett Learning. Photographed by Glen E. Ellman.

For safety's sake, SOPs often prohibit specific actions during a response. As noted previously, all personnel must remain seated with their seat belts securely fastened while the emergency vehicle is in motion. Never unfasten your seat belt to retrieve or don equipment. Do not dismount the apparatus until the vehicle comes to a complete stop. In addition, you should never stand up while riding on apparatus. Do not hold on to the side of a moving vehicle or stand on the rear step. When a vehicle is in motion, everyone aboard must be seated and belted in an approved riding position.

The noise produced by sirens and air horns can have long-term, damaging effects on hearing. For this reason, your department should provide hearing protection for personnel riding on fire apparatus. Some of these devices include radio and intercom capabilities so that firefighters can talk to one another and hear information from the dispatcher or IC.

During response, limit conversation to the exchange of pertinent information. Listen for instructions from the IC, instructions from your company officer, and additional information about the incident over the radio. The vehicle operator's attention should be focused on driving the apparatus safely to the scene of the incident.

The ride to the incident is a good time to consider any relevant factors that could affect the situation. These factors could include the time of day or night, the temperature, the presence of precipitation or wind, the type of occupancy, the type of construction, and the location and type of incident. Using this time to think ahead will help you mentally prepare for the various possibilities that you might encounter at the scene.

When the apparatus arrives at the incident scene, the driver/operator will park it in a location that is both safe and functional. Wait until the vehicle comes to a

SKILL DRILL 2-2

Dismounting a Stopped Apparatus Support Person, NFPA 1010: 5.3.2

1. Become familiar with your riding position and the safest way to dismount.

2. Maintain the one-hand and one-foot placement when leaving the apparatus, especially on wet or potentially icy roadway surfaces.

© Jones & Bartlett Learning. Photographed by Glen E. Ellman.

complete stop before dismounting. Always check for traffic before opening the doors or stepping out of the apparatus. During the dismount, watch for other hazards such as ice and snow, downed power lines, uneven terrain, or hazardous materials that could be present.

Be careful when dismounting apparatus. The increased weight of PPE and adverse conditions can contribute to slips, strains, and sprains. Follow the steps in **SKILL DRILL 2-2** to dismount an apparatus safely.

Traffic Safety on the Scene

An emergency incident scene presents several risks to firefighters in addition to the hazards of fighting fires and performing other duties. One of these dangers is traffic, particularly when the incident scene is on a street or highway. Traffic safety should be a major concern for the first-arriving units because approaching drivers might not see emergency workers or realize how much room firefighters need to work safely.

The first unit or units to arrive at the incident scene have a dual responsibility. Not only must the firefighters focus on the emergency situation facing them, but they also must consider approaching traffic, including other emergency vehicles, and other, less obvious hazards. Always check for traffic before opening doors and dismounting the apparatus, and watch out for traffic when working in the street.

One of the most dangerous work areas for members of the fire service is on a roadway, where traffic may be approaching at high speeds. Follow departmental SOPs to close streets quickly and block access to areas where operations are being conducted. Place traffic cones, flares, emergency scene signage, and other warning devices far enough away from the incident to allow inbound traffic to slow down and be directed away from the work area (**FIGURE 2-9**). Police officers should assist by diverting traffic at a safe distance outside the hazard area. It is not uncommon to set large control zones at the onset of an incident, only to discover that the zones may have been established too liberally. At the same time, control zones should not be defined too narrowly. As the IC gets more information about the incident, the control zones may be expanded or reduced. Wind shifts are a common reason why control zones are modified during the incident. If there is a prevailing wind pattern in your area, then factor that information into your decision-making process when it comes to control zones.

Keep in mind that even in a work area that is properly marked and blocked by apparatus, the work area

FIGURE 2-9 The scene of a crash should be marked properly, and traffic should be diverted so that responders have enough room to work.

© Mike Legeros. Used with permission.

remains a dangerous place. Firefighters must maintain situational awareness for those who may present hazards to an emergency scene, especially drivers who are distracted, drowsy, or intoxicated.

Placement of emergency vehicles on the scene is also critical. With proper placement, such vehicles can act as a barrier between oncoming traffic and the scene. All firefighters working at highway incidents should wear high-visibility safety vests in addition to their normal PPE. These vests should meet the American National Standards Institute (ANSI) 207 standard for public safety vests. Many fire departments have specific SOPs covering required safety procedures for these incidents.

SAFETY TIP

- Do not attempt to mount or dismount a moving vehicle.
- Do not remove your seat belt until the apparatus comes to a complete stop.
- Do not stand directly behind an apparatus that is backing up. Instead, stand off to one side, where the driver can see you in the rearview mirror. All fire apparatus should have working, audible alarms when in reverse gear.

Safe Driving Practices

In 2020, the NFPA estimated there were 15,675 collisions involving fire department emergency vehicles responding to or returning from incidents (Campbell and Hall 2023). As discussed earlier, in 2020, 15 firefighters died in vehicle-related incidents. In addition, 4,975 firefighters were injured when responding to or returning from incidents (Campbell and Hall 2023).

Drivers of fire apparatus have a great responsibility to get the fire apparatus and the crew members to the emergency scene without having or causing a traffic accident en route. Drivers must always exercise caution when driving to an incident. Fire apparatus can be large, heavy, and difficult to maneuver, and drivers must know the streets in their first-due area and any target hazards. They must be able to operate the vehicle skillfully and keep it under control at all times. In addition, they must anticipate all responses from other drivers who might not see or hear an approaching emergency vehicle or know what to do when confronted by one. Operating an emergency vehicle without the proper regard for safety can endanger the lives of both the personnel on the vehicle and any civilian drivers or pedestrians encountered along the way.

Prompt response is a goal, but safe response is a much higher priority. The attitude and ability of the vehicle operator are major factors in crashes. A competent emergency vehicle operator needs to have a confident, but not arrogant, attitude. Aggressive driving has no place in the fire service. The emergency vehicle operator should have good judgment and reactions, mental fitness, maturity, physical fitness, alertness, and good driving habits. It is important to maintain a clean driving record because a person who has been cited for multiple moving violations in their personal vehicle will usually be an increased risk as the driver of an emergency vehicle. To be a good driver, it is important to know the state and local laws relating to motor vehicle operations. In addition, an emergency vehicle operator needs to understand the reaction time, braking distance, and stopping distance of the vehicle.

Emergency driving, even when properly performed, increases risks. Impaired driving dramatically increases those risks. Impairment can result from many different sources—for example, using some prescribed medicines, using some over-the-counter medicines, and being overly fatigued. Anyone who has been drinking alcoholic beverages should not drive. Driving while eating, texting, or talking on a communications device are all forms of distracted driving. Even devices with navigation technology should not be used in such a way that the driver is distracted. Distracted driving during routine driving or when responding to an emergency is to be avoided—it is the cause of many collisions and deaths. In some cases, this has led to criminal prosecution and conviction.

Firefighters who drive emergency vehicles must have special driver training and know the laws and regulations that apply to emergency responses. Many jurisdictions require a special driver's license to operate fire apparatus. The rules that apply to emergency vehicles are specific, and the driver/operator is legally responsible for the safe operation of the vehicle and the safety of the vehicle occupants at all times. These driving skills are not required for the support person, Firefighter I, and Firefighter II designations and are beyond the scope of this resource.

Although most states and provinces permit drivers of emergency vehicles to take exception to specific traffic regulations when responding to emergency incidents, drivers/operators must always consider the potential actions of other drivers before making such a decision. For example, traffic laws require other drivers to yield the right of way to an emergency vehicle. There is no assurance, however, that other drivers will do so when an emergency vehicle approaches. The driver/operator also must anticipate which routes other units responding to the same incident will take. All passengers and drivers of emergency vehicles should wear seat belts on routine and emergency responses.

Many collisions occur when the motor vehicle operator loses control of the vehicle. These crashes may be caused by driving too fast for the prevailing conditions, braking inappropriately, changing directions too abruptly, or tracking around a curve too fast. In addition, many collisions involving emergency vehicles occur on open roads, where excessive speed is often cited as a primary cause. Intersections are common sites of collisions involving emergency vehicles; most of these collisions occur when the emergency vehicle operator fails to stop at an intersection to ensure that other traffic has stopped before proceeding through the intersection.

Finally, a motor vehicle collision itself consists of a series of separate collision events. The first collision occurs when the vehicle collides with a second vehicle or with a stationary object; the second collision occurs when the occupants of the vehicle collide with the interior of the vehicle.

SAFETY TIP

Always fasten your seat belt each and every time you get into a motor vehicle. It is vital that you follow this rule each and every time you climb into a fire apparatus.

Laws and Regulations Governing Emergency Vehicle Operation

Four general principles govern emergency vehicle operation:

- Emergency vehicle operators are subject to all traffic regulations unless a specific exemption is made. A specific exemption is a statement that appears in a statute, such as, "The driver of an authorized vehicle may exceed the maximum speed limits so long as he or she does not endanger life or property."
- Exemptions are legal only when the vehicle is operating in emergency mode.
- Even with an exemption, the emergency vehicle operator can be found criminally or civilly liable if involved in a crash.
- An exemption does not relieve the operator of an authorized emergency vehicle from the duty to drive with reasonable care for all persons using the highway.

Laws governing emergency vehicle operation vary from one state to another. You must follow the laws and regulations of your state regarding emergency vehicle operations. Some states permit private vehicles to operate as emergency vehicles when responding to a fire station or to the scene of an emergency; other states outline a limited use of emergency equipment and dictate certain restrictions. You must understand and follow the specific laws and regulations of your state as well as your department's SOPs.

In many cases, one of the best predictors of future performance is past behavior. Your driving record is an important consideration in your career in a fire department. A person who has past moving violations for speeding, reckless operation, aggressive driving, chargeable crashes, or driving under the influence of alcohol or drugs may be excluded from driving any emergency vehicles. Your driving record, both on duty and off duty, is important to your career in the fire service.

Standard Operating Procedures for Personal Vehicles

In some departments where firefighters are not on duty at the fire station, it may be necessary for members to respond to the emergency scene in their private vehicles. The use of your personal vehicle in a situation requiring an emergency response must be permitted by state laws and regulations, and this operation must also be permitted by your local fire department. If your local fire department does not permit the use of personal vehicles for emergency responses, then you are not permitted to engage in this behavior regardless of what is permitted by your state's laws.

Fire service personnel who respond to emergency incidents in their personal vehicles must follow the specific laws and regulations of their state or province and follow departmental SOPs related to this issue. Some jurisdictions require emergency responders to equip their privately owned vehicles with warning devices and to operate them as emergency vehicles; others grant no special status to privately owned vehicles driven during an emergency response. In some areas, volunteer firefighters responding to an emergency incident use colored lights to request the right of way from other drivers.

Your fire department should have SOPs that spell out whether private vehicles can be used for emergency response. These procedures typically address which kind of training you must complete before you are permitted to use your personal vehicle for emergency response. This training usually includes a course in defensive driving and functioning as an emergency vehicle operator.

Vehicle Collision Prevention

Safe driving practices will prevent most vehicle collisions. If you are using your personal vehicle to respond to an emergency, it is important to take the characteristics of your vehicle into account while driving. For example, four-wheel-drive pickup trucks and large sport utility vehicles have a higher center of gravity than sleek sports cars, and sudden changes in direction are more likely to result in roll-over crashes when a vehicle has a high center of gravity. For this reason, it is critical to learn the characteristics and limitations of your vehicle.

Anticipate the road and road conditions. If you regularly travel the same roads to report to your fire station, learn the characteristics of that strip of roadway so that you are able to anticipate when you need to slow down or stop. Always drive at a safe speed for that road; traveling on a limited-access highway is much different from traveling through a school zone, for example. Observe the traffic conditions, and slow down when traffic congestion is present. Expect that motorists around you may do anything at any time. Upon the approach of an emergency vehicle, for example, some motorists will slow down, some will stop, some will speed up, some will pull to the right, and others will turn left in front of you. Expect the unexpected.

Make allowances for weather conditions. On a clear day, it might seem that you can see forever, but really, you cannot. In conditions of rain, snow, fog, dust, or darkness, the distance over which you can see is greatly reduced; reduce your speed to compensate for the limited visibility. There also may be vehicle collisions, downed trees or power lines, or debris in the road. At night, your vision is limited by the distance your headlights reach, so reduce your speed accordingly. Recognize that you cannot see objects outside the projection of the headlight beams. Understand that the goal of an emergency response is not to drive as fast as you can but rather to arrive on the scene as quickly as you can while maintaining safety.

Adjust your speed of response to accommodate any storm conditions. Rainstorms reduce visibility and produce slippery road surfaces. When water collects on a roadway, your vehicle can hydroplane on the thin film of water that separates your tires and the road surface. You have no control over your vehicle when hydroplaning occurs. Be aware of high winds from tornadoes, hurricanes, or other windstorms that can topple trees onto the roadway. During heavy rainstorms, watch for water flowing over the road. You cannot see how deep the water is, and you cannot tell whether the floodwater has washed out part of the roadway. Slow down when driving after an earthquake or tornado because you will not know where the roadway, bridges, and overpasses have been blocked, damaged, or destroyed.

When operating an emergency vehicle, you are not exempt from the laws of physics. The laws of physics tell us that when the speed of a vehicle doubles, the force exerted by that vehicle increases by a factor of four. Thus, when the speed of a vehicle increases from 20 miles per hour to 40 miles per hour (32 kilometers per hour to 64 kilometers per hour), the force exerted by that vehicle is increased by a factor of four. Higher speeds require more braking power and a longer distance to bring the vehicle to a stop.

To drive safely, you need to understand where the greatest risks are located. As mentioned earlier, studies have shown that many crashes involving emergency vehicles occur at intersections (**FIGURE 2-10**). Most fire departments require all emergency vehicles to come to a full stop when the emergency vehicle operator encounters a stop sign or a red traffic light. If you enter an intersection without stopping, you may not be able to see approaching traffic, and many motorists do not drive defensively. They assume that if there is no stop sign or red light, they can proceed through the intersection without looking for oncoming vehicles.

In addition, many emergency vehicle crashes occur on open stretches of straight roads during daylight hours. The primary cause of these crashes is excessive speed. Most of us have limited experience with

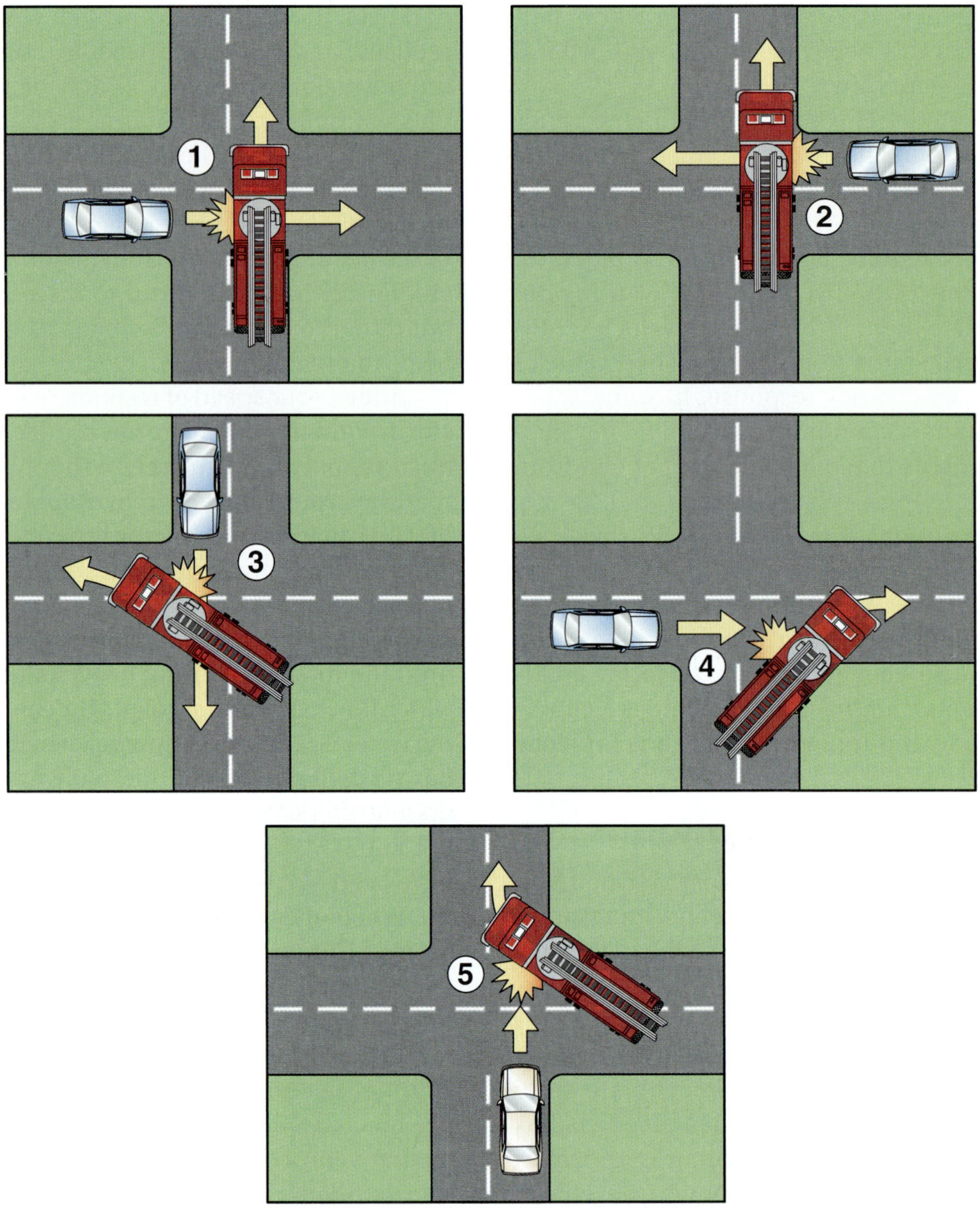

FIGURE 2-10 Every two-lane intersection has five potential crash points, making intersections very dangerous.

© Jones & Bartlett Learning

high-speed driving, so it is important to keep within the limits of your driving skills and within the limits of the vehicle you are operating. Many departments limit the speed during emergency responses to 5 miles per hour (8 kilometers per hour) over the posted speed limit. Multiple studies have shown that little time is saved by driving at high rates of speed.

During an emergency response, be alert for the presence of other emergency vehicles. These vehicles may carry other firefighters responding from their homes or fire, EMS, or law enforcement personnel responding to the same incident. It is difficult to hear the sounds of other apparatus when you are on an emergency response.

Drive with a cushion of safety around you—that is, in such a manner that even if a motorist does the unexpected, you still have enough time and distance to avoid a collision. Remember that the sound of a siren carries a limited distance. Sometimes in an urban environment, motorists may not be able to determine the location of an approaching emergency vehicle because of surrounding buildings, which distort the source of the siren. Similarly, emergency lights have limitations and cannot be depended on to clear a path of travel for you.

TABLE 2-4 Guidelines for Safe Emergency Vehicle Response

1. Drive defensively.
2. Follow agency policies regarding posted speed limits.
3. Always maintain a safe distance. Use the "4-second rule." Stay at least 4 seconds behind another vehicle in the same lane.
4. Maintain an open space or cushion in the lane next to you as an escape route in case the vehicle in front of you stops suddenly.
5. Always assume that other drivers will not hear your siren or see your emergency lights.
6. Select the shortest and least congested route to the scene at the time of dispatch.
7. Visually clear all directions of an intersection before proceeding.
8. Go with the flow of traffic.
9. Watch carefully for bystanders and pedestrians. They may not move out of your way or could move the wrong way.

© Jones & Bartlett Learning

TABLE 2-4 summarizes the guidelines to follow when driving to respond to an emergency call.

The Importance of Vehicle Maintenance

Do not underestimate the importance of proper vehicle maintenance for both fire department vehicles and your private vehicle. It is important to perform regular maintenance of the engine, transmission, and other equipment to ensure dependable starting and running. Keep the brakes and tires in proper condition to ensure dependable stopping. In addition, note that tire tread depth for emergency vehicles should exceed the minimum required for nonemergency vehicles. The suspension system and steering system need to be properly maintained to ensure safe handling of the vehicle. Likewise, the windshield wipers and windshield washers must be kept in good condition to ensure clear vision. Regularly check all headlights, taillights, and turn signals for proper operation. Private vehicles that are used for emergency response should be maintained at a higher level than vehicles that are not subject to the stresses of emergency operation.

Safety at the Incident

At the emergency scene, responders should never charge blindly into action. From the moment that firefighters arrive at an emergency incident scene, their department's SOPs and the structured ICS must guide all of their actions. The officer in command will size up the situation, carefully evaluating the conditions to determine whether the burning building is safe to enter or what needs to be done to make the scene safe for action to take place. Your job is to pay attention to your surroundings and report things that do not match your expectations. For example, observe the time of day, the weather conditions, the people on the scene, and fire and smoke conditions. Wait for instructions and follow directions for the specific tasks to be performed. Firefighters will always work in assigned teams (companies or crews) and be guided by a strategic plan for the incident. Teamwork and disciplined action are essential to provide for the safety of all firefighters and the effective, efficient conduct of operations.

Freelancing, whether done by individual firefighters, groups of firefighters, support personnel, or full companies, is unacceptable. Freelancing cannot be tolerated at any emergency incident. The safety of every person on the scene can be compromised by firefighters who do not work within the system.

For example, a firefighter who enters a burning structure without informing a superior may be trapped by rapidly changing conditions. By the time the firefighter is missed, it may be too late to perform a rescue. Searching for a missing firefighter exposes others to unnecessary risk.

Personnel must not respond to an emergency incident scene unless they have been dispatched or have an assigned duty to respond. Unassigned units and individual personnel arriving on the scene can overload the IC's ability to manage the incident effectively. Individuals who simply show up and find something to do are likely to compromise their own safety and create more problems for the command staff. All personnel must operate within the established system, reporting to a designated supervisor under the direction of the IC.

SAFETY TIP

Teamwork is a hallmark of effective fireground operations. Training and teamwork produce effective, coordinated operations. Unassigned individual efforts and freelancing can disrupt the operations and endanger the lives of both firefighters and civilians.

Teamwork

On the fireground or emergency scene, a firefighting team should always consist of at least two firefighters who work together and are in constant communication with each other. Some departments call this scheme the *buddy system* (**FIGURE 2-11**). In some cases, firefighters work directly with the company officer, and all crew members function as a team. In other situations, two individual firefighters may form a team that is assigned to carry out a specific task. In either case, the company officer must always know where teams are and what they are doing.

Partners or assigned team members should enter together, work together, and leave together. If one member of a team must leave the fire building for any reason, the entire team must leave together, regardless of whether it is a two-person team or an entire crew working as a team.

FIGURE 2-11 A firefighting team should consist of at least two members who work together.

© Jones & Bartlett Learning

Before entering a burning building to perform interior search and rescue or fire-suppression operations, firefighters must be properly equipped with approved and appropriate PPE. Partners should check each other's PPE to ensure it has been donned and is working correctly before they enter a hazardous area. Team members should maintain visual, vocal, or physical contact with one another at all times. At least one member of each team should have a portable radio to maintain contact with the IC or a designated individual in the chain of command who remains outside the hazardous area. Two firefighters must remain outside the hazardous area, properly equipped to respond immediately if the entry team needs to be rescued. This rapid intervention crew/company (also known as a *rapid intervention team*) must be able to communicate with the entry team, either by sight or by radio, and should be ready to provide assistance.

Personnel Accountability

Every fire department must have a **personnel accountability system** to track personnel and assignments on the emergency scene. This system should record the individuals assigned to each company, crew, or entry team; the assignments for each team; and the team's current activities. If any firefighters are reported missing, lost, or injured, the accountability system can identify who is missing and what their last assignment was. Several kinds of accountability systems are acceptable, ranging from paper assignments or display boards to laptop computers and electronic tracking devices.

Accountability systems provide an up-to-date accounting of everyone who is working at the incident and how they are organized. At set intervals, an accountability check, or personnel accountability report, is completed to account for everyone. A **personnel accountability report (PAR)**—an accountability check or roll call taken by a supervisor at an emergency incident that may be conducted as part of the personnel accountability system—is also completed when the operational strategy changes or when a situation occurs that could endanger firefighters. Usually, a company officer reports on the status of each crew. The company officer should always know exactly where each crew is and what it is doing. If a crew splits into two or more teams, the company officer should be in contact with at least one member of each team. A list of personnel and assignments should be readily available at the command post.

All personnel must learn which kind of accountability system their department uses, how to work within it, and how this system works within the ICS. They are responsible for complying with this system and staying in contact with a company officer or assigned supervisor at all times. Most importantly, members of the fire service must always remember that teams must stay together.

Hazard Indicators

Hazardous conditions are any conditions that may directly or indirectly harm first responders in the course of their duties. Direct hazards are hazards that cause immediate harm. Some examples of direct hazards are heat, smoke, unstable structures, unstable surfaces, exposed electrical elements, sharp objects, and hazardous chemicals. Indirect hazards are conditions that could potentially cause harm and include stress, fatigue, and exhaustion. Hazardous conditions may not always be evident by simple observation. For example, smoke and flames may be hidden in a wall and detectable only if searched for. Firefighters train to identify, understand, and properly respond to both visible and hidden hazards. Three observable factors a firefighter should evaluate are the type of construction, type of occupancy, and weather conditions.

The occupancy and use of a building can give clues to potential hazards. Common commercial buildings such as hardware stores and home improvement centers contain huge amounts of highly flammable and hazardous materials. A name such as "Central Plating Company" on a building should cause firefighters to anticipate that hazardous materials could be present. A warning placard on the outside of a building is a more specific hazardous materials indicator. For example, NFPA 704, *Standard System for the Identification of the Hazards of Materials for Emergency Response, 2022 Edition,* describes a system for identifying specific hazards in a building or a container by using a combination of colors and numbers (**FIGURE 2-12**).

Weather is also a factor when evaluating hazards. Although inclement weather can be easy to spot, its effects are not always obvious. For example, it is obvious that there is snow on the ground, but it is easy to forget that snow may add enough weight to a roof to cause it to collapse. Similarly, it is obvious when it is windy, but it is not as obvious that wind can rapidly change a fire's flow path or rapidly increase the intensity of a fire, often with deadly results.

These are just a few examples of observable factors that might indicate a hazard. At every incident, firefighters should look for critical indications of a hazard.

FIGURE 2-12 The NFPA 704 diamond indicates that hazardous materials are present. This diamond indicates that under emergency conditions, the hazardous materials in the building or container are potentially hazardous to health (blue) and could cause significant irritation (1), are flammable (red) under almost all ambient conditions (3), and are normally stable (yellow and 0). The absence of a symbol on the white diamond indicates that there are no special hazards—materials that react violently with water, materials that can oxidize, or gases that can asphyxiate a person—present.

© Jones & Bartlett Learning. Photographed by Glen E. Ellman.

Scene Hazards

Fire service personnel must be aware of their surroundings when performing their assigned tasks at an emergency scene. At an incident, make a safe exit from the fire apparatus and look at the building or situation for safety hazards such as traffic, downed utility wires, and adverse environmental conditions. The introduction of new technologies requires all crew members to be familiar with their run areas and observant of new hazards. This includes the increasingly common photovoltaic power systems (solar power systems) and battery energy storage systems within electrical power grids. An incident on a street or highway must first be secured with proper traffic- and scene-control devices. A variety of traffic incident management techniques may be appropriate. Flares, traffic cones, and barrier tape are all measures that can help keep the public at a safe distance from the scene and divert traffic around the area. Emergency vehicles can be placed to block traffic and

SKILL DRILL 2-3

Establish and Operate Within a Protected Work Area Support Person, NFPA 1010 5.3.1, 5.3.2, 5.3.3

1. Don a high-visibility safety vest in addition to normal PPE.
2. Check for traffic before opening the doors or stepping out of the apparatus.
3. Place traffic cones, flares, emergency scene signage, and other warning devices far enough away from the incident to allow traffic to slow down and be directed away from the work area.
4. Operate as directed within the protected work area, maintaining situational awareness for those who may present hazards to an emergency scene.

© Jones & Bartlett Learning

protect the incident scene. Always operate within established boundaries and protected work areas.

Changing fire conditions also will affect safety. Through situational awareness, firefighters monitor the changing conditions and maintain safety. This should continue even during the overhaul phase and while picking up equipment—watch out for falling debris, smoldering areas of fire, and sharp objects. If a safety officer is not on the scene, another qualified person should be assigned to monitor the atmosphere to ensure it is safe to remove SCBA or enter the area without SCBA. Because the chance of injury increases when you are tired, do not let down your guard even though the main part of the firefighting operation is over.

For the steps to safely establish and work within a protected work area, see **SKILL DRILL 2-3**.

Utilities

Controlling utilities is one of the first tasks that must be accomplished at many working structure fires. Once utilities have been controlled, this information should be communicated to the IC. If additional utility hazards are found while conducting operations, the IC should be notified immediately. Everyone working on the fireground should know the status of the utilities. If control of the utilities cannot be completed, the IC should determine the safest course of action. Most departments have written SOPs that define when the utilities are to be shut off. Although this responsibility is often assigned to a particular company or crew, all firefighters should know how to shut off a building's electrical, gas, and water service.

FIGURE 2-13 An example of a residential gas meter.

© Shadowspeeder/Shutterstock

Because most structures use electricity or gaseous fuel to operate lighting, heating, cooking, and manufacturing equipment, it is important to shut these utilities off as early as possible during fire-suppression activities (**FIGURE 2-13**). The IC should identify which utilities are present during size-up and request that the appropriate utility companies respond to disconnect service to the building. If it is safe to do so, fire service personnel can often shut off these services to the structure before the utility company arrives.

Controlling utilities is particularly important if firefighters need to open walls or ceilings to look for fire in the void spaces, to cut ventilation holes in roofs, or to penetrate through floors. Because these spaces often contain both electrical lines and gas pipes, the danger of electrocution or explosion exists unless these utilities are disconnected.

Electrical, gas, and water utilities at a burning building may be disconnected for any of several reasons. For example, if faulty electrical equipment or a gas appliance caused the fire, shutting off the supply will help alleviate the problem. It also will reduce the risk of injury to firefighters.

Electrical Service

Electricity is present in most structure fires, many outdoor emergencies, and many traffic collisions. Disconnecting electrical service reduces the risk of electrical injury to firefighters. For example, a firefighter using a saw to open a void space could be electrocuted if the tool contacts energized electrical wires. Having the electrical service disconnected can prevent problems such as short circuits and electrical arcing that could result from fire or water damage. In addition, disconnecting electrical service eliminates potential ignition sources that might cause an explosion if leaking gas has accumulated inside the building. The electrical supply must be disconnected at a location outside the area where gas might be present. Interrupting the power from a remote location alleviates the risk of an electrical arc that could cause a gas explosion.

The type of electrical service delivery depends on the utility company providing that service and the age of the system. The most common installation in older areas is a service drop from aboveground utility wires to the electric meter, which is typically mounted on the outside of the building (**FIGURE 2-14**). Some newer developments include underground distribution systems that include underground cables and transformers. Electrical meters may be located inside or outside of a building. There may be separate meters for each tenant or one meter for multiple occupancies.

When a fire occurs in a building, the electrical service should be turned off quickly to reduce the risk of injury or death to firefighters, even if there is no involvement of electrical equipment in the fire. If possible, this shut-off should take place at the main circuit breaker box. Attach a lockout tag to prevent someone from accidentally turning the electrical current back on. Some buildings have an exterior main disconnect switch that shuts off electrical power to the building. These should be identified during preincident planning.

If it is not possible to turn the electricity off at the breaker box or the main disconnect switch, firefighters must notify the electrical utility to send a representative to the fire scene to disconnect the service from a location outside the building. In many cases, the local utility company is automatically notified of any working fire.

FIGURE 2-14 The electrical service often has an exterior meter connected to aboveground utility wires.

© Jones & Bartlett Learning

All electrical equipment should be considered as potentially energized until the power company or a qualified electrical professional confirms that the power is off. To shut off the electrical utilities, follow the steps in **SKILL DRILL 2-4**.

Electrical wires may be energized yet nearly invisible when lying on the ground or when dangling from a pole or a building. This is especially true in conditions of limited visibility, such as darkness, fog, and smoke. Always check carefully for overhead power lines before raising ladders. During any major fire, the electrical power supply to the building should be turned off as part of the fire-ground task called "controlling the utilities."

In addition to structure fires, fire departments are often called to electrical emergencies, such as those involving downed power lines, fires, arcing wires, and transformer fires. Whenever there is a possibility that there are live power lines, have the power disconnected or turn off the power if you are trained and permitted by department policy.

Fire apparatus should be parked outside the area and away from power lines when responding to a call for an electrical emergency. A downed power line should always be considered energized until the power company confirms that it is dead. Secure the area

SKILL DRILL 2-4

Shutting Off Electrical Utilities Support Person, NFPA 1010: 5.3.7

1. Don PPE and enter the personnel accountability system.
2. Acknowledge the assignment.
3. Locate the appropriate disconnect switch or main breaker, and then shut the power off.
4. Attach a lockout tag and lock if required.
5. Notify command that the electricity is shut off.

© Jones & Bartlett Learning. Photographed by Glen E. Ellman.

around the power line and keep the public at a safe distance. Never drive fire apparatus over a downed line or attempt to move it using tools. If a sparking power line causes a brush fire, and it is safe to do so, attempt to contain the fire from a safe distance, but do not use water near the power line.

Fire departments must work with their local electrical utility companies to identify the different types of electrical services and arrangements used in the area and to determine the proper procedures for dealing with each type. Often, a main disconnect switch can be operated to interrupt the power. In general, to turn off an electrical service to a structure, you must first identify the circuit breaker box that supplies the building or part of the building in which you need to disconnect the power. Then you can turn off the main circuit breaker that supplies the whole breaker box. It is important to leave a firefighter at the circuit breaker box or to use a lockout system that tags the box to ensure that the power does not get turned on accidentally. Most utility companies will train firefighters to identify and operate shut-off devices on typical installations. Other utility companies request that the fire department call a company representative to shut off the power. Shutting off large systems that involve high-voltage equipment should be done by a company technician or a trained individual from the premises.

Sometimes a utility company representative needs to be called to interrupt power from a remote location, such as a utility pole. This step may be necessary if the outside wires have been damaged by fire, if firefighters are working with ladders or aerial apparatus, or if an explosion is possible. In urban areas, the electrical utility company can usually dispatch a qualified technician or crew to respond quickly to a fire or emergency incident. In many cases, qualified technicians are dispatched automatically to all working fires. Some utility companies can interrupt the service to an area from a remote location in response to a fire department request.

Gas Service

Many structures use either natural gas or propane for heating and cooking. These two energy sources also have many industrial applications. If a gas line inside a structure becomes compromised during a fire, the escaping gas can add fuel to the fire. The means by which the gas is supplied to the structure must be located to stop the flow. Shutting down gas service eliminates the potential for explosion as a result of damaged, leaking, or ruptured gas piping. For example, a power saw slicing through a gas pipe can cause a rapid release of gas and create the potential for an explosion. A structural collapse can rupture gas lines or create leaks in the stressed piping.

Both natural gas and liquefied petroleum gas (LPG) are used for heating and cooking. Generally, natural gas is delivered through a network of underground pipes. By contrast, LPG is usually delivered by a tank truck and stored in a container on the premises. In some areas, LPG is distributed from one large storage tank through a local network of underground pipes to several customers.

A single valve usually controls the natural gas supply to a building. This valve is generally located outside the building at the entry point of the gas piping; natural gas service has a distinctive piping arrangement. In older buildings, the shut-off valve for a natural gas system may be located in the basement. The shut-off valve for a natural gas system is usually a quarter-turn valve with a locking device so that it can be secured in the off position (**FIGURE 2-15**). When the handle is in line with the pipe, the valve is open; when the handle is at a right angle to the pipe, the valve is closed. A special key, an adjustable wrench, or a spanner wrench can be used to turn the handle. If the gas is supplied by an underground distribution system, the flow can be stopped by closing a quarter-turn valve on the gas meter. If the gas is supplied from an outside LPG storage cylinder, closing the cylinder valve will stop the flow. After the gas service has been shut off, attach a lockout or shut-off tag or lock to ensure it is not turned back on. To shut off gas utilities, follow the steps in **SKILL DRILL 2-5**.

FIGURE 2-15 The natural gas shut-off valve is generally located outside the building at the entry point of the gas piping. In older buildings, the shut-off valve for a natural gas system may be located in the basement.

© lucag_g/Shutterstock

The shut-off valve for an LPG (propane) system is usually located at the storage tank (**FIGURE 2-16**). The more common type of LPG valve, however, has a distinctive handle that indicates the proper rotation direction to open or close the valve. To shut off the flow of gas, rotate the handle to the fully closed position.

A gas valve that has been shut off must not be reopened until the system piping has been inspected by a qualified person. Air must be purged from the system, and any pilot lights must be reignited to prevent gas leaks.

Water Service

If a serious water leak has occurred inside the building, shutting off the water supply may help to minimize additional water damage to the structure and contents.

Water service to a building can usually be shut off by closing one valve at the entry point. Many communities permit the water service to be turned off at the connection between the utility pipes and the building's system. This underground valve, which is often accessed through a curb box, is located outside the building and can be operated with a special wrench or key. In most cases, another valve is found inside the building, usually in the basement (if there is one), where the water line enters. In warmer climates, water supply valves are sometimes located above ground.

Lifting and Moving

Lifting and moving objects are part of a firefighter's daily duties. Do not try to move something that is too heavy alone—ask for help. Never bend at the waist to lift an object; instead, bend at the knees, and use your legs to lift the weight. If you must move objects over a long distance, use equipment such as handcarts, hand trucks, and wheelbarrows.

In addition to moving inanimate objects, firefighters must often move sick or injured victims. Do not move an injured person, unless their life is in danger, until the appropriate medical personnel are on scene to stabilize the victim. Discuss and evaluate the options before moving a victim, and then proceed carefully. If necessary, request help. Never be afraid to call for

SKILL DRILL 2-5

Shutting Off Gas Utilities Support Person, NFPA 1010: 5.3.7

1. Don PPE and enter the personnel accountability system.
2. Acknowledge the assignment.
3. Locate the exterior gas shut-off valve and close it.
4. Attach a shut-off or lockout tag and lock if required.
5. Notify command that the gas is shut off.

© Jones & Bartlett Learning. Photographed by Glen E. Ellman.

FIGURE 2-16 The shut-off valve for an LPG system is usually located at the storage tank.

© Jones & Bartlett Learning. Photographed by Amanda Mitchell.

additional resources, such as another engine or truck company, to assist in lifting and moving a heavy victim.

Adverse Weather Conditions

In adverse weather conditions, fire personnel must dress appropriately. Firefighter PPE, consisting of a structural firefighting protective coat, pants, boots, and helmet, can keep you warm and dry in rain, snow, or ice. Firefighting gloves and knit caps also help retain body heat and keep you warm. If conditions are icy, make smaller movements, watch your step, and keep your balance. During the hot summer months, it is difficult to stay cool in PPE. Many departments have SOPs covering appropriate PPE to use in summer months. Check your department's SOPs for guidance on acceptable PPE use in adverse weather.

Rehabilitation

Never be afraid or embarrassed to admit you need a break when on the emergency scene. Typically, breaks are in the form of recycling and rehabilitation. Recycling is brief and consists of taking a few minutes to rest, hydrate, and change the SCBA bottle before getting back to work.

Rehabilitation is a systematic process that provides longer periods of rest and recovery for emergency workers during an incident. It is usually conducted in

FIGURE 2-17 In the rehabilitation area, firefighters can rest and rehydrate.

© Jones & Bartlett Learning. Photographed by Glen E. Ellman.

a designated area away from the hazards of the emergency scene. The rehabilitation area, or "rehab," is usually staffed by fire and EMS personnel.

While in rehabilitation, personnel should take advantage of the opportunity to rest, rehydrate, have their vital signs checked, and receive treatment for minor injuries (**FIGURE 2-17**). Rehabilitation gives firefighters the chance to cool off in hot weather and warm up in cold weather. Rehabilitation time can also be used to replace SCBA cylinders, obtain new batteries for portable radios, and make repairs or adjustments to tools or equipment. Firefighting teams can discuss recently completed assignments and plan their next work cycle. When a crew is released from rehabilitation, its members should be rested, refreshed, and ready for another work cycle. If members of the crew are too exhausted or unable to return to work or have not been medically cleared to return to work, they should be replaced and released from the incident. More information on firefighter rehabilitation is presented in Chapter 20, *Firefighter Rehabilitation.*

Violence

The safety of fire service personnel can be jeopardized by violence. Emergency responders are sometimes dispatched to calls involving domestic disputes, active shooters, injuries from an assault, or other violent scenes. In such incidents, the staging area and the fire apparatus should be located at some distance from the scene until the police arrive, investigate, and declare the scene safe. Only then should firefighters proceed to enter the emergency scene. There have been violent incidents where firefighters have been targeted, and in some cases, they have been mistaken for intruders. This increases the need for situational awareness. A firefighter's personal safety should always be paramount. If there is any threat to your personal safety, retreat as quickly as possible from the emergency scene to a safe distance and request the police to secure the scene. Do not become a victim.

If you are confronted with a potentially violent situation, do not respond violently. Remain calm, speak quietly, attempt to gain the person's trust, and call for police assistance. You might consider taking additional classes to increase your understanding and develop appropriate skills for these situations.

Safety at the Fire Station

The fire station is just as much a workplace as the fireground. Indeed, firefighters spend much of their time during a shift at their fire station. The length of each shift varies, from 12 to 24, 48, or 72 hours and, in some cases, to a 96-hour shift. Some may be longer. The need for safety around the fire station is as important as it is on the fireground or at other emergency scenes. While in the fire station, be careful when working with power tools and equipment, ladders, electrical appliances, pressurized cylinders, and hot surfaces. Practice using tools and equipment properly and safely before attempting to use them at an actual emergency incident. Follow the proper procedures and safety precautions both in training and at an incident scene.

SAFETY TIP

Remember these guidelines to stay safe—both on and off the job:

- You are personally responsible for safety. Keep yourself safe. Keep your teammates safe. Keep citizens—your customers—safe.
- Work as a team. The safety of the entire firefighting unit depends on the efforts of each member. Become a dependable member of the team.
- Follow orders. Freelancing can endanger other firefighters as well as yourself.
- Think! Before you act, think about what you are doing. Many people are depending on you.

Equipment should always be in excellent condition and ready for use. Proper maintenance of tools includes sharpening, lubricating, and cleaning each tool. All firefighters should be able to do basic repairs such as changing a saw blade or a hand-light battery. Practice using and maintaining tools and equipment at the fire station until you can perform these tasks quickly and safely. Remember, injuries that occur at the firehouse can be just as devastating as those that occur at an emergency incident scene.

Safety Outside Your Workplace

Continue to follow safe practices when you are off duty. An accident or injury, regardless of where it happens, can end your career in the fire service. For example, if you are using a ladder while off duty, follow the same safety practices that you would use while on duty. Keep your seat belt fastened in your personal vehicle, just as you are required to do when you are on duty.

CASE STUDY

You Are the Support Person CONCLUSION

The firefighters in your crew are at a large fire in an apartment building that started on the third floor, extended into the attic, and is now venting through the roof. Your captain gave the IC the accountability tags, and a crew was assigned to confine the fire on the third floor. The crew members gather their gear, and as they approach the apartment building, you consider the ways firefighters can accomplish their mission while avoiding preventable injury or death.

1. **What are the leading causes of firefighter injuries and deaths?**

   Answer: The leading causes of firefighter injuries and deaths are stress, overexertion, and medical issues. Before the incident, proper nutrition, hydration, physical conditioning, sleep, and stress management will reduce these risks. During the incident, rehabilitation and medical monitoring can further reduce the risk of firefighter injury and death. In addition, rehabilitation allows a fire crew to ensure that they are working together and helps reduce freelancing. After the incident, decontamination and, where possible, a return to proper nutrition, hydration, physical conditioning, sleep, and stress management can further reduce these risks.

2. **What are the potential hazards during this incident?**

   Answer: Potential hazards found on the fireground include the changing conditions of the fire itself as well as the effect of the fire on the building. In addition, utilities, poorly maintained equipment, poorly trained and/or freelancing firefighters, extreme environmental conditions, uneven terrain, hazardous materials, and possibly roadway traffic near the operational area all pose significant fire-ground hazards.

3. **What safety measures should be taken during this incident?**

   Answer: Hazards that can be safely mitigated should be immediately addressed. For example, a ladder with a broken rung should be immediately removed from service and replaced with a functional ladder. Hazards that cannot be mitigated should be marked off and brought to the attention of the next officer in your chain of command or to the incident safety officer. For example, traffic cones and banner tape should be used to mark off an area that has live broken electrical lines.

WRAP-UP

SUMMARY

- Firefighting is inherently dangerous. The goal of the fire service is to reduce preventable deaths and injuries to firefighters by reducing unnecessary risks. This requires the dedicated efforts of every member of the fire community working together.
- Adequate training, equipment, physical fitness, and a healthy lifestyle are all essential for keeping firefighters safe.
- Burnout and compassion fatigue are common, stress-induced conditions in the firefighting community. It is critical to recognize signs of cumulative stress in yourself and others in the community.
- It is crucial to know and follow all SOPs and SOGs when riding the fire apparatus, including donning PPE either prior to mounting the apparatus or after you've dismounted; wearing your seat belt while the apparatus is in motion; and properly mounting, stowing, or securing all equipment aboard the apparatus.
- It is crucial to know and follow all SOPs and SOGs when operating a fire apparatus or personal vehicle. Maintaining safe driving practices on and off the job is required of all fire service personnel.
- Maintaining safety at an emergency incident scene requires strict adherence to department SOPs and SOGs. Wait for instructions and follow directions for the specific tasks to be performed. Teamwork and disciplined action are essential to provide for the safety of firefighters and the effective, efficient conduct of operations.
- Maintaining safety at an emergency incident scene requires controlling traffic in the vicinity of the scene. Follow departmental SOPs to close streets quickly and block access to areas where operations are being conducted.
- Personal accountability systems are used to track personnel and assignments on the emergency scene and perform intermittent personal accountability reports to determine if anyone is missing or injured.
- It is crucial to control utilities, including gas, electricity, and water service, on a structural fire scene. Everyone working the scene must know the status of these utilities. If control of the utilities cannot be completed, the IC should determine the safest course of action.

KEY TERMS

Code of Federal Regulations (CFR) A collection of permanent rules published in the *Federal Register* by the executive departments and agencies of the U.S. federal government. Its 50 titles represent broad areas of interest that are governed by federal regulation. Each volume of the CFR is updated annually.

compassion fatigue A disorder characterized by a gradual lessening of compassion over time

critical incident stress debriefing (CISD) A postincident meeting designed to assist rescue personnel in dealing with psychological trauma resulting from an emergency. (NFPA 1006)

critical incident stress management (CISM) A program designed to reduce acute and chronic effects of stress related to job functions. (NFPA 450)

employee assistance programs (EAPs) An employer-sponsored service designed for personal or family problems, including mental health, substance abuse, various addictions, marital problems, parenting problems, emotional problems, or financial or legal concerns. (NFPA 450)

freelancing The dangerous practice of acting independently of command instructions.

Occupational Safety and Health Administration (OSHA) The U.S. federal agency that regulates worker safety and, in some cases, responder safety. OSHA is part of the U.S. Department of Labor.

personnel accountability report (PAR) Periodic reports verifying the status of responders assigned to an incident or planned event. (NFPA 1026)

personnel accountability system A system that readily identifies both the location and function of all members operating at an incident scene. (NFPA 1550)

posttraumatic stress disorder (PTSD) A behavioral disorder that develops after a person has experienced a critical incident; characterized by reexperiencing

KEY TERMS CONTINUED

the event and overresponding to stimuli that recall the event; symptoms include depression, startle reactions, flashback phenomena, and dissociative episodes (e.g., amnesia of the event).

response Immediate and ongoing activities, tasks, programs, and systems to manage the effects of an incident that threaten life, property, operations, or the environment. (NFPA 1600)

structural firefighting protective clothing All of the clothing elements of the structural firefighting protective ensemble.

REVIEW QUESTIONS

1. What should fire service personnel's approach to safety be?
2. What is the top cause of firefighter deaths by cause of injury?
3. What is the cause of most firefighter injuries and deaths?
4. What should a firefighter incorporate into their daily routine?
5. A firefighter injured during training should not return to duty until ____________.
6. What actions should a firefighter perform during an inspection of their equipment to be ready for assignment?
7. What is the highest priority when responding in an emergency apparatus?
8. What should firefighters be able to do to promote safety at the fire station?
9. What should firefighters do to stay safe while off duty?

DISCUSSION QUESTIONS

1. Does a culture of safety mean a firefighter should never do anything dangerous?
2. How can a firefighter be "safe" in an inherently dangerous profession?
3. How can fire service personnel reduce their risk of dying or being injured as a result of stress and overexertion?
4. What steps can be taken to increase safety and effectiveness overall in the fire service?
5. How does working in teams increase firefighter safety and effectiveness?
6. How does freelancing affect the safety and effectiveness of emergency operations?

REFERENCES

American Cancer Society. 2020, June 9. "Alcohol Use and Cancer." Accessed August 23, 2024. www.cancer.org/cancer/risk-prevention/diet-physical-activity/alcohol-use-and-cancer.html.

Campbell, Richard B., and Shelly Hall. 2023. "United States Firefighter Injuries." National Fire Protection Association. www.nfpa.org/education-and-research/research/nfpa-research/fire-statistical-reports/firefighter-injuries-in-the-united-states.

Campbell, Richard, and Jay T. Petrillo. 2023. "Fatal Firefighter Injuries in the United States in 2022." National Fire Protection Association. www.nfpa.org/-/media/files/news-and-research/fire-statistics-and-reports/emergency-responders/osfff.pdf.

Centers for Disease Control. 2023, April 20. "Firefighter Resources, Cancer and Other Illnesses." Accessed August 2, 2024. www.cdc.gov/niosh/firefighters/health.html.

Firefighter Behavioral Health Alliance. 2017. "Firefighter Behavioral Health Alliance." Accessed October 30, 2023. www.ffbha.org/.

Firefighter Cancer Support Network. 2013. "Taking Action Against Cancer in the Fire Service." Accessed November 15, 2023. https://firefightercancersupport.org/wp-content/uploads/2017/11/taking-action-against-cancer-in-the-fire-service-pdf.pdf.

Firefighter Cancer Support Network. 2022. "Firefighter Cancer Support Network: Together We Can." Accessed October 30, 2023. https://firefightercancersupport.org/.

International Association of Firefighters. 2019, March 27. "Taking Action against Occupational Cancer." Accessed October 30, 2023. www.iaff.org/news/taking-action-against-occupational-cancer/.

National Fallen Firefighters Foundation. 2022. "Everyone Goes Home—Firefighter Life Safety Initiatives." Accessed October 30, 2023. www.everyonegoeshome.com/.

National Fire Protection Association. 2017. *NFPA 1451, Standard for a Fire and Emergency Service Vehicle Operations Training Program.* 2018 Edition. Quincy, MA: National Fire Protection Association.

National Fire Protection Association. 2018. *NFPA 1600, Standard on Continuity, Emergency, and Crisis Management.* 2019 Edition. Quincy, MA: National Fire Protection Association.

National Fire Protection Association. 2019. *NFPA 1951, Standard on Protective Ensembles for Technical Rescue Incidents.* 2020 Edition. Quincy, MA: National Fire Protection Association.

National Fire Protection Association. 2020. *NFPA 450, Emergency Medical Services and Systems.* 2021 Edition. Quincy, MA: National Fire Protection Association.

National Fire Protection Association. 2020. *NFPA 704: Standard System for the Identification of the Hazards of Materials for Emergency Respons*e. 2022 Edition. Quincy, MA: National Fire Protection Association.

National Fire Protection Association. 2020. *NFPA 1006, Standard for Technical Rescue Personnel Professional Qualifications.* 2021 Edition. Quincy, MA: National Fire Protection Association.

National Fire Protection Association. 2020. *NFPA 1250. Recommended Practice in Fire and Emergency Service Organization Risk Management.* 2020 Edition. Quincy, MA: National Fire Protection Association.

National Fire Protection Association. 2021. *NFPA 1581, Standard on Fire Department Infection Control Program.* 2022 Edition. Quincy, MA: National Fire Protection Association.

National Fire Protection Association. 2021. *NFPA 1582, Standard on Comprehensive Occupational Medical Program for Fire Departments.* 2022 Edition. Quincy, MA: National Fire Protection Association.

National Fire Protection Association. 2023. *NFPA 1010, Standard on Professional Qualifications for Firefighters.* 2024 Edition. Quincy, MA: National Fire Protection Association.

National Fire Protection Association. 2023. *NFPA 1550, Standard for Emergency Responder Health and Safety.* 2024 Edition. Quincy, MA: National Fire Protection Association.

National Fire Protection Association. 2023. *NFPA 1026, Standard for Incident Management Personnel Professional Qualifications.* 2024 Edition. Quincy, MA: National Fire Protection Association.

National Volunteer Fire Council. 2022. *Share the Load.* Accessed October 30, 2023. www.nvfc.org/programs/share-the-load-program/.

Chapter Opener: © Eric Scruggs

CHAPTER

3

Support Person

Personal Protective Equipment

KNOWLEDGE OBJECTIVES

After studying this chapter, you will be able to:

- List protective equipment available for use when operating at an emergency scene.
- List the conditions that require respiratory protection.
- List the respiratory hazards posed by smoke and fire.
- Describe the types of breathing apparatus.
- Describe the differences between open- and closed-circuit breathing apparatus.
- Describe the limitations of self-contained breathing apparatus (SCBA).
- Describe the physical and psychological limitations of an SCBA user.
- List and describe the major components of SCBA cylinders.
- Explain the procedures for returning an SCBA air cylinder to service.
- Explain the procedures for refilling SCBA air cylinders.

SKILLS OBJECTIVES

After studying this chapter, you will be able to perform the following skills:

- Demonstrate the ability to don and doff a protective ensemble.
- Demonstrate how to properly wear a protective ensemble.
- Perform field reduction of contaminants from a protective ensemble.
- Demonstrate how to clean and inspect an SCBA.
- Properly exchange/replace an SCBA air cylinder from a rack or apparatus.
- Replace an SCBA air cylinder on a firefighter.
- Perform a visible and operational inspection of an SCBA.
- Replace an SCBA air cylinder.
- Refill an SCBA air cylinder from a cascade or compressor system.
- Clean an SCBA.

ADDITIONAL NFPA STANDARDS

- **NFPA 853**, *Standard for the Installation of Stationary Fuel Cell Power Systems, 2020 Edition*
- **NFPA 1550**, *Standard for Emergency Responder Health and Safety, 2024 Edition*
- **NFPA 1700**, *Guide for Structural Fire Fighting, 2021 Edition*
- **NFPA 1851**, *Standard on Selection, Care, and Maintenance of Protective Ensembles for Structural Fire Fighting and Proximity Fire Fighting, 2020 Edition*
- **NFPA 1852**, *Standard on Selection, Care, and Maintenance of Open-Circuit Self-Contained Breathing Apparatus (SCBA), 2019 Edition*
- **NFPA 1970**, *Standard on Protective Ensembles for Structural and Proximity Firefighting, Work Apparel and Open-Circuit Self-Contained Breathing Apparatus (SCBA) for Emergency Services, and Personal Alert Safety Systems (PASS), 2025 Edition*
- **NFPA 1977**, *Standard on Protective Clothing and Equipment for Wildland Fire Fighting and Urban Interface Fire Fighting, 2022 Edition*
- **NFPA 1984**, *Standard on Respirators for Wildland Fire Fighting Operations and Wildland Urban Interface Operations, 2022 Edition*
- **NFPA 1990**, *Standard for Protective Ensembles for Hazardous Materials and CBRN Operations, 2022 Edition*

CASE STUDY

You Are the Support Person

One cool fall evening, you are dispatched for a report of a smoke odor in the area. These calls are normal for this time of year as residents start to use their fireplaces for the first time. As the engine drives slowly through the neighborhood, you notice the distinctive smell of a working house fire rather than that of firewood. When you turn the corner, you see smoke drifting from the basement window of a house. Your captain completes a quick size-up and tells two firefighters to pull a cross-lay and attack the fire.

1. Which personal protective equipment is needed for this situation?
2. Once the fire is over, what is the proper way to care for the protective equipment?

Introduction

Two safety components used by fire service personnel require special consideration: personal protective equipment and self-contained breathing apparatus. **Personal protective equipment (PPE)** is the broad term that includes clothing and equipment worn to protect the wearer from the hazards they encounter as they do their job. **Self-contained breathing apparatus (SCBA)** is a component of PPE and is a personal respirator worn by a firefighter that provides an air supply to the firefighter. To remain safe and do your job effectively, you need to understand the purposes for which this equipment is made, what it can do, and what it cannot do. Exceeding the protection offered by PPE can result in severe injuries and death.

Personal Protective Equipment

PPE is an essential component of a firefighter's safety system. It enables a person to survive under conditions that might otherwise result in death or serious injury. Different PPE ensembles are designed for specific hazardous conditions, such as structural firefighting, wildland firefighting, airport rescue and firefighting, hazardous materials operations, and emergency medical operations. The various PPE ensembles provide specific protections, so an understanding of their designs, applications, and limitations is critical. It is important for fire service personnel to understand what protection PPE can provide so that you can adequately choose which is best for each situation.

The complete set of clothing and equipment that a firefighter wears when attacking a structure fire (**structural firefighting**) is referred to as the **structural firefighting protective ensemble**. This ensemble consists of **structural firefighting protective clothing**—personal protective clothing that provides full-body coverage that covers every inch of the body and provides protection from heat and fire, keeps water away from the body, and helps reduce injuries from cuts or falls—SCBA for respiratory protection, and a **personal alert safety system (PASS)** device, which is an electronic device that emits a loud, audible signal when a firefighter becomes trapped or injured (**FIGURE 3-1**). Some departments have their firefighters carry additional equipment, such as hand lights, radios, and so forth, to provide more complete protection.

To be effective, the entire ensemble must be worn whenever potential exposure to those hazards exists. Structural firefighting protective ensembles allow firefighters to enter burning buildings and work in areas with elevated temperatures and concentrations of

FIGURE 3-1 The structural firefighting protective ensemble provides protection from multiple hazards.

Courtesy of Central County Fire & Rescue.

FIGURE 3-2 The structural firefighting ensemble consists of a helmet, protective hood, protective coat, protective pants, boots, gloves, SCBA, and a PASS device.

© Jones & Bartlett Learning. Photographed by Glen E. Ellman.

toxic gases. Without PPE, firefighters would be unable to conduct search and rescue operations or perform fire-suppression activities. The structural firefighting ensemble consists of the following (**FIGURE 3-2**):

- Personal protective clothing
  - Helmet
  - Hood
  - Coat
  - Trousers (pants or coveralls)
  - Boots
  - Gloves
- SCBA
- PASS device

All of these elements must be correctly worn together to provide the necessary level of protection.

Personal Protective Clothing

Personal protective clothing is the first layer of PPE for firefighters. Without personal protective clothing, firefighters would not be able to enter a burning building to conduct an interior attack or rescue people from toxic environments.

Work Uniforms

The lowest level of personal protective clothing is normal street clothing or work uniforms. Firefighters who are not responding to a call, police officers, and emergency medical services (EMS) providers typically wear this as a part of their daily uniform. In fact, at this level, the personal protective clothing is the PPE system because personnel wearing these uniforms do not wear devices to protect their airways or alert others if they are injured, disabled, or killed.

Certain synthetic fabrics, such as nylon and polyester, melt at relatively low temperatures, even when worn under a complete personal protective ensemble. This can cause severe burns because as the material melts, it sticks to the skin, and when that material is removed, it removes layers of skin. Therefore, clothing containing nylon or polyester, even if these materials are blended with natural fibers, should not be worn in a firefighting environment. Clothing made of natural fibers, such as cotton or wool, may char or burn if exposed to flames, but it does not melt. When worn under a protective ensemble, these materials will not be exposed to flame and are generally safer to wear. NFPA 1970, *Standard on Protective Ensembles for Structural and Proximity Firefighting, Work Apparel and Open-Circuit Self-Contained Breathing Apparatus (SCBA) for Emergency Services, and Personal Alert Safety Systems (PASS), 2025 Edition*, defines criteria for selecting appropriate fabrics for work uniforms.

SAFETY TIP

Do not mix and match different brands and styles of personal protective clothing. Some styles are not compatible with others and may leave gaps that will expose the wearer.

Protective Clothing for Structural Firefighting

The next level of protection is provided by structural firefighting protective clothing, which is personal protective clothing that provides full-body coverage to protect from heat and fire, keep water away from the body, and help reduce injuries from cuts or falls. Structural firefighting protective clothing is often referred to as **turnout gear** or **bunker gear**.

TIP

Special synthetic fibers such as Nomex and polybenzimidazole (PBI), which are used in structural firefighting protective clothing, have excellent resistance to high temperatures.

SAFETY TIP

The structural firefighting protective ensemble protects the wearer from the wide range of interior and exterior fires to which a fire department responds. It is not designed to protect personnel against the hazards encountered when responding to hazardous materials, water rescue, high-angle rescue, or wildland fire incidents. Instead, each of these specialized functions requires a specific set of PPE. If your department performs specialized operations, you need the appropriate PPE for these activities.

A structural firefighting protective ensemble provides full-body coverage and offers the following:

- Provides thermal protection
- Repels water
- Provides impact protection
- Protects against cuts and abrasions
- Furnishes padding against injury
- Allows firefighters to be more easily seen by others in smoky conditions

For example, a moisture barrier between different layers of PPE keeps liquids and vapors, such as hot water or steam, from reaching the skin. The knees in the pants may be reinforced with pads for greater protection when crawling. Reflective trim adds visibility in dark or smoky environments.

Each item of protective clothing is designed to overlap from one item to the next. For example, boots overlap the pants, coats overlap the pants, and coats overlap the gloves. This ensures that the entire body is protected at all times. It is vitally important that the PPE a firefighter wears is sized appropriately and is not too tight or too loose.

There are a number of different manufacturers and styles of structural firefighting protective ensembles. All protective clothing for firefighters must be manufactured according to exacting standards. The requirements for protective clothing for firefighters are outlined in two standards:

- NFPA 1970, *Standard on Protective Ensembles for Structural and Proximity Firefighting, Work Apparel and Open-Circuit Self-Contained Breathing Apparatus (SCBA) for Emergency Services, and Personal Alert Safety Systems (PASS), 2025 Edition*
- NFPA 1977, *Standard on Protective Clothing and Equipment for Wildland Fire Fighting and Urban Interface Fire Fighting, 2022 Edition*

According to NFPA 1970, each item must have a permanent label verifying that the particular item meets the requirements of a certification organization (**FIGURE 3-3**). Labels also include the appropriate term for the piece of PPE and manufacturer information, including date of manufacture, model name or number, size or size range, and principal material of construction. Usage limitations, as well as cleaning and maintenance instructions, should also be provided.

TIP

Most manufacturers measure firefighters for the appropriate-sized coats and pants. It is important to be sure that PPE is adequately sized for the firefighter to whom it is assigned.

Structural firefighting PPE will protect firefighters under limited conditions. The longer a firefighter remains in an environment with elevated temperatures, the more heat energy the clothing absorbs. After a certain period of time in a high-temperature environment, the gear will start to transfer more heat through the coat, and the wearer will start to feel this stored heat energy transmitted to their body. This can happen quickly in a fire situation. Once the firefighter starts to feel this heat transfer take place and it becomes uncomfortable, they should start planning to exit the environment.

Helmet

The **structural firefighting protective helmet** provides protection to the firefighter's head from falling debris, and the attached face shield provides eye protection (**FIGURE 3-4**). Helmets are manufactured in several designs and shapes using a variety of materials. Every design must meet the requirements specified in NFPA 1970.

The components of a fire helmet are as follows:

- Hard outer shell: This must be lined with energy-absorbing material and have a suspension system to provide impact protection against falling objects. The outer shell also repels water, protects against steam, and creates a thermal barrier against heat and cold. Its shape helps to deflect water away from the head and neck.
- Face and eye protection: The face shield is attached to the helmet and provides face and eye protection. Some helmets also have goggles attached to provide additional eye protection. The face shield (and goggles) is used when an SCBA is not needed or when the SCBA face piece is not in place.

Garment Safety Label

⚠ DANGER

Please read to confirm your understanding of these warnings and instructions. **1234**
Serious injury or death can result from a failure to heed these warnings.

This garment should be worn for FIREFIGHTING DUTIES ONLY.
THIS GARMENT CANNOT AND DOES NOT PROVIDE PROTECTION AGAINST CBRN TERRORISM AGENTS.

Before wearing this garment, please read to confirm your understanding of the *Instruction, Safety and Training Guide for Users* accompanying this garment.

The Guide for Users provides:
- Serious safety information and protective clothing restrictions.
- Proper sizing.
- Measures for dressing and removing protective clothing.
- Cleaning, decontamination, inspection and storing directives.
- NFPA 1500-compatible usage.
- Useful information limitations and procedures for retiring garments.

Only those who are properly trained in firefighting techniques should wear this garment, and the wearer should have knowledge of the proper selection, fit, use, care and limitations of protective clothing and equipment.
- PLEASE BE AWARE that this garment provides only limited protection against heat and flames.
- The wearer should minimize exposure to heat. Burns without warning may occur, even without receiving damage to garment. Avoid contact with hot objects.
- Burns to skin happen when skin reaches a temperature of 118°F. And fires can burn at temperatures up to 2000°F.
- Protection can be diminished by even a small amount of dampness or moisture and/or compression in your garment.
- Exertion in hot conditions may result in heat fatigue or poor decision-making. If you feel dizzy, dehydrated, unfocused, or short of breath, PLEASE find a safe area, remove this garment, and find medical attention.
- If this clothing is soiled or damaged in any way, DO NOT USE. Soiled or damaged garments will NOT offer the desired protection. ALWAYS follow manufacturer's cleaning instructions.
- This garment has limited useful life. Inspect this garment regularly and discard when appropriate, paying full attention to the *Guide for Users*. See also NFPA 1700.

DO NOT REMOVE OR WRITE ON THIS LABEL! REV. 4.0 12/31

Garment Cleaning Label

Call immediately with any questions. **1234**
J&B
1-800-555-1212

CLEANING AND STORAGE INSTRUCTIONS
- Use the *Guide for Users* for cleaning, inspection, storing, and altering instructions.
- Never use chlorine bleach or the protection used in the formation of this garment will be severely compromised.
- For coats, remove DRD and launder using mild detergent and warm water by hand only.
- Fasten all hooks and D-rings, turn garment inside out, and launder in a laundry bag, machine washing in warm water, with mild detergent and only when necessary, liquid chlorine bleach. Rinse twice in cool water but DO NOT USE FABRIC SOFTENERS.
- DO NOT DRY CLEAN.
- Hang garment to dry, away from sunlight and/or fluorescent light, and store similarly.
- THIS PROTECTIVE FIRE FIGHTING GARMENT MEETS THE REQUIREMENTS OF NFPA 1971, 2018 EDITION. PROTECTIVE GARMENT FOR STRUCTURAL FIRE FIGHTING IN ACCORDANCE WITH NFPA 1971-2018.
- When worn with the inner liner and outer shell assembled together, this garment meets the PPE criteria of US Dept. of Labor OSHA Bloodborne Pathogens Standard, Title 29 CFR, Part 1910, 1030, and CAL-OSHA Standard Title 3 Section 3406.

Rev. 4.0 12/31 **DO NOT REMOVE OR WRITE ON THIS LABEL**

Garment Information Label

TECHSQUARE MOISTURE BARRIER (TEPR)
GLOSS 4G GUDERE F-92 (G) THERM.LINER
NOMEX F-92 QUILT
REQ:091772
MFG DATE:10/15/2018
CUT:01202012
MODEL:SRDM
LINER:0120SRDM
SIZE:051711T

Garment Liner Attachment Safety Label

⚠ WARNING

The inner liner on its own WILL NOT protect against heat, flame, chemical or biological hazards. DO NOT wear the inner liner without the SAME SIZE AND MODEL outer shell, as identified on labels located on each detachable component. Please assemble and wear together ALL of the following items in order to avoid the risk of injury or death: 1. protective coat and pant with outer shell, attached inner liner and DRD installed in coat 2. gloves 3. boots 4. helmet with eye protection 5. protective hood 6. SCBA 7. PASS device

ALWAYS confirm that all ensemble layers have the suitable overlap and all layers fit with satisfactory slackness. A tight fit will lower insulation protection and will limit mobility.

I

MADE IN THE U.S.A.
DO NOT REMOVE OR WRITE ON THIS LABEL!

FI# 0120
REV. 4.0 12/31

FIGURE 3-3 Mandatory labels provide important information about each item of protective clothing.

© Jones & Bartlett Learning

- Chin strap: This strap maintains the helmet in the proper position and helps keep the helmet on the head during an impact.
- Inner liner: The inner layer provides added thermal protection and protects the wearer's neck. These liners have ear tabs that the firefighter pulls down before entering a burning building to provide maximum protection.
- Adjustable inner suspension system: The suspension system holds the shell away from the head and cushions the head against impacts. To ensure proper fit, put the helmet on and adjust it with the protective hood in place. Then make further adjustments as needed.

In many departments, helmet shells are color-coded according to the firefighter's rank and function. Some fire departments include a shield or badge on the front of the helmet that identifies the firefighter's rank and company. Each department is unique, but it is fairly standard for chief officers to wear a white helmet. Bright, fluorescent reflective materials or decals are applied to helmets to make personnel more visible in all

FIGURE 3-4 A helmet is constructed with multiple layers and components.

© Jones & Bartlett Learning. Photographed by Glen E. Ellman.

FIGURE 3-5 A protective hood.

© Jones & Bartlett Learning. Photographed by Glen E. Ellman.

types of lighting conditions. Some decals are luminescent (glow in the dark).

Hood

Although the helmet's ear tabs cover the ears and neck, this area is still at risk for burns when the head is turned or the neck is flexed. A **structural firefighting protective hood** is constructed of a flame-resistant material such as Nomex, PBI, or carbon fiber, and it covers the whole head and neck (except for that part of the face that would be protected by the SCBA face piece) to provide additional thermal protection for these areas (**FIGURE 3-5**). The lower part of the protective hood, which is called the **bib**, drapes down inside the protective coat. Because of the danger from the contaminants in smoke itself, some protective hoods provide particulate protection to reduce the amount of contaminants that reach the skin. Protective hoods are available in various styles, lengths, materials, and colors. One type is a particulate barrier hood, which filters out many of the contaminants firefighters encounter to reduce their exposure to carcinogens. Per NFPA 1970, a particulate hood must block a minimum of 90 percent of particulates ranging in size from 0.1 to 1.0 microns.

FIGURE 3-6 A structural firefighting protective coat.

© Jones & Bartlett Learning. Photographed by Glen E. Ellman.

Coat

The **structural firefighting protective coat**, sometimes called a **turnout coat** or **bunker coat**, protects the arms and body (**FIGURE 3-6**). Structural firefighting protective coats consist of three layers: the outer shell, the moisture barrier, and the thermal barrier.

The outer layer or shell is constructed of a sturdy, flame-resistant material, such as Nomex, Kevlar, or PBI. Fluorescent reflective material applied to the outer shell makes the firefighter more visible in smoky conditions and at night. When light-colored fabrics are used for protective coats, it makes it easier to identify contaminants, such as hydrocarbons, blood, and body fluids on the coat.

The second layer of the protective coat is the moisture barrier, which usually consists of a flexible membrane attached to the thermal barrier material. The moisture barrier helps prevent the transfer of water, steam, and other fluids to the skin. It is critical because water applied to a fire generates large amounts of superheated steam, which can engulf firefighters and burn unprotected skin.

The third layer of the protective coat creates a thermal barrier, which is made of a multilayered or quilted material that insulates the body from external temperatures. It enables firefighters to operate in the elevated temperatures generated by a fire. It also keeps the body warm during cold weather.

The front of the protective coat has an overlapping flap to provide a secure seal. The inner closure is secured first, and then the outer flap is secured, creating a double seal. Several different combinations of zippers, Velcro, snaps, and D-rings can be used to secure the inner and outer closures.

The collar of the protective coat works with the protective hood to protect the neck. The collar has snaps or a Velcro closure system in front to keep it in a raised position.

The coat's sleeves include thumb wristlets (straps or reinforced openings). Wristlets prevent the sleeves from riding up on the wrists so that no skin is exposed between the gloves and the sleeves, which could result in wrist burns. It is important that the wearer use the thumb wristlets.

Protective coats have an additional integrated safety component. The **drag rescue device (DRD)** is webbing integrated into the back of the protective coat with a handle that is accessible just below the collar. A rescuer can grab this handle and use it to drag the incapacitated firefighter to safety.

Pockets in the coat can be used for carrying small tools or extra gloves. Additional pockets or loops can be installed to hold radios, microphones, flashlights, or other accessories.

Trousers

Structural firefighting protective trousers, also called **bunker pants** or **turnout pants**, are constructed in a waist-length design (pants) or a bib-overall configuration (**FIGURE 3-7**). Like protective coats, protective pants must satisfy the conditions specified in NFPA 1970.

Protective pants are constructed with the same three layers as a protective coat. The outer shell resists abrasions and repels water. The second layer is a moisture barrier that protects the skin from liquids and steam burns. The third, inner layer is a quilted, thermal barrier that protects the body from elevated temperatures. Protective pants are reinforced around the ankles and knees with leather or extra padding to protect those areas from injury, especially if the firefighter needs to crawl to stay below hot gases and smoke.

Protective pants are manufactured with a double-fastener system at the waist, similar to the front flap of the protective coat. Suspenders hold the pants up. Florescent or reflective stripes around the ankles provide added visibility. Pants should be large enough that you can put them on quickly. They also need to be roomy enough to allow you to bend your knees and crawl easily, but they should not be bigger than necessary.

Footwear

Structural firefighting protective footwear (boots) protects the feet and ankles from heat sources, keeps them dry, prevents puncture injuries, and protects the toes from crushing injuries. Structural firefighting boots can be constructed of rubber or leather, and they are available in a variety of styles. Taller boots provide shin protection. Rubber firefighting boots come in a step-in style without laces (**FIGURE 3-8**). Leather firefighting boots are available in pull-on style or in a shorter version with laces (**FIGURE 3-9**). Many firefighters install a zipper on the laced boots to make it faster to put on and take off this component of PPE.

FIGURE 3-7 Structural firefighting protective pants in a waist-length design.

© Jones & Bartlett Learning. Photographed by Glen E. Ellman.

FIGURE 3-8 Rubber firefighting boots.

© Jones & Bartlett Learning. Photographed by Glen E. Ellman.

FIGURE 3-9 Leather firefighting boots.

© Jones & Bartlett Learning. Photographed by Glen E. Ellman.

FIGURE 3-10 Firefighting gloves.

© Jones & Bartlett Learning. Photographed by Glen E. Ellman.

Both rubber and leather styles must meet the requirements specified in NFPA 1970, and some boots are dual certified to meet the requirements of NFPA 1970 and the more stringent requirements of NFPA 1990, *Standard for Protective Ensembles for Hazardous Materials and CBRN Operations, 2022* Edition, which requires that boots provide liquid splash protection to guard against hazardous materials. The outer layer of boots repels water and is flame and cut resistant. Boots also must have a heavy sole with a slip-resistant design, a puncture-resistant sole, and a reinforced toe to prevent injury from falling objects. An inner liner constructed of a material such as Nomex or Kevlar adds thermal protection.

Like all protective clothing, boots must be the correct size. The foot should be secure within the boot to prevent ankle injuries and provide secure footing on ladders or uneven surfaces. Improperly sized boots cause blisters and other problems.

Gloves

A **structural firefighting protective glove** protects hands from heat, cuts, and abrasions (**FIGURE 3-10**). Gloves are an important part of the firefighting ensemble because most fire-suppression tasks require the use of the hands. Gloves must provide adequate protection while permitting the manual dexterity needed to accomplish tasks at an emergency scene. NFPA 1970 specifies that gloves must be resistant to heat, liquid absorption, vapors, cuts, and penetration. A liner adds thermal protection and serves as a moisture barrier.

Many firefighters carry a second set of gloves in their gear or on the apparatus so that they can change them when the first pair becomes dirty, wet, or damaged. Do not wring or twist wet gloves because this motion can tear or damage the inner liners.

Although gloves furnish needed protection, they inevitably reduce manual dexterity. For this reason, manual skills should be practiced while wearing gloves to become accustomed to working with them on and to learn to adjust their movements accordingly.

SAFETY TIP

Plain leather work gloves or plastic-coated gloves should never be used for structural firefighting. Gloves used for firefighting must be labeled as such and meet the specifications given by NFPA 1970. Latex or nitrile gloves are required for rendering emergency medical care.

Donning and Doffing Personal Protective Clothing

The process of **donning** (putting on) the structural firefighting protective ensemble must be performed properly and quickly. Protective clothing must be donned in a specific order to obtain maximum protection. Following a set pattern of donning PPE reduces the time it takes to dress. This exercise does not include donning an SCBA, which will be discussed later. First become proficient in donning personal protective clothing, and then add the SCBA. Protective clothing is donned either in the firehouse before mounting the apparatus or on the scene after you arrive. *Do not* don protective clothing inside the apparatus while en route to an emergency incident. Instead, stay in your assigned seat, properly secured by a seat belt or safety harness, while the vehicle is in motion.

To **doff** (remove) the structural firefighting protective ensemble, reverse the procedure you used when donning it. Doffing should be done at the scene so that the interior of the apparatus is not contaminated with anything from the emergency scene.

Current NFPA standards require decontaminated gear to be stored in a separate room of the fire station that has good ventilation and limited sunlight. The separation and ventilation aid in keeping cancer-causing agents away from personnel, and reduced sunlight helps improve the life of PPE by preventing unnecessary exposure to ultraviolet (UV) rays, which can break down materials. Even after it is thoroughly cleaned, PPE should not be kept in living areas, sleeping areas, or vehicles because even when cleaned, it may still contain some carcinogens. Wherever it is stored, PPE must be properly maintained, organized, and ready for the next emergency response. It should be laid out in a logical order for donning quickly, and the boots should be inside the legs of the protective pants so that you can step into the boot and pants at the same time.

Follow the steps in **SKILL DRILL 3-1** to don the structural firefighting protective ensemble. Follow the steps in **SKILL DRILL 3-2** to doff the structural firefighting protective ensemble.

> **SAFETY TIP**
>
> Do not ride back to the station wearing or sitting next to contaminated PPE.

SKILL DRILL 3-1

Donning Personal Protective Clothing Support Person, NFPA 1010: 5.1.2

1. Place your protective hood over your head and down around your neck.

2. Step into your boots and protective pants, and then pull up the pants. Place the suspenders over your shoulders, and then secure the front of the pants.

Continues.

SKILL DRILL 3-1 CONTINUED

Donning Personal Protective Clothing Support Person, NFPA 1010: 5.1.2

3. Put on your protective coat and close the front of it.

4. Place your helmet on your head with the ear tabs extended and adjust the chin strap securely. Turn up your coat collar and secure it in front.

5. Put on your gloves and secure the wristlet.

6. Have your partner check your clothing.

© Jones & Bartlett Learning. Photographed by Glen E. Ellman.

SKILL DRILL 3-2

Doffing Personal Protective Clothing Support Person, NFPA 1010: 5.1.2

1. Remove your gloves.

2. Open the collar of your protective coat.

3. Release the helmet chin strap, and then remove your helmet.

Continues.

SKILL DRILL 3-2 CONTINUED

Doffing Personal Protective Clothing Support Person, NFPA 1010: 5.1.2

4. Remove your protective coat.

5. Remove your protective pants and boots.

6. Remove your protective hood and prepare gear for reuse.

© Jones & Bartlett Learning. Photographed by Glen E. Ellman.

Additional Personal Equipment

All fire service personnel should always carry a **hand light**. A good working hand light can illuminate your surroundings, mark your location, and help you find your way under difficult conditions. In addition to carrying a hand light, hands-free options such as helmet-mounted lights or other lights that clip to clothing or SCBA may be useful.

Two-way radios link the members of a firefighting team. It is ideal for all personnel to have a radio when working inside a burning building or in a hazardous atmosphere. If radios are not available for everyone, at least one team member should have a radio. Some fire departments provide a radio for on-duty personnel. Follow your department's standard operating procedure (SOP) on radio use. A radio should be considered part of PPE and always carried with you.

Whenever you are operating in a roadway or are close to traffic, you are required to wear a brightly colored reflective safety vest that meets the latest iteration of the American National Standards Institute (ANSI) and the International Safety Equipment Association (ISEA) *High Visibility Safety Apparel and Accessories* standard (ANSI/ISEA 107-2020; ANSI and ISEA 2020).

Some eye protection is provided by the face shield mounted on the fire helmet. When additional eye protection is needed, such as when using power saws or hydraulic rescue tools, and if goggles are not part of the helmet, approved goggles or safety glasses can be used. Goggles can be carried easily in the pocket of the protective coat.

First responders are often exposed to loud noises, such as sirens and engines. Because hearing loss is cumulative, it is important to limit exposure to loud sounds. Headsets with a boom microphone and an intercom system on the apparatus provide hearing protection by reducing engine and siren noise, permit communications between crew members, and enable everyone to hear radio communications (**FIGURE 3-11**). A small speaker is incorporated into large earmuffs located at each riding and operating position. Flexible earplugs are useful in other situations involving loud sounds. All personnel should use the hearing protection supplied by their departments to prevent hearing loss when using power equipment or other equipment that produces noise.

Everyone will have personal preferences about the additional tools and equipment they carry in their pockets. Observe which items seasoned members of your department carry in their pockets. Their choices will help you decide whether you need to carry similar equipment. The following are some common items:

FIGURE 3-11 Some fire apparatus are equipped with a headset and intercom systems. These systems can provide hearing protection, permit communications between crew members, and enable everyone to hear radio communications.

© Jones & Bartlett Learning. Photographed by Glen E. Ellman.

- Spare flashlight
- Multi-tool
- Lineman's cutters
- Wedges
- Webbing
- Screwdriver with an interchangeable head
- Shove knife
- Folding knife

Inspection and Maintenance of Personal Protective Clothing

Approved personal protective clothing is built to rigorous standards. It requires proper care if it is to continue to afford maximum protection. A complete set of approved clothing (excluding SCBA) costs more than $3,000. The gear is expensive, so it should be properly maintained to ensure adequate protection for those wearing it. Avoid unnecessary cuts or abrasions on the outer material. This material already meets NFPA standards—do not look for new opportunities to test its effectiveness.

Inspecting PPE

NFPA 1851, *Standard on Selection, Care, and Maintenance of Protective Ensembles for Structural Fire Fighting and Proximity Fire Fighting, 2020 Edition*, recommends a routine inspection before each shift and after each time structural firefighting gear is worn. In addition to the routine inspection, an advanced inspection should take place annually or when damage is noticed during the routine inspection. The advanced inspection can be performed by a third party or by trained personnel within the department. During the routine inspection, look for excessive wear, contamination, dirt, and damage. If the fabric is damaged, the item must be taken out of service and properly repaired to retain its protective qualities. Clothing that is worn or damaged beyond repair must be replaced immediately because it will not be able to protect you. Follow the manufacturer's instructions for repairing or replacing the protective clothing.

The chin strap and suspension system must be properly adjusted, and all parts of the helmet must be kept in good repair. Never modify the inside of the helmet basket because it can reduce the protection level the helmet offers.

> **SAFETY TIP**
>
> It is recommended that all personnel have access to an extra set of personal protective clothing for times when their gear is dirty or has become damaged and needs repair.

Cleaning PPE

PPE should be cleaned whenever it becomes dirty—for example, after being exposed to smoke, blood or body fluids, or chemicals—to ensure your safety and health. Smoke contains carcinogens, body fluids can transmit diseases, and chemicals can cause burns. It is equally important to remove all of your clothing and thoroughly shower as soon as possible to reduce your exposure to these hazardous materials. In the field, you can use wet wipes to remove dirt from your face and neck so that you are not breathing it in on your way back to the firehouse.

Protective clothing must be cleaned in accordance with the manufacturer's instructions as soon as possible after it has been exposed to a fire or other situation involving any hazardous materials. Dirt builds up in the clothing fibers during routine use and exposure to fire environments. Smoke particles also become embedded in the outer shell and continue to off-gas if not properly cleaned. The interior layers will frequently be soaked with perspiration, which is another reason to clean it regularly. Allowing fire by-products to stay embedded in protective clothing will break down the protective components of the gear, causing the potential for the material to fail.

Other contaminants are formed from the by-products of burned plastics, melted tar, petroleum products, other synthetic products, or other contaminants. These residues, which are flammable, can become trapped between the fibers or build up on the outside of protective clothing, damaging the materials and reducing their protective qualities. A firefighter who is wearing protective clothing contaminated with flammable residues is, in essence, bringing additional fuel into the fire on the clothing.

Keeping personal protective clothing clean reduces your exposure to carcinogens and maintains the protective properties of the clothing. Regular and routine cleaning should remove most of these contaminants. Items that have been exposed to other chemicals or hazardous materials may need to be cleaned by a professional cleaning company or be impounded for decontamination or disposal.

Field reduction of contaminates is performed at the incident scene. This is the first step in cleaning protective clothing that has been exposed to hazardous conditions. While the firefighter is still wearing the exposed protective clothing, the firefighter should be sprayed down using a small-diameter hose to remove surface contaminants (**FIGURE 3-12**). Next, the firefighter should doff the PPE at the scene and then use a brush or water to remove as much dirt and debris as possible before placing the PPE into a bag or container. When the firefighter returns to the station, they should thoroughly clean the PPE following the manufacturer's instructions. Contaminated PPE should never be placed or transported in the passenger area of a vehicle.

Fire departments should have special washing machines that are approved for cleaning protective clothing. Do not launder protective clothing in personal or public washing machines. This could cause cross-contamination with home laundry. If a department does not have approved cleaning devices, they can contract with an outside firm to clean and repair protective clothing. Follow your fire department's policies in

FIGURE 3-12 Perform a field reduction of contaminants to remove dirt and debris before returning to the station.

© Jones & Bartlett Learning. Photographed by Glen E. Ellman.

regard to cleaning contaminated PPE. Your fire department may require that the PPE be bagged in a biohazard bag and professionally cleaned. In all cases, the manufacturer's instructions for cleaning and maintaining the garment must be followed. These instructions can be found on a label affixed to the inside of the garments by the manufacturer. Always follow the manufacturer's cleaning instructions and use only an approved detergent. If you use other methods or cleaning solutions, you may reduce the effectiveness of the garment and create an unsafe situation for the wearer. According to NFPA 1851, advanced cleaning should be done every 6 months. Advanced cleaning is performed by someone who has received additional training to do so; sometimes this is an outside agency. If you wash your protective clothing at the station, follow the manufacturer's instructions for drying these garments properly.

SAFETY TIP

Make sure your PPE is dry before using it on the fireground. If wet protective clothing is exposed to the high temperatures of a structural fire, the water trapped in the liner materials will turn into steam and be trapped inside the moisture barrier, which could cause painful steam burns.

Helmets also require regular cleaning and maintenance. The outer shell should be cleaned with a mild soap as recommended by the manufacturer. The inner parts of helmets should be removed and cleaned according to the manufacturer's instructions.

Protective hoods and gloves get dirty quickly and should be cleaned according to the manufacturer's instructions. Dirty protective hoods may contain large quantities of carcinogens after a fire. Most protective hoods can be washed with appropriate soaps or detergents. Do not use gloves or protective hoods that have holes in them; repair or discard them. Even a small cut or opening in these items can result in a burn injury.

Boots should be maintained according to the manufacturer's instructions. Do not store these items or any other PPE in direct sunlight. Leather boots must be properly maintained to keep them supple and in good repair. All types of boots need to be repaired or replaced if the outer shell is damaged. Exposure of the metal safety toe or boot shank could allow the body to conduct electricity if used in an electrified environment.

As discussed, correctly storing your PPE is another important step in the proper care and maintenance of your PPE. PPE should be stored only in the proper PPE storage areas.

SAFETY TIP

When doffing PPE, *do not* leave a contaminated protective hood or SCBA mask around your neck and below your nose and mouth. You will absorb or inhale additional toxins.

Respiratory Protection

Respiratory protection equipment is an essential component of the structural firefighting PPE. Respiratory protection is both expensive and complicated; using this equipment confidently requires practice. A firefighter must be proficient in using it before they engage in fire-suppression activities.

The atmosphere of a burning building is considered to be an environment that is **immediately dangerous to life and health (IDLH)** because of the presence of airborne contaminants. An IDLH environment is one in which conditions pose an immediate threat to life or could cause irreversible or delayed adverse health effects. Attempting to work in such an environment without proper respiratory protection can cause serious injury or death. A firefighter should never enter or operate in a hazardous atmosphere without first ensuring they have appropriate respiratory protection.

More fire deaths are caused by smoke inhalation than burns. Clearly, adequate respiratory protection is essential for firefighter safety. In fact, the products of combustion from house fires and commercial fires may be so toxic that just a few breaths can result in death. This is why respiratory protection precautions must always be taken before entering the fire.

When arriving at the scene of a fire, there is no way to measure the immediate danger to a firefighter's life and health posed by that fire. The firefighter must don full PPE and approved breathing apparatus before entering and operating within this atmosphere. This includes response to outside fires, such as vehicle and dumpster fires. Respiratory protection and PPE also must be used in any situation where there is a possibility of toxic gases or oxygen deficiency, such as in a confined space. Always assume that the atmosphere is hazardous until it has been tested and proven to be safe.

The firefighter must continue to wear full PPE and approved breathing apparatus during **overhaul**—the process of finding, exposing, and suppressing any smoldering or hidden pockets of fire in an area that has been burned—until the air has been tested and proven to be safe. The firefighter should not remove their respiratory protection even if it appears the fire has been suppressed. They should wait for the signal from command that overhaul has been completed and it is safe to do so.

Respiratory Hazards of Fires

Firefighters need respiratory protection because a fire involves a complex series of chemical reactions that can rapidly affect the atmosphere in unpredictable ways.

Overview of Smoke and Products of Combustion

The most readily evident by-product of a fire is **smoke**, which is the airborne product of combustion consisting of solids (particles), liquids (vapors), and gases. Think of smoke as unburned fuel. Even though smoke is the constant companion of the firefighter, there is not always a thorough understanding of the link between the components of smoke and the reason people get sick or die when they breathe smoke or why firefighters may contract heart disease or experience neurological dysfunction, cancer, or other illnesses related to a single or repeated exposure to smoke.

Essentially, smoke is a collection of gaseous products from burning materials—especially organic materials, such as plastic, nylon, rubber products, wood, and paper—made visible by the presence of small particles of incomplete combustion (carbon). Some of the harmful substances found in smoke include soot, carbon monoxide, cyanide compounds, formaldehyde, and the oxides of nitrogen. These substances are dangerous if inhaled, and repeated exposures to these materials may have negative health effects. Cancer is prevalent in the fire service, more so than in most any other profession, and this high rate should come as no surprise once the by-products of combustion are understood. Many known and suspected human carcinogens are liberated as fire gases and particulates.

Most people identify carbon monoxide as the main harmful component of fire smoke but struggle to list more than a handful of substances beyond that. Less often acknowledged are compounds, such as ammonia, hydrogen chloride, sulfur dioxide, hydrogen sulfide, hydrogen cyanide, carbon dioxide, the oxides of nitrogen, formaldehyde, acrolein, polycyclic aromatic hydrocarbons, and soot.

Smoke production depends on several factors, including the chemical makeup of the burning material, the temperature of the combustion process, and the influence of ventilation (oxygenation). Make no mistake: Fire (combustion) is a complex process, and the smoke produced during a fire event is an intricate collection of particulates, superheated air, and gaseous chemical compounds.

The extensive use of synthetic manufacturing and construction materials (for example, plastics, nylon, rubber products, fire retardants, laminates, and foams, such as Styrofoam and polystyrene) has significant effects on fire behavior and smoke production. Synthetic substances ignite and burn quickly, causing rapidly developing fires and toxic smoke and making structural firefighting today more dangerous than ever before.

As an example, consider polyurethane foam, the most predominant substance in a typical mattress. It consists of many different chemicals, including polyol (an organic alcohol molecule and the majority of the polyurethane compound), toluene diisocyanate (TDI), methylene chloride, and ammonia-based catalysts. When polyurethane foam is exposed to heat, the parent substances break down and bond with each other, creating many new compounds. Some of those compounds are irritants, such as hydrogen chloride and ammonia, which may cause eye irritation or airway problems in smoke exposures. Other compounds, such as carbon monoxide and cyanide compounds, are acutely toxic when inhaled.

The thermal decomposition of polyurethane foam in the mattress fire scenario is broadly representative of the way gases are liberated by a working fire. (Keep in mind this is a limited representation of the process; many more substances are also liberated.) Certainly, each substance is present in varying levels depending on the material(s) involved, the heat of combustion, and the available oxygen (ventilation of the fire). Aside from the toxic gases produced, it is equally important to recognize the visible part of combustion (aside from the flames), which consists of carbon particles of varying sizes, ranging from large embers to particles you cannot see with the naked eye (**FIGURE 3-13**).

This collection of particulates—what people traditionally call smoke—is actually **soot**. Soot is a known human carcinogen, just like benzene and formaldehyde (also by-products of combustion). With that point in mind, take a moment and think about your perception of smoke and revise it to this: The combustion process liberates things you can see (particulates) and things you cannot see (gases). **TABLE 3-1** breaks smoke down into these two main categories to make it easier to understand and separate the hazards. This matrix is not inclusive of all potentially harmful substances, but it does illustrate the point that many properties of smoke can cause acute and chronic health effects. It also shows that smoke is not just *one thing* but rather a dynamic, multifaceted mixture of gases and particulates that changes from minute to minute at any fire.

FIGURE 3-13 Notice how the beam of light is visible because of the particulates in the air.

© Jones & Bartlett Learning. Courtesy of Rob Schnepp.

TABLE 3-1 Smoke Matrix

	Harm You Now	Harm You Later
What you see	Air	Soot, particulates
What you don't see	Carbon monoxide (CO), hydrogen cyanide (HCN), oxides of nitrogen (NOx), sulfur dioxide (SO_2), hydrogen chloride (HCl), hydrogen sulfide (H_2S)	Aldehydes, benzene

© Jones & Bartlett Learning. Courtesy of Rob Schnepp.

Another group of compounds commonly found in fire smoke comprises the polycyclic aromatic hydrocarbons (PAHs), which are classified as probable or possible human carcinogens. More than 100 PAHs have been identified and categorized by various regulatory agencies. This group of substances occurs naturally in materials, such as coal and crude oil. These substances are also generated during the combustion of organic materials and can be found in vehicle exhaust; tobacco smoke; and the smoke generated from structure fires, vehicle fires, wildland fires, or any other type of fire. PAHs can exist as particles or gases. Chances are that you have been exposed to PAHs throughout your life from a common and perhaps surprising activity—grilling food (**FIGURE 3-14**).

When PAHs are generated during a structure fire, they may bind with the soot, resulting in dermal and inhalation exposures. PAHs are believed to be immunosuppressants, a property perhaps contributing to the mechanism by which they are suspected to cause cancer. Examples of PAHs include the following chemicals:

- Anthracene
- Benzopyrene
- Methylchrysene
- Phenanthrene
- Pyrene

Responders not only have to be concerned about acute exposure to PAHs and other materials but also need to consider this group of chemicals (and many others) as potential contaminants of their structural firefighting protective clothing. NFPA 1851

FIGURE 3-14 The thermal decomposition of muscle meats such as chicken, beef, and pork generates PAHs and heterocyclic amines (HCAs) when they are cooked over an open flame.

© Jones & Bartlett Learning. Courtesy of Rob Schnepp.

offers detailed guidance on many aspects of cleaning and maintaining structural firefighting protective clothing.

Products of Combustion

Smoke particles consist of unburned, partially burned, and completely burned substances. Many smoke particles are so small that they can pass through the natural protective mechanisms of the respiratory system and enter the lungs or pass through the skin, move into the bloodstream, and be transported throughout the body. Some are toxic to the body and can result in severe injuries or death if they are inhaled. These particles can also be extremely irritating to the eyes and digestive system.

Smoke vapors are finely suspended liquids—that is, aerosols. This is similar to fog, which consists of small water droplets suspended in the air. When oil-based compounds burn, they produce small hydrocarbon droplets that become part of the smoke. If the droplets are inhaled or ingested, the hydrocarbons can affect the respiratory and circulatory systems. In addition, some toxic droplets in smoke vapor can cause poisoning if they are absorbed through the skin.

A fire also produces several types of gases. The amount of oxygen available to the fire and the type of fuel being burned determine which gases are produced. Many of the gases produced by residential or commercial fires are highly toxic. Carbon monoxide (CO), hydrogen cyanide, and phosgene are three gases commonly present in smoke. These gases are discussed in more detail in Chapter 5, *Fire Behavior.*

The harmful products in smoke require firefighters to use respiratory protection in all fire environments, regardless of whether the environment is known to be contaminated, suspected of being contaminated, or could possibly become contaminated without warning. The use of respiratory protection allows firefighters to enter and work in a hazardous atmosphere with a safe, independent air supply.

SAFETY TIP

Research conducted by Underwriters Laboratories (UL) demonstrated that residual smoke on gear retains many heavy metals, including arsenic, cobalt, chromium, lead, mercury, phosphorus, and phthalates. Exposure to these chemicals has been linked to an increased risk of cancer. Dirty PPE is a danger to your health!

Carbon Monoxide and Hydrogen Cyanide: Silent Killers

Smoke is one of the first observable signs of a working fire. Firefighters note the volume, color, and force of smoke as it exits a fire building—all good indicators of what the fire is doing inside—and use that information to implement appropriate fireground tactics. Firefighters may aggressively enter smoky buildings to search for victims but rarely perform a conscious evaluation of the toxic substances lurking in the smoke. This could be a significant oversight in terms of treating smoke inhalation victims, according to studies performed in Paris, France, and Dallas County, Texas. These studies, although conducted many years ago, focused on CO and hydrogen cyanide (HCN) specifically because they are acutely toxic, present to some degree in nearly all fires, and have clinical interventions available to reverse the adverse health effects of the exposure. Clearly, many toxic substances are generated during a typical structure fire, and smoke inhalation is a complicated illness. It is also clear that successful medical treatment cannot be accomplished by a single drug or single action and that smoke inhalation victims are quite sick.

The Paris and Dallas County studies were done more than 20 years ago, which underscores the fact that the presence of HCN and CO in smoke is not a new revelation; however, the fire service and other disciplines have recently been connecting the dots to understand that smoke is a bigger health and safety issue than previously imagined. In short, these studies identify and evaluate the impact of HCN and CO on victims of smoke inhalation. The discussion here does not focus on the fact that those two substances make smoke toxic; rather, the important point is that both are toxic substances that are common by-products of combustion (present at some level at most fires), and their effects are treatable with clinical interventions. Again, many other toxins can be found in smoke—the goal here is to highlight two fire gases found in nearly any kind of fire.

The Paris study was designed to prospectively assess the role of cyanide in smoke-related morbidity and mortality. *Morbidity* is the incidence of injury or disease within a population; *mortality* is the incidence of death in a population. In the Paris study, blood samples were drawn from fire survivors as well as fatalities (at the time of exposure), and the cyanide and carbon monoxide levels were measured. In several deaths, cyanide levels registered in the lethal range, whereas carbon monoxide levels were in nontoxic concentrations. This suggests cyanide toxicity as the primary cause of death. The study also revealed another bit of interesting information: Deaths also occurred in victims with both cyanide *and* carbon monoxide levels in the nontoxic range, perhaps revealing a relationship between carbon monoxide and cyanide in victims of smoke inhalation. The results of the Paris study can be summarized as follows:

- Cyanide and carbon monoxide are both important determinants of smoke inhalation–associated morbidity and mortality.
- Cyanide concentrations are directly related to the probability of death.
- Cyanide poisoning may be more predominant than carbon monoxide poisoning as a cause of death in certain fire victims.
- Cyanide and carbon monoxide may potentiate each other's harmful effects.

The Dallas County study measured blood cyanide levels in victims after exposure to fire smoke and, in many respects, echoed the findings of the Paris study (Silverman et al. 1988). Over a 2-year period, researchers collected blood samples from a total of 187 smoke inhalation patients within 8 hours of exposure. There were 144 viable patients at the University of Texas Health Sciences emergency department; 43 victims were dead on arrival at the Dallas County Medical Examiner's office.

Of the 144 living patients who reached the emergency room, 12 had blood cyanide concentrations exceeding 1 mg/L. Of these 12 patients, 8 eventually died. None had blood carboxyhemoglobin (COHb) concentrations suggesting carbon monoxide as the cause of death (i.e., ≥50 percent). Although some of the patients had extensive burns that may have contributed to their death, three of them (patients 2, 3, and 9) had total body surface area burns of 4 percent or more. According to the study, blood cyanide levels greater than 1 mg/L had a significant impact on patient outcome. More importantly, the study found that elevated cyanide levels were pervasive in victims of smoke inhalation, and cyanide concentrations were directly related to the probability of death.

Lastly, both studies dispelled a long-held belief in the fire service—that carbon monoxide is the sole (or at least predominant) killer in fire smoke. In fact, it appears that cyanide plays a role in smoke-related deaths and injuries, perhaps more often than we think. Therefore, any victim(s) exposed to significant amounts of smoke or rescued from a closed-space structure fire may be suffering from cyanide toxicity. However, remember that smoke has many properties that are capable of causing acute illness and leaving an indelible impression on the body.

Depending on the dose, hydrogen cyanide has the ability to incapacitate a victim, preventing escape from the fire environment and thereby increasing the exposure to more cyanide, carbon monoxide, and other toxic by-products of combustion. Although this theory is currently unsupported with human data related to smoke exposures, there is information to substantiate the "knock-down" potential of cyanide. In the mid-1980s, studies were conducted on monkeys exposed to the fumes of heated polyacrylonitrile. When polyacrylonitrile is broken down by pyrolysis, cyanide is liberated. Cyanide-exposed monkeys first hyperventilated and then rapidly lost consciousness at a dose-dependent concentration. A concentration of 200 parts per million (ppm) was associated with rapid incapacitation but not with elevated blood cyanide concentrations measured hours after exposure. The direct correlation to human data is unknown at present but could be interpreted in the following way: Hydrogen cyanide could be partly responsible for rendering firefighters and civilians incapable of self-rescue when exposed to smoke.

In summary, you can look at carbon monoxide and cyanide exposures in this way: Carbon monoxide reduces the amount of oxygen carried to the cells; cyanide renders the cells incapable of using whatever oxygen is present.

Oxygen Deficiency

Normal outside or room air contains approximately 21 percent oxygen. A decrease in the amount of oxygen in the air drastically affects an individual's ability to function (**TABLE 3-2**). An atmosphere with an oxygen concentration of less than 19.5 percent is considered oxygen

TABLE 3-2 Physiological Effects of Reduced Oxygen Concentration

Oxygen Concentration	Effect
21%	Normal breathing air
17%	Judgment and coordination impaired; lack of muscle control
12%	Headache, dizziness, nausea, fatigue
9%	Unconsciousness
6%	Respiratory arrest, cardiac arrest, death

© Jones & Bartlett Learning

deficient by the **Occupational Safety and Health Administration** (OSHA 1970). If the oxygen level drops below 19.5 percent, people can experience disorientation, an inability to control their muscles, and irrational thinking, which can make escaping a fire much more difficult.

When a fire is burning within an enclosed area, oxygen depletion occurs in two ways. First, the fire consumes large quantities of the available oxygen, decreasing the concentration of oxygen in the atmosphere. Second, the fire produces large quantities of other gases, which displace the oxygen that would otherwise be present inside the enclosed area.

Increased Temperature

Heat is also a respiratory hazard of fire. The temperature of the smoke varies, depending on the fire conditions and the distance the smoke has traveled. Inhaling superheated smoke can cause severe burns of the respiratory tract immediately. If the smoke is hot enough, a single inhalation can cause fatal respiratory burns. More information about fire behavior and products of combustion appears in Chapter 5, *Fire Behavior.*

Respiratory Hazards in Other Toxic Environments

Not all hazardous atmospheric conditions are caused by fires. Indeed, fire service personnel may encounter oxygen-deficient atmospheres in numerous types of emergency situations. For example, toxic gases may be released in hazardous materials incidents from leaking storage containers or industrial equipment, from chemical reactions, or from the normal decay of organic materials. Internal combustion engines or improperly operating heating appliances can produce carbon monoxide. Respiratory protection is just as important in these situations as it is in a fire-suppression operation.

Types of Breathing Apparatus

There are two basic categories of breathing apparatus: atmosphere-supplying respirators and air-purifying respirators.

An **atmosphere-supplying respirator (ASR)** uses air that is supplied from a source independent of the ambient (room) air. There are two types of ASRs. One type uses air carried in a compressed gas cylinder, such as the fire service SCBA. The construction and operation of SCBA will be more fully described in the following sections.

The second type of ASR is a supplied-air respirator. A **supplied-air respirator (SAR)**, also known as an **air-line respirator**, uses an external source for the breathing air (**FIGURE 3-15**). In this type of device, a hose line is connected to a breathing-air compressor or to compressed air cylinders located outside the hazardous area. The user breathes air through the line and exhales through a one-way valve. SARs are more frequently used in industrial settings and in some hazardous materials situations.

FIGURE 3-15 An SAR may be needed for special rescue operations.

Courtesy of 3M.

The second major type of breathing apparatus is an air-purifying respirator. An **air-purifying respirator (APR)** has an air-purifying filter, cartridge, or canister that removes specific air contaminants by passing ambient (room) air through the air-purifying element (**FIGURE 3-16**). APRs remove particulate matter or specific gases and vapors. They are used in some industrial settings where the air is monitored carefully to determine the precise quantity of contaminants that are present. APRs are not suitable for firefighting operations because you need to know the type and concentration of contaminants present so that you can use the correct filter. This is not possible at a fire scene. Furthermore, APRs do not supply oxygen in an oxygen-deficient atmosphere.

A **powered air-purifying respirator (PAPR)** is similar to the standard APR, but it includes a small fan to

FIGURE 3-16 APRs offer specific degrees of protection if the hazard present is known and the appropriate filter canister is used.

Courtesy of Sperian Respiratory Protection.

help circulate air into the mask (**FIGURE 3-17**). The fan draws outside air through the filters and into the mask via a low-pressure hose. PAPRs make it easier for the wearer to breathe, help reduce fogging in the mask, and provide a constant flow of cool air across the face. PAPRs are not considered to be true positive-pressure units—that is, a unit that creates higher pressure inside the mask than the atmospheric pressure outside the mask. This is because it is possible for the wearer to "outbreathe" the flow of supplied air—that is, inhale more air than is being supplied via the low-pressure hose. This creates negative pressure—pressure less than the atmospheric pressure outside the mask—inside the mask. When this happens, the seal with the face is compromised, which will possibly allow contaminants to enter the face mask.

APRs and PAPRs are typically worn to respond to a hazardous materials situation where the type and quantity of contaminants are known. These respirators should not be used in structural firefighting situations.

Self-Contained Breathing Apparatus and Personal Alert Safety System

The SCBA is an essential component of the PPE used for structural firefighting. An SCBA provides clean breathing air (the same as environmental air) to firefighters

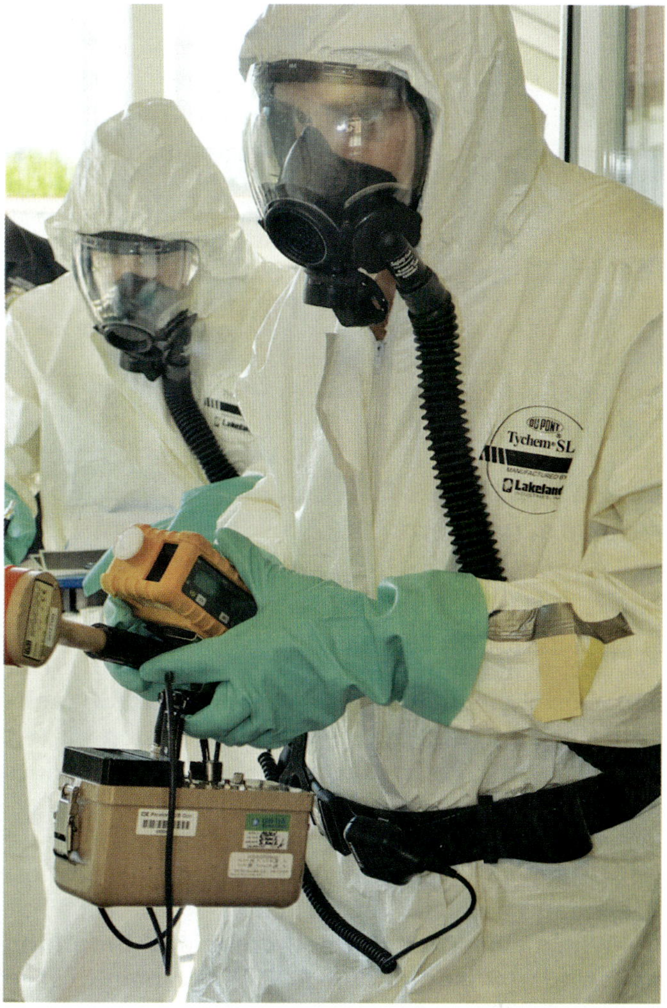

FIGURE 3-17 PAPRs have a small fan to help circulate air into the mask.

Courtesy of Chris Hawley.

who are working in the hostile environment of a fire. Without adequate respiratory protection, firefighters would not be able to enter toxic environments. The SCBA must meet rigid manufacturing specifications so that it can function in the increased temperatures and smoke-filled environments that firefighters encounter. When properly maintained, this equipment will provide enough air to enable firefighters to perform rigorous tasks.

TIP

Do not confuse SCBA and SCUBA. An SCBA is used by firefighters during fire-suppression activities. A self-contained underwater breathing apparatus (SCUBA) is used by divers while swimming underwater.

FIGURE 3-18 Open-circuit SCBA.

Courtesy of 3M.

The two main types of SCBA are open-circuit and closed-circuit devices. An **open-circuit self-contained breathing apparatus (open-circuit SCBA)** uses a cylinder of compressed air that provides a limited air supply to the user, and exhaled air is released into the atmosphere through a one-way valve (**FIGURE 3-18**). Open-circuit SCBAs are the type most commonly used in the fire service.

With a **closed-circuit self-contained breathing apparatus (closed-circuit SCBA)**, commonly called a **rebreather**, exhaled air is not released to the outside environment. Instead, the user's exhaled air passes through a mechanism that removes carbon dioxide and chemically generates oxygen. The air is then "rebreathed" by the wearer (**FIGURE 3-19**). Many closed-circuit SCBA units also include a small oxygen cylinder. A closed-circuit SCBA is often used for extended operations, such as mine rescue work and operations in long tunnels where breathing apparatus must be worn for a long time.

Using an SCBA requires that firefighters develop unique skills, including different breathing techniques. Proficiency in the use of SCBA and other PPE requires ongoing training and practice.

Limitations of SCBA

Like any type of equipment, SCBA has its limitations. Some of these limitations apply to the equipment; others apply to the user's physical and psychological abilities. Because an open-circuit SCBA carries its own air supply in a pressurized air cylinder, its use is limited by the amount of air in the cylinder. SCBA air cylinders for structural firefighting must carry enough air for a minimum of 30 minutes. Air cylinders rated for 45 minutes, 60 minutes, and 75 minutes are also available. These duration ratings are based on ideal laboratory conditions. The realistic useful life of an SCBA air cylinder for firefighting operations is usually much less than the rated duration, and actual use time depends on the size of the user, their physical fitness and conditioning, their physical exertion, and how calm they are. An SCBA air cylinder generally has a realistic useful life of no more than 50 percent of the rated time. For example, an SCBA air cylinder rated for 30 minutes can be expected to last for a maximum of 15 minutes during strenuous firefighting.

FIGURE 3-19 A closed-circuit SCBA is commonly referred to as a *rebreather*.

© Jones & Bartlett Learning. Photographed by Glen E. Ellman.

Firefighters must manage their working time while using SCBA so that they have enough time to exit the hazardous area before exhausting the air supply. To properly manage the air supply, a firefighter must consider the following factors:

- The time and the effort it will take to reach the task destination. Climbing stairs takes more energy and air than walking across a flat floor, for example.
- The amount of air that will be available upon reaching the task destination.

- The amount of time necessary to complete the task and the air that will be used during that period. Some tasks take more energy and air.
- The amount of time it will take to reach a safe area. At the end of this time, the firefighter must have a reserve of air for unexpected emergencies.

An SCBA provides a limited window of time for firefighting and a safe exit from the hazardous conditions of the fire. It is essential that firefighters have a margin of safety built into their air supply for the unexpected. In some cases, a firefighter may need to begin exiting from the fire scene before half of their air supply is exhausted.

The weight of an SCBA varies, based on the manufacturer and the type and size of the air cylinder. Generally, an SCBA weighs at least 25 lb (11 kg). The added weight and bulk of the SCBA decrease the user's flexibility and mobility and shift the user's center of gravity.

This equipment limits normal sensory awareness—the senses of smell, hearing, and sight are all affected by the apparatus. As previously noted, the design of the SCBA face piece limits the firefighter's vision—particularly their peripheral vision. The face-piece lens may fog up under some conditions, further limiting visibility. Face pieces also can fail if the temperature and radiant heat are too high (**FIGURE 3-20**). NIST studies show that the face-piece lens can soften at approximately 302°F (150°C), with the lens starting to "craze" or bubble at higher temperatures and then actually fail at temperatures above 419°F (215°C) (Mensch, Braga, and Bryner 2011, 1). SCBA may affect the user's ability to communicate, depending on the type of face piece and any additional hardware provided, such as voice amplification and radio microphones. The equipment can be noisy during inhalation and exhalation, which may limit the user's hearing as well, especially when coupled with a protective hood and helmet.

FIGURE 3-20 This SCBA face piece shows damage from heat.

© Jones & Bartlett Learning. Photographed by Dave Casey.

Altogether, the protective clothing and SCBA that must be worn when fighting fires can weigh more than 50 lb (23 kg). Moving with this extra weight requires additional energy, which, in turn, increases air consumption and body temperature. Taken collectively, this activity places additional stress on a firefighter's body.

The weight and bulk of the complete PPE ensemble limit a firefighter's ability to walk, climb ladders, lift objects, and crawl through restricted spaces. Firefighters must become accustomed to these limitations and learn to alter their movements accordingly. Practice and conditioning are key to becoming proficient in wearing and using PPE while fighting fires.

Physical Limitations of the SCBA User

Conditioning is important for SCBA users. A person in good physical condition will be able to perform more work per cylinder of air than a person who is overweight or out of shape. An out-of-shape or obese firefighter will consume the air supply from this equipment more quickly and will have to exit the fire building long before a well-conditioned firefighter must do so. Overweight or poorly conditioned firefighters also are at greater risk for heart attack resulting from physical stress.

Psychological Limitations of the SCBA User

In addition to the physical limitations, the user must make mental adjustments when wearing an SCBA. Breathing through an SCBA is different from normal breathing, and it can be stressful. Covering your face with a face piece, the noises of the air rushing in and valves opening and closing, and exhaling against positive pressure are all foreign sensations. The surrounding environment, which is often dark and filled with smoke, is foreign as well.

Components of SCBA

As with protective clothing, the NFPA specifies the minimum requirements for SCBA and PASS devices, as outlined in the following standards:

- NFPA 1852, *Standard on Selection, Care, and Maintenance of Open-Circuit Self-Contained Breathing Apparatus (SCBA), 2019 Edition*
- NFPA 1970, *Standard on Protective Ensembles for Structural and Proximity Firefighting, Work Apparel and Open-Circuit Self-Contained Breathing Apparatus (SCBA) for Emergency Services, and Personal Alert Safety Systems (PASS), 2025 Edition*
- NFPA 1984, *Standard on Respirators for Wildland Fire Fighting Operations and Wildland Urban Interface Operations, 2022 Edition*

An SCBA consists of four main parts: the harness, the air cylinder, the regulator assembly, and the face-piece assembly (**FIGURE 3-21**). Although the basic features and operations of all models are similar, you need to become familiar with the specific model of SCBA used by your department.

Harness

The **SCBA harness** consists of the backpack or frame for mounting the working parts of the SCBA and the straps and fasteners used to attach the SCBA to the firefighter (**FIGURE 3-22**). It is usually constructed of a lightweight metal or composite material. Most SCBA harnesses have two adjustable shoulder straps and a waist belt. These must be constructed of a material, such as Kevlar, that is strong and able to withstand elevated temperatures. Depending on the specific model of SCBA, the waist belt and shoulder straps carry different proportions of the pack's weight. The procedures for tightening and adjusting the straps also vary based on the model. The SCBA harness must be secure enough to keep the SCBA firmly fastened to the user but not so tight that it interferes with breathing or movements. The waist belt must be tight enough to keep the SCBA from moving from side to side or getting caught on obstructions. Some SCBA models are equipped with a reinforced harness or hand loop, similar to the DRD on a protective coat, that can be used to help drag a fallen firefighter out of danger.

FIGURE 3-21 The components of SCBA.

Courtesy of 3M.

Breathing Air Cylinder

The **breathing air cylinder** (or **air cylinder**) on open-circuit SCBA holds the compressed breathing air for an SCBA. This removable air cylinder is attached to the SCBA harness and can be swapped with another cylinder quickly in the field.

All fire service personnel should be familiar with the type of air cylinders used in their departments. Air cylinders are marked with the materials used in their construction, the working pressure, and the rated duration.

The air pressure in filled SCBA air cylinders ranges from 2200 to 5500 pounds per square inch (psi; 15,168 to 37,920 kilopascals [kPa]). The greater the air pressure, the more air that can be stored in the cylinder. Low-pressure air cylinders, which are pressurized at

FIGURE 3-22 SCBA harnesses come in a variety of models.

© Jones & Bartlett Learning. Photographed by Glen E. Ellman.

2200 psi (15,168 kPa), can be constructed of steel or aluminum and are usually rated for 30 minutes of use. Composite air cylinders are generally constructed of an aluminum shell wrapped with carbon, Kevlar, or glass fibers. They are significantly lighter in weight; can be pressurized up to 5500 psi (37,920 kPa); and are rated for 30, 45, 60, or 75 minutes of use.

As previously noted, the rated duration times are established under laboratory conditions. A working firefighter can quickly use up the air because of exertion, so the ratings should be viewed with caution. Generally, the working time available for a particular air cylinder is half the rated duration.

The neck of an air cylinder is equipped with a hand-operated shut-off valve that controls the flow of air leaving the air cylinder. Be careful not to damage the threads or let any dirt get into the outlet of the air cylinder. The **air-cylinder pressure gauge** located near the air-cylinder valve shows the amount of air currently in the cylinder. This gauge cannot be viewed by the user while wearing the SCBA, so a second pressure gauge, called the **remote pressure gauge**, is located on the shoulder strap or in another location where it can be seen while the SCBA is being used. If the two pressure gauges are working correctly, the readings should be within 10 percent of each other.

Regulator

An **SCBA regulator** controls the flow of air to the user (**FIGURE 3-23**). The regulator may be mounted on the waist belt or shoulder strap of the SCBA harness or attached directly to the face piece. Inhaling decreases the air pressure in the face piece. This change in pressure opens the regulator, which in turn releases air from the cylinder. When inhalation stops, the regulator shuts off the air supply. Exhaling opens a second valve (the exhalation valve), thereby expelling the exhaled air into the atmosphere. Most SCBA units are equipped with a **dual-path pressure reducer**, a feature that automatically provides a backup method for air to be supplied to the regulator if the primary passage malfunctions.

A proper seal between the face piece and the face must always be maintained. In addition to making sure your face piece fits well and is correctly donned, SCBA regulators maintain a slightly higher air pressure to the face piece in relation to the ambient air pressure outside the face piece. This is called *positive pressure*. This helps prevent the hazardous atmosphere outside the face piece from leaking into the face piece during inhalation. If any leakage occurs in the area where the face piece and the face make the seal, the positive-pressure breathing air inside the face piece keeps the hazardous atmosphere from entering the device. Breathing with this slight positive pressure may require some practice. New firefighters often report that it takes more energy to breathe when first using positive-pressure SCBA, but this sensation gradually decreases.

FIGURE 3-23 SCBA regulator.

Courtesy of 3M.

An SCBA also has a **regulator purge/bypass valve**, which allows the air from the air cylinder to enter the face piece without passing through the regulator. This will create a constant flow of air into the face piece; however, this action will rapidly deplete the remaining air supply in the air cylinder. Firefighters use this valve if the air supply is partially or completely cut off during use. In the event that it is necessary to open the regulator purge/bypass valve, they should immediately exit from the IDLH area. During normal use, the regulator purge/bypass valve can be momentarily opened to remove condensation or residual air from the respirator after the air cylinder valve is turned off.

The regulator may have an air saver/donning switch that prevents the rapid loss of the air supply if the cylinder valve is open and the face piece is removed from the face or the regulator is removed from the face piece. In other words, if you open the air-cylinder valve, air will not flow unless you depress the air saver/donning switch.

Many models of SCBA regulators exist. Firefighters must learn how to operate the particular model that is used in their department. They should be able to operate the regulator in the dark and while wearing firefighting gloves.

FIGURE 3-24 SCBA face pieces come in several sizes.

© Jones & Bartlett Learning. Photographed by Glen E. Ellman.

Face Piece

The **face piece** delivers breathing air to the wearer and protects the face from high temperatures and smoke (**FIGURE 3-24**). It consists of a face mask with a clear lens and an exhalation valve. On SCBA models with a harness-mounted regulator, the face piece also includes a flexible low-pressure supply hose. In some models, the regulator is attached directly to the face piece.

Full face pieces cover the nose, mouth, and eyes. Half face pieces cover just the nose and mouth. The part that comes in contact with the skin is made of special rubber or silicone because these materials provide for a tight seal. Exhaled air is expelled from the face piece through the one-way exhalation valve, which has a spring mechanism to maintain positive pressure inside the face piece. Because it is difficult to communicate through a face piece, a voice-amplification device or mechanical diaphragm is used to facilitate communication. A mechanical diaphragm is a vibrating airtight membrane that transmits the firefighter's voice without the use of electricity.

Face pieces are equipped with **nose cups**, an insert inside the face piece that fits over the user's mouth and nose. The nose cup has two uses. It prevents the build-up of carbon dioxide (CO_2) by directing exhaled air toward the exhalation valve. It also helps prevent fogging of the clear lens. Fogging occurs because the compressed air you breathe is dry, but the air you exhale is moist. If fogging occurs, the regulator purge/bypass valve can be opened slightly for a second or two to clear the condensation.

A leak in the face-piece seal may result from an improperly sized face piece, an improper donning procedure, or facial hair around the edge of the face piece. In particular, the following factors can affect the seal on an SCBA face piece:

- Facial hair, sideburns, or beard
- A low hairline that interferes with the sealing surface
- Ponytails or buns that interfere with the smooth and close fit on the head harness
- A skull cap that projects under the face piece or temple pieces
- The absence of teeth
- An improperly sized face piece

> **TIP**
>
> Fogging is a greater problem in colder climates.

An improperly fitted face piece may lead to exposure to a hazardous environment. Such leaks are dangerous for two reasons. First, a large leak could overcome the positive pressure in the face piece and allow contaminated air to enter the face piece. Second, a leak of any size will deplete the breathing air and reduce the amount of time available for firefighting.

When wearing SCBA, the protective hood is worn over the face piece and under the helmet. After securing the face-piece straps, the firefighter should carefully fit the protective hood around the face piece so that no areas of bare skin are left exposed. The protective hood must fit snugly around the clear area of the face piece so that vision is not compromised and hot gases cannot leak between the face piece and the protective hood.

Face pieces are manufactured in several sizes. NFPA 1550 requires that all firefighters have their face pieces fit-tested annually to ensure that they are wearing the proper size. Some departments issue individual face pieces to each firefighter; others provide a selection of sizes on each apparatus. NFPA 1550 also requires that the sealing surface of the face piece be in direct contact with the user's skin; that is, hair or a beard cannot be in the seal area.

Additional Features

An SCBA is also required to have a **heads-up display (HUD)**, which must be visible to the user while wearing the face piece, enabling the user to constantly monitor the amount of air in the air cylinder. Most SCBA face pieces contain **light-emitting diodes (LEDs)** that indicate the amount of air remaining in the cylinder. These indicate whether the air cylinder is 100, 75, 50, or 35 percent full. Other LEDs on the display may provide

FIGURE 3-25 An HUD enables the user to constantly monitor the amount of air in the air cylinder.

© Jones & Bartlett Learning. Photographed by Glen E. Ellman.

FIGURE 3-26 The RIC uses the RIC UAC to refill the SCBA cylinder of a trapped firefighter who is running out of air. This connection is used only for the emergency refilling of an air cylinder.

Courtesy of MSA.

additional information—for example, low batteries or other problems (**FIGURE 3-25**).

NFPA standards require that SCBA include two **end-of-service-time-indicator (EOSTI)** or low-air alarms, which operate independently of each other and activate different senses. For example, one alarm might ring a bell, whereas the second alarm might vibrate, whistle, or flash an LED.

This warning device tells the user that the end of the breathing-air supply is approaching. Currently, the EOSTI or low-air warning alarm is constructed to sound when the pressure in the air cylinder is down to 34 percent (+/− 2%) rated service pressure for a 2216 pounds per square inch gauge (psig) SCBA; 33 percent (+/− 2%) for a 3000 psig SCBA; 31 percent (+/− 2%) for a 4,500 psig SCBA; 29 percent (+/- 2%) for a 5,500 SCBA; 28 percent (+/− 2%) for a 6,000 psig SCBA percent of capacity (NFPA 1970 2025). At this pressure level, the firefighter must begin to exit from the IDLH environment. Most fire departments require firefighters to exit from the IDLH area before the EOSTI alarm sounds because the low-air alarm does not sound until two-thirds of the air supply has been exhausted. Per NFPA 1970, this reserve of one-third of the air supply is not always adequate for a safe escape.

Communications between firefighters wearing SCBA are difficult. To facilitate communication, an SCBA is required to be equipped with a voice communication system. This functionality may be as simple as a mechanical voice diaphragm, or it may be as sophisticated as an electronic system. Some SCBA devices are also equipped with voice communications systems that use wireless communication interoperability to interface with mobile radios. Recent changes have been made in the NFPA 1970 standards to improve the voice communication intelligibility and volume of face pieces.

If an SCBA is equipped with an **universal emergency breathing safety system (UEBSS)** meets the performance requirements listed in NFPA 1970, the SCBA can be used to "buddy breathe,' or share their available air supply when one is low or out of air. An SCBA is required to be equipped with a **rapid intervention crew/company universal air connection (RIC UAC)** (**FIGURE 3-26**). A **rapid intervention crew/company (RIC)**, sometimes called a **rapid intervention team (RIT)**, is a team of firefighters who stand by, fully dressed in PPE and SCBA, during emergency operations so that they can be quickly deployed to rescue a trapped firefighter. The RIC uses this connection to refill the SCBA cylinder of a trapped firefighter who is running out of air. A universal air connection (UAC) is attached to a hose from a full air cylinder and brought to the downed firefighter by an RIC, who refills the downed firefighter's cylinder. This connection should not be used for the routine refilling of a cylinder. The RIC UAC is designed only to be used for emergency refilling of an air cylinder. It is important that the dust cap is in place over the UAC to prevent damage. This universal connection can be used between SCBA packs of any manufacturer.

Fire department SCBA devices are required to be certified to provide protection against certain chemical, biological, radiological, and nuclear agents—that is, agents that could be released as a result of a terrorist attack. An SCBA provides protection against these agents by preventing chemical fumes, disease-causing biological organisms, and radioactive particles from being inhaled and entering the firefighter's respiratory system. However, an SCBA does not protect from contamination by other means of transmission.

Per NFPA 1970, many accessories are available for SCBA, including data logging, unit IDs, tracking devices, corrective eye lenses for face pieces, thermal imaging capabilities, and electronic communications devices. Data logging and unit IDs aid in keeping track of the use of each SCBA and checking the proper functioning of each unit. Some of these tracking devices work by sending out sounds of varying pitch. Corrective lenses for face pieces enable SCBA users to use their face pieces as corrective eyewear. Electronic communications systems integrate radio communications between SCBA users. Some SCBA devices offer a thermal imaging device that can be added to the unit to provide each firefighter with the ability to detect heat sources. If your SCBA is equipped with any of these accessories, you must become competent in the operation of these extra devices. To ensure proper functioning of your SCBA, do not add any devices or accessories that are not approved by the manufacturer of the equipment.

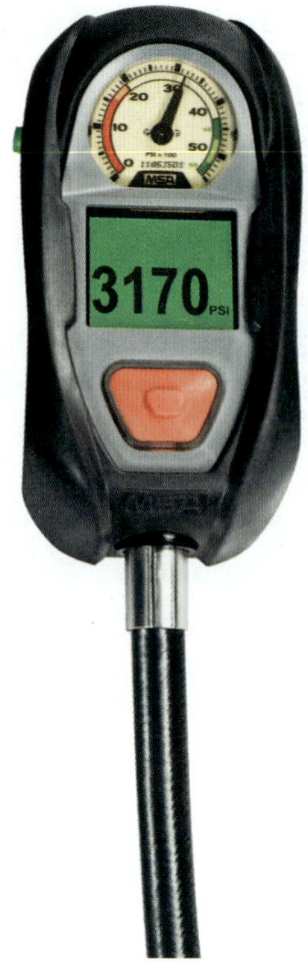

FIGURE 3-27 An integrated PASS device can save a firefighter's life.

Courtesy of MSA.

Pathway of Air through an SCBA

In an open-circuit SCBA, the breathing air is stored under pressure in the air cylinder. This air passes through the cylinder valve into the high-pressure air line (hose), which then carries it to the regulator. The regulator opens when the user inhales, reducing the air pressure on the downstream side. In an SCBA unit with a face-piece–mounted regulator, this low-pressure air goes directly into the face piece. In units with a harness-mounted regulator, the low-pressure air travels from the regulator through a low-pressure hose into the face piece. From the face piece, the air is inhaled through the user's air passages and into the lungs. When the user exhales, the used air is returned to the face piece. The exhaled air is exhausted from the face piece through the exhalation valve. As the pressure in the face piece drops, the exhalation valve closes, and the regulator opens. This cycle repeats with every inhalation.

Personal Alert Safety System

As described earlier, a PASS is an electronic device that emits a loud audible signal when a firefighter becomes trapped or injured to help colleagues locate the downed firefighter. Newer PASS devices may include a radio transmitter that sends a signal to the command post when the alarm sounds. Many SCBA models are manufactured with an integrated PASS device (**FIGURE 3-27**).

The PASS device combines an electronic motion sensor with an alarm system. If the user remains motionless for 30 seconds, it produces a low warning tone before sounding a full alarm. The user can reset the device by moving during this warning period. A firefighter in distress can manually activate this device.

Turning on the air supply in an SCBA with an integrated PASS device automatically activates the PASS device. This ensures that a firefighter does not forget to turn on the PASS device when entering a hazardous area. Nonintegrated PASS devices, found with older SCBA devices, must be turned on manually. Some PASS devices may be used by firefighters not wearing SCBA. All firefighters must confirm that their PASS devices are on and working properly before they enter a burning building or hazardous area. It is important to learn how to activate and deactivate the PASS device on your SCBA.

Limitations of the Structural Firefighting Protective Ensemble

Unfortunately, even today's advanced PPE has drawbacks and limitations. Understanding those limitations will help the wearer avoid situations that could result in serious injury or death.

FIGURE 3-28 It takes practice to don the full structural firefighting protective ensemble.

© Jones & Bartlett Learning. Photographed by Glen E. Ellman.

First, the structural firefighting protective ensemble is not easy to don. It includes several components, all of which must be put on in the proper order and correctly secured. You must be able to don your equipment quickly and correctly, either at the fire station before you respond to an emergency or after you arrive at the scene. Practice donning your protective clothing ensemble until you can do so quickly and smoothly (**FIGURE 3-28**).

The ensemble is heavy, weighing nearly 50 lb (23 kg). This increased weight means that every activity—even walking—requires more energy and strength. It also limits mobility. Full bunker gear not only limits the range of motion, but it also makes movements awkward and difficult. Tasks such as advancing an attack line up a stairway or using an axe to ventilate a roof can be difficult, even for a firefighter in excellent physical condition.

Because PPE retains body heat and perspiration, it is difficult for the body to cool itself when wearing this equipment. Perspiration is retained inside the protective clothing rather than evaporating to cool you. As a consequence, firefighters in full protective gear can rapidly develop elevated body temperatures, even when the ambient temperature is cool. The problem of overheating is more acute when surrounding temperatures are high, which is one reason why firefighters must undergo regular rehabilitation and adequate fluid replacement. Removing your ensemble will help you to cool down quickly.

Wearing PPE also decreases normal sensory abilities. Wearing heavy gloves, for example, reduces the sense of touch. Structural firefighting protective coats and pants protect skin but reduce its ability to determine the temperature of hot air. Sight is restricted when wearing an SCBA. The plastic face piece reduces peripheral vision, and the helmet, protective hood, and coat make turning the head difficult. Both the helmet ear tabs and the protective hood over the ears limit hearing. Speaking becomes muffled and distorted by the SCBA face piece, even if it is equipped with a special voice amplification system.

PPE absorbs smoke particles that continue to **off-gas** (emit harmful chemicals in the form of a gas) toxins from the hazardous conditions from which they just came. Remove as much of your PPE as possible when you arrive at rehabilitation, and then place the PPE in an area away from direct contact with yourself and others. Much of the structural firefighting gear worn by firefighters contains a chemical known as *PFAS*. PFAS is a known carcinogen and has been linked to several health problems. Research is being conducted to find an alternative. This is just another reason to always decontaminate your gear and keep it stored separately from living quarters and passenger vehicles.

For these reasons, it is important to become accustomed to wearing and using PPE. Practicing skills while wearing PPE will help you become comfortable with its operation and limitations.

SCBA Inspection and Maintenance

An SCBA must be properly cleaned, inspected, and prepared for the next use each time it is used, whether in an actual emergency incident or as part of a training exercise. The air cylinder must be changed or refilled; the face piece and regulator must be sanitized according to the manufacturer's instructions; and the unit must be cleaned, inspected, and checked for proper operation.

After operating in a fire environment, an SCBA will be coated with dirt and soot, which contains many dangerous and carcinogenic substances. It is important to thoroughly clean all parts of the SCBA as soon as possible. Cleaning SCBAs is just as important as cleaning firefighting gear to rid it of any carcinogens or other harmful substances. Follow the instructions of the SCBA manufacturer and of your department. It is the user's responsibility to ensure that the SCBA is in good working order and in ready-to-use condition before it is returned to the fire apparatus.

The U.S. Department of Transportation (DOT) limits the number of years that a cylinder can be used. In addition, the DOT requires that SCBA cylinders be tested on a periodic basis to identify any defects or damage that might render the pressurized cylinder unsafe. This test is called a **hydrostatic test**. The date and results of hydrostatic tests are listed on a label on the cylinder. Hydrostatic testing is usually handled by certified third parties. NFPA 1852 acknowledges the requirements from the DOT within the standard.

Cylinders constructed of different materials have different testing requirements. Aluminum, steel, and carbon-fiber cylinders must be hydrostatically tested every 5 years (DOT 2017). Cylinders constructed of composite materials such as Kevlar or fiberglass fibers must be tested every 3 years (DOT 2017). All personnel should know which types of cylinders are used by their departments and must check each cylinder for a current hydrostatic test date before filling it.

Inspection

In addition, each SCBA must be inspected and tested on a regular basis to ensure that it will function properly at an emergency scene. In career departments, inspection and testing are done at the beginning of each shift. In volunteer departments, this step is commonly performed on a weekly schedule. A complete annual inspection and maintenance procedure must also be performed on each SCBA unit. The annual inspection must be performed by a certified manufacturer's representative or a person who has been trained and certified to perform this work.

If an SCBA inspection reveals any problems that cannot be remedied by routine maintenance, the SCBA must be removed from service for repair. Only properly trained and certified personnel are authorized to repair SCBA.

The purpose of a visual SCBA inspection is to identify any parts of the SCBA that are visibly damaged and need to be repaired or replaced to ensure continued safe operation. This visual inspection can be done in conjunction with the operational testing sequence discussed next. Follow the steps in **SKILL DRILL 3-3** to conduct a visual inspection of an SCBA. A more detailed inspection is required if a cylinder has been exposed to excessive heat, has come into contact with flame, has been exposed to chemicals, or has been dropped.

SKILL DRILL 3-3

Visual SCBA Inspection Support Person, NFPA 1010: 5.5.2

1. Visually inspect the air cylinder and valve assembly for dents and gouges. Look for black or discolored areas that indicate exposure to flame.

2. Check the cylinder for the current hydrostatic test date and date of manufacture. Check the air-cylinder pressure gauge to be sure it is full.

SKILL DRILL 3-3 CONTINUED

Visual SCBA Inspection Support Person, NFPA 1010: 5.5.2

3. Inspect hose and rubber parts for damage or deterioration.

4. Inspect the SCBA harness, webbing, buckles, fasteners, and cylinder retention system for damage.

5. Verify that the SCBA has been cleaned according to the manufacturer's and department's recommendations. Inspect the regulator for intact gaskets and visible damage.

6. Inspect the face piece for damage and worn components. Look for damage to the lenses, and check for the presence of a nose cup.

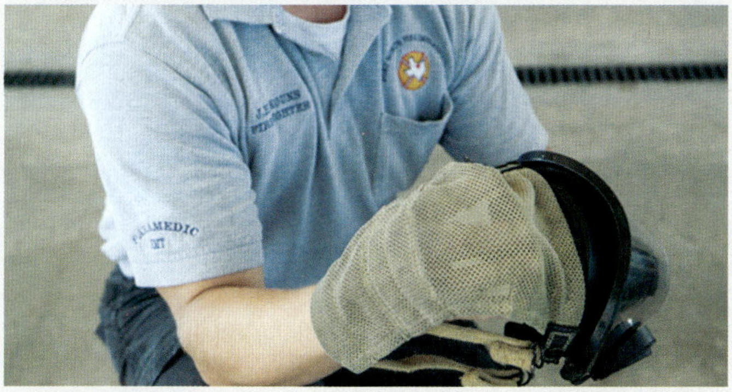

7. Inspect the head harness to confirm that all parts are present and working properly.

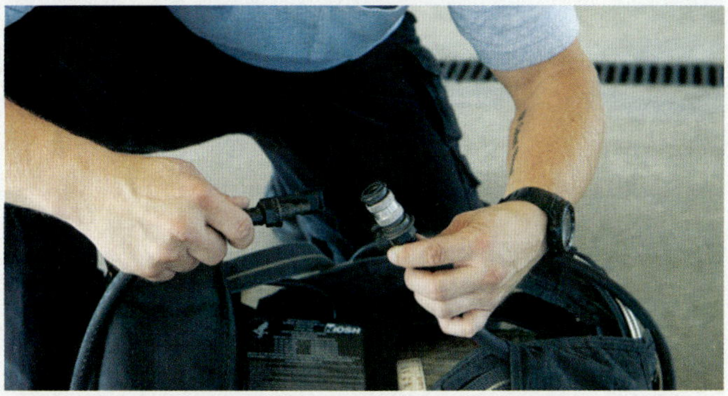

8. Check the quick disconnects and the RIC UAC to make sure that they are not damaged, that they are operating properly, and that the dust cap is in place.

© Jones & Bartlett Learning. Photographed by Glen E. Ellman.

Operational Testing

A pressurized SCBA cylinder contains a tremendous amount of potential energy. Not only does the air within the cylinder exert considerable pressure on its walls, but the cylinder itself is used under extreme conditions on the fireground. If the cylinder ruptures and suddenly releases this energy, it can cause serious injury or death. For this reason, cylinders must be regularly inspected and tested to ensure they are safe. The operational testing sequence is designed to check the function of the many parts of the SCBA to ensure safe use of the device. It concentrates on the working parts of the SCBA. Follow the steps in **SKILL DRILL 3-4** for operational testing of an SCBA.

SKILL DRILL 3-4

SCBA Operational Inspection Support Person, NFPA 1010: 5.5.2

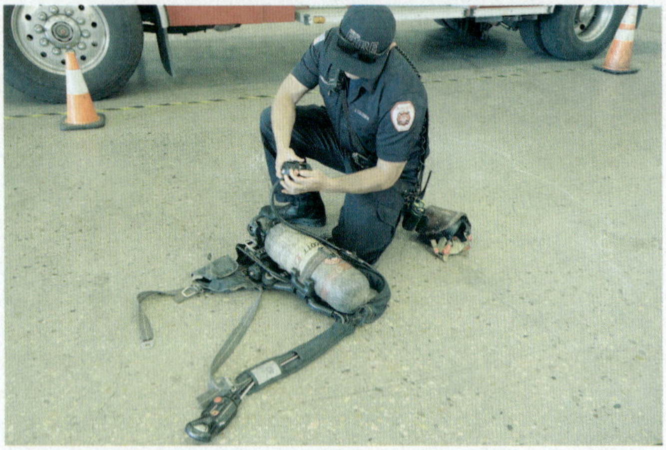

1. Check the regulator purge/bypass valve to be sure it is closed.

2. Depress the air saver/donning switch, if present, to stop the flow of air.

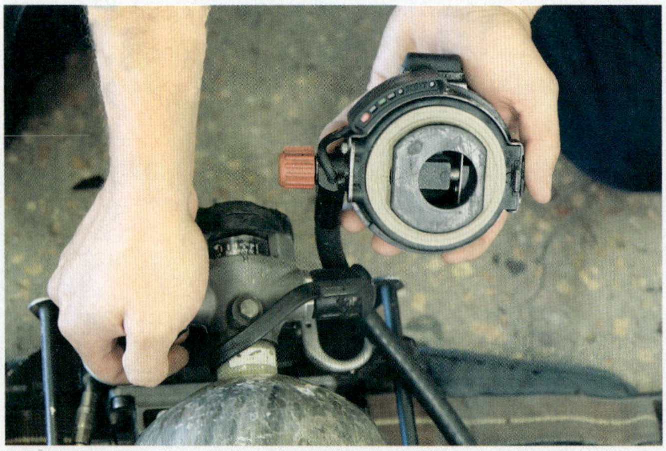

3. Slowly open the air-cylinder valve. Check for proper operation of the HUD and of the low-battery indicator. Confirm that the low-air alarm and the PASS device are working.

4. Check the remote pressure gauge for proper operation.

SKILL DRILL 3-4 CONTINUED

SCBA Operational Inspection Support Person, NFPA 1010: 5.5.2

5. Don the face piece. Adjust it to obtain a good seal. Inhale sharply to start the flow of air. Breathe normally to check for proper operation.

6. Remove the regulator or face piece; air should flow freely.

7. Depress the air saver/donning switch to stop the flow of air.

8. Open the regulator purge/bypass valve to check for air flow.

Continues.

SKILL DRILL 3-4 CONTINUED

SCBA Operational Inspection Support Person, NFPA 1010: 5.5.2

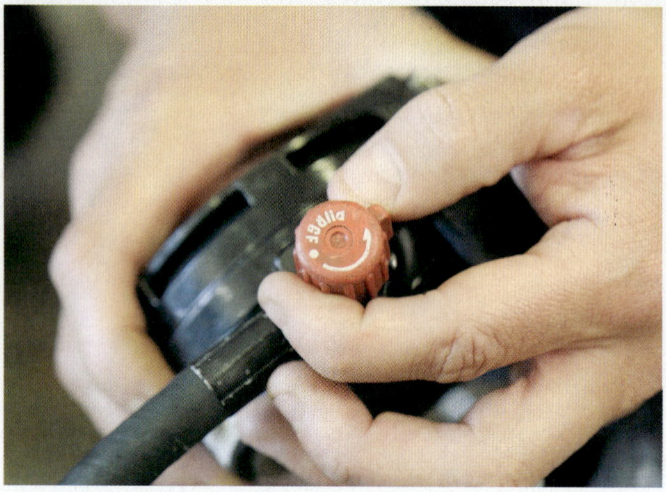

9. Close the regulator purge/bypass valve to stop the flow of air.

10. Rotate the air-cylinder valve to close it.

11. Open the regulator purge/bypass valve slightly to vent residual air pressure from the system. Watch the HUD to verify its proper operation as the air pressure is exhausted.

12. Once the air flow stops, close the regulator purge/bypass valve. Complete any reporting that is required.

© Jones & Bartlett Learning. Photographed by Brandon Fryman/JRM Creative.

Replacing SCBA Cylinders

A used air cylinder can be quickly replaced with a full cylinder in the field to enable a firefighter to continue firefighting activities. It's important for a firefighter to be sure they are physically able to fight a second round with the fire. It is better for the firefighter to allow themselves a few minutes in rehab than to get back into the fire immediately and require rescue from other crew members. Firefighters should not overtax themselves by replacing the air cylinder and going back to work without adequate rest when they need it.

A firefighter who is working alone must doff their SCBA harness to replace the air cylinder. Two firefighters who are working together can change each other's cylinders without removing their SCBA harness. This procedure may vary slightly depending on the model of SCBA being used. Follow the procedure recommended by the SCBA manufacturer and by your department's SOPs.

Practice changing air cylinders until you become proficient at this task. All personnel should be able to change an air cylinder in the dark and while wearing gloves if necessary. Follow the steps in **SKILL DRILL 3-5** to replace an SCBA air cylinder when you are working alone.

SKILL DRILL 3-5

Replacing an SCBA Cylinder Support Person, NFPA 1010: 5.5.1

1. Place the SCBA on the floor or a bench.

2. Close the air-cylinder valve.

3. Open the regulator purge/bypass valve to bleed off the pressure.

4. Disconnect the high-pressure supply hose. Keep the ends clean.

Continues.

SKILL DRILL 3-5 CONTINUED

Replacing an SCBA Cylinder Support Person, NFPA 1010: 5.5.1

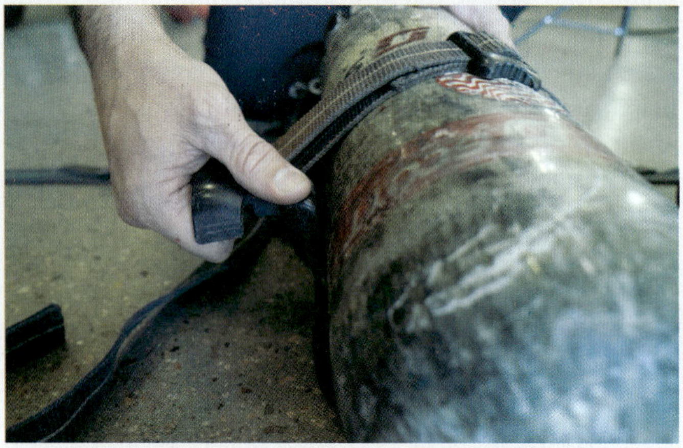

5. Release the air cylinder from the SCBA harness, and then remove the depleted air cylinder.

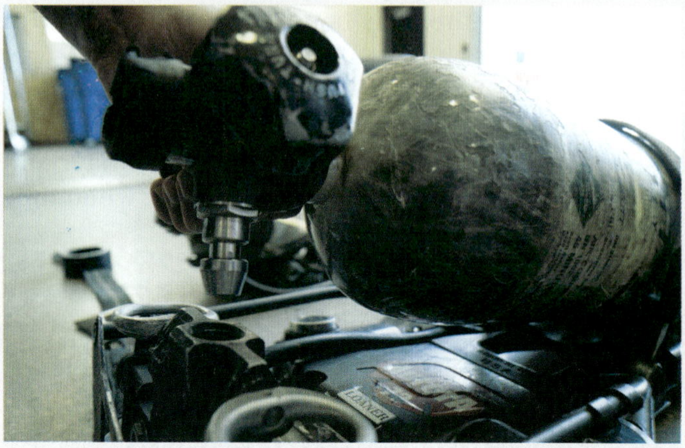

6. Slide a full air cylinder into the SCBA harness. Align the outlet to the supply hose. Lock the air cylinder in place.

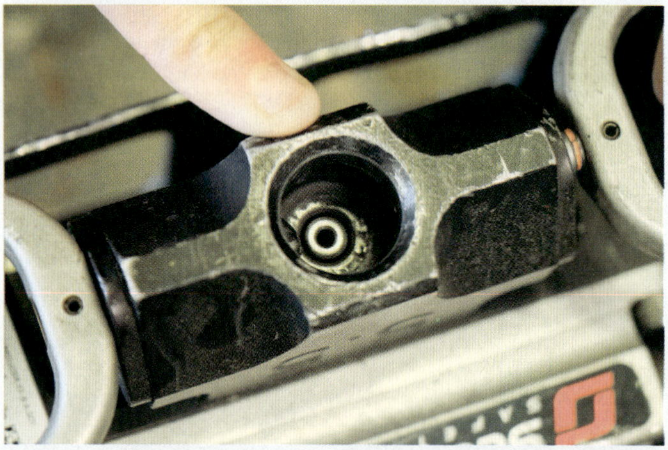

7. Check that the O-ring is present and in good shape.

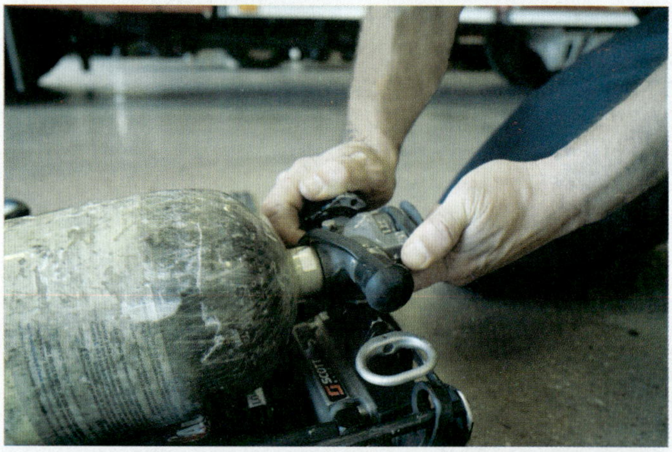

8. Connect the high-pressure hose to the air cylinder. Hand-tighten only.

9. Open the air-cylinder valve. Check the air-cylinder pressure gauge and the remote pressure gauge.

© Jones & Bartlett Learning. Photographed by Brandon Fryman/JRM Creative.

If a firefighter needs to quickly reenter a fire scene, a support person can replace their air cylinder while they continue to wear their SCBA harness. Follow the steps in **SKILL DRILL 3-6** to replace an SCBA air cylinder for a firefighter.

Refilling SCBA Cylinders

SCBA air cylinders can be refilled by equipment permanently located at a maintenance facility or at a firehouse or mounted on a truck or a trailer for mobile use.

SKILL DRILL 3-6

Replacing an SCBA Cylinder on a Firefighter Support Person, NFPA 1010: 5.5.1

1. Remove the regulator from the face piece or remove the face piece so that the firefighter can breathe ambient air.

2. Close the air-cylinder valve on the used air cylinder.

3. Open the regulator purge/bypass valve to bleed off pressure from the high-pressure supply line.

4. Disconnect the high-pressure supply line.

Continues.

SKILL DRILL 3-6 CONTINUED

Replacing an SCBA Cylinder on a Firefighter Support Person, NFPA 1010: 5.5.1

5. Release the SCBA air cylinder from the SCBA harness and set it aside.

6. Slide the full SCBA air cylinder into the SCBA harness.

7. Check for the presence and satisfactory condition of the O-ring.

8. Lock the SCBA air cylinder into place.

9. Connect the high-pressure line to the SCBA cylinder.

10. Open the air-cylinder valve, and then notify the firefighter that the SCBA cylinder change is complete. State the pressure reading of the SCBA air cylinder to the firefighter.

© Jones & Bartlett Learning. Photographed by Brandon Fryman/JRM Creative.

FIGURE 3-29 Compressors filter and compress atmospheric air before transferring it to an SCBA air cylinder.

© Jones & Bartlett Learning. Photographed by Glen E. Ellman.

FIGURE 3-30 Cascade systems store filtered and compressed breathing air in storage cylinders and transfer it to SCBA air cylinders.

Courtesy of BAUER COMPRESSORS INC.

Mobile filling units are often brought to the scene of a large fire. Before an air cylinder is refilled, the hydrostatic test date must be checked to ensure that its certification has not expired.

Two types of systems are used to refill air cylinders. A **compressor** filters atmospheric air, compresses it to a high pressure, and transfers it to the SCBA air cylinders (**FIGURE 3-29**). A **cascade system** has several large storage cylinders of compressed breathing air with pressure ranging from lower pressure to higher pressure connected by a high-pressure manifold system (**FIGURE 3-30**). The empty SCBA air cylinder is connected to the cascade system, and compressed air is transferred from the storage cylinders to the SCBA air cylinder. The specific steps for filling SCBA air cylinders vary with different systems.

Proper training is required to fill SCBA cylinders. Refilling SCBA cylinders requires special precautions because of the high pressures involved in this procedure. While it is being refilled, the SCBA cylinder must be placed in a shielded container designed to prevent injury if the cylinder ruptures. In addition, special procedures must be followed to ensure that the air used to fill the SCBA cylinder is not contaminated. Only personnel who have been trained on the safe use of your department's equipment should refill air cylinders. Follow the manufacturer's recommendations. Follow the steps in **SKILL DRILL 3-7** to safely fill SCBA air cylinders.

Cleaning and Sanitizing SCBA

An SCBA must be cleaned after each use. The first step in cleaning the SCBA is to rinse the entire unit with clean water using a hose. It is important to clean the outside of the SCBA *before* taking it apart. This prevents foreign substances from getting into SCBA connectors and internal parts. The SCBA harness assembly and air cylinder can be cleaned with a mild detergent or soap-and-water solution. If additional cleaning is needed, the unit can be scrubbed with a stiff brush. After scrubbing, the SCBA harness and air cylinder should be rinsed with clean water. Most SCBA manufacturers provide specific instructions for the care and cleaning of their models. Follow the

SKILL DRILL 3-7

Refilling SCBA Cylinders Support Person, NFPA 1010: 5.5.1

1. Complete the fill record form, including the date, hydrostatic test date, and air-cylinder serial number.

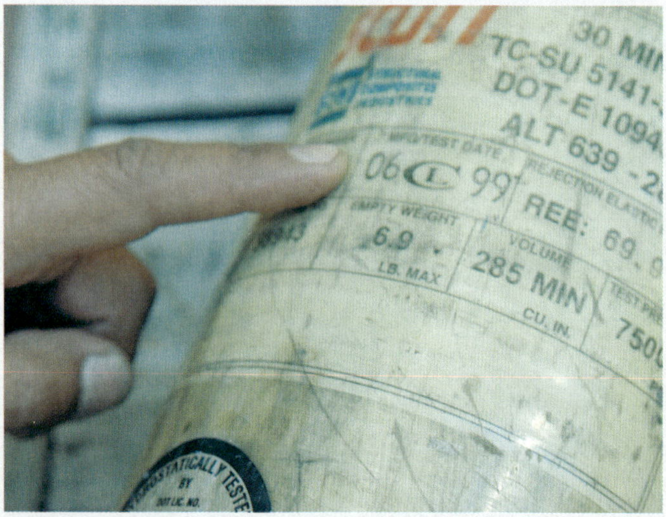

2. Ensure the air cylinder is safe to fill by checking its date of manufacture and the hydrostatic test date.

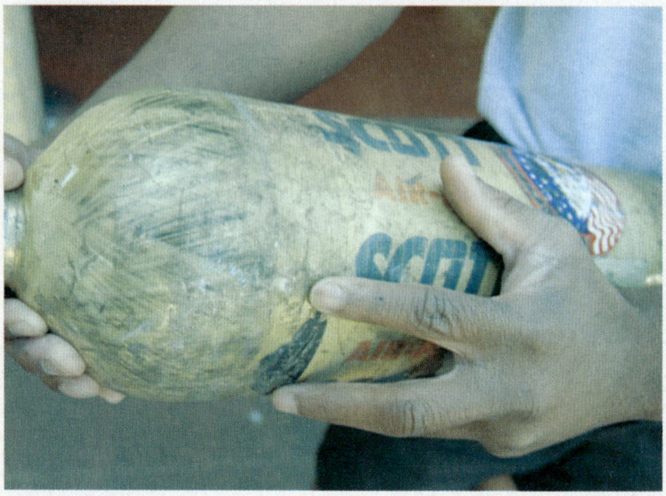

3. Check the air cylinder for visible damage. Follow the compressor or cascade system filling procedures. Secure the system after use according to system procedures.

© Jones & Bartlett Learning. Photographed by Glen E. Ellman.

manufacturer's cleaning instructions and the protocols of your department.

After a fire, face pieces and regulators can be cleaned with a mild detergent or soap-and-water solution or with a disinfectant cleaning solution. The face piece should be fully submerged in the cleaning solution. If additional cleaning is needed, a soft brush can be used to scrub the face piece. During the cleaning process, avoid scratching the lens or damaging the exhalation valve. The regulator can be cleaned with the same solution, but it should not be submerged. The face piece and regulator should then be rinsed with clean water. Follow the manufacturer's recommendations, and avoid the use of aerosol cleaners or any alcohol-containing cleaner because these materials degrade the rubber material in the face piece. Face pieces should be air dried or wiped with a soft, nonabrasive cloth to avoid scratching the lens.

Allow the SCBA time to dry completely before returning it to service. Also check for any damage before returning the equipment to service. Follow the steps in **SKILL DRILL 3-8** to clean and sanitize an SCBA.

SKILL DRILL 3-8

Cleaning an SCBA Support Person, NFPA 1010: 5.5.2

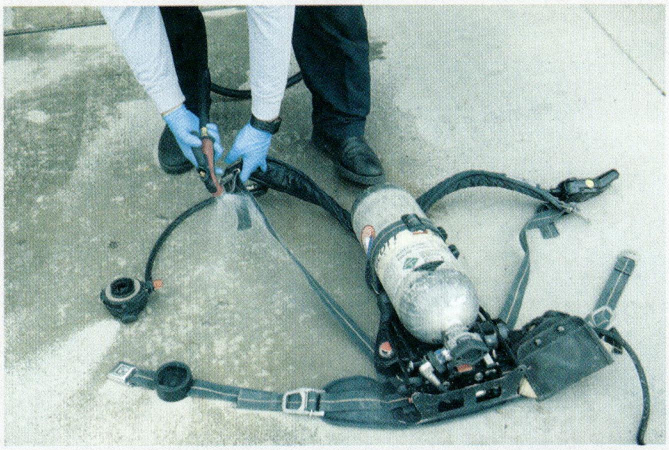

1. Rinse the entire unit using a hose with clean water. Inspect the SCBA for any damage that might have occurred before cleaning.

2. Detach the SCBA air cylinder from the harness. On some models, the regulator can be removed from the SCBA harness.

Continues.

SKILL DRILL 3-8 CONTINUED

Cleaning an SCBA Support Person, NFPA 1010: 5.5.2

3. Using a stiff brush, along with a mild detergent or soap-and-water solution, scrub the SCBA air cylinder and harness. Rinse and set these pieces aside to dry.

4. In a 5-gallon (19-liter) bucket, make a mild detergent-and-water or soap-and-water solution or prepare the manufacturer's recommended cleaning and disinfecting solution and water. Submerge the SCBA face piece in the cleaning solution. For heavier cleaning, allow the face piece to soak.

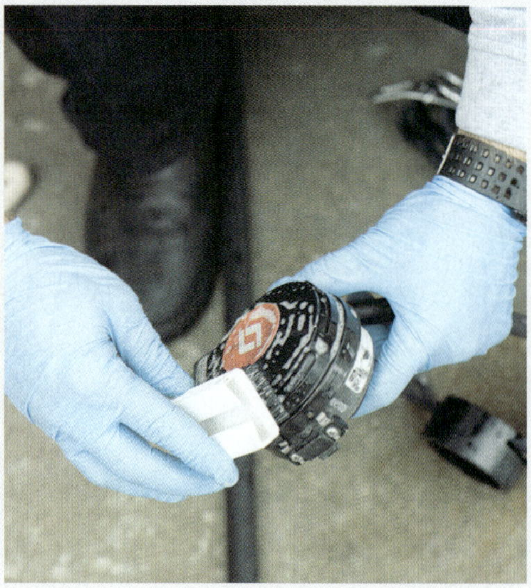

5. Clean the regulator with the soapy water or cleaning solution, following the manufacturer's instructions. Use a soft brush, if necessary, to scrub contaminants from the face piece and regulator.

6. Completely rinse the face piece and the regulator with clean water. Do *not* submerge the regulator. Set them aside and allow them to dry. Reassemble and inspect the entire SCBA before placing it back in service.

© Jones & Bartlett Learning. Photographed by Glen E. Ellman.

CASE STUDY

You Are the Support Person CONCLUSION

One cool fall evening, you are dispatched for a report of a smoke odor in the area. These calls are normal for this time of year as residents start to use their fireplaces for the first time. As the engine drives slowly through the neighborhood, you notice the distinctive smell of a working house fire rather than that of firewood. When you turn the corner, you see smoke drifting from the basement window of a house. Your captain completes a quick size-up and tells two firefighters to pull a cross-lay and attack the fire.

1. **Which personal protective equipment is needed for this situation?**

   Answer: When participating in a fire attack, a firefighter needs to put on the entire structural firefighting ensemble and wear an SCBA.

2. **Once the fire is over, what is the proper way to care for the protective equipment?**

   Answer: First, another member of the team needs to spray the firefighter down with a small-diameter hose to remove surface contaminants. Then the firefighter must doff their PPE and use a brush and water to remove as much dirt and debris as possible. Finally, they must transport the PPE to the station according to the department's SOPs and then thoroughly clean it following the manufacturer's instructions.

WRAP-UP

SUMMARY

- PPE is an essential component of firefighter safety. To be effective, the entire structural firefighting protective ensemble must be worn when attacking a structure fire.
- Smoke from a fire can create highly toxic and unpredictable atmospheric changes. SCBA is an essential component of structural firefighting that provides clean breathing air to firefighters.
- Firefighters must consider the limitations of the SCBA (limited windows of use relative to the amount of breathable air, weight of the apparatus, size and shape of the apparatus, and the reduction of normal sensory awareness while wearing the apparatus), as well as their own physical and psychological limitations.
- Support personnel must be proficient in assisting firefighters with their SCBAs, including cleaning, inspections, and exchanging and replacing air cylinders.

KEY TERMS

air cylinder See *breathing air cylinder.*

air-cylinder pressure gauge The device on an SCBA that measures and displays pressure readings to indicate the quantity of breathing air available.

air-line respirator See *supplied-air respirator.*

air-purifying respirator (APR) A respirator that removes specific air contaminants by passing ambient air through one or more air purification components. (NFPA 1984)

atmosphere-supplying respirator (ASR) A respirator that supplies the respirator user with breathing air from a source independent of the ambient atmosphere and includes self-contained breathing apparatus (SCBA) and supplied-air respirators (SARs). (NFPA 1970)

KEY TERMS CONTINUED

bib The lower part of the protective hood that is part of the structural firefighting ensemble.

breathing air cylinder The pressure vessel or vessels that are an integral part of the self-contained breathing apparatus (SCBA) and that contain the breathing gas supply; can be configured as a single cylinder or other pressure vessel or as multiple cylinders or pressure vessels. (NFPA 1970)

bunker coat See *structural firefighting protective coat.*

bunker gear See *structural firefighting protective clothing.*

bunker pants See *structural firefighting protective trousers.*

cascade system A method of piping air tanks together to allow air to be supplied to the self-contained breathing apparatus (SCBA) fill station using a progressive selection of tanks, each with a higher pressure level. (NFPA 1900)

closed-circuit self-contained breathing apparatus (closed-circuit SCBA) A recirculation-type self-contained breathing apparatus (SCBA) in which the exhaled gas is rebreathed by the wearer after the carbon dioxide has been removed from the exhalation gas and the oxygen content within the system has been restored from sources, such as compressed breathing air, chemical oxygen, liquid oxygen, or compressed gaseous oxygen. Also called *rebreather.* (NFPA 1970)

compressor A device used for increasing the pressure and density of a gas. (NFPA 853)

doff The process of properly removing a member's personal protective equipment (PPE) and respiratory protection to limit additional contamination and exposure. (NFPA 1700)

don The process of properly dressing in full personal protective equipment (PPE), ensuring all exposed skin and the airway are protected. (NFPA 1700)

drag rescue device (DRD) A fabric handle integrated just below the collar at the back of the protective coat that a rescuer can grab to drag an incapacitated firefighter to safety.

dual-path pressure reducer A feature that automatically provides a backup method for air to be supplied to the regulator of a self-contained breathing apparatus (SCBA) if the primary passage malfunctions.

end-of-service-time-indicator (EOSTI) A warning device on a self-contained breathing apparatus (SCBA) that alerts the user that the reserved air supply is being utilized. (NFPA 1970)

face piece The part of the SCBA consisting of the face mask and exhalation valve that delivers breathing air to the firefighter and protects the face from high temperatures and smoke.

hand light A small, portable light carried by firefighters to improve visibility at emergency scenes; it is often powered by rechargeable batteries.

heads-up display (HUD) Visual display of information and system conditions status that is visible to the wearer. (NFPA 1970)

hydrostatic test A test performed by filling pressure-containing components completely with water or other incompressible fluid while expelling all contained air, closing or capping all open ports of the pressure-containing components, and then raising and maintaining the contained pressure to pressurize the pressure-containing components to a prescribed value through an externally supplied pressure-generating device. (NFPA 1900)

immediately dangerous to life and health (IDLH) Any condition that would pose an immediate or delayed threat to life, cause irreversible adverse health effects, or interfere with an individual's ability to escape unaided from a hazardous environment. (NFPA 1700)

light-emitting diodes (LEDs) Electronic semiconductors that emit a single-color light when activated. LEDs are used for operational displays in self-contained breathing apparatus (SCBA).

nose cups An insert inside the face piece of a self-contained breathing apparatus (SCBA) that fits over the user's mouth and nose.

Occupational Safety and Health Administration (OSHA) The U.S. federal agency that regulates worker safety and, in some cases, responder safety. It is part of the U.S. Department of Labor.

off-gas To emit harmful chemicals in the form of a gas.

open-circuit self-contained breathing apparatus (open-circuit SCBA) An SCBA in which the exhaled air is released into the atmosphere and is not reused.

overhaul A firefighting term involving the process of final extinguishment after the main body of the fire has been knocked down. All traces of fire must be extinguished at this time. (NFPA 1700)

personal alert safety system (PASS) A device that continually monitors for lack of movement of the wearer and automatically activates an alarm signal, indicating the wearer is in need of assistance; can also be manually activated to trigger the alarm signal. (NFPA 1970)

personal protective equipment (PPE) The full complement of garments firefighters are required to wear while on an emergency scene, including turnout coat, protective trousers, firefighting boots, firefighting gloves, a protective hood, self-contained breathing apparatus (SCBA), a personal alert safety system (PASS) device, and a helmet with eye protection. (NFPA 1010)

powered air-purifying respirator (PAPR) An air-purifying respirator that uses a powered blower to force the ambient air through one or more air-purifying components to the respiratory inlet covering. (NFPA 1984)

rapid intervention crew/company (RIC) A dedicated crew of at least one officer and three members, positioned outside the immediately dangerous to life and health (IDLH) area, trained and equipped as specified in NFPA 1407, who are assigned for rapid deployment to rescue lost or trapped members. Also called *rapid intervention team (RIT)*. (NFPA 1550)

rapid intervention crew/company universal air connection (RIC UAC) A system that allows emergency replenishment of breathing air to the self-contained breathing apparatus (SCBA) of disabled or entrapped fire or emergency services personnel. (NFPA 1970)

rapid intervention team (RIT) See *rapid intervention crew/company (RIC)*.

rebreather See *closed-circuit self-contained breathing apparatus (closed-circuit SCBA)*.

regulator purge/bypass valve A device or devices designed to bypass a regulator.

remote pressure gauge The device on a self-contained breathing apparatus (SCBA) that measures and displays pressure readings to indicate the quantity of breathing air available and that is located on the shoulder strap or in another location where it can be seen by the user while the SCBA is in use.

SCBA harness The backpack or frame for mounting the working parts of the self-contained breathing apparatus (SCBA) and the straps and fasteners used to attach the SCBA to the firefighter.

SCBA regulator The part of the self-contained breathing apparatus (SCBA) that reduces the high pressure in the cylinder to a usable lower pressure and controls the flow of air to the user.

self-contained breathing apparatus (SCBA) An atmosphere-supplying respirator that supplies a respirable air atmosphere to the user from a breathing-air source that is independent of the ambient environment and designed to be carried by the user. (NFPA 1970)

smoke The airborne solid and liquid particulates and gases evolved when a material undergoes pyrolysis or combustion, together with the quantity of air that is entrained or otherwise mixed into the mass. (NFPA 1700)

soot Black particles of carbon produced in a flame. (NFPA 1700)

structural firefighting The activities of rescue, fire suppression, and property conservation in buildings or other structures, vehicles, railcars, marine vessels, aircraft, or like properties. (NFPA 1010)

structural firefighting protective clothing All of the clothing elements of the structural firefighting protective ensemble.

structural firefighting protective coat The element of the protective ensemble that provides protection to the upper torso and arms, excluding the hands and head. Also called *bunker coat* or *turnout coat*. (NFPA 1970)

structural firefighting protective ensemble Multiple elements of compliant protective clothing and equipment that, when worn together, provide protection from some risks, but not all risks, of emergency incident operations. (NFPA 1970)

structural firefighting protective footwear The element of the protective ensemble that provides protection to the foot, ankle, and lower leg. (NFPA 1970)

structural firefighting protective glove The element of the protective ensemble that provides protection to the hand and wrist. (NFPA 1970)

structural firefighting protective helmet The element of the protective ensemble that provides protection to the head. (NFPA 1970)

structural firefighting protective hood The interface element of the protective ensemble that provides limited protection to the coat/helmet/self-contained breathing apparatus (SCBA) fac-piece interface area. (NFPA 1970)

structural firefighting protective trousers The element of the protective ensemble that provides protection to the lower torso and legs, excluding the ankles and feet. Also called *bunker pants* or *turnout pants*. (NFPA 1970)

supplied-air respirator (SAR) An atmosphere-supplying respirator for which the source of breathing

KEY TERMS CONTINUED

air is not designed to be carried by the user. Also called *air-line respirator.* (NFPA 1852)

turnout coat See *structural firefighting protective coat.*

turnout gear See *structural firefighting protective clothing.*

turnout pants See *structural firefighting protective trousers.*

two-way radios Portable communication devices used by firefighters. Every firefighting team should carry at least one radio to communicate distress, progress, changes in fire conditions, and other pertinent information.

universal emergency breathing safety system (EBSS) A device on a self-contained breathing apparatus (SCBA) that allows users to share their available air supply in an emergency situation.

REVIEW QUESTIONS

1. What are the six clothing items that make up the set of protective clothing for structural firefighting?
2. Which NFPA standard do structural firefighting protective coats need to meet?
3. How many layers are structural firefighting protective coats and pants constructed with, and what do they consist of?
4. How often should a routine inspection be performed on structural firefighting gear?
5. What percentage of oxygen does outside or room air contain? At what percentage of oxygen do people start experiencing disorientation and other side effects?
6. What are the two categories of breathing apparatus?
7. Which type of SCBA is typically used for structural firefighting?
8. What is the realistic useful life of an SCBA air cylinder?
9. What are the four main parts of an SCBA?
10. What is the range of air pressure in filled SCBA air cylinders?
11. What does the EOSTI alarm indicate?
12. What components make up the complete PPE ensemble?
13. Which agency requires periodic hydrostatic testing of SCBA air cylinders?
14. How often should an SCBA be cleaned?

DISCUSSION QUESTIONS

1. What are some of the reasons you should practice donning all the components of the structural firefighting protective ensemble quickly and correctly?
2. Discuss some of the psychological limitations of the SCBA user and explain why they might be stressful.
3. Explain why it is important to clean and store your firefighting gear appropriately.

REFERENCES

American National Standards Institute and the International Safety Equipment Association. 2020, September. *American National Standard for High-Visibility Safety Apparel* (ANSI/ISEA 107-2020). Arlington, VA: ISEA.

Centers for Disease Control and Prevention (CDC). 2021. "Respirators." Last reviewed August 22, 2021. Accessed June 23, 2023. www.cdc.gov/niosh/topics/respirators/default.html.

Mensch, Amy, George Braga, and Nelson Bryner. 2011. "Fire Exposures of Fire Fighter Self-Contained Breathing Apparatus Facepiece Lenses" (NIST Technical Note 1724). National Institute of Standards and Technology, November 2011. Accessed June 22, 2023. http://ws680.nist.gov/publication/get_pdf.cfm?pub_id=909917.

National Fire Protection Association. 2017. *NFPA 1404, Standard for Fire Service Respiratory Protection Training.* 2018 Edition. Quincy, MA: National Fire Protection Association.

National Fire Protection Association. 2018. *NFPA 1852, Standard on Selection, Care, and Maintenance of Open-Circuit Self-Contained Breathing Apparatus (SCBA).* 2019 Edition. Quincy, MA: National Fire Protection Association.

National Fire Protection Association. 2019. *NFPA 853, Standard for the Installation of Stationary Fuel Cell Power Systems.* 2020 Edition. Quincy, MA: National Fire Protection Association.

National Fire Protection Association. 2019. *NFPA 1851, Standard on Selection, Care, and Maintenance of Protective Ensembles for Structural Fire Fighting and Proximity Fire Fighting.* 2020 Edition. Quincy, MA: National Fire Protection Association.

National Fire Protection Association. 2020. *NFPA 99, Health Care Facilities Code.* 2021 Edition. Quincy, MA: National Fire Protection Association.

National Fire Protection Association. 2020. *NFPA 1700, Guide for Structural Fire Fighting.* 2021 Edition. Quincy, MA: National Fire Protection Association.

National Fire Protection Association. 2021. *NFPA 1977, Standard on Protective Clothing and Equipment for Wildland Fire Fighting.* 2022 Edition. Quincy, MA: National Fire Protection Association.

National Fire Protection Association. 2021. *NFPA 1984, Standard on Respirators for Wildland Fire Fighting Operations and Wildland Urban Interface Operations.* 2022 Edition. Quincy, MA: National Fire Protection Association.

National Fire Protection Association. 2021. *NFPA 1990, Standard for Protective Ensembles for Hazardous Materials and CBRN Operations.* 2022 Edition. Quincy, MA: National Fire Protection Association.

National Fire Protection Association. 2023. *NFPA 1010, Standard on Professional Qualifications for Firefighters.* 2024 Edition. Quincy, MA: National Fire Protection Association.

National Fire Protection Association. 2023. *NFPA 1550, Standard for Emergency Responder Health and Safety.* 2024 Edition. Quincy, MA: National Fire Protection Association.

National Fire Protection Association. 2023. *NFPA 1900, Standard for Aircraft Rescue and Firefighting Vehicles, Automotive Fire Apparatus, Wildland Fire Apparatus, and Automotive Ambulances.* 2024 Edition. Quincy, MA: National Fire Protection Association.

National Fire Protection Association. 2023. *NFPA 1970, Standard on Protective Ensembles for Structural and Proximity Firefighting, Work Apparel and Open-Circuit Self-Contained Breathing Apparatus (SCBA) for Emergency Services, and Personal Alert Safety Systems (PASS).* 2024 Edition. Quincy, MA: National Fire Protection Association.

Occupational Safety and Health Administration (OSHA). n.d. "Oxygen Concentration Efforts." Accessed July 27, 2023. www.osha.gov/sites/default/files/2018-12/fy15_sh-27664-sh5_Confined_Space_Handout_Effects_of_Oxygen.pdf.

Occupational Safety and Health Administration. 1970. "Occupational Safety and Health Standards: Personal Protective Equipment, Respiratory Protection." C.F.R. 1910.124, updated September 26, 2019. Accessed July 26, 2023. www.osha.gov/laws-regs/regulations/standardnumber/1910/1910.134.

Silverman, Steven H., Gary F. Purdue, John L. Gary F., Hunt, and Robert O. John L., Bost., Robert O. 1988. "Cyanide Toxicity in Burned Patients." *Journal of Trauma: Injury, Infection, and Critical Care* 28, no. 2, 171–176.

U.S. Department of Transportation. 2017. "General Requirements for Requalification of Specification Cylinders." 49 C.F.R. 180.205. Accessed June 18, 2018. www.ecfr.gov/cgi-bin/text-idx?SID=475de645c98062d4043077d9794a583f&mc=true&node=se49.3.180_1205&rgn=div8.

Chapter Opener: © Eric Scruggs

CHAPTER

4

Support Person

Fire Service Communications

KNOWLEDGE OBJECTIVES

After studying this chapter, you will be able to:

- Explain methods of receiving emergency and nonemergency fire department communications.
- Identify other modes of fire service communication.
- Describe the procedures for reporting an emergency.
- Describe the procedures for handling emergency calls.
- Explain the procedures for transmitting emergency information to a dispatch center.
- Explain the importance of following departmental standard operating procedures (SOPs) for receiving and processing communications.
- Identify the information to be obtained when taking a report of an emergency to enable necessary assistance to be dispatched.
- Identify radio departmental procedures and codes for using fire department radios.
- Describe the basic procedures and etiquette of effective radio communications.
- Describe when to use plain language and how ten-codes are implemented in communications.
- Outline the information provided in size-up and progress reports.
- Recognize routine traffic, emergency traffic, and emergency evacuation signals.
- Outline the information provided in size-up and progress reports.

SKILLS OBJECTIVES

After studying this chapter, you will be able to perform the following skills:

- Receive a phone call and obtain, route, and document information according to department procedures.
- Demonstrate the relay of information to the dispatch center.
- Demonstrate the ability to record information according to department standard operating procedures (SOPs).
- Demonstrate the ability to operate fire department communications equipment and technology.
- Demonstrate how to send and receive messages over the fire department radio.

ADDITIONAL NFPA STANDARDS

- **NFPA 72**, *National Fire Alarm and Signaling Code, 2022 Edition*
- **NFPA 450**, *Standard for Emergency Services Communications, 2021 Edition*
- **NFPA 1225**, *Standard for Emergency Services Communications, 2022 Edition*
- **NFPA 1410**, *Standard on Training for Emergency Scene Operations, 2020 Edition*
- **NFPA 1900**, *Standard for Aircraft Rescue and Firefighting Vehicles, Automotive Fire Apparatus, Wildland Fire Apparatus, and Automotive Ambulances, 2024 Edition*

CASE STUDY

You Are the Support Person

At 0304 hours, your crew is dispatched to a report of a fire at 3256 West Madison Street. While en route, the dispatcher advises that the caller reported that smoke was coming from the eaves of the building and then hung up. With this new information, the crew prepares for a possible house fire. The dispatcher radios back and says that this may be a false report because the caller used a cell phone that was identified as being on the east side of town. It seems like you may have been woken up for another false call, but then the dispatcher comes back on the radio with a frantic sound in her voice and says that a home security company just reported a fire alarm at 3256 East Madison Street.

1. What is the process for receiving 911 calls in an emergency communications center?
2. How does the dispatcher know where the cell phone call originated?
3. How are alarm systems monitored and then reported to the communications center?

Introduction

Fighting a fire requires the coordination of numerous resources and people. Different crews and equipment fill various functions on the fire ground, many of which occur simultaneously or must not start until something else has been accomplished. Ensuring a smooth interaction between all of the firefighting tasks being performed requires good communication among fire suppression crews, officers, and members of the Incident Command System (ICS). In addition, fire departments need to have good communication protocols, techniques, and equipment to manage incoming calls for emergency assistance. Finally, fire service communications also encompass postincident reporting because sharing data and information about the cause of the fire and activities undertaken to suppress a fire is important for preventing fires and being able to suppress them as safely and efficiently as possible.

Every fire department depends on a functional **public safety communications center**—otherwise known simply as a **communications center**—which is the location where 911 calls for that community or jurisdiction are directed. When someone requests assistance or an alarm sounds, the communications center dispatches the appropriate units to the incident. **Dispatch** means to select units to respond to a reported incident, alert those units, and then communicate the request for service by quickly and accurately transmitting the information to them. The communications center maintains communication with those units during the incident. It also tracks the location and status of every other fire department unit. The communications center must always know which units are available to be dispatched to an incident, and it must be able to contact those units promptly. The communications center is responsible for redeploying units to maintain adequate coverage for all areas. It is the link between firefighters on the scene, the rest of the organization, and others. The communications center monitors communications from the incident scene and processes all requests for assistance or special resources.

At the scene, fire service personnel need to communicate with one another so that the incident commander (IC) can manage the operation efficiently based on progress reports or requests for assistance from firefighters. The ICS depends on the presence of a functional onsite communications system. During incidents, fire service personnel must be able to communicate not only with one another but also with other emergency response agencies.

In addition to these special communications requirements, a fire department must have a communications infrastructure that allows it to function as an effective organization. Basic administration and day-to-day management require an efficient communications network, including telephone and data links with every fire station and work site. Rapidly developing technology and advanced communications systems are improving fire department communications. As a support person, you must be familiar with the communications systems, equipment, and procedures used in your department.

Communications Centers

A large percentage of the United States has access to some type of 911 system to report an emergency. These calls are directed to a **public safety answering**

point (PSAP), also known as a 911 call center. PSAPs are often co-located within the jurisdiction's *public safety communications center* or *communications center*. There are several types of communications centers. A communications center may be a **stand-alone communications center** serving a single specific fire department, or it may be a regional center serving many fire departments. In addition, it may be co-located with other communications centers of various public safety agencies, or it may be integrated, bringing police, fire, and emergency medical services (EMS) into one location. The primary responsibilities of the communications center are to process 911 calls, as well as other nonemergency public safety calls.

When fire, EMS, and law enforcement communications are in the same facility, calls can be answered, processed, and dispatched immediately. A communications center may have independent personnel and systems for each emergency responding agency, or all employees may be trained to receive calls and dispatch responders to any type of emergency incident. If each of these agencies has a stand-alone communications center that operates in separate facilities, each call must be transferred to the appropriate communications center. Some agencies have a mobile communications center that allows dispatchers to be onsite and run communications for a larger incident.

Regardless of how or where the call is transferred, the public safety communications center is the hub of the emergency response system. It serves as the central processing point for all information relating to an emergency incident and all of the information relating to the location, status, and activities of responding units.

The size and complexity of the communications center vary depending on the needs of the department. For example, the communications center for a small, rural department might be a small room in a fire station staffed by one person at a time. The communications center for a large department with multiple stations in an urban area might be a specially designed, highly sophisticated facility with advanced technological equipment that requires special training and multiple personnel to be onsite. Regardless of their size, location, and configuration, all communications centers perform the same basic functions (**FIGURE 4-1**).

FIGURE 4-1 All communications centers perform the same basic functions.

© Jones & Bartlett Learning. Photographed by Glen E. Ellman.

Dispatchers

The employee who staffs a communications center is called a **telecommunicator**, more commonly called a **dispatcher**. Dispatchers receive, process, and disseminate information; understand and follow complicated procedures; perform multiple tasks effectively; memorize information; and make decisions quickly. Dispatchers should be trained to work in a public safety communications environment and have completed advanced training and professional certification programs to ensure that skilled, competent individuals can fulfill these critical roles in the public safety system.

The job of a dispatcher can be complicated, demanding, and extremely stressful. A dispatcher must possess certain qualities to be successful. For example, one of the dispatcher's most important skills is the ability to communicate effectively to obtain critical information, even when the caller is highly stressed or in extreme personal danger. If an emotional caller criticizes or insults the dispatcher, the dispatcher must still respond professionally and focus on obtaining the essential information. Voice control and the ability to maintain composure under pressure are important qualities for dispatchers. They must always be clear, calm, and in control.

Dispatchers must be skilled in operating the systems and equipment in the communications center. They must understand and follow the fire department's operational procedures, particularly those relating to dispatch policies and protocols, radio communications, and incident management. The dispatcher must always keep track of the status and location of each unit and must monitor the overall deployment and availability of resources throughout the system. NFPA 1225, *Standard for Emergency Services Communications, 2022 Edition*, contains a complete list of qualifications for dispatcher candidates.

Communications Facility Center Requirements

The communications center should be designed and operated to ensure that its critical mission can be performed with a high degree of reliability. The performance requirements in NFPA 1225 govern the design and construction of public safety communications centers. These requirements apply whether the communications center serves a small community with only one or two fire stations or a metropolitan area with dozens of stations.

The communications center must be well protected against natural and human-made threats, and it must be able to withstand predictable damaging forces, such as floods, earthquakes, snowstorms, tornadoes, hurricanes, and other severe storms. It must be located so that it can function during times of civil unrest. The communications center must be able to operate at maximum capacity, without interruption, even when other community services are severely affected. The building must be equipped with emergency generators and other systems so that it can continue to operate for several days in even the most challenging conditions.

NFPA 1225 also requires backup systems for all of the critical equipment in a communications center so that the failure of a single component or system will not disable the entire operation. For example, there must be more than one way of transmitting a message from the communications center to each agency. A backup radio transmitter should be available, and the telephone system must be able to receive calls even if part of the system is damaged. The design of the facility must minimize its vulnerability to a fire originating inside the building as well as to nearby fires. The center also must be secured to prevent unauthorized entry.

A backup communications center at a different location must be established. If some unanticipated situation makes it impossible to operate from the primary location, this backup location can be activated to ensure ongoing operation of the communications function.

Plans need to be in place describing the procedure for reporting emergencies if the emergency communications center fails. Some communities locate fire department and law enforcement vehicles at strategic locations in the community so that citizens can report emergencies. Agencies may plan to staff a watch desk to take reports of emergencies from citizens who walk into the building.

Communications Center Equipment

Depending on the size of the communications center and the specific role of each, most communications centers have the following equipment:

- Phone lines dedicated to receiving 911 calls
- Phone lines that connect directly to other agencies
- Nonemergency phone lines
- Equipment to receive alarms from public or private fire alarm systems
- Teletype (TTY)/telecommunications device for the deaf (TDD) equipment
- Computers that use a **geographic information system (GIS)**, which is an application that layers a wide variety of data about a community on top of a map of the community. For example, users can see the location of gas lines and fire hydrants; demographic data; topographical data, including which areas are prone to flooding; and much more.
- Equipment for alerting and dispatching units to emergency calls
- Two-way radio system(s)
- Recording systems to record phone calls and radio communications
- System for maintaining and managing records
- Backup generators to supply electricity in case of an electrical outage

Computer-Aided Dispatch

Many communications centers have **computer-aided dispatch (CAD)**, which is a combination of hardware and software that assists dispatchers by performing specific functions more quickly and efficiently than doing them manually (**FIGURE 4-2**). For example, once the address of the incident is determined and the incident description is in the CAD system, the system can make a recommendation on the appropriate units to dispatch very quickly. Data links that provide the location of the caller and automatically enter the address can also save time. If duplicate addresses exist or if the address entered is not a valid location, the CAD system prompts the dispatcher to ask the caller for more information.

The most advanced type of CAD system utilizes **global positioning system (GPS)** data devices to confirm the location of the fire department units. The CAD system determines which units are available to respond to a call and the closest fire stations in order of response

FIGURE 4-2 A CAD system enables a dispatcher to work more quickly and efficiently.

© Jones & Bartlett Learning. Photographed by Glen E. Ellman.

FIGURE 4-3 Some CAD systems transmit dispatch information directly to terminals in fire stations and MDTs in the apparatus.

© Jones & Bartlett Learning. Photographed by Glen E. Ellman.

for any location, and it keeps track of units assigned to incidents and units that are temporarily assigned to cover different areas. The CAD system then indicates the units that can respond quickly to an alarm, even if some of the units that would normally respond to the call are currently unavailable.

A major advantage of a CAD system is that it automatically captures and stores every event as it occurs. Before such systems were developed, someone had to write down and timestamp every response, including every radio transmission, and all of these hard-copy logs had to be retained for future reference.

Some CAD systems use the radio system or cellular signals to transmit dispatch information directly to **mobile data terminals (MDTs)** on an apparatus or to computers in the fire station (**FIGURE 4-3**). In addition, CAD systems can also provide immediate access to information such as preincident plans, hazardous materials lists, lockbox locations, and information such as whether persons with limited mobility reside at the address of the emergency call. By linking the CAD system to other data files, fire service crews gain access to even more useful information, such as travel route instructions, maps, and reference materials.

Voice Recorders and Activity Logs

Almost everything that happens in a communications center is recorded, 24 hours a day, by a voice recording system or an activity logging system. A **voice recording system** is a system that records communications over the phone and the radio. Digital recording systems, many of which are cloud based, allow a recording to be recalled at any time. Older recording systems have an instant-playback unit that allows the dispatcher to replay conversations only from the previous 10 to 15 minutes. Replaying calls is valuable if the caller speaks quickly, has a challenging-to-understand manner of speaking, hangs up, or is disconnected.

An **activity logging system** is a computer system that keeps a detailed record of every incident and activity that occurs. These records include every call received or made; every unit dispatched; and every significant event related to emergency incidents, such as reports of when a unit is en route and when it has arrived at the scene, when the incident is under control, and when the last unit leaves the scene.

Voice recorders and activity logs are maintained for several reasons. First, they serve as legal records of the official delivery of a government service by a public agency. These records may be required for legal proceedings, sometimes years after the incident occurred. They may be needed to defend a department's actions when questions are raised about an unfortunate outcome. These records document the events and often demonstrate that the organization and its employees performed ethically, responsibly, and professionally. They also make it difficult to hide an error if a mistake was made.

Second, records are valuable for reviewing and analyzing information about department operations. Good record-keeping enables a fire department to examine what happened on a particular call, as well as to measure workloads, system performance, activity trends, and other factors as part of its planning and budget preparation. For example, analysts and planners can use data from CAD systems to study deployment strategies and make the most efficient use of fire department resources.

Communications Center Operations

Basic functions that a dispatcher performs in a communications center include the following:

- Receiving calls reporting emergency incidents and dispatching the appropriate units
- Supporting the operations of the fire department and other response units delivering emergency services
- Coordinating operations among responding agencies
- Always keeping track of the status of each response unit
- Monitoring the level of coverage and dispatching additional units as needed
- Notifying designated individuals, such as on-call chiefs, and agencies, such as utility companies or critical incident stress debriefing teams, of particular events and situations
- Maintaining information required for dispatch purposes

All of these activities must be performed accurately and efficiently, even in the most challenging circumstances. For example, if a disastrous event that involves many members of the community occurs, the activity level in a communications center will increase rapidly as calls come in. Dispatchers must not let the chaos outside affect the operations inside. If the communications center fails to perform its mission, the emergency responders will not be able to deliver much-needed emergency services.

Most emergency responses begin when a call comes into the communications center, which then dispatches appropriate units. When a unit goes to an incident, it is called going on a **run**. When the dispatcher dispatches units, it is called *generating* or *creating a run.* If a citizen reports an emergency directly to a fire station or if a crew discovers a situation, the information and the request for help must immediately be relayed to the communications center. The communications center can then initiate the emergency response process.

The dispatcher's first responsibility is to obtain the information that is required to dispatch the appropriate units to the correct location (**FIGURE 4-4**). Each jurisdiction will have slightly different standard operating procedures (SOPs), but the dispatcher will usually use the CAD system information to confirm which agencies or units should respond and transmit the necessary information to them. As stated in NFPA 1225, the generally accepted performance objective is to dispatch units to the highest-priority calls within 60 seconds from the time a call reaches the communications center. Responding to a reported emergency without using the proper dispatch procedures can result in confusion and delay in the necessary resources reaching the emergency.

FIGURE 4-4 Dispatchers must obtain the needed information and relay it accurately to the appropriate responders.

© Jones & Bartlett Learning. Photographed by Glen E. Ellman.

The communications center must have both a primary and a backup method of transmitting alarms to stations. Although radio, telephone, and public address systems are often used to transmit information to fire stations, some communications centers use computers to transmit dispatch messages to MDTs in apparatus or to printers at fire stations. Some fire departments still use a system of bells or tones to transmit alarms. Volunteers or rural departments may use outdoor sirens or horns to summon firefighters to an emergency. Dispatch messages can also be transmitted over pagers or cell phones.

The major steps in processing an emergency incident include the following:

1. Call receipt
2. Location validation
3. Classification and prioritization
4. Unit selection
5. Dispatch

Call Receipt

Call receipt is the process of receiving a call for service and obtaining the information necessary to initiate a response. Calls come from the following sources:

- Telephone calls from the general public or other municipal agencies
- Municipal fire alarm systems

- Private automatic fire alarm systems
- Calls reported directly to fire stations
- Calls initiated via radio from fire or police units that are already "on the road"

Telephone Communication

The 911 calls received by a PSAP or communications center are generated mostly from a person calling on a phone, such as a cell phone, home or business landline, or a roadside emergency telephone. According to the National Emergency Number Association (NENA 2021), 80 percent of 911 calls come from wireless devices. Some calls are made from a phone using **Voice over Internet Protocol (VoIP)**. VoIP converts a person's voice into a digital signal that is sent via the Internet instead of over traditional phone lines. In most communities, phone calls to 911 connect the caller with a PSAP. The PSAP telecommunicator either takes the information directly from the caller or transfers the call to the appropriate agency, based on the nature of the emergency. In some systems, calling 911 connects the caller directly with a dispatcher, who obtains the required information and assigns the appropriate units.

Each communications center also has a normal 10-digit, nonemergency phone number. Some people use that number to report an emergency because they are not sure that their situation is serious enough to be considered a true emergency. Many communities have implemented the 311 system to handle nonemergency calls. The purpose of the 311 system is to reduce the number of nonemergency calls to 911. This system often links callers with community resources that can assist with a problem when it is not life threatening. Any number, whether it is listed as an emergency or nonemergency number, that is published as a fire department phone number should always be answered, including on nights and weekends.

The dispatcher who takes the call must conduct a telephone interview, asking the caller questions to obtain the required information. The first pieces of information the dispatcher needs to know are the location of the emergency and the nature of the problem. The dispatcher needs to remember that the caller thinks the situation is an emergency, and they must treat every call as an emergency until it is determined that no emergency exists. The caller may be distressed, scared, angry, or excited and unable to organize their thoughts to communicate the problem clearly. If English is not the caller's first language, the caller might not know the right words to explain the situation and might not understand any questions the dispatcher asks. Dispatchers must not allow a caller's strong emotions to affect their ability to do their job. They must speak calmly, remain professional, and use active listening techniques, following the agency's SOPs, to interpret the information. Many communications centers provide dispatchers with a structured set of questions designed to obtain accurate information on the nature of the situation.

Dispatchers should not allow gaps of silence to occur while questioning the caller. If the caller is suddenly silent, something may have happened to them, or the caller may be in personal danger. Conversely, if the dispatcher is silent, the caller might think that the dispatcher is no longer on the line or no longer listening.

Disconnects are another problem dispatchers need to be prepared to handle. Callers to 911 might hang up accidentally or be disconnected after reaching the communications center and before describing the reason for the call. The dispatcher should first attempt to call the person back using the automatic number identification from the initial call. If the dispatcher cannot reach the caller and the caller's location is already known, the dispatcher can dispatch a police officer to the location to determine if more help is needed.

With just two critical pieces of information—the location and nature of the problem—the dispatcher can initiate a response. Local guidelines or SOPs, however, may require the dispatcher to obtain additional information. Getting the caller's name and confirming their phone number is useful in case it is necessary to call them back to obtain additional information.

If the caller is in danger or distress, the dispatcher should try to keep the line open and remain in contact with the caller until help arrives, or in some cases, the dispatcher can advise the caller what to do next. For example, if a building fire is being reported, the dispatcher might advise the occupants to evacuate and wait outside. If the 911 call is for a medical incident, it will be routed to a dispatcher trained to provide medical self-help instructions to callers based on the caller's description of the patient's symptoms. These specially trained dispatchers are called **emergency medical dispatchers (EMDs)**.

On occasion, a communications center may receive calls about issues that cannot be handled by the fire department or other agencies participating in the communications system. If this happens, the dispatcher should make every effort to accommodate the caller and, if possible, transfer the caller to the proper agency and provide the caller with the proper contact information in case the transfer fails. The caller will appreciate the assistance and retain a positive image of the department's service.

FIGURE 4-5 TTY/TDD systems enable people with speech or hearing impairments to communicate via text using a phone.

© Jones & Bartlett Learning

In addition to receiving calls from different types of phones, communications centers receive calls from the following:

- TTY/TDD systems
- Text-to-911
- Direct-line phones
- Multi-line telephone systems

TTY/TDD Systems. People with a speech or hearing impairment can communicate by phone using a **TTY/TDD system**, which has a keyboard and a screen for displaying text rather than transmitting audio, physically connected to a phone or a computer (**FIGURE 4-5**). The Americans with Disabilities Act (ADA) requires that every call-taking position within a PSAP must have its own TTY or TTY-compatible equipment and be able to receive calls via TTY/TDD systems (ADA.gov 2016, 35.161 and 35.162). Dispatchers must know how to operate this equipment. A communications center might have a dedicated telephone line with its own 10-digit number, but the **Federal Communications Commission (FCC)** states that where 911 is not available and a PSAP provides emergency services via a 7- or 10-digit number, it still must provide direct, equal access to TTY callers.

Text-to-911. Text-to-911 is currently available only in certain locations. The FCC, which regulates interstate and international communications between the United States and other countries by radio, television, wire, satellite, and cable, encourages emergency call centers to begin accepting texts, but it is up to each call center to decide the particular method in which to implement and deploy text-to-911 technology (FCC 2015).

Direct-Line Phones. It is important to have a second means to communicate with emergency agencies and responding units if the cellular network or radio system becomes inoperable. A **direct-line phone** (or **ring-down phone**) connects two predetermined points. When there is a phone at either end of the direct line, picking up the phone at one end causes the phone at the other end to ring. Direct lines may connect the communications center to each fire station in its jurisdiction. They also often link police and fire communications centers or two fire communications centers that serve adjacent areas. Direct lines also may connect hospitals, private alarm companies, utility companies, airports, and similar facilities with the communications center and each other.

Multi-Line Telephone Systems. A multi-line telephone system (MLTS) is a phone system that allows multiple users throughout an organization, such as a business, a hotel, or a university, to make and receive calls at the same time. In most cases, when someone calls 911 from a phone connected to an MLTS, the caller's location within the organization, such as a floor or room number, is transmitted along with the street address.

Municipal Fire Alarm Systems

Some communities still have a **municipal fire alarm system**, which is a network of fire alarm boxes and emergency telephones on street corners or in public places. Although these have largely been replaced by landlines and cell phones, they still are used in areas where there are no landlines and cell phone service is unreliable. A **fire alarm box** transmits a coded signal to the communications center and, in some cases, directly to individual fire departments. A manual fire alarm box, often called a "pull box," is activated when the lever is pulled (**FIGURE 4-6**). The coded signal identifies the location of the box, but it does not indicate the type of emergency that is occurring. Fire alarm boxes in municipal fire alarm systems may also be connected to automatic fire alarm systems in private commercial, industrial, and residential buildings. When the building's fire alarm system is activated, the connected municipal fire alarm box transmits the alarm to the communications center. Municipal fire alarm boxes are usually connected by networks of dedicated cables or through a dedicated radio frequency.

If a community chooses to keep its municipal fire alarm system, in many cases, they replace the boxes connected to the system by underground cables with radio-operated fire alarm boxes, which use radio waves to transmit the signal (**FIGURE 4-7**).

FIGURE 4-6 Fire alarm boxes are still in use in some parts of the country.

© Beschi Mauro/Shutterstock

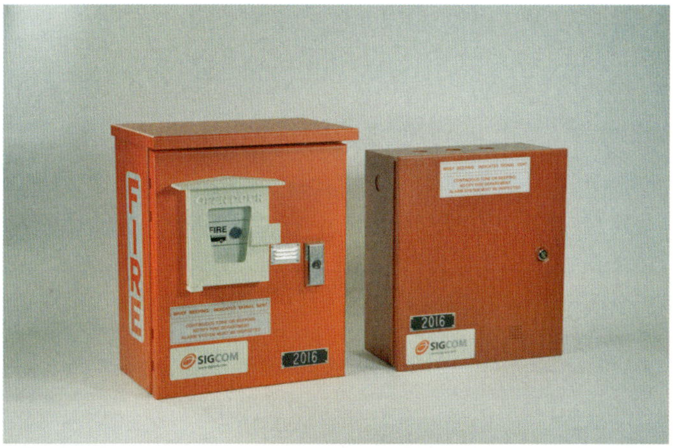

FIGURE 4-7 A radio-operated fire alarm box.

Courtesy of Signal Communications Corporation (Sigcom).

Some communities have converted their municipal fire alarm systems into call box systems. A **call box** is a direct-line phone or a radio that connects a caller directly to a dispatcher. The caller can request a full range of emergency assistance—from police, fire, or emergency medical assistance to a tow truck. Emergency call boxes are often located along major highways in

FIGURE 4-8 Wireless call boxes are sometimes found in places without other kinds of telephone service.

© Jones & Bartlett Learning. Photographed by Glen E. Ellman.

remote areas and near bridges, tunnels, subways, large complexes such as universities, and other locations where nearby phones are not available and where cell phones do not work. Call boxes may be solar powered so that they can be installed in locations without electric or telephone service (**FIGURE 4-8**).

Private Automatic Fire Alarm Systems

Many private, commercial, industrial, school, and residential buildings have automatic fire alarm systems. These systems use heat detectors, smoke detectors, water flow detectors, or other devices to initiate an alarm.

The connection used to transmit an alarm from a private fire alarm system to a communications center depends on many factors. These systems may be monitored by a privately operated alarm monitoring service. When the monitoring service detects that an alarm has been initiated at a location, staff at the monitoring service alerts the communications center, sometimes via a direct line. Some communications centers provide monitoring services. No matter how the alarm reaches the communications center, the result is the creation of an incident report and the dispatch of fire department resources.

Walk-Ins

Most emergencies are reported by telephone, but sometimes people come to a fire station seeking assistance. When a **walk-in**—also known as a **door banger**—arrives, someone at the station should immediately contact the communications center and describe the situation, even if the units at the station can handle the situation without assistance. If the walk-in did not clearly provide information about the incident, the personnel interacting with the walk-in should consider putting the walk-in on the phone to convey information directly to the dispatcher. The dispatcher needs to generate a run and create an incident report, as well as dispatch any needed additional units.

People should be able to come to a fire station and report an emergency at any time, even when the station is unoccupied. Many departments install a direct-line phone to the communications center just outside each fire station. These phones should be labeled with a sign instructing the public that if the station is vacant, they can pick up the phone in the red box to report an emergency.

Radio Calls from Fire or Police Units

A fire or police unit responding to a call or on patrol may come upon or be notified of an emergency situation by a bystander. This is referred to as a "flag down." The fire or police unit then contacts the communications center via radio to report the emergency and the need to assign units to the call. At other times, a change in status at an emergency incident may require additional or specialty units. For example, if a unit arrives at a fire scene and finds that the fire is much larger than the original report, they would request additional units to be dispatched to the scene. Likewise, if additional units are en route but the units on the scene have the situation under control, the IC on the scene will notify the communications center so that the additional units can be recalled.

Location Validation

To process the request for service and before dispatching units, the dispatcher must confirm that the location information received is adequate and there is no confusion about the address. For example, two streets in the same town or city might have similar names, such as 123 East Main Street and 123 West Main Street or 456 Spring Street and 456 Spring Avenue. Some larger cities even have two streets with the same name. This process of elimination is done through the GIS system in CAD. The information provided must point to a valid location on a map or in a street index system, and that location must be within the geographic jurisdiction of the potential dispatch units.

Enhanced 911

To help identify the location of an incident, Enhanced 911 (E911) was developed. If the call is made with a landline, E911 provides the phone number of the phone making the call and the location where the call originated. Most 911 systems currently in operation are E911 systems.

In E911 systems, **automatic number identification (ANI)** displays the phone number of the phone making the call. **Automatic location identification (ALI)** shows the location of the telephone making the call, the subscriber's name, and other details (**FIGURE 4-9**). If the call comes from a cell phone, the signal is picked up by the closest cell tower, which may not be in the same town or city. If the cell user has location services activated, the caller can be located using GPS. Otherwise, cell carriers can locate a phone through triangulation by calculating the distance from multiple cell phone towers to which the cell phone connected.

If the call is made using VoIP, the user must have provided the VoIP provider with their address in order for E911 to be able to retrieve that location. However, because VoIP calls can be made from anywhere with a connection to the Internet, that address might not be the location where the call originated. Efforts to provide location validation for calls received from cell phones and VoIP are advancing; however, in some areas, the dispatcher cannot determine the location of 911 calls made from these devices (911.gov 2023).

If the ANI and ALI are provided, the dispatcher should do their best to confirm that the information

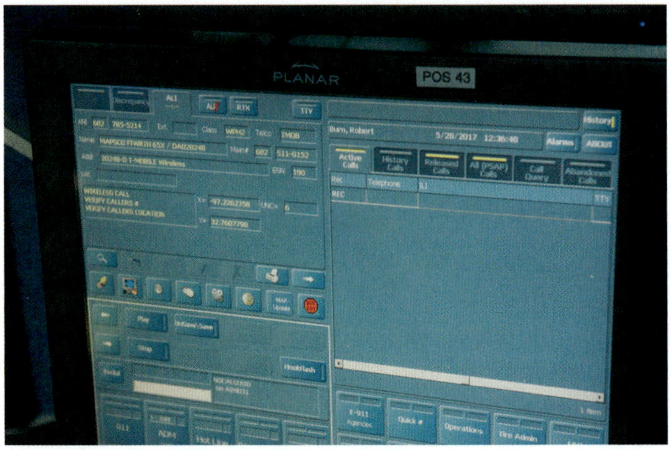

FIGURE 4-9 An ALI system provides information about the origin of a 911 call.

© Jones & Bartlett Learning. Photographed by Glen E. Ellman.

that is generated by E911 is correct and identifies the actual location of the emergency. The provided information might be inaccurate for a variety of reasons, such as the following:

- If the caller recently moved, the database may contain the old address.
- Sometimes a billing address for the phone service is provided rather than the physical location of the telephone.
- If the call was made from a location other than the site of the emergency incident, the location provided by the E911 system would not be the correct location to send responders.
- If the call was made from a cell phone or VoIP, the ANI and the ALI may not display.
- Many people calling 911 may not know their exact location.
- Numerous calls from cell phones may come in for the same incident, making it hard to determine the exact location of the emergency and overloading the communications center.

Next Generation 911

Next Generation 911 (NG911) systems allow all types of digital information, including voice, text messages, photos, and video, to be transmitted to a PSAP or stand-alone communications center. In addition to being able to receive digital data, PSAPs with NG911 are also able to receive data and alerts from safety devices, medical devices, and sensors, and they will be able to issue emergency alerts to wireless devices and highway alert systems. NG911 relies on a GIS to determine the location from which a call is made. Next-generation communications centers also provide CAD to CAD services; permit dispatch centers to receive data, texts, video clips, and other communications from the general public; incorporate CAD and GPS; and automate 911 hang-ups.

Call Classification and Prioritization

Call classification and prioritization is the process of assigning a response category based on the nature of the reported problem. Response categories in a department might be as follows:

- Critical/severe
- Major/high
- Medium
- Minor/low

Some calls qualify for emergency response (lights and siren); others are classified as nonemergency. Fire departments respond to many types of situations, ranging from a fire in a dumpster to a fire in a high-rise building, someone who collapsed, or an incident with multiple casualties. The nature of the call dictates its urgency and priority. High-priority calls are dispatched immediately. Sometimes it is necessary to delay dispatch of a lower-priority call if a more urgent incident is reported or if several calls come in at the same time.

Unit Selection

The call classification and prioritization help the dispatcher select the appropriate units to respond to the incident. **Unit selection** is the process of determining exactly which unit or units should be dispatched based on the location and classification of the incident. The department's SOPs and dispatch protocols dictate the type and number of units to dispatch to each category of incident. The response can range from assigning a single unit to assigning multiple units and dozens of firefighters. Generally, the standard assignment to each type of incident in each geographic response area is stored on **run cards**. Run cards list units in the order of response based on response distance or estimated response time. They often specify the units that should be dispatched if there are multiple alarms. Run cards are prepared in advance. At communications centers with a CAD system, the information is entered into the CAD, but the cards are retained so that they can be used if the CAD fails.

Most CAD systems are programmed to recommend which units to dispatch to an incident based on the location of the incident, the call type and classification, and the status of all units. The dispatcher can accept the recommendation or adjust based on the circumstances or additional information. When additional units are needed at an incident, the same process is used.

Selecting units requires quick decision-making skills, even when the CAD system is programmed to follow set policies for various situations. Usually, the policy is to dispatch the closest available unit that can provide the needed assistance. When all units are available and in their fire stations, the dispatcher can make this determination quickly based on information already programmed into the CAD. Unit selection becomes more complicated when units are not in their fire stations, are on an assignment, or are unavailable. Some communications centers have vehicle locator systems that track apparatus using GPS devices. These systems allow dispatchers to quickly identify the closest units in service. Departments sometimes enter into automatic mutual aid agreements with other departments to dispatch the closest available units, even if those units are not from the same jurisdiction as the incident.

Dispatch

After the dispatcher selects the unit or units to respond to the incident, the next step is dispatch. Fire departments use a variety of dispatch systems. The communications center must have at least two ways to send a dispatch message from the communications center to each fire station.

Fire station alerting systems receive a tone or series of tones to get the attention of personnel in the fire station. This is followed by a verbal or automated message providing the call type and units to respond. The communications center and the fire station alerting system are connected for dispatch via one of the following:

- Hard-wired circuit that directly connects the communications center to the alerting system and usually used when the communications center and the firehouse are located in the same building
- Phone line, usually a fiber-optic line
- Computer-based data link
- Microwave transmission system that transmits call information to a printer located in the firehouse
- Radio system that sends a signal to the station alerting system when firehouses are located great distances from the communications center

Regardless of how the dispatch message is relayed, it is broadcast over speakers in the fire station so that everyone in the station knows the location and nature of the incident and which units are being dispatched. If a unit is out of the station and available, the dispatcher contacts the unit via the radio system. The fire station or each unit must confirm that the message was received and the unit is responding. If the communications center does not receive confirmation within a set time, it must dispatch substitute units.

CAD systems can be programmed to alert the appropriate fire station automatically. A CAD system can send dispatch information to computer terminals or printers, sound distinctive tones, turn on lights and public address speakers, turn off the stove, and open the garage bay doors. The dispatch message can be sent directly to each individual vehicle that has an MDT. The CAD notification is often accompanied by a verbal announcement over the fire station speakers and the radio.

Fire departments with volunteer responders must be able to reach members with a dispatch message. Some departments issue pagers, use cell phones, or have online or mobile apps to reach volunteers. Some CAD systems send text messages to cell phones. Some volunteer fire departments rely on outdoor sirens, horns, or whistles to notify their members of an emergency. These audible devices usually can be activated by remote control from the communications center. Volunteers then contact the communications center by phone or radio to receive specific instructions.

Operational Support and Coordination

After the communications center dispatches the units, it begins providing operational support and coordination. A dispatcher in the communications center must remain in contact with the responding units throughout the entire incident. They need to confirm that the dispatched units received the alarm, record the en route and on-scene arrival times of the units, receive updates from the units on the scene, and provide additional and updated information to the units.

Generally, the IC uses the radio to communicate with a dispatcher operating a radio in the communications center. The communications center closely monitors the radio and provides needed support for the incident. Communications from the IC to the communications center include the following:

- Progress and incident status reports
- Requests for additional units or the release of extra units when they are no longer needed
- Requests for outside resources, such as a utility company's vacuum truck or equipment from a local construction company
- Notifications to municipal officials and on-call personnel
- Requests for information such as building plans

Each part of the public safety network—fire, EMS, and police—needs to know what other agencies are doing at an incident. The communications center coordinates the fire department's activities and requirements with other agencies and resources. For example, the dispatcher might need to notify a nonemergency resource, such as a utility company, about the incident and request that they take certain actions, such as shutting off a gas line.

> **TIP**
>
> The communications center should have accurate, current telephone numbers and contact information for every relevant agency.

Status Tracking and Deployment Management

The communications center must always know the location and status of every fire department unit. Units should never get "lost" in the system. As conditions change at an incident, units may need to be reassigned, or additional units may need to be requested from outside the normal response area or from other districts.

Tracking the status of units can be difficult if the dispatcher must rely on radio reports and colored magnets or tags on a map or status board. CAD systems that use GPS data make this job much easier because as status changes are entered, the map and status descriptions are automatically updated.

Communications centers continually monitor the availability of units in each geographic area and redeploy units to balance coverage when necessary, even when no major incidents are in progress. This might mean requesting coverage from surrounding jurisdictions. Many fire departments list both unit relocations and multiple alarm units—that is, units to be dispatched if an incident requires additional responders—on the run cards. Usually, the supervisor in a communications center is responsible for determining when and where to redeploy units, as well as requesting coverage from surrounding jurisdictions. Sometimes companies from the same department are relocated to another station; other times, units from a different jurisdiction may be redeployed to respond for coverage if a large-scale incident depletes local resources. This redeployment plan is often described in a formal memorandum of understanding (MOU) under regional or statewide mutual aid plans. These plans must include a system for tracking every unit and a designated communications center for maintaining contact with all units.

Taking Calls at the Fire Station

One of the first things you should learn when you are assigned to a fire station is how to use the telephone, intercom, and radio systems. You must be able to use them to answer a call or to initiate a response. When answering a call at the fire station, similar to what a dispatcher does, you should solicit needed information from the caller and then, if needed, initiate a response to an emergency by contacting the communications center or transferring the call to the communications center.

The crew member who answers the telephone in a fire station, fire department facility, or communications center is a representative of the fire department. Use your department's standard greeting when you answer the phone: "Good afternoon, Pleasant Town Fire Department. How may I help you?" Be prompt, polite, professional, and concise. Based on the caller's response, decide whether it is an emergency or a nonemergency call. Keep nonemergency calls as short as possible so that incoming phone lines are available to receive emergency calls.

An emergency call can come in on any fire department phone line, for example, if someone calls a fire station directly instead of dialing 911. If this happens, the crew member who answers the call is responsible for ensuring that the caller receives the appropriate emergency assistance. Detailed SOPs for obtaining information and processing calls should be provided to the crew member assigned to answer incoming emergency telephone lines. The department's SOPs should outline the exact steps to take, such as whether the crew member who answered the call should take the information from the caller or connect the caller directly to the communications center. Even though fire service personnel are not specifically trained in **telephone interrogation** (the phase in a 911 call during which the telecommunicator asks questions to obtain vital information, such as the location of the emergency), the SOPs and understanding the nature of the information needed to dispatch the appropriate response should guide the crew member as they communicate with the caller.

Follow the steps in **SKILL DRILL 4-1** to receive a call, obtain the essential information from a caller, and initiate a response to an emergency.

New personnel should tour the communications center for their department. Observing the operation of the center provides a much better understanding of the role of the dispatcher.

Radio Systems

Fire department communications systems rely on two-way radios. As discussed earlier in the chapter, radios are one of the ways a communications center transmits dispatch information to fire stations. Radios are also used in many other situations, including the following:

- Communication between the communications center and individual units
- Communication between units at an incident scene
- To page firefighters
- To link MDTs

A radio system is an integral component of the ICS because it links the IC or unified command with all the

SKILL DRILL 4-1

Receiving a Call at a Fire Station and Initiating a Response to an Emergency Support Person, NFPA 1010: 5.2.1 and 5.2.2

1. Answer the phone promptly and professionally. Identify yourself, your agency, and your company. Determine immediately whether the caller is reporting an emergency, and if they are, follow your department's SOPs. Obtain the following information from the caller:
   - Incident location (including cross streets and identifying landmarks)
   - Type of incident or situation
   - Safety concerns at the scene
   - When the incident occurred
   - Caller's name
   - Location of the caller if it is different from the incident location
   - Caller's callback number

   Always terminate the call in a courteous manner, and let the caller hang up first.

2. Record the date and time of the call and the responses to your questions. Initiate a response following the department SOPs. The SOPs in your department may vary from the steps listed here. Follow the SOPs of the authority having jurisdiction (AHJ) for your department's communications.

© Jones & Bartlett Learning. Photographed by Glen E. Ellman.

units at an incident, up and down the chain of command. A radio may be a firefighter's only means to call for help in a dangerous situation.

Fire departments use many types of radios and radio systems, and technological advances are rapidly adding new features and system configurations. Although it would be impossible to describe all of the possible features and principles of operation of every radio system here, this section describes some common systems and operating features. Be sure you learn how to operate the radio assigned to you and learn your department's radio SOPs.

Radio Equipment

The fire service primarily uses three types of radios: portable radios, mobile radios, and a base station.

FIGURE 4-10 A portable radio should be carried by all department personnel.

© Jones & Bartlett Learning. Photographed by Glen E. Ellman.

FIGURE 4-11 A mobile radio is permanently mounted in a vehicle.

© Jones & Bartlett Learning. Photographed by Glen E. Ellman.

A **portable radio** is a hand-held, two-way radio that is small enough for a crew member to always carry (**FIGURE 4-10**). The body of a portable radio contains a speaker and microphone, on/off switch or knob, volume control, channel-select switch, and push-to-talk (PTT) button. At least one—if not every—member of a team carries a portable radio during an emergency incident.

Portable radios have an antenna to receive and transmit signals. A popular optional attachment is an extension microphone/speaker unit that can be clipped to a collar or shoulder strap while the radio remains in a pocket or pouch.

Portable radios are usually powered by a rechargeable battery. Battery-operated portable radios also have limited transmitting power, usually 1 to 5 watts (w). This means the signal can be transmitted only within a certain range and is easily blocked or overpowered by a stronger signal.

A **mobile radio** is permanently mounted in the vehicle and powered by the electrical system in the vehicle. They are more powerful than portable radios, usually 25 to 50 w, so they have a greater range than portable radios (**FIGURE 4-11**). In most departments, every fire department vehicle has a mobile radio.

Mobile radios usually have a fixed speaker and an attached, hand-held microphone on a coiled cord. The PTT button is on the microphone, and the antenna is usually mounted on the exterior of the vehicle. Fire apparatus often include protective headsets with a combined intercom/radio system, enabling crew members to talk to each other in the apparatus and hear any announcements over the radio.

A **base station** is a radio permanently mounted in a building, such as a fire station, communications center, or remote transmitter site. Base station radios are more powerful than either portable or mobile radios, usually 50 to 100 w. The antenna of the base station is often mounted on a radio tower, so communications are transmitted over the widest possible coverage area (**FIGURE 4-12**). Public safety radio systems for a large geographic area often have additional transmitters installed at different locations to extend the range of the transmission.

MDTs transmit data using both the radio system and cellular signals. Crew members can press buttons

FIGURE 4-12 The antenna for a base station is often mounted on a radio tower to provide maximum coverage.

© Jones & Bartlett Learning. Photographed by Glen E. Ellman.

FIGURE 4-13 SCBA with a voice amplifier.

Courtesy of 3M.

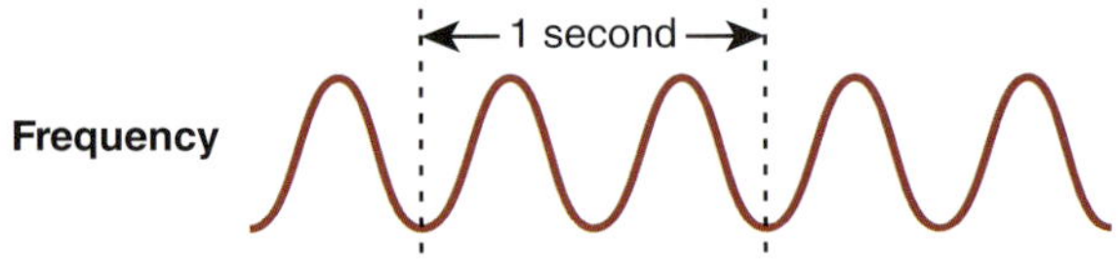

FIGURE 4-14 The frequency of a radio wave is the number of times the wave oscillates per second.

Courtesy of the National Aeronautics and Space Administration (NASA).

on the MDT to communicate to dispatch that they are en route or on the scene. MDTs also have GPS technology so that satellites can track the location of the apparatus responding to and arriving at the incident scene. MDTs also allow responders to verify the location of an incident. For example, instead of having to radio the dispatcher to ask whether they said 11345 Main Street or 11354 Main Street, they can look at the MDT where the address is displayed. Some departments use MDTs to track unit status, transmit dispatch messages, and exchange many other types of information.

To enable firefighters to communicate more effectively while wearing self-contained breathing apparatus (SCBA), some manufacturers provide accessories for SCBA face pieces that amplify a firefighter's voice for face-to-face communication. Another accessory connects the SCBA face mask to a firefighter's portable radio via Bluetooth (**FIGURE 4-13**). These devices enable firefighters to hear messages more clearly while wearing SCBA and to transmit information over their portable radios.

Radio Operation

In the United States, the FCC has established strict limitations governing the assignment of radio channels (frequencies) to ensure that all users have adequate access. Every radio system must be licensed and operated within these guidelines. To do this, the FCC licenses an agency to operate on one or more specific frequencies called *bands*. The most commonly used bands for public safety communications are the **very high-frequency (VHF) band** and the **ultrahigh-frequency (UHF) band**. The frequencies available to the U.S. fire service are in several different ranges, including 33 to 46 megahertz (MHz; VHF low band), 150 to 174 MHz (VHF high band), 450 to 460 MHz (UHF band), 700 MHz, and 800 to 900 MHz (FCC 2022) (**FIGURE 4-14**).

The FCC allocates additional groups of frequencies in specific geographic areas that have a high demand for public safety agency radio channels. Each band has certain advantages and disadvantages relating to geographic coverage, topography (hills and valleys), and penetration into structures. Some bands have many different users, which can cause interference problems, particularly in densely populated metropolitan areas.

Generally, a radio can be programmed by a qualified technician to operate on several frequencies in a particular band. Communication may be affected if neighboring police, fire, and EMS systems within the same jurisdiction operate on different bands. When different agencies working at the same incident communicate via different bands, they must make complicated arrangements so that they can communicate with one another. Being able to communicate across different radio bands and between different agencies is called **interoperability**. Public safety agencies rely on three main types of radio systems to overcome interoperability barriers:

- Simplex or line-of-sight systems
- Conventional radio repeater systems
- Trunked radio repeater systems

A **simplex communication system**, sometimes referred to as a **line-of-sight system**, allows communication to flow in only one direction at a time (**FIGURE 4-15**). A drawback of a simplex or line-of-sight system is that it has limited range, providing radio coverage to a small area, such as one city block. This limited coverage is not adequate for dispatching units or communicating between the communications center and apparatus assigned to a run. It is used by fire departments when they need their signals to penetrate a specific building or when operational security is critical. Some fire departments use a simplex communication system for on-scene communications. This configuration is sometimes called a **talk-around channel** because the radios transmit and receive on the same frequency, and the signal is not sent over a **repeater channel**, which is a channel that transmits to a repeater. A **repeater** is a combination of a radio receiver and a transmitter that receives a signal and retransmits—repeats—it to a wider geographic location. A talk-around channel often works well for short-distance communications, such as from the location of the IC to crews inside a house fire or from one unit to another inside a building. Conversations on a talk-around channel are not routinely monitored by the communications center. If the IC wants to maintain contact with the communications center, they must use a radio channel designated to transmit to a repeater.

Radio messages can only be transmitted over a limited distance for two reasons. First, the signal weakens as it travels farther from the source and eventually becomes undetectable. Second, buildings, tunnels, and topography create interference and block the signal from traveling. These problems are more significant for hand-held portable radios, which have limited transmitting power. Mobile radios that operate in systems that cover large geographic areas face similar problems. To compensate for these shortcomings, fire departments use radio repeater systems, which use a repeater installed in a location that has a line of sight to a large geographic area, greatly extending the operating range of radios. This location could be the top of a building or a hill. A radio repeater system can be digital or analog,

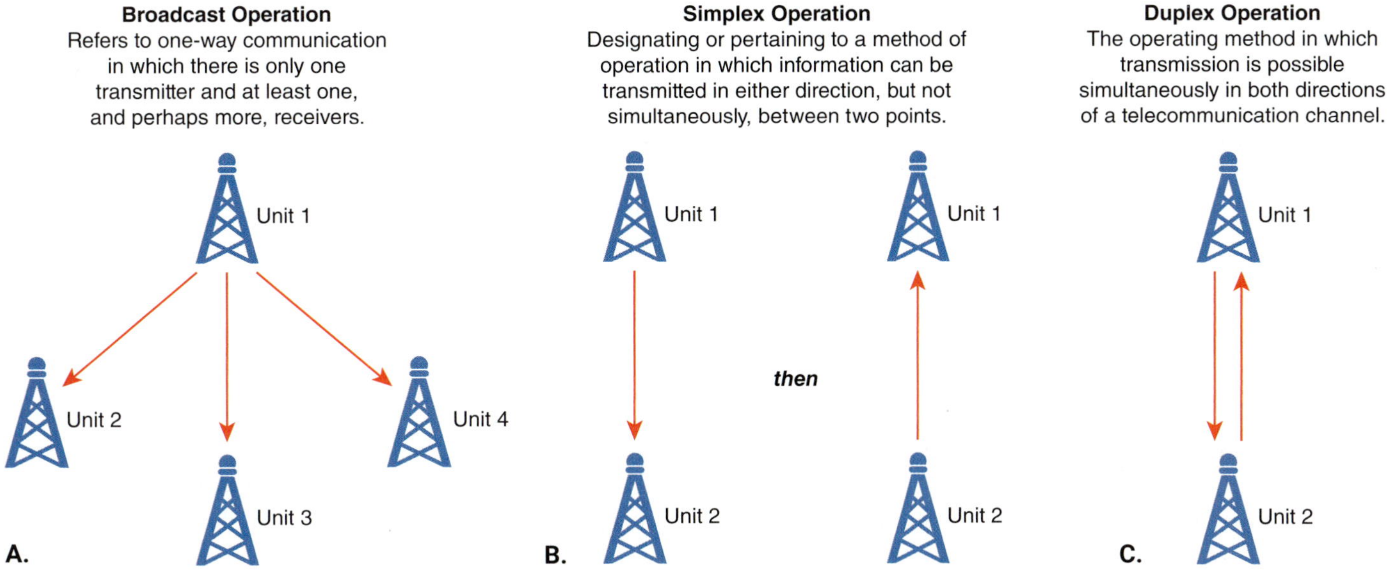

FIGURE 4-15 **A.** On a broadcast channel, transmission goes in only one direction. **B.** On a simplex channel, transmission goes in both directions, but only one party can transmit at a time. **C.** On a duplex channel, transmission goes in both directions, and both parties can transmit at the same time.

© Jones & Bartlett Learning

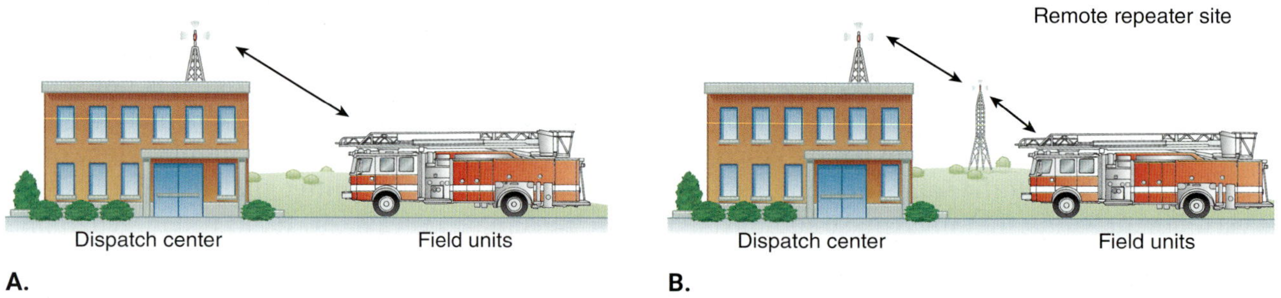

FIGURE 4-16 **A.** A transmission from a portable or mobile radio may not be strong enough to travel to everyone it needs to reach. **B.** A repeater intercepts the transmissions and retransmits them with more power so that the message travels to a wider geographic area.

© Jones & Bartlett Learning

and it typically can receive and send communications over multiple channels. A drawback to repeater systems is that they are inefficient when used by large numbers of public safety personnel.

A repeater has a large antenna that can receive lower-power signals, such as those from a portable radio. That signal is then retransmitted to all the radios set on the designated channel with all the power of a base station. Communications systems that use repeaters usually have outstanding systemwide communications and can get the best signal from portable and mobile radios. This approach enables the transmission to reach a wider coverage area (**FIGURE 4-16**).

Many public safety repeater systems have multiple receivers called *voting receivers*, which are geographically distributed over the service area to capture weak signals. A **voting receiver** receives radio signals and forwards the strongest signal to the repeater. If the original signal is strong enough to reach one of the voting receivers, the system is effective. If the signal does not reach a repeater, no one will receive it.

A **mobile repeater** boosts signals at an incident scene by capturing the weak signals from portable radios and rebroadcasting them. It can be permanently mounted in a vehicle or set up near the IC. Some fire departments use a **cross-band repeater**, which boosts the power of a weak signal and then retransmits it over a different radio band. For example, a cross-band repeater allows a firefighter using a UHF portable radio inside a building to communicate with an IC who is outside on a VHF radio. Similar systems are often used in large buildings or underground structures so that crews working inside can communicate with units on the outside or with the communications center.

A **trunked radio system** is a repeater system that operates on the same principles as a conventional radio repeater system, except that a computer connected to a control channel sets the operating frequencies. When users press the PTT button, the radio transmits a "key" that signals to the control channel that a message to the trunk group is being transmitted. The system automatically directs that signal to the channel assigned to the trunk group. Trunked radio systems support many more users than conventional repeater systems.

A **digital radio** allows the transmission of digital signals to transmit data or analog signals to transmit voices that have been digitized and compressed by a computer. Using digital technology is more efficient than using analog technology, and it also enables more users to communicate at the same time in a limited portion of the radio spectrum. When digital radios are used in a trunked radio system, instead of the trunk group being assigned to only one or two frequencies, many frequencies are assigned to a group. These can be thought of as virtual channels that appear as conversations occur and disappear when they are completed. As a radio conversation in a trunk group begins, the computer scans for the next open frequency and directs all the radios in the trunk group to receive the message on that frequency. As the conversation continues, the frequency may change automatically because the computer constantly monitors the frequency load and reassigns the transmissions to unused frequencies.

Trunked systems make the use of available frequencies more efficient and can provide coverage over multiple trunk groups. Because trunked systems allow different radios to be tied together for a given incident, it is possible to link different agencies on the same system. For instance, public works, the highway department, the fire department, and the police department could all be linked for a specific incident if they needed to share radio communications.

A **multiplex channel** combines analog and digital signals so that they can simultaneously transmit two or more different types of information, such as voice and data, in both directions over the same frequency.

FirstNet

Cell phones are used by emergency responders to assist them in daily operations. The First Responder Network Authority (FirstNet Authority), part of the U.S. Department of Commerce, is a nationwide broadband network for communication among first responders and communications centers to provide a national coordinated response to emergencies. Officially launched in 2018, the mission of the FirstNet Authority is "to ensure the building, deployment, and operation of the nationwide broadband network that equips first responders to save lives and protect U.S. communities" (FirstNet Authority n.d.). That network is called FirstNet. FirstNet was designed to work with NG911 so that responders can get the same information that is sent to the communications center. The FirstNet dedicated network allows first responders to have priority cell service.

Using a Radio

As a support person, you must learn how to operate a radio assigned to you as well as how to use the radio systems used by your department. You must know when to identify the channel you are using, such as "Channel 3," "Tac 5," or "Charlie 2"; which buttons to press; and which knobs to turn. Study the training materials and SOPs provided by your department to learn how to properly use and maintain the radio, and don't be afraid to ask questions if there is something you don't understand.

When a radio is assigned to you, make sure you receive an explanation on how to operate and maintain it. Learn the protocol for recharging the battery or replacing the batteries if the device is not rechargeable. Whether the radio assigned to you has a rechargeable battery or one that needs to be replaced, check it at the beginning of each shift to make sure you have full power. Because the battery has a limited capacity, it must be recharged or replaced after extended operations, so check it again when you return to the station after a run.

If you hold a portable radio right side up, perpendicular to the ground, with the antenna pointing toward the sky, you will get better transmission and reception. Range and transmission quality also improve if you remove the radio from the radio pocket or belt clip before you use it.

All radio transmissions should be pertinent to the situation. Avoid unnecessary radio traffic. Keep in mind that most radio communications are automatically recorded. These recordings provide a complete record of everything that is said and can be admissible as evidence in a legal case.

TIP

When using a radio, always maintain a professional demeanor. In addition to the fact that radio communications are recorded, remember that anyone in your department and in the communications center, as well as anyone with a radio scanner, including the news media and the public, can hear everything you say. Don't say anything of a sensitive nature or anything you might regret later.

A common way to communicate by radio is the "Hey you, it's me" method. This method helps ensure that radio traffic reaches the appropriate person and is properly understood. In the following example of the "Hey you, it's me" method, E1 stands for the crew leader of Engine 1:

IC: Engine 1 (Hey you), Elm Street Command (it's me).

E1: Go ahead, Elm Street Command (you); this is Engine 1 (me).

IC: Engine 1, I want you to exit the building and report upon exiting.

E1: Engine 1 copies. We will exit the building and report upon exiting.

In this example, the IC was clear about which unit they were trying to reach. That crew leader then acknowledged that they heard the communication. Once the IC gave their orders, the unit crew leader repeated the orders to confirm that they accurately understood the command.

The National Incident Management System (NIMS) and the NFPA recommend that fire departments use plain English for radio communications, as in the previous example. Using this method helps reduce miscommunications. However, some departments use codes for standard messages. A few departments have developed intricate systems of numeric codes to fit any situation. **Ten-codes**, a system of coded messages that begin with the number 10, were once widely used and are still commonly used by law enforcement. Ten-codes, however, are not standardized across the country, with a few exceptions—for example, almost everyone knows that "10-4" means acknowledgment, or OK. To be effective, codes must be understood by all parties and mean the same thing to all users. Ten-code systems are problematic when units from different jurisdictions need to communicate with one another or when multiple agencies need to communicate at a large emergency incident.

If your department uses radio codes, you must learn all the codes and understand how and when they are used. If your department uses plain English, be aware that even plain English may include coded terminology, such as "code blue" for a cardiac arrest or "MVC" for motor vehicle collision. The objective is to be clearly and easily understood without having to explain every message in detail. If your department has a few codes or special words for special situations, they should be specified in the SOPs.

To use a radio, follow the steps in **SKILL DRILL 4-2**.

SKILL DRILL 4-2

Using a Radio Support Person, NFPA 1010: 5.2.2

1. If you are holding the radio, position it right side up, perpendicular to the ground, with the antenna pointing toward the sky. If the radio is in your pocket or a pouch, try to position it in the same manner.

2. Listen to determine if a radio communication is routine or emergency traffic. Ensure that the channel is clear of any other traffic. Press and hold the PTT button, then wait at least 2 seconds before speaking to allow the system to capture the channel without cutting off the first part of the message. Some systems sound a distinctive tone when the channel is ready after the PTT button is pressed.

3. Hold the microphone 1 to 2 inches (2.5 to 5 centimeters) from your mouth, then speak across the microphone at a 45-degree angle. Do not have something in your mouth when speaking on the radio. Know what you are going to say before you start talking. Speak clearly and keep the message brief and to the point.

4. Release the PTT button only after you have finished speaking so that the message does not get cut off.

© Jones & Bartlett Learning. Photographed by Brandon Fryman/JRM Creative.

Incident Command System and Fire Department Communications

According to the National Institute for Occupational Safety and Health (NIOSH), firefighter injuries and deaths are often caused by poor communication (Ludwig 2019). Effective communication is essential to reduce complications at incident scenes and in training drills and to be able to successfully complete an incident operation. The ICS, which should be utilized at every incident, helps ICs, fire officers, and firefighters communicate more effectively. **Unity of command**—the concept that each person reports to only one direct supervisor, each supervisor answers to only one boss, and so on—minimizes confusion by ensuring that each individual at an incident reports to only one supervisor so that important messages are communicated to the correct person. Unity of command also reduces occurrences of freelancing. Yonkers Fire Department Chief Chris Kiernan used to say that a good command structure created an environment of "controlled chaos." One of the benefits of the ICS is that it allows ICs to create sections to manage different aspects of the incident, and within those sections, smaller groups can be created. As early as possible, the IC should assign a communications unit leader.

During an incident, the IC or the communications unit leader should identify and assign specific radio channels to each company, crew, or unit as the incident determines. When multiple agencies, especially multiple agencies from different jurisdictions, respond to the same incident, it is essential that a means of interoperable communication is established as soon as possible. Some agencies have radio systems that allow for interoperability. Other agencies have adopted VoIP communication systems that allow transmission between radios and cell phones by adding an accessory device to portable radios (**FIGURE 4-17**).

In disaster situations, the larger cellular network providers can deploy 5G-capable, portable cell signal towers called *cell on wheels* (COWS). One major cellular network provider has flying COWS (cell on wings), which are 4G- and 5G-capable drones.

Size-Up and Progress Reports

A **size-up** is the rapid evaluation and analysis of an incident by an officer or the IC to determine which resources need to be deployed and which actions can be undertaken safely. The first-arriving unit at an incident—whether a firefighter, a company officer, or a chief—needs to give a brief initial size-up report and establish command. This initial report should convey a preliminary assessment of the situation and give other units a sense of what is happening so that they can anticipate what their assignments may be when they arrive at the scene. The communications center should repeat the initial size-up information. For example, on arrival at the scene, an officer, or an engine company firefighter if there is no officer currently on the scene, might make the following report:

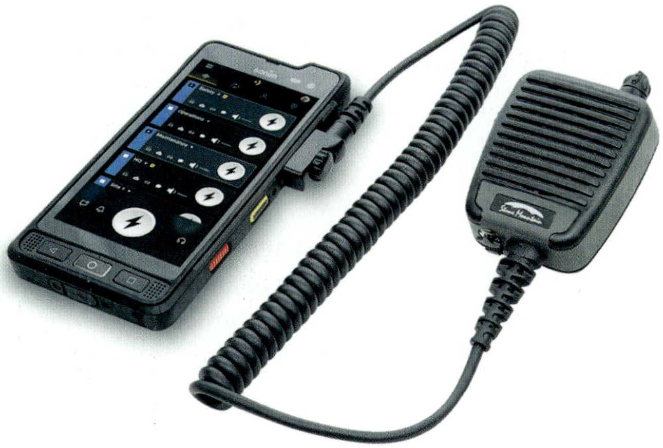

FIGURE 4-17 PTT accessories for cell phones allow transmission between cell phones and radios.
Courtesy of Instant Connect Software LLC.

> Communications, this is Engine 410. Engine 410 is on location at the intersection of High Street and Chambersburg Road. We have an electric vehicle fully involved in fire. Engine 410 is assuming High Street Command. This will be an offensive operation.

The communications center would respond by announcing:

> Engine 410, I have you on the scene assuming High Street Command. Electric vehicle fire fully involved. Offensive operation.

As the incident progresses, and for the duration of an incident, the IC should provide regular progress reports or updates to the communications center. The communications center should prompt the IC to report at set intervals, known as **time marks**. Time marking allows the IC to assess the progress of the incident and decide whether the strategy or tactics should be changed. All progress reports are noted in the activity log to document the incident.

If additional resources, such as apparatus, equipment, or personnel, are needed, the IC transmits these requests to the communications center. The communications center will advise the IC of additional resources that are dispatched.

Emergency Messages

Sometimes the dispatcher must interrupt normal radio transmissions for emergency traffic. **Emergency traffic** is an urgent message that takes priority over all other communications. When a unit needs to transmit emergency traffic, the dispatcher generates a distinctive alert tone to notify everyone on the frequency to stand by so that the channel is available for the emergency communication. Once the emergency message is complete, the dispatcher notifies all units to resume normal radio traffic.

To alert the communications center that emergency traffic needs to be transmitted, some radio systems have emergency buttons located on portable and mobile units that a firefighter in trouble can push to transmit an emergency signal to the dispatcher. This button alerts the communications center that the ID assigned to that radio needs to transmit emergency traffic. Some radio systems can also identify the location of the unit in trouble. If your radios are equipped with this feature, learn the procedure for using it.

Mayday

The most important emergency traffic is a firefighter's call for help. Most departments use the word **mayday** to indicate that a firefighter is in imminent danger and requires immediate assistance, is lost, or is missing. If a mayday call is transmitted, all other radio traffic should stop immediately. The firefighter making the mayday call must describe the situation, identify their location, and describe the help needed.

Some agencies use the acronym LUNAR to report a mayday, which stands for the following:

- **Location**, your location in the building/incident
- **Unit**, the unit to which you are assigned
- **Name**, who you are
- **Air**, the amount of air you have in your cylinder, or **Assignment**, where you were last assigned
- **Resources**, what you need to get you out of the mayday situation.

In other agencies, the person reporting the mayday reports **who** (they are), **what** (the situation is), and **where** (they are located) instead of providing a LUNAR report. An example of a mayday call is as follows:

Firefighter: MAYDAY, MAYDAY, MAYDAY.

[All radio traffic stops.]

IC: Unit calling MAYDAY, go ahead.

Firefighter: This is Engine 403. We are on the second floor and running out of air. Fire has cut off our escape route. We request a ladder to the window on the Charlie side (rear of building) of the building so we can evacuate.

IC: Command copied. Engine 403, your escape route is cut off by fire. I am sending the rapid intervention crew to Charlie side (rear of building) with a ladder.

Evacuation Message

Another emergency traffic message is "evacuate the building," which directs all units inside a structure to abandon the building immediately. Most fire departments have specified a standard **evacuation signal** to warn all personnel to pull back to a safe location. The evacuation signal that is commonly used is a sequence of three blasts on an apparatus air horn, repeated several times, or sirens sounded "high-low" for 15 seconds. An evacuation warning should be announced at least three times to ensure that everyone hears it; the IC also should announce the warning on the portable radio. Because no universal evacuation signal has been established, you need to learn your department's SOP for emergency evacuation.

After the evacuation, normal radio traffic should not begin again until the IC conducts a roll call of all units. This step ensures that all personnel have safely exited the building. After the roll call, the IC will allow the resumption of normal radio traffic.

CASE STUDY

You Are the Support Person CONCLUSION

At 0304 hours, your crew is dispatched to a report of a fire at 3256 West Madison Street. While en route, the dispatcher advises that the caller reported that smoke was coming from the eaves of the building and then hung up. With this new information, the crew prepares for a possible house fire. The dispatcher radios back and says that this may be a false report because the caller used a cell phone that was identified as being on the east side of town. It seems like you may have been woken up for another false call, but then the dispatcher comes back on the radio with a frantic sound in her voice and says that a home security company just reported a fire alarm at 3256 East Madison Street.

1. **What is the process for receiving 911 calls in an emergency communications center?**

   Answer: Upon receiving a call, the dispatcher determines the type of emergency that exists; the location of the emergency; and the type of resources required, such as police, fire, and ambulance. The dispatcher then assigns the call to all needed resources using the jurisdiction's SOPs. Once the units are dispatched, the dispatcher continues to interview the caller for additional information. If the dispatcher is trained as an EMD, they provide the caller with instructions for emergency medical care over the phone.

2. **How does the dispatcher know where the cell phone call originated?**

   Answer: The best way for a dispatcher to know the location of a cell phone caller is to have the caller tell them the address directly. Current technology does not provide an exact location of a caller from a cell phone, although because many cell phones have GPS, cell phone carriers may be able to provide the latitude and longitude to the communications center. Otherwise, the approximate location of the caller is provided by the cellular carrier by using cell tower triangulation.

3. **How are alarm systems monitored and then reported to the communications center?**

   Answer: When heat, fire, or smoke is detected by an alarm system, the communications center needs to be notified. Notification may occur via a direct call from someone at the location; a transmitter at the location dialing the communications center and playing a prerecorded message; or a central monitoring station receiving notification of the alarm, causing monitoring personnel to contact the communications center.

WRAP-UP

SUMMARY

- Coordination and communication between fire service personnel and emergency responders is vital for effectively responding to fire emergencies. Different crews and equipment fill various functions on the fire ground, many of which occur simultaneously or must occur in a particular order.
- Communications centers serve as central processing points for all information relating to emergency incidents and all of the information relating to the location, status, and activities of responding units. Their primary responsibility is to process 911 calls, as well as other nonemergency public safety calls.
- Although most emergencies are reported by telephone, the public must be able to come to a fire station and report an emergency at any time, even when the station is unoccupied. Many departments install a direct-line phone to the communications center just outside each fire station. These phones should be labeled with a sign instructing the public that if the station is vacant, they can pick up the phone in the red box to report an emergency.
- When a fire or police unit comes upon an emergency or is flagged down by a bystander, the unit contacts the communications center via radio to report the emergency and the need to assign units to the call. Units may also report the need for additional units to be dispatched once they arrive on the scene.
- Call classification and prioritization is the process of assigning a response category of severity based on the nature of the reported problem. Some calls qualify for emergency response (lights and siren); others are classified as nonemergency. High-priority calls are dispatched immediately. Sometimes it is necessary to delay dispatch of a lower-priority call if a more urgent incident is reported or if several calls come in at the same time.
- A dispatcher in the communications center must remain in contact with the responding units throughout the entire incident. They need to confirm that the dispatched units received the alarm, record the en route and on-scene arrival times of the units, receive updates from the units on the scene, and

SUMMARY CONTINUED

provide additional and updated information to the units.

- When answering a call at the fire station, you should solicit needed information from the caller and then, if needed, initiate a response to an emergency by contacting the communications center or transferring the call to the communications center.
- A radio system is an integral component of the ICS because it links the IC or unified command with all the units at an incident, up and down the chain of command. A radio may be a firefighter's only means to call for help in a dangerous situation. Be sure you learn how to operate the radio assigned to you and learn your department's radio SOPs.
- As a support person, you must learn how to operate a radio assigned to you, as well as how to use the radio systems used by your department. Study the training materials and SOPs provided by your department to learn how to properly use and maintain the radio, and don't be afraid to ask questions if there is something you don't understand.
- The first-arriving unit at an incident—whether a firefighter, a company officer, or a chief—needs to give a brief initial size-up report and establish command. This initial report should convey a preliminary assessment of the situation and give other units a sense of what is happening so that they can anticipate what their assignments may be when they arrive at the scene.
- When a unit needs to transmit emergency traffic, the dispatcher generates an alert tone to notify everyone on the frequency to stand by until the emergency message is complete. The most important emergency traffic is a firefighter's call for help (mayday). If a mayday call is transmitted, all other radio traffic should stop immediately. The firefighter making the mayday call must describe the situation, identify their location, and describe the help they need. Another important emergency traffic message is "evacuate the building," which directs all units inside a structure to abandon the building immediately.

KEY TERMS

activity logging system A device that keeps a detailed record of every incident and activity that occurs.

automatic location identification (ALI) A series of data elements that informs the recipient of the location of the alarm. (NFPA 1225)

automatic number identification (ANI) A series of alphanumeric characters that informs the recipient of the source of the alarm. (NFPA 1225)

base station A stationary radio transceiver with an integral alternating-current (AC) power supply. (NFPA 1225)

call box A system of telephones connected by phone lines, radio equipment, or cellular technology to a communications center or fire department.

call classification and prioritization The process of assigning a response category based on the nature of the reported problem. (NFPA 1225)

call receipt The process of receiving a call for service and obtaining the information necessary to initiate a response.

communications center See *public safety communications center.*

computer-aided dispatch (CAD) A combination of hardware and software that provides data entry; makes resource recommendations; and notifies and tracks those resources before, during, and after fire service alarms, preserving records of those alarms and status changes for later analysis. (NFPA 1225)

cross-band repeater A repeater that boosts a weak signal and then retransmits it over a different radio band.

digital radio A radio that transmits information via radio waves using digital data or analog (voice) signals that have been converted to a digital signal and compressed.

direct-line phone A phone that connects two predetermined points and does not require the user on either end to dial to cause the phone at the other end to ring. Also called *ring-down phone.*

dispatch To send out emergency response resources promptly to an address or incident location for a specific purpose. (NFPA 450)

dispatcher See *telecommunicator.*

door banger See *walk-in.*

emergency medical dispatcher (EMD) Personnel specifically trained and certified in interviewing techniques, prearrival instructions, and call prioritization. (NFPA 450)

emergency traffic An urgent message, such as a call for help or evacuation, transmitted over a radio that takes precedence over all normal radio traffic.

evacuation signal A distinctive signal intended to be recognized by the occupants as requiring evacuation of the building. (NFPA 1900)

Federal Communications Commission (FCC) The federal regulatory authority that oversees radio communications in the United States.

fire alarm box A device connected via an underground cable to a municipal fire alarm system.

geographic information systems (GIS) A system of computer software, hardware, data, and personnel to describe information tied to a spatial location. (NFPA 450)

global positioning system (GPS) A satellite-based radio navigation system composed of three segments: space, control, and user. (NFPA 1900)

interoperability The ability to communicate across different radio bands and between different agencies.

line-of-sight system See *simplex communication system.*

mayday A verbal declaration indicating that a firefighter is lost, missing, or trapped and requires immediate assistance.

mobile data terminal (MDT) Technology that allows firefighters to receive data while in the fire apparatus or at the station.

mobile radio A two-way radio that is permanently mounted in a fire apparatus.

mobile repeater A repeater used at an incident scene to boost signals at the scene by capturing the weak signals from portable radios and rebroadcasting them.

multiplex channel Simultaneous transmission of multiple data streams, most often voice signals, in either or both directions over the same frequency.

municipal fire alarm system A network of fire alarm boxes and emergency telephones on street corners or in public places.

Next Generation 911 (NG911) A system designed to replicate traditional Enhanced 911 (E911) features and functions and provide additional capabilities to provide access to emergency services from all connected communications sources and provide multimedia data capabilities for public safety answering points (PSAPs) and other emergency service organizations. (NFPA 1225)

portable radio A battery-operated, hand-held transceiver. (NFPA 1225)

public safety answering point (PSAP) A facility equipped and staffed to receive emergency and nonemergency calls requesting public safety services via telephone and other communication devices. (NFPA 1225)

public safety communications center A building or portion of a building that is specifically configured for the primary purpose of providing emergency communications services or public safety answering point (PSAP) services to one or more public safety agencies under the authority or authorities having jurisdiction. Also called *communications center.* (NFPA 1225)

repeater A combination of a radio receiver and transmitter that receives radio signals, usually over multiple channels, and retransmits—repeats—them to a wider geographic location.

repeater channel A radio channel that transmits to a repeater.

ring-down phone See *direct-line phone.*

run The act of a unit traveling to an incident.

run cards Information prepared in advance and stored that describes a predetermined response to an emergency.

simplex communication system A radio system that allows communication to flow in only one direction at a time. Also called *line-of-sight system* or *talk-around channel.*

size-up The process of gathering and analyzing information to help fire officers make decisions regarding the deployment of resources and the implementation of tactics. (NFPA 1410)

stand-alone communications center A public safety communications center that serves and dispatches a single emergency response agency.

talk-around channel See *simplex communication system.*

telecommunicator An individual whose primary responsibility is to receive, process, or disseminate information of a public safety nature via telecommunication devices. Also called *dispatcher.* (NFPA 1225)

KEY TERMS CONTINUED

telephone interrogation The phase in a 911 call during which the telecommunicator asks questions to obtain vital information, such as the location of the emergency.

ten-codes A system of predetermined coded messages, such as "What is your 10-20?" used by responders over the radio.

time marks Status updates provided to the communications center every 10 to 20 minutes. Such an update should include the type of operation, the progress of the incident, the anticipated actions, and the need for additional resources.

trunked radio system A repeater system that uses a computerized shared bank of frequencies to make the most efficient use of radio resources.

TTY/TDD systems User devices that allow speech- and/or hearing-impaired citizens to communicate over a telephone system. TTY stands for teletype, and TDD stands for telecommunications device for the deaf; the displayed text is the equivalent of a verbal conversation between two hearing persons.

ultrahigh-frequency (UHF) band Radio frequencies between 300 and 3000 MHz.

unit selection The process of determining exactly which unit or units should be dispatched after a call to 911 is received by a communications center, based on the location and classification of the incident.

unity of command The concept that each person reports to only one direct supervisor.

very high-frequency (VHF) band Radio frequencies between 30 and 300 MHz; the VHF spectrum is further divided into high and low bands.

Voice over Internet Protocol (VoIP) Technology that converts a person's voice into a digital signal that can be sent via the Internet to another device.

voice recording system A system that records communications over the phones and the radio.

voting receiver A device in a repeater system that receives signals from radio transmissions and then sends the strongest signal to the repeater.

walk-in A person who comes to a fire station seeking assistance rather than calling 911. Also called *door banger*.

REVIEW QUESTIONS

1. What does *dispatch* mean?
2. List 11 types of equipment that are in most communications centers.
3. Describe E911.
4. What is VoIP?
5. What is an EMD?
6. What happens if the dispatcher does not receive confirmation that a unit is responding within a set time?
7. What are the three types of radios that fire service personnel primarily use?
8. In what type of system are radio signals transmitted between the sender and the receiver on the same frequency?
9. What is a base station that uses two separate frequencies—one to receive messages and one to transmit messages—to retransmit the message to a wider coverage area?

DISCUSSION QUESTIONS

1. Describe why the job of dispatcher is so stressful.
2. A walk-in arrives at your fire station and tells you that scaffolding collapsed two blocks from the station, and there are many injuries. What actions should you take?
3. If you answer the phone at the fire station and the caller is reporting an incident, what information do you need to obtain from the caller?

REFERENCES

911.gov. 2023. "Frequently Asked Questions." Updated March 8, 2023. Accessed May 30, 2023. www.911.gov/calling-911/frequently-asked-questions.

ADA.gov. 2016. "Americans with Disabilities Act Title II Regulations." ADA.gov, October 11, 2016. Accessed March 12, 2023. www.ada.gov/law-and-regs/title-ii-2010-regulations/#-35161-telecommunications.

Dolcourt, Jessica. 2014. "Text to 9-1-1: What You Need to Know (FAQ)." CNET, May 5, 2014. Accessed May 27, 2023. www.cnet.com/news/text-to-911-what-you-need-to-know-faq.

Federal Communications Commission. 2015. "Best Practices for Implementing Text-to-911." Updated December 22, 2015. Accessed May 30, 2023. www.fcc.gov/general/best-practices-implementing-text-911.

Federal Communications Commission (FCC). 2016. "911 Wireless Services." Accessed May 27, 2023. www.fcc.gov/consumers/guides/911-wireless-services.

Federal Communications Commission (FCC). 2020. "Text to 911: What You Need to Know." Updated January 6, 2020. Accessed June 29, 2023. www.fcc.gov/consumers/guides/what-you-need-know-about-text-911.

Federal Communications Commission. 2022. "Public Safety Spectrum." Updated March 4, 2022. Accessed May 30, 2023. www.fcc.gov/public-safety/public-safety-and-homeland-security/policy-and-licensing-division/public-safety-spectrum.

FirstNet Authority. n.d. "Home Page." Accessed March 12, 2023. www.firstnet.gov.

Hawkins, Dan. 2007. *Communications in the Incident Command System*. Centers for Disease Control and Protection, March 2007. Accessed September 28, 2022. www.cdc.gov/niosh/erhms/pdf/cops-interoperable-communications-technology-program.pdf.

Ludwig, Gary. 2019. "Examining the Role of Culture in Firefighter Deaths." FireRescue1, June 10, 2019. Accessed May 30, 2023. www.firerescue1.com/fire-chief/articles/examining-the-role-of-culture-in-firefighter-deaths-8ndTaFl986zrxm81.

National Emergency Number Association. 2021. "9-1-1 Statistics." February 2021. Accessed March 9, 2023. www.nena.org/page/911Statistics.

National Fire Protection Association. 2019. *NFPA 1410, Standard on Training for Emergency Scene Operations*. 2020 Edition. Quincy, MA: National Fire Protection Association.

National Fire Protection Association. 2020. *NFPA 450, Emergency Medical Services and Systems*. 2021 Edition. Quincy, MA: National Fire Protection Association.

National Fire Protection Association. 2021. *NFPA 72, National Fire Alarm and Signaling Code*. 2022 Edition. Quincy, MA: National Fire Protection Association.

National Fire Protection Association. 2021. *NFPA 1225, Emergency Services Communications*. 2022 Edition. Quincy, MA: National Fire Protection Association.

National Fire Protection Association. 2023. *NFPA 1010, Standard on Professional Qualifications for Firefighters*. 2024 Edition. Quincy, MA: National Fire Protection Association.

National Fire Protection Association. 2023. *NFPA 1900, Standard for Aircraft Rescue and Fire Fighting Vehicles, Automotive Fire Apparatus, Wildland Fire Apparatus, and Automotive Ambulances*. 2024 Edition. Quincy, MA: National Fire Protection Association

U. S. Department of Transportation (DOT). *NG911 & FirstNet*. Accessed May 28, 2023. www.npstc.org/download.jsp?tableId=37&column=217&id=3974&file=NG911_FirstNet_Guide_170820.pdf.

Chapter Opener: © Eric Scruggs

CHAPTER

5

Support Person

Fire Behavior

KNOWLEDGE OBJECTIVES

After studying this chapter, you will be able to:

- Understand the basic principles of fire dynamics.
- Explain how fires are spread by conduction, convection, and radiation.
- Describe the four methods of extinguishing fires.
- Define Class A fires and the risks associated with them.
- Define Class B fires and the risks associated with them.
- Define Class C fires and the risks associated with them.
- Define Class D fires and the risks associated with them.
- Define Class K fires and the risks associated with them.

SKILLS OBJECTIVES

There are no skills objectives for Support Person candidates. NFPA 1010 contains no Support Person Performance Requirements for this chapter.

ADDITIONAL NFPA STANDARDS

- **NFPA 1**, *Fire Code, 2024 Edition*
- **NFPA 10**, *Standard for Portable Fire Extinguishers, 2022 Edition*
- **NFPA 24**, *Standard for the Installation of Private Fire Service Mains and Their Appurtenances, 2022 Edition*
- **NFPA 53**, *Recommended Practice on Materials, Equipment, and Systems Used in Oxygen-Enriched Atmospheres, 2021 Edition*
- **NFPA 67**, *Guide on Explosion Protection for Gaseous Mixtures in Pipe Systems, 2019 Edition*
- **NFPA 99**, *Health Care Facilities Code, 2024 Edition*
- **NFPA 115**, *Standard for Laser Fire Protection, 2020 Edition*
- **NFPA 268**, *Standard Test Method for Determining Ignitibility of Exterior Wall Assemblies Using a Radiant Heat Energy Source, 2022 Edition*
- **NFPA 291**, *Recommended Practice for Water Flow Testing and Marking of Hydrants, 2022 Edition*
- **NFPA 1403**, *Standard for Live Fire Training, 2018 Edition*
- **NFPA 1410**, *Standard on Training for Emergency Scene Operations, 2020 Edition*

CASE STUDY

You Are the Support Person

It is 0346 and your crew arrives at a modern-looking two-story house for the report of a garage fire called in by a neighbor. Through the smoke and flames, you observe what appears to be a popular model of an electric vehicle parked inside the garage and involved in the fire. As firefighters begin applying water to the body of the fire, you witness fire conditions growing rapidly and extending into the house. The hair on the back of your neck rises as you get an unsettled feeling.

1. Why would this situation make you uneasy?
2. What elements of this scene may make extinguishment of the fire challenging?

Introduction

Since prehistoric times, when humans first discovered how to use fire for cooking food and keeping warm, fire has fueled our lives. Although we have progressed from using open fires for cooking and for warmth, we continue to depend on fires to provide the energy for modern society. Burning fossil fuels creates most of our electric power. Our gasoline- and diesel-powered vehicles depend on small explosions (internal combustion) to propel them. Open flames in well-designed furnaces heat most of our homes.

Unfortunately, destruction of lives and property by uncontrolled fires has also been occurring since ancient times. Despite our advanced technology for detecting and combating fires, we continue to wage an ongoing battle with fire. In the United States, the fire death rate steadily increased between 2012 and 2021 (U.S. Fire Administration [USFA] 2023). According to the National Fire Protection Association (NFPA), in the United States in 2021, there were 3,800 civilian fire deaths and 14,700 civilian fire injuries. Fire departments responded to 1,353,500 fires. These fires resulted in a property loss of $15.9 billion. Seventy-nine percent of the civilian fire deaths and 86 percent of the civilian fire injuries were caused by home structure fires (Hall and Evarts 2021, 1–2). Understanding how fires start, the factors that cause them to grow, the characteristics of different types of fires, and the information provided by smoke is a good start to learning how to prevent unintended fires and how to extinguish them safely when they occur.

What Is Fire?

Fire is the visible result of combustion. **Combustion** is a rapid chemical chain reaction between a material or a substance and oxygen that produces heat and usually light. For the purposes of the fire service, the terms *combustion* and *fire* can be used interchangeably. Materials and substances that are capable of catching fire are **combustible**. Fire is characterized by the production of a flame, which can be many different colors, depending on what is burning and how much heat is produced. The location where a fire starts is the **area of origin**. This is sometimes called the *fire compartment* when the fire is in a structure, the **fire seat**, or the **seat of the fire** in casual conversation.

Matter

Matter is anything that has mass and volume—in simple terms, it weighs something, and it takes up space. Matter is made up of elements; an **element** is a substance that cannot be chemically broken down into simpler substances. Elements are often referred to by their symbol. For example, oxygen is an element, and its symbol is O. Hydrogen is another element, and its symbol is H. Elements are made up of atoms; an **atom** is the smallest unit of an element that retains the properties of the element. Some atoms, such as oxygen, exist in pairs because they are more stable that way. Therefore, the symbol for stable oxygen is O_2, which essentially means two oxygen atoms bonded together. A **molecule** is made up of atoms that are chemically bonded together. This type of bond is called a **molecular bond**. For example, oxygen and hydrogen are elements. When two atoms of hydrogen (H) bond with one atom of oxygen (O), one molecule of water (H_2O) is formed.

The physical form of a material is referred to as its **state of matter** (**FIGURE 5-1**). There are three states of matter: solid, liquid, and gas.

A **solid** has a specific size and shape. Cold makes most solids more brittle, whereas heat makes them more flexible. Most of the molecules that make up a solid are inside of it—that is, they are cushioned or

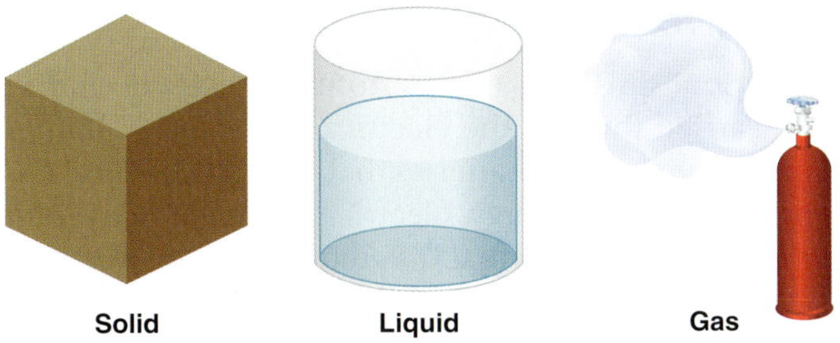

FIGURE 5-1 The three states of matter are solid, liquid, and gas.

© Jones & Bartlett Learning

insulated by the outer surface of the solid. Only a limited number of the molecules that make up the solid are present on its surface.

A **liquid** has a specific volume but does not have a specific size or shape. A liquid assumes the shape of the container in which it is placed. Most liquids expand when heated, and if heated enough, they turn into gases. Liquids cannot be compressed. This means that when you apply pressure to one end of a container, the liquid will move. This characteristic allows firefighters to pump water for long distances through pipelines or hose.

A **gas**, sometimes called a **vapor**, does not have a specific size, shape, or volume. Instead, a gas expands to fill the volume of the container into which it is released. The gas we most commonly encounter is air, the mixture of invisible, odorless, tasteless gases that surrounds the earth. Air is composed of 21 percent oxygen; 78 percent nitrogen; 1 percent argon; and trace amounts of other gases such as carbon dioxide, neon, methane, helium, krypton, hydrogen, and xenon.

Matter can change from one physical state to another under specific conditions. For example, water is a liquid at room temperature. When it reaches 32°F (0°C), it freezes—that is, it changes to a solid. When it reaches 212°F (100°C), it vaporizes—that is, it changes to a gas. The water molecules do not change—the molecules are all still H_2O—only the physical state of the water changes.

Energy

Energy is the ability to do work. The law of conservation of energy states that energy cannot be created or destroyed. Instead, energy is converted from one form to another. Think of an automobile. Chemical energy in the gasoline is converted to mechanical energy when the car moves down the road. When you apply the brakes to stop the car, the mechanical energy used to do this is converted to heat energy by the friction between the wheel rotors and the brake pads. Electrical energy can be converted to heat or to light. Mechanical energy can be converted to electrical energy through a generator. In a house fire, the stored chemical energy in the wood structure and plastic-based contents of a house is converted into heat and light energy during the fire. Energy exists in many forms, including chemical, mechanical, electrical, light, nuclear, and heat.

Energy is kinetic or potential. **Kinetic energy** is the energy possessed by an object because of its motion. **Potential energy** is energy stored by an object as a result of its position or condition. For example, if you hold a ball, it has potential energy. If you drop the ball, the potential energy changes to kinetic energy, and the ball falls to the ground.

Heat Energy

Heat energy (also referred to as **thermal energy**) is the potential energy of a combustible material and the kinetic energy released when heat is applied to the material. Heat can be measured using different units, including the following:

- **British thermal unit (BTU)**: A BTU is the amount of heat energy required to raise 1 pound of water at sea level by 1°F. BTU is a common measure of heat used in the United States. For example, architects designing a new fire station might select a commercial heating system based on its BTU rating to ensure it has the capacity to adequately heat the building.
- **Calorie**: A calorie is the amount of heat energy required to raise 1 gram of water at sea level by 1°C .
- **Joule**: A joule is a measure of heat energy equal to 0.4 calorie. One calorie is equal to about 4.2 joules, and 1 BTU is equal to approximately 1,055 joules.

Most people use the terms *heat* and *temperature* interchangeably; however, they have different meanings. Heat is a form of energy. As heat is applied to a material, the material's molecules increasingly move and vibrate, causing its temperature to rise. **Temperature** is the measurement of this movement and tells us how hot or cold something is. The two most common units of measurement for temperature are degrees Fahrenheit (°F) and degrees Celsius (°C). Residential structure fires can reach well over 1000°F (538°C).

Changing the temperature of a material can change its physical state. For example, when water is heated, the molecules move and vibrate until the water is boiling, and when it reaches 212°F (100°C), the physical state of the water molecules changes to gas—steam.

When the temperature of water is lowered—as heat leaves or is removed from the water—the movement of the molecules slows. When the temperature reaches 32°F (0°C), it changes to a solid—ice.

Chemical Energy

Chemical energy is the stored potential energy in molecular bonds and the kinetic energy created by a chemical reaction. Some chemical reactions are **exothermic**, which means they produce, or give off, heat. Others are **endothermic**, which means they absorb heat. Combustion is an example of an exothermic reaction. The reaction between water and ammonium salts in an instant ice pack is an example of an endothermic reaction—when combined, the two rapidly absorb heat. Most chemical reactions occur because the molecular bonds established between elements are broken when energy is applied.

Oxidation is the process during which oxygen combines chemically with another substance to create a new compound. For example, steel that is exposed to oxygen results in rust. The process of oxidation can be extremely slow; indeed, it can take years for oxidation to become evident. Slow oxidation does not produce easily measurable heat. Combustion is a type of fast oxidation.

Heat is produced whenever oxygen combines with a combustible material. If the reaction occurs slowly in a well-ventilated area, the heat is released harmlessly into the air. If the reaction occurs rapidly or within an enclosed space, the material can be heated to its ignition temperature. **Ignition temperature** is the minimum temperature at which a material ignites in the presence of oxygen. This process is called **autoignition**. The fire that results is energy released as a result of the chemical reaction between oxygen and the fuel. An example of this occurs when a bundle of rags soaked with linseed oil releases enough heat through oxidization, causing the rags to ignite spontaneously.

Mechanical Energy

Mechanical energy is the potential energy stored because of the position of an object (for example, a ball you hold) or the kinetic energy of an object in motion. Water falling over a dam is an example of mechanical energy. Mechanical energy is converted to heat when two materials rub against each other and cause friction. For example, a fan belt rubbing against a seized pulley or vehicle tires spinning on pavement produce heat. Heat also is produced when mechanical energy is used to compress air in a compressor.

Electrical Energy

Electrical energy is energy from an electrical charge. It is carried through the electrical wires inside homes and can be stored in batteries that convert chemical energy to electrical energy. Electrical energy is converted to heat energy in several different ways. For example, electricity produces heat when it flows through a wire or any other conductive material. The greater the flow of electricity and the greater the resistance of the material, the greater the amount of heat produced. Examples of electrical energy that can produce enough heat to start a fire include electric heating elements, overloaded wires, electrical arcs, batteries and other energy storage systems, and lightning.

Light Energy

Light energy is produced by electromagnetic waves packaged in discrete bundles called *photons*. This energy travels as thermal radiation, a form of heat. When light energy is hot enough, it can sometimes be seen in the form of visible light. One example of light energy is the radiant energy we receive from the sun. We think of candles, fires, light bulbs, and lasers as forms of light energy. We should recognize that although these objects do produce light energy, they also produce heat. If light energy is of a frequency that we cannot see, the energy may be felt as heat but not seen as visible light.

Nuclear Energy

Nuclear energy is the potential energy stored within the nucleus of an atom or the kinetic energy released by splitting the nucleus of an atom into two smaller nuclei (fission) or by combining two small nuclei into one large nucleus (fusion). Nuclear energy is stored in radioactive materials. Nuclear reactions release large amounts of energy in the form of heat. These reactions can be controlled, as in a nuclear power plant, or uncontrolled, as in an atomic bomb explosion. In a nuclear power plant, the nuclear reaction releases carefully controlled amounts of heat. This is used to heat water to produce steam, which powers a steam turbine generator. The mechanical energy from the generator is converted to electrical energy. Both uncontrolled explosions and controlled reactions release radioactive material, which can cause injury or death.

Fuels

Fuel is matter in any state—solid, liquid, or gas—that stores energy and is combustible. When the energy in fuel is released, for example, when it is burning, it is converted into another type of energy. Think of the vast

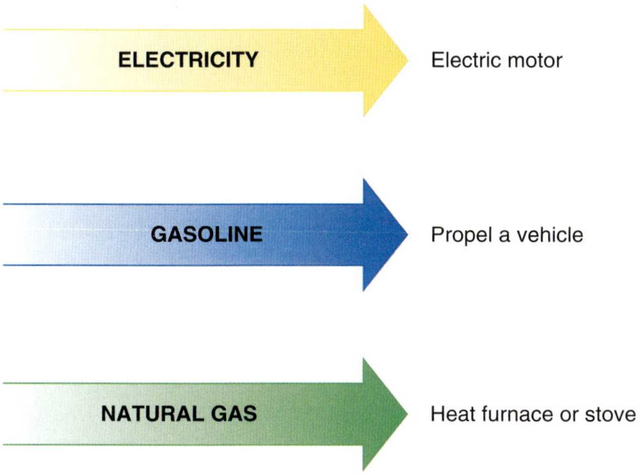

FIGURE 5-2 Energy being converted to work.

© Jones & Bartlett Learning

amount of kinetic energy in the form of heat and light that is released during a large fire. This energy released in the form of heat and light is stored in the fuel as potential energy before it is burned. The release of the potential energy in a gallon of gasoline, for example, can move a car many miles down the road (**FIGURE 5-2**).

Fire requires fuel that is in the form of combustible gas in order for the chemical process of combustion to occur. When solid and liquid fuels are heated, they produce a gas (they **off-gas**), and that is what ultimately ignites. Materials that remain in a solid or liquid state will not burn while in that state. Solid fuels such as wood, liquid fuels such as gasoline, and gaseous fuels such as propane all burn after they have changed into a gas.

Conditions Needed for Fire

To understand the behavior of fire, you need to consider the three basic elements needed for combustion to occur: fuel, oxygen, and heat. First, a combustible fuel must be present. Second, oxygen must be available in sufficient quantities. Third, a source of ignition (heat) must be present. These three components together form the **fire triangle** (**FIGURE 5-3**).

The fourth factor is combustion, the chemical chain reaction that occurs when the fuel, oxygen, and heat interact. One way of visualizing this process is to show the chemical chain reaction joining the elements of fuel, oxygen, and heat. This depiction reflects the central role that the chemical reaction plays in maintaining the process of flaming combustion. This relationship is sometimes characterized as the **fire tetrahedron**. A tetrahedron is a four-sided, three-dimensional figure—a four-sided pyramid. Each

FIGURE 5-3 A fire needs all three components of the fire triangle: fuel, oxygen, and heat.

© Jones & Bartlett Learning

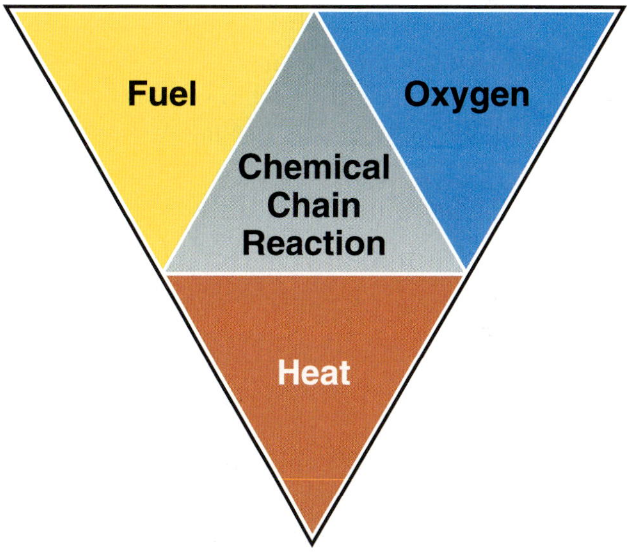

FIGURE 5-4 A chemical chain reaction unites fuel, oxygen, and heat in the fire tetrahedron.

© Jones & Bartlett Learning

side of the fire tetrahedron represents one of the four elements needed for a fire to occur (**FIGURE 5-4**).

The key point to remember is that fuel, oxygen, and heat must all be present for a fire to start and continue burning. If you remove any of these elements, the fire will go out. A **fuel-limited fire** is a fire that has sufficient oxygen for fire growth but has a limited amount of fuel available for burning. When all of the fuel is consumed, the fire stops burning. One example of a fuel-limited fire occurs in cases where an outdoor fire consumes the fuel supply. It also can occur in a room or compartment fire when the fuel supplied by the compartment contents is being consumed and the fire has not extended beyond the compartment because of fire-resistant construction. A **ventilation-limited fire** is a fire that has sufficient fuel for fire growth but has a limited amount of oxygen available, so it doesn't burn as rapidly as it would with an unlimited supply of oxygen. A ventilation-limited fire can grow quickly if more oxygen is added.

Chemistry of Combustion

Almost all fuels consist of hydrogen (H) and carbon (C) atoms; hence, they are called *hydrocarbons.* Fuels may contain a wide variety of other elements, but for our purpose, we can describe the self-sustaining chemical reaction of combustion simply by considering the hydrogen and carbon in the fuel. When a hydrocarbon (H–C) combines with oxygen (O_2), it produces two by-products, water (H_2O) and carbon dioxide (CO_2). This can be written as follows:

$$H\text{–}C + O_2 = H_2O + CO_2$$

In addition, the production of large quantities of heat and light is an integral part of combustion. Thus, the reaction for combustion should be rewritten as follows:

$$H\text{–}C \text{ (hydrocarbon fuels)} + O_2 = H_2O + CO_2 + \text{Light} + \text{Heat}$$

Products of Combustion

Of course, the reactions that occur in most actual fires are not as simple as this basic reaction that results in only water and carbon dioxide suggests. First, hydrocarbon fuels are complex molecules that contain complex chains of atoms in addition to hydrogen and carbon. As a consequence, the combustion process produces numerous toxic by-products in addition to water and carbon dioxide. Second, most structure fires encountered by fire crews today are ventilation-limited fires. This results in **incomplete combustion**, which is a combustion process during which the fuel is not completely consumed. Incomplete combustion produces significant quantities of deadly gases and compounds and a variety of other by-products, which are released into the atmosphere. All of these by-products are collectively called **smoke**. Because smoke is the product of incomplete combustion and it contains unburned hydrocarbons, you need to remember that it is a form of fuel. A column of hot black smoke coming in contact with an adequate supply of oxygen and an ignition source can ignite suddenly and violently. Smoke is composed of three components: solids in the form of **smoke particles**, liquids in the form of an **aerosol**—tiny water droplets suspended in the air similar to fog—and gases.

The particles in smoke include unburned, partially burned, and completely burned materials. The unburned particles are usually readily visible. Partially burned particles become part of the smoke because inadequate oxygen is available to allow for their complete combustion. Completely burned particles are primarily ash. The unburned and partially burned particles include a variety of substances. Some smoke particles are small enough that they get past the protective mechanisms of the respiratory system and enter the lungs. Most smoke particles are toxic. The concentration of the three types of particles depends on the amount of oxygen that was available to fuel the fire.

Smoke aerosol is formed when water is applied to a fire and tiny water droplets become suspended in the smoke or haze that forms. The vaporization of these water droplets may produce steam contained within the smoke. That steam may not be visible, and thus firefighters risk steam burns if not adequately protected. If the fuel contains oil-based compounds, small oil-based droplets become part of the smoke. Oil-based or lipid compounds can cause great harm when their smoke is inhaled. Some types of toxic aerosols cause poisoning if absorbed through the skin.

Smoke contains a wide variety of gases. The composition of gases in smoke varies depending on the amount of oxygen available to the fire at any instant. The composition of the gases produced by a burning substance influences the composition of the gases in the smoke. In other words, a fire fueled by wood produces a different composition of gases than a fire fueled by a petroleum-based fuel such as plastic.

Almost all gases produced by a fire are toxic, including the following:

- **Carbon monoxide (CO)**, produced by incomplete combustion, is deadly in fairly small quantities. Hemoglobin molecules in the blood normally carry oxygen (O_2) to the body's cells. When inhaled, CO quickly replaces the oxygen in the bloodstream because it binds with the hemoglobin molecules 200 times more readily than oxygen does. Even a small concentration of CO can quickly cause disability and death.
- **Hydrogen cyanide (HCN)** is formed from the incomplete combustion of plastic and foam products, such as furniture and polyvinyl chloride (PVC) piping used in residential construction. HCN is quickly absorbed by the blood and interferes with cellular respiration. Low-level exposure can cause cyanosis, headache, dizziness, unsteady gait, and

nausea. Just a small amount of HCN can render a person unconscious. Recent studies have indicated that cyanide poisoning in firefighters may be far more common than previously recognized.

- **Phosgene** is a gas formed from incomplete combustion of many common household products, including vinyl materials. This gas affects the body in several ways. At low levels, it causes itchy eyes, a sore throat, and a burning cough. At higher levels, phosgene gas can cause pulmonary edema (fluid retention in the lungs) and death.
- Fires produce many other gases that may affect the human body in harmful ways. Among these toxic gases are hydrogen chloride (HCl) and several compounds containing different combinations of nitrogen (N) and oxygen (O_2). Many of the compounds cause cancer, years after exposure. Higher rates of exposure to smoke increase the possibility of firefighters contracting cancer.

Carbon dioxide (CO_2), which is produced when sufficient oxygen is available for complete combustion, is not toxic, but it displaces oxygen in the atmosphere and can cause hypoxia and asphyxiation. Small amounts of CO_2 do exist naturally in our atmosphere, so an increase may seem entirely normal, but in fact, CO_2 in amounts above normal is immediately dangerous to life and health (IDLH) because of the lack of sufficient oxygen. Because CO_2 is heavier than air, it may pool in low-lying areas such as a basement or confined space. Firefighters must use a self-contained breathing apparatus (SCBA) in such a setting.

A discussion of the by-products of combustion would be incomplete without considering heat. Because smoke is the result of fire, it is hot. The temperature of smoke varies depending on the conditions of the fire and the distance the smoke travels from the fire. Injuries from smoke occur when the particles, droplets, and gases that make up smoke are inhaled. In addition, inhaling the hot gases in smoke can cause severe burns to the respiratory tract. In addition, all of the components in smoke can cause severe burns to the skin.

Fire Spread and Heat Transfer

The exchange of heat energy between materials is known as **heat transfer**. When there is a difference in temperature between two objects, heat transfers from a hotter object to a cooler object until the objects reach equal temperatures. When two objects have the same temperature, heat transfer does not occur.

The **heat release rate (HRR)** is the rate at which heat energy is generated, and it is measured in watts (1 watt is equivalent to 1 joule per second). **Heat flux** is the measure of the heat transfer to or from one surface to another. For example, if a sofa is actively burning and heat is moving to the ceiling, the heat flux would indicate how much heat was being transferred to the ceiling. Heat flux is expressed as kilowatts per square meter (kW/m^2), watts per square centimeter (W/cm^2), or BTUs per square foot (BTU/ft^2). Heat flux depends on two factors: the difference in temperature between the two substances and the ability of each material to conduct heat. The higher the difference in temperature between the two substances, the faster the rate of transfer.

Heat is transferred in three ways: conduction, convection, and radiation.

Conduction

Conduction is the process of transferring heat between solids that are touching each other (**FIGURE 5-5**). This heat transfer occurs because of the kinetic energy of the molecules within the solid materials. Conduction transfers energy directly from one molecule to another, much the same way energy is transferred between billiard balls when struck.

The ability of a material to conduct heat is known as **thermal conductivity**. Objects vary in their ability to conduct energy. The thermal conductivity of an object depends on its composition, its moisture content, and how porous it is; that is, objects that have more tightly packed molecules are more efficient in conducting heat than objects that are less densely constructed. For example, inorganic materials such as metals are good conductors of heat; organic materials such as plastic, rubber, and wood are weak conductors of heat. If one end of a copper pipe is heated, the heat will be readily conducted along the pipe.

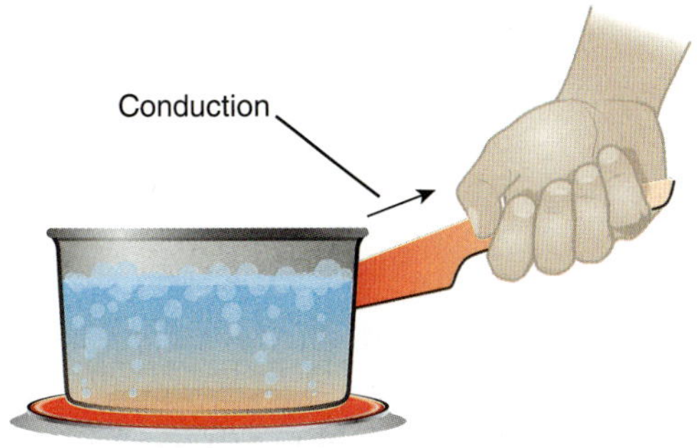

FIGURE 5-5 Conduction through a solid object.

© Jones & Bartlett Learning

Heat transfer by conduction is dependent on three factors. The first factor is the thermal conductivity of the material being heated—better conductors transfer more heat. The second factor is the size of the area being heated. If all other factors remain the same, heating a small area results in less heat transfer than heating a larger area. The third factor is the difference in temperature between the heated object and the object that is being heated. As discussed previously, the greater the temperature difference or energy difference between the two materials, the greater the rate of heat transfer.

Insulating materials limit the transfer of heat by conduction. Air or other gas is trapped in small pockets within the insulating material, and gases are not efficient conductors of heat from one solid to another because the gas molecules are farther apart than the molecules in denser materials. It is important to note, however, that although some insulating materials, such as polyurethane foam, are poor conductors, they can be highly combustible.

Convection

Convection is the transfer of heat in a gas or liquid by circulating the material from hotter areas to cooler areas (**FIGURE 5-6**). During a fire, the smoke and hot gases generated by the fire move by convection. The heat of the fire warms the gases and particles in the smoke. The hotter and less dense column of gases rises and displaces cooler, denser gases downward. A large fire burning in the open can generate a **plume**, also called a **thermal column**, which is an elongated column of heated gases and smoke that rises high in the air by convection. This convection stream can carry smoke and large bands of burning fuel for several stories before the gases cool and fall back to the earth.

When a fire occurs in a building, the convection currents generated by the fire rise in the room and travel along the ceiling. This is known as a **ceiling jet**. The hot gases carried by these currents may ultimately heat combustible room surfaces and contents to their ignition temperatures, and those materials will ignite. Firefighters can take advantage of convection currents by using the cooler, lower layer to advance a hose line or conduct a search for occupants.

As the fire grows, the volume of hot gases and smoke increases. If the pressure in the area of origin, or **fire compartment**—the room in which the fire started—is sufficient, the hot gases push laterally outside the fire compartment. Because the pressure in the lower parts of the fire compartment is lower, cooler air is drawn into the lower levels of the fire compartment. This route along which the heat, smoke, and hot gases move from the higher-pressure area inside the fire compartment to the lower-pressure area outside the fire compartment is the **flow path**. A flow path needs at least one intake vent, such as the front opening of a fireplace, and at least one exhaust port, such as the chimney above the fireplace. In other words, the flow path transfers heat, hot gases, and smoke from the higher pressure within the fire area toward the lower-pressure areas through doorways, window openings, and roof openings. Fire growth progresses along the flow path toward the exhaust vent. Convection currents, and therefore the flow path, are influenced by the layout of the building. For example, air flows quickly through a structure with an open floor plan and less quickly through a structure made up of small, closed rooms.

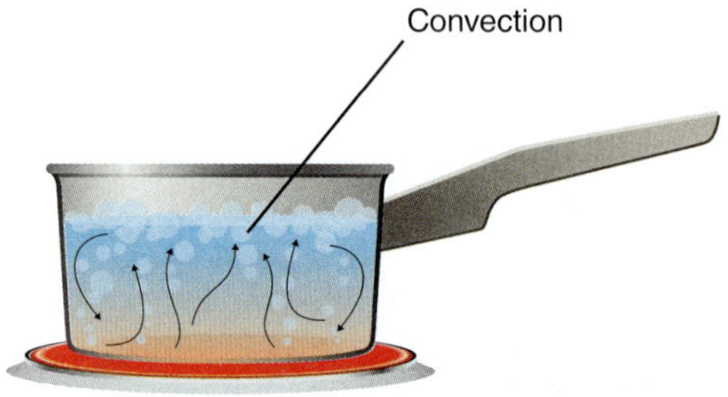

FIGURE 5-6 Convection transfers heat by the flow of gases or fluids from hotter areas to cooler areas.

© Jones & Bartlett Learning

Thermal Radiation

Thermal radiation is the transfer of heat through the emission of energy in the form of invisible electromagnetic waves. If you hover your hand over a burner on a stove, you feel the radiant heat from that burner (**FIGURE 5-7**). The sun radiates energy to the earth. When this energy is absorbed—for instance, as the sun's energy touches your body on a warm day—it is converted to heat. The direction in which the thermal radiation travels can be changed or redirected, as when a mirror reflects the sun's rays and bounces the energy in another direction. For example, heat from a fire is

FIGURE 5-7 Thermal radiation.

© Jones & Bartlett Learning

transferred in a direct line away from the source object and absorbed by cooler objects, including liquids and gases. If not properly protected, vehicles and structures next to a house on fire may become exposed to thermal radiant heat and ignite into additional fire problems. Thermal radiation also occurs in a fire compartment when heat from the fire is radiated back from the surfaces in the area of origin.

Thermal radiation from a heat source travels in all directions. The effect of thermal radiation, however, is not seen or felt until the radiation strikes an object and heats the surface of the object. Thermal radiation is a significant factor in the growth of a campfire from a small flicker of flame to a fire hot enough to ignite large logs. The growth of a small fire in a wastebasket to a full-blown room-and-contents fire is due in part to the effect of thermal radiation. A building that is fully involved in fire radiates a tremendous amount of energy in all directions. Indeed, the radiant heat from a large building fire can travel several hundred feet to ignite an unattached building.

The solid fuel in modern structures contains large amounts of petroleum-based materials, and this burning material produces a large amount of heat energy. This increased HRR creates conditions in which firefighters are exposed to higher levels of energy in a shorter amount of time, potentially saturating their personal protective equipment (PPE). If this happens, that heat energy is then transferred to the firefighter's body, potentially leading to dangerous conditions.

Methods of Extinguishment

There are four methods of extinguishing fires (**FIGURE 5-8**):

- Cool the burning material (remove heat from the fire tetrahedron).
- Exclude oxygen from the fire (remove oxygen from the fire tetrahedron).
- Remove fuel from the fire (remove fuel from the fire tetrahedron).
- Interrupt the chemical reaction with a flame inhibitor (remove heat from the fire tetrahedron).

There are many variations in the way that these methods can be implemented. Sometimes a combination of these methods is used to achieve suppression of fires.

The most common method of extinguishing a fire is to cool the burning material. Setting up a water supply,

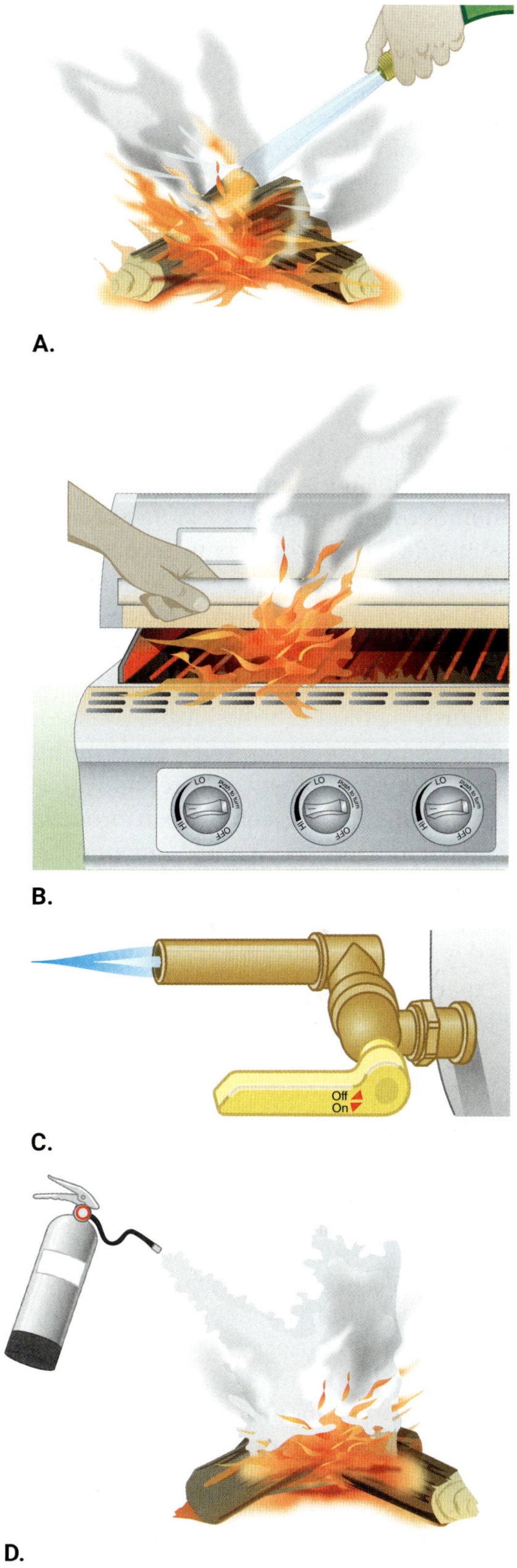

FIGURE 5-8 The four basic methods of fire extinguishment. **A.** Cool the burning material. **B.** Exclude oxygen from the fire. **C.** Remove fuel from the fire. **D.** Interrupt the chemical reaction with a flame inhibitor.

© Jones & Bartlett Learning

laying hose lines to the fire, and applying water to the fire are all steps in implementing this method of fire extinguishment.

A second method of extinguishing a fire is to exclude oxygen from the fire. A simple way to do this is to place the lid on an unvented charcoal grill or to close the door to a fire compartment. Reducing the amount of air reaching a fire will retard fire growth or extinguish the fire. To extinguish petroleum fires, firefighters use foam instead of water. They spray the foam in such a way that it creates a blanket that smothers the fire by separating the flammable gas from the oxygen in the air.

A third method of extinguishing a fire is to remove the fuel. For example, if a fire is being fed by a supply of natural gas, shutting off the supply extinguishes that fire. Remember that smoke is a product of incomplete combustion and is fuel, so reducing or cooling hot gases and smoke also diminishes the supply of fuel available to a fire. In wildland fires, a **firebreak**—a swath where the fuel (trees and brush) is removed—cut around a fire puts further fuel out of reach of the fire.

The fourth method of extinguishing a fire is to interrupt the chemical reaction with a **flame inhibitor**, a chemical extinguishing agent that reacts with the fuel to chemically disrupt the combustion process. These fire-extinguishing agents are applied with portable extinguishers or through a fixed suppression system designed to flood an enclosed space. They leave no residue and are often used to protect electronics and computer systems.

FIGURE 5-9 A Class A fire involves wood, paper, or other ordinary combustibles.

© schankz/Shutterstock

FIGURE 5-10 A Class B fire involves flammable liquids such as gasoline.

© thaloengsak/iStock/Getty Images

Classes of Fire

Fires are categorized into one of five classes based on the type of fuel: Class A, Class B, Class C, Class D, and Class K.

Class A Fires

A **Class A fire** involves ordinary solid combustible materials such as wood, paper, plastics, and cloth (**FIGURE 5-9**). Natural vegetation, such as the grass that burns in ground cover fires, is also considered to be part of this group of materials. The methods most commonly used to extinguish Class A fires are cooling the fuel with water to a temperature that is below the ignition temperature or using a combination of limiting ventilation and applying water.

Class B Fires

A **Class B fire** involves flammable or combustible liquids, such as gasoline, kerosene, diesel fuel, grease, tar, lacquer, oil-based paints, and motor oil, or gases such as propane and natural gas (**FIGURE 5-10**). These fires can be extinguished by shutting off the supply of fuel or by using foam to separate the oxygen in the air from the fuel.

Class C Fires

A **Class C fire** involves energized electrical equipment (**FIGURE 5-11**). A Class C fire could involve building wiring and outlets, fuse boxes, circuit breakers, transformers, generators, or electric motors. Power tools; lighting fixtures; household appliances; and electronic devices such as televisions, radios, and computers could be involved in Class C fires as well. Class C fires are extinguished with an extinguishing agent that does not conduct electricity. Incorrectly attacking a Class C fire with an extinguishing agent that conducts electricity can result in injury or death. Once the power is cut to a Class C fire, the fire is treated as a Class A or Class B fire, depending on the type of material that is burning.

FIGURE 5-11 A Class C fire involves energized electrical equipment.

© PhotoStock-Israel/Alamy Stock Photo

Class D Fires

A **Class D fire** involves combustible metals such as sodium, magnesium, zirconium, lithium, potassium, and titanium (**FIGURE 5-12**). Class D fires cannot be extinguished with water. In fact, applying water to fires involving these metals will result in violent explosions. This is because when these metals are heated and water is applied to them, the water molecules (H_2O) react with the heated metal and produce hydrogen gas (H_2), which burns with an explosive force. Instead, these fires must be attacked with extinguishing agents that prevent explosions, smother the fire, and snuff out the supply of oxygen. Many automobile manufacturers use magnesium components to reduce the weight and increase the strength of vehicles, so firefighters must take care when extinguishing a vehicle fire.

FIGURE 5-12 A Class D fire involves metals such as magnesium, sodium, or titanium.

© Andrew Lambert Photography/Science Source

Class K Fires

A **Class K fire** involves combustible cooking oils and fats and often occurs in kitchens (**FIGURE 5-13**). Heating vegetable fats, animal fats, or oils in appliances such as deep-fat fryers can result in serious fires that are difficult to fight with ordinary fire extinguishers. Class K extinguishing agents contain a wet agent that combines with cooking oils to produce a soap-like substance. The resulting soapy foam absorbs heat from the fire.

FIGURE 5-13 A Class K fire involves combustible cooking oils and fats.

© Kathie Nichols/Shutterstock

Fires Involving Mixed Materials

Some fires fit into more than one class. For example, a fire involving a wood building, Class A, could also involve petroleum-based contents, Class B. Likewise, a fire involving energized electrical circuits—a Class C fire—also might involve Class A or Class B materials. A vehicle fire may involve Class A, B, C, and even D materials, depending on the components used to construct the vehicle. If a fire involves live electrical sources, it should be treated as a Class C fire until the source of electricity has been disconnected to isolate the ignition source and protect firefighters. Once isolated, firefighters may treat the fire based on the classes of materials involved.

Characteristics of Solid-Fuel Fires

Most building fires that fire crews encounter are fed by solid fuels. In structure fires, the building and most of the contents exist as solids. A variety of solid fuels are found in most buildings. These include wood and wood-based products, fabrics, paper, and carpeting, as well as petroleum-based fuels, such as plastics and petroleum-based foam cushions. Wood is a commonly encountered building material. Older furniture was commonly constructed of solid-wood frames, and natural materials such as cotton were used for padding and upholstery. Modern furniture is more frequently constructed from petroleum-based foams and fabrics and is made with plastic frames and engineered wood that includes glue or adhesive that makes the wood more flammable. Wall coverings in older houses are often paper. Today, wall coverings are more likely to be latex-based paint or other petroleum-based materials.

Solid fuels burn when they are heated sufficiently to change them into flammable vapors. This process of liberating gaseous fuel vapors as a result of the heating of a solid fuel is called **pyrolysis**. Pyrolysis is evident when wood, heated sufficiently, breaks down into vapors and char (**FIGURE 5-14**). Some fuels change from a solid form directly into a vapor. Other solid fuels first change into a liquid before becoming a vapor. Plastics and petroleum-based materials pyrolyze faster than wood-based products. This means that it requires less heat to decompose and ignite these fuels than it does to ignite wood.

Wood contains varying amounts of moisture. When wood is heated, the first change that occurs is that the water is vaporized and escapes. This causes the wood to begin to char. Charring results in a number of protective mechanisms, including burning off the softer exterior cellulose and exposing a tougher inner layer that requires higher heat to ignite. Also, the carbon in the charring creates a protective layer. As the wood chars, it begins to lose mass, which can compromise its load-carrying capabilities. When the wood is heated to about 425°F (218°C), pyrolysis begins. As the wood is heated, the energy required to vaporize the water delays the beginning of pyrolysis.

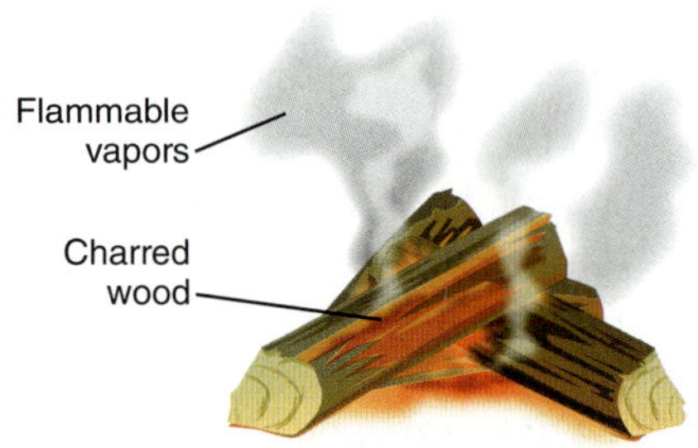

FIGURE 5-14 Pyrolysis is evident when wood, heated sufficiently, breaks down into vapors and char.

© Jones & Bartlett Learning

The characteristics of solid fuels that influence their combustion can be broken into three major categories: the composition of the fuel, the amount of the fuel, and the configuration of the fuel.

Composition of Solid Fuel

The chemical composition of the fuel has a significant impact on how the fuel burns. For example, wood is composed primarily of a natural, combustible fiber called *cellulose*. Most of the contents of modern structures, including synthetic foams, upholstery coverings, wall coverings, and plastic furniture, are manufactured from petroleum products. Petroleum-based products generally contain more potential heat energy than products made from natural products. This means that fires fueled by modern petroleum-based products have a much higher HRR than fires fueled by natural products such as wood and cotton.

The amount of moisture content in the fuel is another factor that influences how the fire burns and the amount of heat that is released. Fuel with more moisture takes longer to ignite than fuel with less moisture. The water contained in a fuel serves as a cooling agent. It requires heat to vaporize the water from the fuel before the fuel can be pyrolyzed to create a flammable vapor. For example, green firewood from a recently cut tree contains a lot of moisture and is difficult to burn in a fireplace, whereas firewood that has been allowed to dry will ignite more easily and burn faster, with a higher HRR.

Amount of Fuel

The second characteristic of fuel that determines how it burns is the amount of fuel available to the fire. If all other factors are the same, when more fuel is available, the HRR is higher than when less fuel is available. For example, under identical conditions, a fire fueled with four wooden pallets will have a much lower HRR than a fire fueled with 40 wood pallets.

Configuration of Fuel

The third characteristic of fuel that determines how it burns is the configuration of the fuel. Three factors contribute to this: the surface-to-mass ratio, the orientation of the fuel, and the continuity of the fuel.

Surface-to-Mass Ratio

The size and shape of a solid fuel greatly affect the ability of the fuel to ignite, the time it takes the fuel to be consumed, and the HRR of the burning fuel. For example, consider a large wood log that is 10 feet (3 meters) long. This log has a large mass or weight, but the surface area of the log is relatively small compared to its mass. If we cut the log into beams of 4 by 4 inches (in.; 10 by 10 centimeters [cm]), the total mass remains the same, but the surface area is much greater than it was when in the form of a single log. If the original log were cut into thin shingles, it would again have the same original mass as the log but would have a far greater total surface area than the original log or the 4-by-4-in. (10-by-10-cm) beams. The energy required to ignite the low surface-to-mass ratio log is much greater than the energy required to ignite the higher surface-to-mass ratio shingles. The higher the surface-to-mass ratio, the less energy that is required to ignite the fuel. This is because less energy is required to heat the material to a temperature sufficient to burn. For example, holding a match to a large log usually will not result in ignition of the log, but holding a match to a thin piece of shingle from the same source usually will result in ignition of the shingle.

Orientation of the Fuel

The second configuration factor that affects how a fuel burns is the orientation of the fuel. For example, a board in a horizontal position burns more slowly than the same size board in a vertical position. A board in a horizontal position that is ignited on one corner will produce hot fire gases that rise away from the surface of the board. As convection currents carry the heat from the fire upward, that heat is carried away from the surface of the board. If the same board is in a vertical position and it is ignited on one of the lower corners, the convection created by the hot fire gases will transfer some of the heat along the vertical surface of the board. The heat absorbed by convective heat transfer on the vertical board will be much greater than the heat that is absorbed by the board in a horizontal position.

Continuity of the Fuel

The third configuration factor that affects how a fuel burns is the continuity of the fuel, or how close one piece of fuel is to the next piece of fuel. Fuel that is in contact with other fuel will ignite faster and reach a peak HRR more quickly. The closer the fuel is, the easier and more quickly it will ignite. This rate of growth is partly due to spread by radiation, convection, and in some cases, close proximity conduction. Continuity can occur in a horizontal direction along floors, ceilings, or horizontal surfaces of building contents. It also can occur in a vertical direction along walls or the vertical surfaces of building contents.

Solid-Fuel Fire Development

A ventilated, fuel-limited, solid-fuel fire progresses through four classic stages of growth: the incipient stage, the growth stage, the fully developed stage, and the decay stage (**FIGURE 5-15**).

1. Incipient Stage

The first stage of fire development is the **incipient stage**. The incipient stage occurs when there is an adequate supply of fuel, oxygen, and heat or ignition (**FIGURE 5-16**). At the incipient stage, a fire is small and confined to the initial fuel that was ignited. Because the fire is small at this stage, the initial growth is largely dependent on the type of fuel and how much of it can be pyrolyzed into a gas. Radiant heat from the fire begins to pyrolyze increasing amounts of fuel into flammable gases. A fire at the incipient stage consumes relatively small amounts of oxygen, so at this stage, the fire is fuel dependent.

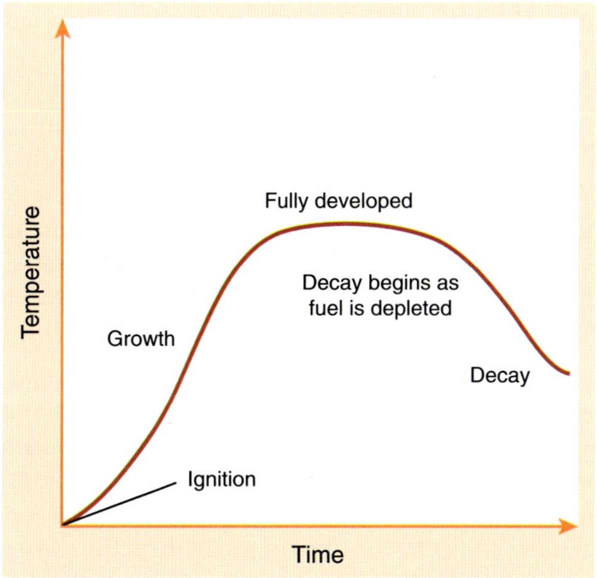

FIGURE 5-15 Illustration of a traditional fire growth curve with a fuel-limited fire.

Modified from National Institute of Standards and Technology. "Fire dynamics." Updated June 2, 2021. www.nist.gov/el/fire-research-division-73300/firegov-fire-service/fire-dynamics

FIGURE 5-16 The incipient stage of a fire.

Courtesy of the National Institute of Standards and Technology.

FIGURE 5-17 The growth stage of a fire.

Courtesy of the National Institute of Standards and Technology.

As the fire grows and produces more heat, a plume of hot gases and smoke begins to rise from the fire. As this plume rises, it is cooled by the surrounding cooler air. As increasing amounts of heat are generated, the plume carries more heat and hot gases upward. If it reaches the ceiling, it will spread out in a circle that increases in size. These hot gases begin to heat the surfaces with which they make contact.

In the incipient stage, because the amount of heat is limited, there is little increase in the temperature of the room. The amount of smoke and combustion gases in the compartment is usually not high enough to pose a serious threat to occupants if they exit quickly. Many fires in the incipient stage can be extinguished safely with a portable fire extinguisher.

It is important to understand, however, that the stages of fire growth do not have a clear dividing line between them. A fire can progress rapidly from the incipient stage to the growth stage. Because fire can grow rapidly, building occupants and fire service personnel who are not equipped with a complete ensemble of PPE, including SCBA, should exit from even a small fire.

2. Growth Stage

The second stage of fire development is the **growth stage**. A fire in the growth stage produces more interaction and is more dependent on the environment in the compartment around it than a fire in the incipient stage (**FIGURE 5-17**). Especially important are the composition of the compartment surfaces and contents and the placement and configuration of the compartment contents. The amount of ventilation is also an important factor.

As a fire begins to grow, it produces larger amounts of heat, producing a higher HRR. The hot gases and smoke increase in density and turbulence as additional fuel is consumed. The location of the fuel in a compartment influences the fire development. If the fuel is in the center of the compartment, it may **entrain**, that is, encircle, draw along, and transport air from four sides, and the hot gases and smoke in the plume mix with the cooler air. This cooling reduces the rate of growth, the length of the flame, and the length of the plume, thereby reducing the vertical extension of the fire. These factors help to slow the rate of fire growth. If the burning fuel is located close to a corner of the compartment, the plume can entrain air from only two sides. As a result, the rate of growth will be faster and the length of the flame and the plume will be longer than if the burning fuel were in the middle of the room. If the fuel is located close to a wall, it entrains air on three sides, resulting in a growth rate slower than that of a fire located in a corner but faster than that of a fire with fuel in the middle of the room. A higher fire plume increases the hot gases at the ceiling of the compartment. More hot gases at the ceiling will radiate more heat back to the room surfaces, the room contents, and the burning fuel, further increasing the rate of fire growth.

Specific types of fire conditions are described next. These include thermal layering, rollover, flashover, backdraft, the behavior of ventilation-limited fires, and rapid fire growth. These conditions are not limited to the growth stage of fire development but are introduced here because they are likely to occur in the growth stage.

Thermal Layering

As a fire continues to grow, more hot gases are generated. Hot gases are lighter and buoyant and therefore tend to seek a higher level in the compartment. Because cooler gases are heavier, they settle at lower levels in the compartment. The phenomenon of gases forming into layers according to temperature is called **thermal layering**, also known as **heat stratification**. Whenever the ventilation of the fire changes, it changes the thermal layering.

As the hot fire gases rise, they increase the rate of heat transfer back to the contents and walls of the fire compartment. The layer of hot gases in the upper level of the compartment creates an increase in pressure. This layer of hot gases tends to push out of the upper level through any available openings, such as doors or windows. Because the cooler gases at the bottom of the room exert less pressure than the hotter gases, cooler air is drawn into the fire compartment, usually at lower levels.

As the hot gases exit the fire compartment at the upper levels and the cooler air enters the fire compartment at a lower level, there is a level in the compartment where the pressure exerted by the lighter hot gases flowing out and the pressure of the cooler air flowing in is equal. This is called the **neutral plane**. In order for a neutral plane to exist, there must be a flow of cooler air entering the compartment and a flow of hot gases exiting the compartment.

As the fire continues to grow and the amount of heat generated increases, small flames "dance" in the hot gas layer. These isolated flames are an indication that the gases in the hot layer are near their ignition temperature. Often, these flames appear around the edges of the hot plume because this area has sufficient oxygen for combustion. These isolated flames may indicate that the fire is a ventilation-limited fire. This also may indicate that conditions are approaching a point where the room and contents will autoignite.

Rollover

Rollover, also called **flameover**, is the spontaneous ignition of hot gases in the upper levels of a room or compartment (**FIGURE 5-18**). During the growth stage of the fire, the hottest gases rise to the top of the room. If any of these gases reach their ignition temperature from the radiation of heat coming from the flaming fire or from direct contact with open flames, they will ignite. In other words, the upper layer of flammable gases catches fire. These flames can flicker across the ceiling and then go out. Alternatively, when they get a few degrees hotter, they can extend throughout the room at ceiling level. Rollover is a sign that the temperature is rising, and if it continues to rise, the temperature throughout the room will reach the point of flashover, when the room and contents spontaneously and rapidly ignite.

Flashover

Flashover is the rapid transition from a fire that is growing by igniting one type of fuel to a fire where all of the exposed surfaces have ignited. It is a rapid change or transition from the growth stage to the fully developed stage. During a fire, it may be hard to determine which stage a fire is in. Fires do not always follow the four classic stages of development. A flashover can occur whenever heat and sufficient oxygen are available to support combustion (**FIGURE 5-19**). Flashovers are less likely to occur if the fire is in a large room or a compartment that supplies more air during fire development. Larger areas entrain more air in the fire plume, which cools the plume and reduces the radiation and convection of heat to the room surfaces and contents.

The critical temperature for a flashover to occur is approximately 1000°F (540°C). Once this temperature is reached, all fuels in the room are involved in the fire,

FIGURE 5-18 Rollover is a sign that flashover is imminent unless actions are taken to change the fire conditions.

© Jones & Bartlett Learning. Photographed by Glen E. Ellman.

FIGURE 5-19 Flashover is the rapid change or transition from the growth stage to the fully developed stage.

© Jones & Bartlett Learning. Courtesy of Dave Casey.

including the floor coverings. This means that the temperature at the floor level may be as high as the temperature of the ceiling just before flashover. Firefighters, even with full PPE, cannot survive for more than a few seconds in a flashover, so it is important to be able to recognize the signs that may indicate a flashover is imminent. These signs include off-gassing or smoke from furniture and carpeting, a sudden increase in heat, and zero visibility. Firefighters must consider the potential risk level of entering a structure showing signs of an impending flashover and consider the possible benefits. If flashover has occurred, there is extreme risk for firefighters and no benefit because any victims exposed to the flashover will be deceased. There is no sense in firefighters risking their lives to try to save someone who is already dead.

Flashover may not occur if the fire remains ventilation limited. A ventilation-limited fire produces a limited amount of heat energy. If the supply of oxygen is not increased, the fire may enter a decay stage. During this decay stage, heat continues to pyrolyze fuels in the compartment, producing additional vaporized fuel in the form of smoke. A fire in the decay stage will flash over quickly, however, if temperatures are maintained and sufficient oxygen is added (**FIGURE 5-20**). Keeping doors and windows to the fire compartment closed helps to reduce the amount of oxygen available to the fire and may help prevent a flashover. It is important to note that even with increased ventilation, many compartment fires remain ventilation limited rather than advancing to a fully developed fire because there is insufficient oxygen to permit the fire to freely burn.

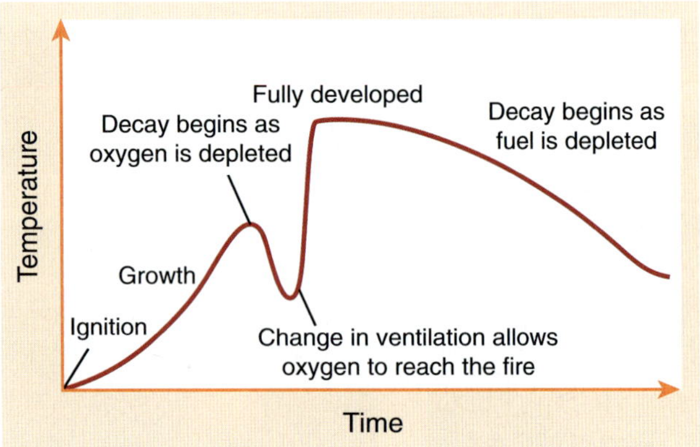

FIGURE 5-20 Illustration of a ventilation-limited fire, which often occurs in houses with lightweight construction and petroleum-based contents.

Modified from National Institute of Standards and Technology. "Fire dynamics." Updated June 2, 2021. www.nist.gov/el/fire-research-division-73300/firegov-fire-service/fire-dynamics

Backdraft

A **backdraft** is an explosion caused by the introduction of oxygen into a compartment where superheated gases and contents are hot enough for ignition but do not have sufficient oxygen to combust. Superheated gases are gases heated above their ignition temperatures. Backdrafts require a "closed box"—that is, a room, compartment, or building that is ventilation limited. When the fire compartment has a limited amount of oxygen, combustion is reduced, yet superheated fuel in the form of smoke still fills the compartment (**FIGURE 5-21**). If a supply of oxygen is introduced into the room by ventilation, such as opening the front door or an accidental event such as a window breaking because of excess heat, sudden and explosive combustion

FIGURE 5-21 These photos show the rapid transition to a backdraft. The photos were taken only seconds after the compartment door was opened.

© Jones & Bartlett Learning. Photographed by Glen E. Ellman.

can occur as a result of the presence of the superheated flammable gases. This explosive combustion may exert enough force to cause severe injury or death to firefighters.

Signs and symptoms of an impending backdraft include the following:

- A confined fire with a large heat build-up
- Little or no flame visible from the exterior of the building
- A "living fire," where the building appears to be breathing as a result of smoke puffing out of and then being sucked back into the building
- Smoke that seems to be ventilating out under pressure
- Smoke-stained windows, which indicates a significant fire
- No visible smoke
- **Turbulent smoke flow**
- Thick yellowish smoke, which indicates that it contains sulfur compounds

It is important for responders to stay alert for the conditions that signal a possible backdraft and to reduce the possibility of a backdraft developing.

Behavior of Ventilation-Limited Fires

Experiments conducted by Underwriters Laboratories (UL), the National Institute of Standards and Technology (NIST), and the New York City Fire Department (FDNY) have consistently demonstrated that many building fires become ventilation limited because of a limited supply of oxygen. Newer houses are tightly sealed, with added insulation and caulking, double-paned windows, and storm windows. A fire in a building during the growth stage can consume enough oxygen to reduce the concentration of oxygen below the level needed to maintain a growing fire. When this occurs, the fire becomes ventilation limited. Assessing a ventilation-limited fire presents a challenge to firefighters. A fire in this condition may be misleading and incorrectly appear to be a small fire in the incipient stage, even though it contains a large amount of energy in the form of hot gases and smoke. Research indicates that fires in modern residential occupancies are likely to enter a ventilation-limited decay stage prior to the arrival of the first-due engine company (Kerber 2009).

Introducing air into a ventilation-limited fire can result in explosive, rapid fire growth. Research has demonstrated that firefighters making entry through the front door can introduce enough air into the fire area to produce rapid fire growth and flashover. This change can occur so fast that it is not possible to escape this deadly environment. Everyone needs to consider carefully what constitutes ventilation. Opening any door, window, skylight, or roof introduces oxygen into a burning building. Repeated experiments have been conducted that produced rapid fire growth after the front door was opened. Rapid fire growth can be prevented by cooling the fuel, removing the fuel, or controlling the amount of oxygen present. During rapid growth, a fire can progress to its maximum HRR unless actions are taken to limit the supply of oxygen or apply water to the fire.

From the outside, a ventilation-limited fire may be deceiving—it may look like a small fire. Often flames are not visible. All that is needed is more oxygen to produce explosive fire growth. Be aware that a fire that appears to be "small and relatively harmless" may in fact be a ventilation-limited fire containing enormous amounts of energy that just needs a healthy dose of oxygen to grow rapidly.

3. Fully Developed Stage

The third stage of fire development is the **fully developed stage**. During this stage, the fire is consuming the maximum amount of fuel possible, and it is achieving the maximum HRR possible for the fuel supply and oxygen present (**FIGURE 5-22**). During the fully developed stage, the fire may be ventilation limited or fuel limited. When a fire has an unlimited supply of oxygen, the fire is fuel limited. Consider a building under construction that has been framed with

FIGURE 5-22 The fully developed stage of a fire.

Courtesy of the National Institute of Standards and Technology.

wood and combustible sheathing but lacks windows and doors. If this building catches fire, the fully developed fire may be fuel limited because it has an abundant supply of oxygen. In contrast, most structure fires that fire crews encounter have limited openings from the fire compartment to the outside. Because there are limited openings to admit oxygen and exhaust the hot fire gases and smoke, these fires are ventilation limited.

During the fully developed stage, as the fire releases large quantities of heat, the energy pyrolyzes large amounts of fuel, generating large quantities of smoke and fire gases. The smoke and fire gases increase pressure in the fire compartment and force hot fire gases from the fire compartment. The HRR will be dependent on the size of the compartment openings, which provide a supply of oxygen and openings for hot fire gases to exit.

When the hot fire gases are exhausted from the fire compartment, if they are above the ignition temperature of the gases, they may ignite upon mixing with a fresh supply of oxygen. This is what produces flames from compartment openings during this active burning state. If these hot fire gases travel to another room or compartment adjacent to the fire compartment, they can transfer enough heat to spread the fire to those rooms or compartments. As the hot vaporized fuel travels to adjacent areas, it can ignite when it reaches a new supply of oxygen. The large quantities of heat transferred to adjacent areas will pyrolyze fuel in these areas, increasing the supply of vaporized fuel available to the fire. This provides an additional source of fuel for the fire.

Not all fires reach the fully developed stage. If the fire compartment has limited openings that result in the fire being ventilation limited, the fire probably will not reach the conditions needed to achieve the maximum rate of heat production of a fully developed fire that has an unlimited supply of oxygen.

4. Decay Stage

The fourth and final stage in fire development is the **decay stage**. This condition occurs because of a decreasing fuel supply or because of a limited oxygen supply (**FIGURE 5-23**).

During the decay stage of a fuel-limited fire, active flaming combustion decreases or stops. The heat continues to pyrolyze the fuel and create flammable gases and vapors. These flammable fuel products will continue to burn, but the rate of combustion will continue to decrease and eventually stop when the fuel supply is exhausted.

FIGURE 5-23 The decay stage of a fire.
Courtesy of the National Institute of Standards and Technology.

During the decay stage of a ventilation-limited fire, the rate of combustion slows, and visible flames decrease or disappear. Because there is still significant heat in and around the fire and the fire compartment, fuels continue to pyrolyze and create additional flammable vapors and gases. The rate of pyrolysis slows as the rate of combustion slows, but large quantities of flammable fuel may still be present. If additional oxygen is introduced into the fire compartment, rapid or violent fire growth can develop quickly, and the fire will return to the fully developed stage.

During the decay stage, smoke may travel inside the structure some distance from the fire. When it comes in contact with a source of ignition, the flammable mixture will ignite—often in a violent manner. This is known as a **smoke explosion**. Smoke explosions can occur during the decay stage of a fire, but they also sometimes occur during final extinguishment or overhaul. A smoke explosion does not occur because of a change to the ventilation profile, such as an open door or window. Instead, it occurs when smoke travels within the structure to an ignition source.

Characteristics of Liquid-Fuel Fires

Fires involving liquid fuels have some different characteristics from fires that involve solid fuels. Recall that solid fuels do not burn in the solid state but instead must be converted to a gas and mixed with oxygen before they will burn. Liquids share the same characteristic. They must be converted to a gas and mix with oxygen before they will burn. Like solid fuel, heat is what causes this state change. Three conditions must be present for a vapor and air mixture to ignite:

- The fuel and air must be present at a concentration within a flammable range.
- There must be an ignition source with enough energy to start ignition.
- The ignition source and the fuel mixture must make contact for long enough to transfer the energy to the air–fuel mixture.

As liquids are heated, the molecules in the liquid become more active, and the speed of vaporization of the molecules increases. Most liquids eventually reach their boiling point during a fire. **Boiling point** is the temperature at which a liquid continually vaporizes in sustained amounts and, if held at that temperature long enough, turns completely into a gas. As the boiling point is reached, the amount of flammable gas generated increases significantly (**FIGURE 5-24**). Because most liquid fuels are a mixture of compounds (for example, gasoline contains approximately 100 compounds), the fuel does not have a single boiling point. The single compound with the lowest ignition temperature determines the flammability of the mixture—the temperature at which that fuel will spontaneously ignite.

Liquids that have a lower molecular weight tend to vaporize more readily than liquids with a higher molecular weight. Liquids that vaporize more readily are more volatile than liquids that vaporize more slowly. In addition, the higher the temperature, the quicker the liquid will evaporate. As more of the liquid vaporizes, it may reach a point where enough gas is present in the air to create a flammable vapor–air mixture.

Two additional terms are used to describe the flammability of liquids: flash point and fire point. The **flash point** is the lowest temperature at which a liquid or solid produces a flammable gas. It is measured by determining the lowest temperature at which a liquid produces enough vapor to support a small flame or flash fire for a short period of time until the fuel is consumed (the flame may go out quickly) (**TABLE 5-1**). The **fire point** (also known as the **flame point**) is the lowest temperature at which a liquid produces enough vapor to sustain a continuous fire. For most materials, the fire point is only slightly higher than the flash point.

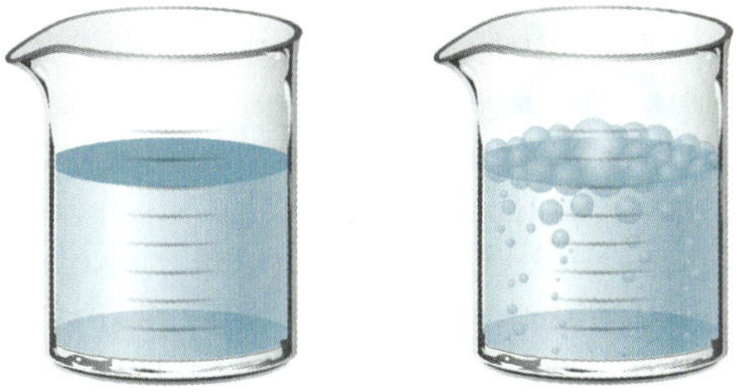

FIGURE 5-24 As liquids are heated, the molecules become more active, and the speed of vaporization increases.

© Jones & Bartlett Learning

Specific Gravity

A significant challenge faced when extinguishing flammable liquid-fuel fires is that many flammable liquids float on water. The term **specific gravity** refers to the density of a liquid compared to water, which has a specific density of 1.0. Gasoline, for example, has a specific gravity ranging from 0.72 to 0.76 and thus floats on top of water. **TABLE 5-2** lists the specific gravity of several liquids.

When flammable liquids have a specific gravity that is lighter than water, they float on water. If those liquids catch fire, the flames will continue to burn on the water's surface, and attempts to extinguish the flames with more water will simply spread the pool of burning flammable liquids across the water's surface.

TABLE 5-1 Flash Point

Liquid	Flash Point
Water	N/A
Gasoline	−45°F/−43°C
Acetone	−4°F/−20°C
#2 grade diesel	125°F/52°C

Modified from Friedman, Raymond. 1998. *Principles of Fire Protection Chemistry and Physics*. 3rd edition, p. 115. National Fire Protection Association.

TABLE 5-2 Specific Gravity Examples

Liquid	Specific Gravity
Water	1.0
Gasoline	0.72–0.76
Acetone	0.79
#2 grade diesel	0.84

Data from Barsan, Michael E., ed. September 2007. NIOSH Pocket Guide to Chemical Hazards. Accessed August 23, 2023. https://stacks.cdc.gov/view/cdc/21265

Alternative extinguishing agents, such as firefighting foam, may need to be applied to the floating pool of flammable liquids. In some cases, it may actually be the most practical to allow the pool of flammable liquids to burn while firefighters protect nearby structures and objects.

Heat-Induced Tears in Containers

Some liquid fuels are stored and transported in nonpressurized containers. A **heat-induced tear (HIT)** occurs when a nonpressurized vessel is exposed to direct **flame impingement**—flames in direct contact with the surface of a material transferring radiant heat—for a prolonged period, causing the structural integrity of the container to fail. When the container reaches a critical temperature, it ruptures and suddenly releases the flammable liquid inside. Once the escaping flammable liquid reaches its flash point and the gas finds an ignition source, the fuel rapidly ignites into a large fireball. A fireball is different from an explosion. A **fireball** is a burst of flames that rapidly ignites available flammable vapors but is not under pressure. An **explosion** is a violent and pressurized release of energy.

For example, when crude oil is transported in tank cars along the railroad, it is normally transported under very little, if any, pressure. If a tank car carrying crude oil derails, it might appear to catch fire and explode, but it is more probable that it experienced an HIT. A key distinction between HITs and explosions is that a HIT does not typically result in fragments of the container being launched outward.

Characteristics of Gas-Fuel Fires

Gas-fuel fires present unique challenges to firefighters. By learning about the characteristics of flammable gas fuels, you can help prevent injuries or deaths in emergency situations and work to mitigate the conditions causing the problem.

Vapor Density

Vapor density is the weight of a gas compared to an equal volume of dry air (**TABLE 5-3**). The weight of air is assigned the value of 1. A gas with a vapor density of less than 1 rises to the top of a confined space or rises in the atmosphere. For example, hydrogen gas, which has a vapor density of 0.07, is a very light gas. Conversely, a gas with a vapor density greater than 1 is heavier than air and settles close to the ground. For example, propane gas, which has a vapor density of 1.55, settles to the ground when it is released from a container. Carbon monoxide has a vapor density of 0.97—almost the same as that of air—so it mixes readily with all layers of the air. In situations where a flammable gas is present, responders need to know the vapor density of the escaping fuel so that they can take action to prevent the ignition of the fuel and allow the gaseous fuel to safely escape into the atmosphere.

TABLE 5-3 Vapor Density of Common Gases

Gaseous Substance	Vapor Density
Carbon monoxide	0.97
Hydrogen	0.07
Methane	0.55
Propane	1.55

Data from Barsan, Michael E., ed. September 2007. NIOSH Pocket Guide to Chemical Hazards. Accessed August 23, 2023. https://stacks.cdc.gov/view/cdc/21265

Flammable Range

Flammable gases need oxygen to burn. Mixtures of flammable gas and air will burn only when they are mixed in specific proportions. If too much fuel is present in the mixture, there will not be enough oxygen to support the combustion process; if too little fuel is present in the mixture, there will not be enough fuel to support the combustion process. The range of air–fuel mixtures that will burn varies from one fuel to another. For example, carbon monoxide will burn when mixed with air in concentrations between 12.5 percent and 74 percent. Natural gas, however, will burn only when it is mixed with air in concentrations between 4.5 percent and 15 percent.

The lower percentage is the **lower explosive limit (LEL)**, also referred to as the **lower flammable limit (LFL)**. The LEL is the minimum percentage of gaseous fuel that must be present in an air–fuel mixture for the mixture to be flammable. For example, the LEL of carbon monoxide is 12.5 percent. The higher percentage is the **upper explosive limit (UEL)**, also referred to as the **upper flammable limit (UFL)**. The UEL is the maximum percentage of gaseous fuel that can be present in an air–fuel mixture for the mixture to be flammable. The UEL of carbon monoxide is 74 percent. The **flammable range**, also called the **explosive limits**, is the range between the lower and upper flammable limits. The terms *flammable range* and *explosive limits* are used interchangeably because under most conditions, if the flammable air–fuel mixture will not explode, it will not ignite. Test instruments are available to measure the percentage of fuels in air–fuel mixtures to determine when an emergency scene is safe.

Boiling Liquid/Expanding Vapor Explosions

One potentially deadly set of circumstances involving fuels that are gases in normal atmospheric conditions but are stored as a liquid under pressure in a container is a **boiling liquid/expanding vapor explosion (BLEVE)**. A BLEVE can occur when a liquid fuel is stored in a closed vessel under pressure. If the vessel is filled with propane, for example, the bottom part of the vessel will contain liquid propane, and the upper part of the vessel will contain gaseous propane (vapor) (**FIGURE 5-25**). If the sealed container is subjected to a source of high heat, the heat causes the liquid fuel to convert to its gaseous form. As the heat causes the liquid fuel to convert to its gaseous form, the vapor pressure inside the tank increases until gas escapes from the pressure relief valve. If the fire is not extinguished quickly, more and more gas escapes, the level of the liquid fuel inside the tank drops, and the surface of the tank begins to weaken due to extreme temperatures. If the internal pressure exceeds the strength of the container, the container can catastrophically rupture, and any propane still in its liquid form will vaporize and ignite in an expanding fireball (**FIGURE 5-26**).

The key to preventing a BLEVE is to cool the top of the tank, which contains the vapor with a high-volume of water from a safe distance. This action will prevent the fuel from building up enough pressure to cause a catastrophic rupture of the container.

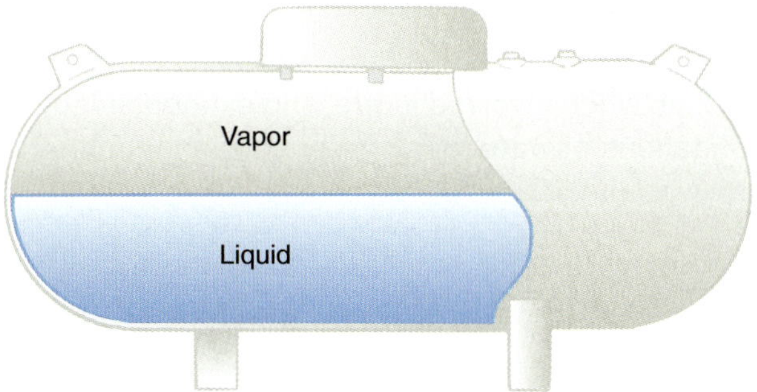

FIGURE 5-25 A propane tank contains both liquid and vapor.

© Jones & Bartlett Learning

Characteristics of Battery Fires

Batteries and stored energy systems are increasingly used to power and propel everything from electronic devices to scooters, bicycles, cars, and even as backup power to buildings (**FIGURE 5-27**). Lithium-ion technology is often used as a rechargeable battery solution

FIGURE 5-26 This photo shows a fireball formed from an ignited BLEVE.

© Ivan Cholakov/Shutterstock

FIGURE 5-27 The power source for electric vehicles is often a collection of numerous battery cells connected together and built into the floor of the car.

© leonello/iStock/Getty Images Plus/Getty Images

because it can deliver sustained amounts of power while charged; with older alkaline battery technology, the power will fade as the amount of charge declines with use.

Lithium-ion batteries are small, lightweight, and have a high energy density. They are fabricated with a positive electrode, called a *cathode*; a negative electrode, called an *anode*; and an electrolyte (a liquid, gel, or paste-like substance that conducts electrical current) that fills the space between the two electrodes. When lithium-ion batteries are charging, lithium ions (positively charged atoms) move from the positive cathode through the electrolyte to the negative anode. When these batteries discharge power, the lithium ions move from the anode to the cathode.

Although designs vary and the technology continues to evolve, electric vehicles are often constructed with a power source that comprises a collection of smaller battery cells connected together. It may be helpful to think of a skateboard when visualizing where the battery cells are positioned in an electric vehicle. In many modern electric vehicles, battery cells form the base of the car, much like a skateboard's deck. If an electric vehicle is involved in a motor vehicle crash and flipped onto its side or top, firefighters may be more directly exposed to the battery cells forming the floor of the car. Batteries in electric vehicles should always be treated as charged and having the ability to energize the car.

One of the dangers that electric vehicles pose is fires. When battery cells are compromised because of damage, extreme heat, or corrosion, the individual cells may self-discharge, overheat, catch fire, and even explode. Because these battery cells are generally constructed within a sealed metal case, overheating may cause the battery to swell from within, potentially causing the battery to explode as a result of overpressurization. When lithium-ion batteries catch fire, they can release immense amounts of heat, toxic gases, and smoke.

CASE STUDY

You Are the Support Person CONCLUSION

It is 0346, and your crew arrives at a modern-looking two-story house for the report of a garage fire called in by a neighbor. Through the smoke and flames, you observe what appears to be a popular model of an electric vehicle parked inside the garage and involved in the fire. As firefighters begin applying water to the body of the fire, you witness fire conditions growing rapidly and extending into the house. The hair on the back of your neck rises as you get an unsettled feeling.

1. **Why would this situation make you uneasy?**

   Answer: A car in the garage suggests that there are occupants inside the burning structure, necessitating efforts to coordinate search and rescue operations while also conducting fire attack operations.

2. **What elements of this scene may make extinguishment of the fire challenging?**

   Answer: Electric vehicles have batteries that pose significant risks when involved in fire, and extinguishing these fires can prove to be very difficult, often requiring thousands of gallons of water or a strategy of isolating the vehicle away from other combustibles to allow it to burn out on its own.

WRAP-UP

SUMMARY

- The four basic methods of extinguishing fires are **cooling the burning material**, **excluding oxygen from the fire**, **removing fuel from the fire**, and **interrupting the chemical reaction with a flame inhibitor**.
- A **Class A fire** involves ordinary solid combustible materials such as wood, paper, plastics, and cloth. The methods most commonly used to extinguish Class A fires are cooling the fuel with water to a temperature that is below the ignition temperature or using a combination of limiting ventilation and applying water.
- A **Class B fire** involves flammable or combustible liquids, such as gasoline, kerosene, diesel fuel, grease, tar, lacquer, oil-based paints, and motor oil, or gases such as propane and natural gas. These fires can be extinguished by shutting off the supply of fuel or by using foam to separate the oxygen in the air from the fuel.
- A **Class C fire** involves energized electrical equipment. Class C fires are extinguished with an extinguishing agent that does not conduct electricity.
- A **Class D fire** involves combustible metals such as sodium, magnesium, zirconium, lithium, potassium, and titanium. Class D fires cannot be extinguished with water. They must be attacked with extinguishing agents that prevent explosions, smother the fire, and snuff out the supply of oxygen.
- A **Class K fire** involves combustible cooking oils and fats and often occurs in kitchens. Class K extinguishing agents contain a wet agent that combines with cooking oils to produce a soap-like substance. The resulting soapy foam absorbs heat from the fire.
- The characteristics of a solid fuel that determine how the fire will burn include the chemical composition, the amount of fuel available, and the configuration of the fuel.
- The configuration factors that affect how a solid fuel burns include the size and shape of the solid fuel, the orientation of the fuel, and the continuity of the fuel (how close one piece of fuel is to the next piece of fuel).
- A ventilated, fuel-limited, solid-fuel fire progresses through four classic stages of growth: **the incipient stage** (a small fire confined to the initial fuel that was ignited), **the growth stage** (a fire that produces more interaction and is more dependent on the environment in the compartment around it), **the fully developed stage** (a fire that is consuming the maximum amount of fuel possible and achieving the maximum HRR possible for the fuel supply and oxygen present), and **the decay stage** (a fire with a decreasing fuel supply or a limited oxygen supply).
- Fires involving liquid fuels have some different characteristics from fires that involve solid fuels. Most liquids eventually reach their boiling point during a fire. As the boiling point is reached, the amount of flammable gas generated increases significantly.
- Fires involving gas fuel present unique challenges to firefighters. In situations where a flammable gas is present, responders need to know the vapor density of the escaping fuel, so they can take action to prevent the ignition of the fuel and allow the gaseous fuel to safely escape into the atmosphere.

KEY TERMS

aerosol An intimate mixture of a liquid or a solid in a gas; the liquid or solid, called the *dispersed phase,* is uniformly distributed in a finely divided state throughout the gas, which is the continuous phase or dispersing medium. (NFPA 99)

area of origin The room or general area where a fire started. Also called *fire compartment, fire seat,* or *seat of the fire.*

atom The smallest particle of an element that retains the properties of that element.

autoignition Initiation of combustion by heat but without a spark or flame. (NFPA 921)

backdraft A deflagration (explosion) resulting from the sudden introduction of air into a confined space containing oxygen-deficient products of incomplete combustion. (NFPA 1403)

KEY TERMS CONTINUED

boiling liquid/expanding vapor explosion (BLEVE) An explosion that occurs when pressurized liquefied materials (for example, propane or butane) in a closed container are exposed to a source of high heat, releasing the fuel, which instantly vaporizes and ignites.

boiling point The temperature at which the vapor pressure of a liquid equals the surrounding atmospheric pressure. (NFPA 1)

British thermal unit (BTU) The amount of heat energy required to raise 1 pound of water at sea level by 1°F.

calorie The amount of heat energy required to raise 1 gram of water (at sea level) by 1°C.

carbon dioxide (CO_2) A nontoxic gas produced when sufficient oxygen is available for complete combustion that can displace oxygen in the atmosphere. Also, a colorless, odorless, electrically nonconductive inert gas that is a suitable medium for extinguishing Class B and Class C fires. (NFPA 10)

carbon monoxide (CO) A toxic gas produced through incomplete combustion.

ceiling jet A strong, turbulent convection current that rose to the ceiling and traveled along it.

chemical energy Potential energy in molecular bonds and the kinetic energy created by a chemical reaction.

Class A fire A fire in ordinary combustible materials, such as wood, cloth, paper, rubber, and many plastics. (NFPA 1)

Class B fire A fire in flammable liquids, combustible liquids, petroleum greases, tars, oils, oil-based paints, solvents, lacquers, alcohols, and flammable gases. (NFPA 1)

Class C fire A fire that involves energized electrical equipment. (NFPA 1)

Class D fire A fire in combustible metals, such as magnesium, titanium, zirconium, sodium, lithium, and potassium. (NFPA 1)

Class K fire A fire in a cooking appliance that involves combustible cooking media (vegetable or animal oils and fats). (NFPA 1)

combustible Capable of undergoing combustion. (NFPA 921)

combustion A chemical process of oxidation that occurs at a rate fast enough to produce heat and usually light in the form of either a glow or a flame. (NFPA 1)

conduction Heat transfer to another body or within a body by direct contact. (NFPA 921)

convection Heat transfer by circulation within a medium such as a gas or a liquid. (NFPA 921)

decay stage The stage of fire development within a structure characterized by either a decrease in the fuel load or available oxygen to support combustion, resulting in lower temperatures and lower pressure in the fire area. (NFPA 1410)

electrical energy Energy is produced by an electrical charge.

element A substance that cannot be chemically broken down into a simpler substance.

endothermic A chemical reaction that absorbs heat.

energy The ability to do work.

entrain To encircle, draw along, and transport.

exothermic A chemical reaction that produces heat.

explosion A violent and pressurized release of energy.

explosive limits See *flammable range.*

fire The visible result of combustion.

fire compartment The area of origin when the fire is in a structure.

fire point The lowest temperature at which a liquid will ignite and achieve sustained burning when exposed to a test flame in accordance with ASTM 92, *Standard Test Method for Flash and Fire Points by Cleveland Open Cup Tester.* Also called *flame point.* (NFPA 1)

fire seat See *area of origin.*

fire tetrahedron A geometric shape used to depict the four components—fuel, oxygen, heat, and chemical chain reactions—required for a fire to occur.

fire triangle The three components—fuel, oxygen, and heat—required for combustion.

fireball A burst of flames that rapidly ignites available flammable vapors but is not under pressure.

firebreak A swath where the fuel (trees and brush) is removed.

flame impingement Flames in direct contact with the surface of a material transferring radiant heat.

flame inhibitor A chemical extinguishing agent that reacts with the fuel to chemically disrupt the combustion process.

flameover See *rollover.*

flame point See *fire point.*

flammable range The range in concentration between the lower and upper flammable limits. Also called *explosive limits.* (NFPA 67)

flashover A transition phase in the development of a compartment fire in which surfaces exposed to thermal radiation reach ignition temperature more or less simultaneously, and fire spreads rapidly throughout the space, resulting in full-room involvement or total involvement of the compartment or enclosed space. (NFPA 921)

flash point The minimum temperature at which a liquid or a solid emits vapor sufficient to form an ignitable mixture with air near the surface of the liquid or the solid. (NFPA 115)

flow path The movement of heat and smoke from the higher pressure within the fire area toward the lower-pressure areas accessible via doors, window openings, and roof structures. (NFPA 1410)

fuel A material that will maintain combustion under specified environmental conditions. (NFPA 53)

fuel-limited fire A fire in which the heat release rate and fire growth are controlled by the characteristics of the fuel because there is adequate oxygen available for combustion. (NFPA 1410)

fully developed stage The stage of fire development where the heat release rate has reached its peak within a compartment. (NFPA 1410)

gas The physical state of a substance that has no shape or volume of its own and will expand to take the shape and volume of the container or enclosure it occupies. (NFPA 921)

growth stage The stage of fire development where the heat release rate from an incipient fire has increased to the point where heat transferred from the fire and the combustion products are pyrolyzing adjacent fuel sources, and the fire begins to spread across the ceiling of the fire compartment (rollover). (NFPA 1410)

heat energy The potential energy of a combustible material and the kinetic energy released when heat is applied to the material. Also called *thermal energy.*

heat flux The measure of the rate of heat transfer to a surface, typically expressed in kilowatts per meter squared (kW/m^2) or British thermal units per square feet (BTU/ft^2). (NFPA 268)

heat-induced tear (HIT) A tear in a nonpressurized container that occurs when the container is exposed to direct or indirect flame impingement for a prolonged period, causing the structural integrity of the container to fail.

heat release rate (HRR) The rate at which heat energy is generated by burning. (NFPA 921)

heat stratification See *thermal layering.*

heat transfer The movement of heat energy from a hotter medium to a cooler medium by conduction, convection, or radiation.

hydrogen cyanide (HCN) An extremely toxic gas produced by the incomplete combustion of many common plastic-based materials.

ignition temperature Minimum temperature a substance should attain in order to ignite under specific test conditions. (NFPA 921)

incipient stage The early stage of fire development where the fire's progression is limited to a fuel source and the thermal hazard is localized to the area of the burning material. (NFPA 1410)

incomplete combustion A combustion process during which the fuel is not completely consumed, usually because of a limited supply of oxygen.

joule A measure of heat energy equal to 0.4 calorie.

kinetic energy Energy possessed by an object because of its motion.

laminar smoke flow The smooth or streamlined movement of smoke.

light energy Energy produced by electromagnetic waves packaged in discrete bundles called *photons.*

liquid Matter that has a specific volume but does not have a specific size or shape.

lower explosive limit (LEL) The minimum concentration of a combustible vapor or combustible gas in a mixture of the vapor or gas and gaseous oxidant, above which propagation of flame will occur on contact with an ignition source. (NFPA 115)

lower flammable limit (LFL) See *lower explosive limit.*

matter Anything that has mass and volume.

mechanical energy The potential energy stored because of the position of an object or the kinetic energy of an object in motion.

molecular bond The connection between atoms in a molecule.

molecule Atoms chemically bonded together.

KEY TERMS CONTINUED

neutral plane The interface at a vent, such as a doorway or a window opening, between the hot gas flowing out of a fire compartment and the cool air flowing into the compartment where the pressure difference between the interior and exterior is equal.

nuclear energy Potential energy stored in the nucleus of an atom or the kinetic energy released by splitting the nucleus of an atom into two smaller nuclei (fission) or by combining two small nuclei into one large nucleus (fusion).

off-gas To emit a gas, often harmful organic vapors.

oxidation Reaction with oxygen, either in the form of the element or in the form of one of its compounds. (NFPA 53)

phosgene A gas formed from incomplete combustion of many common household products.

plume The column of hot gases, flames, and smoke rising above a fire. Also called *convection column, thermal updraft,* or *thermal column.* (NFPA 921)

potential energy Energy stored by an object as a result of its position or condition.

pyrolysis A process in which material is decomposed, or broken down, into simpler molecular compounds by the effects of heat alone; pyrolysis often precedes combustion. (NFPA 921)

rollover The condition in which unburned fuel (pyrolysate) from the originating fire has accumulated in the ceiling layer to a sufficient concentration (i.e., at or above the lower flammable limit) that it ignites and burns. This can occur without ignition of, or prior to the ignition of, other fuels separate from the origin. Also called *flameover.* (NFPA 921)

seat of the fire See *area of origin.*

smoke The airborne solid and liquid particulates and gases evolved when a material undergoes pyrolysis or combustion, together with the quantity of air that is entrained or otherwise mixed into the mass. (NFPA 1404)

smoke explosion A violent release of energy that occurs when smoke travels away from its source to a void area or other area separate from the fire compartment and comes in contact with a source of ignition without any change to the ventilation profile.

smoke particles The unburned, partially burned, and completely burned substances found in smoke.

solid Matter that has a specific size and shape; one of the three states of matter.

specific gravity The density of a liquid compared to water (which is 1.0).

state of matter The physical state of a material (solid, liquid, or gas).

temperature The measurement of the movement of molecules used to describe how hot or cold something is.

thermal column See *plume.*

thermal conductivity The ability of a material to conduct heat.

thermal energy See *heat energy.*

thermal layering The phenomenon of gases forming into layers according to their temperatures. Also called *heat stratification.*

thermal radiation The means by which heat is transferred to other objects.

turbulent smoke flow The agitated, boiling, and angry movement of smoke caused by rapid molecular expansion of the gases within the smoke and the restrictions of the box containing the smoke.

upper explosive limit (UEL) The maximum amount of gaseous fuel that can be present in the air if the air–fuel mixture is flammable or explosive.

upper flammable limit (LFL) See *upper explosive limit.*

vapor See *gas.*

vapor density The weight of a gas compared to an equal volume of dry air.

ventilation-limited fire A fire in which the heat release rate and fire growth are regulated by the available oxygen within the space. (NFPA 1410)

REVIEW QUESTIONS

1. What is the fire triangle? What is the fire tetrahedron?
2. What are the three methods of heat transfer?
3. What are the four methods of extinguishing fire?
4. Describe the five classes of fire.
5. How do solid materials burn?
6. What are the four classic stages of fire development?

7. Describe rollover, flashover, and backdraft.
8. How do modern homes differ from those constructed several decades ago?
9. What three conditions must be present for liquid fuels to ignite?
10. What is vapor density?
11. What type of incident can occur when a liquid fuel is stored in a closed vessel under pressure and the container is exposed to heat?
12. What happens when lithium-ion batteries catch fire?

DISCUSSION QUESTIONS

1. Why is it important to understand the five classes of fire?
2. How do factors such as building construction, building contents, flow paths, and wind affect fire growth?
3. Discuss the risks, safety considerations, and tactical priorities associated with fires involving batteries and stored energy systems.

REFERENCES

Hall, Shelby, and Ben Evarts. 2022. "Fire Loss in the United States during 2021." National Fire Protection Association, September 2022. Accessed July 19, 2023. www.nfpa.org/news-and-research/fire-statistics-and-reports/fire-statistics/fires-in-the-us.

Kerber, Stephen. 2009. "Impact of Ventilation on Fire Behavior in Legacy and Contemporary Residential Construction." Underwriters Laboratories (UL). Accessed July 24, 2023. https://ulfirefightersafety.org/research-projects/impact-of-ventilation-on-fire-behavior-in-legacy-and-contemporary-residential-construction.html.

National Fire Protection Association. 2017. *NFPA 1403, Standard on Live Fire Training Evolutions.* 2018 Edition. Quincy, MA: National Fire Protection Association.

National Fire Protection Association. 2017. *NFPA 1404, Standard for Fire Service Respiratory Protection Training.* 2018 Edition. Quincy, MA: National Fire Protection Association.

National Fire Protection Association. 2018. *NFPA 67, Guide on Explosion Protection for Gaseous Mixtures in Pipe Systems.* 2019 Edition. Quincy, MA: National Fire Protection Association.

National Fire Protection Association. 2019. *NFPA 115, Standard for Laser Fire Protection.* 2020 Edition. Quincy, MA: National Fire Protection Association.

National Fire Protection Association. 2019. *NFPA 1410, Standard on Training for Emergency Scene Operations.* 2020 Edition. Quincy, MA: National Fire Protection Association.

National Fire Protection Association. 2020. *NFPA 53, Recommended Practice on Materials, Equipment, and Systems Used in Oxygen-Enriched Atmospheres.* 2021 Edition. Quincy, MA: National Fire Protection Association.

National Fire Protection Association. 2020. *NFPA 921, Guide for Fire and Explosion Investigations.* 2021 Edition. Quincy, MA: National Fire Protection Association.

National Fire Protection Association. 2021. *NFPA 10, Standard for Portable Fire Extinguishers.* 2022 Edition. Quincy, MA: National Fire Protection Association.

National Fire Protection Association. 2021. *NFPA 268, Standard Test Method for Determining Ignitability of Exterior Wall Assemblies Using a Radiant Heat Energy Source.* 2022 Edition. Quincy, MA: National Fire Protection Association.

National Fire Protection Association. 2021. *NFPA 291, Recommended Practice for Water Flow Testing and Marking of Hydrants.* 2022 Edition. Quincy, MA: National Fire Protection Association.

National Fire Protection Association. 2021. *NFPA 550, Guide to the Fire Safety Concepts Tree.* 2022 Edition. Quincy, MA: National Fire Protection Association.

National Fire Protection Association. 2023. *NFPA 1, Fire Code.* 2024 Edition. Quincy, MA: National Fire Protection Association.

National Fire Protection Association. 2023. *NFPA 24, Standard for the Installation of Private Fire Service Mains and Their Appurtenances.* 2022 Edition. Quincy, MA: National Fire Protection Association.

National Fire Protection Association. 2023. *NFPA 99, HealthCare Facilities Code.* 2024 Edition. Quincy, MA: National Fire Protection Association.

U.S. Fire Administration. 2023. "Fire Death and Injury Risk." Last reviewed August 2, 2023. Accessed August 6, 2023. www.usfa.fema.gov/statistics/deaths-injuries/.

Chapter Opener: © Eric Scruggs

CHAPTER

6

Support Person

Portable Fire Extinguishers

KNOWLEDGE OBJECTIVES

After studying this chapter, you will be able to:

- State the primary purposes of fire extinguishers.
- Explain the considerations used when selecting the proper class of fire extinguisher.
- Describe the four methods of extinguishing fires.
- Explain the classification and rating system for fire extinguishers.
- Explain the labeling system for fire extinguishers.
- Describe the three risk classifications for area hazards.
- Describe the types of agents and operating systems used in fire extinguishers.
- Describe the basic steps of fire extinguisher operation.

SKILLS OBJECTIVES

After studying this chapter, you will be able to perform the following skills:

- Select an appropriate extinguisher based on the size and type of fire, and transport the fire extinguisher to the location of the fire.
- Demonstrate the safe extinguishment of a Class A fire with a stored-pressure water-type fire extinguisher.
- Demonstrate the safe extinguishment of a Class A fire with a multipurpose dry-chemical fire extinguisher.
- Demonstrate the safe extinguishment of a Class B flammable liquid fire with a dry-chemical fire extinguisher.
- Operate a carbon dioxide fire extinguisher.

ADDITIONAL NFPA STANDARDS

- **NFPA 1**, *Fire Code, 2024 Edition*
- **NFPA 10**, *Standard for Portable Fire Extinguishers, 2022 Edition*
- **NFPA 11**, *Standard for Low-, Medium-, and High-Expansion Foam, 2021 Edition*
- **NFPA 13**, *Standard for the Installation of Sprinkler Systems, 2022 Edition*
- **NFPA 408**, *Standard for Aircraft Hand Portable Fire Extinguishers, 2022 Edition*
- **NFPA 440**, *Guide for Aircraft Rescue and Firefighting Operations and Airport/Community Emergency Planning, 2024 Edition*
- **NFPA 557**, *Standard for Determination of Fire Loads for Use in Structural Fire Protection Design, 2023 Edition*
- **NFPA 1660**, *Standard for Emergency, Continuity, and Crisis Management: Preparedness, Response, and Recovery, 2024 Edition*
- **NFPA 1700**, *Guide for Structural Fire Fighting, 2021 Edition*
- **NFPA 1900**, *Standard for Aircraft Rescue and Firefighting Vehicles, Automotive Fire Apparatus, Wildland Fire Apparatus, and Automotive Ambulances, 2024 Edition*

CASE STUDY

You Are the Support Person

You are the designated on-duty cook for the day, and the kitchen is bare. You go into the grocery store while the rest of your crew waits in the apparatus. As you make your way through the produce section, an employee runs up to you and says that there is a fire in the back room. You quickly radio your officer to inform them of the situation, then immediately proceed to the back room. You locate a small fire in a trash can that is just starting to spread. You quickly scan the room, looking for a fire extinguisher. Not seeing one, you ask the employee where the closest extinguisher is located.

1. What type of fire extinguisher would be most effective for this fire? Why?
2. As the fire grows, at what point will using a fire extinguisher become ineffective?
3. How do fire inspectors ensure the fire extinguisher will work when needed?

Introduction

Portable fire extinguishers are required in many types of occupancies as well as in commercial vehicles, boats, aircraft, and various other locations. Fire-prevention efforts encourage citizens to learn how to use and keep fire extinguishers in their homes, particularly in their kitchens. Fire extinguishers have been used successfully to put out small fires, preventing millions of dollars in property damage, as well as saving lives. Most fire extinguishers are easy to operate and can be used effectively by an individual with only basic training. Members of the fire service often provide fire extinguisher training for the public, so it is essential that you understand the characteristics and operations of each type of fire extinguisher. You need to be able to select the most appropriate extinguisher to use for different types of fires; you also need to know which extinguishers must not be used for certain fires. (The selection of the proper fire extinguisher builds on the information presented in Chapter 5, *Fire Behavior*.) You must be able to operate the most common types of portable fire extinguishers correctly and effectively to reduce the risk of personal injury and property damage, and you must know how to inspect and maintain them.

Purpose of a Fire Extinguisher

Fire extinguishers range in size from models that can be operated with one hand to large, wheeled models that contain several hundred pounds of **extinguishing agent** (material used to stop the combustion process) (**FIGURE 6-1**). Extinguishing agents include water, water with additives, dry chemicals, wet chemicals, dry powders, and gaseous agents. Each agent is suitable for a specific type of fire.

Portable fire extinguishers are used to extinguish incipient-stage fires and to control fires where traditional methods of fire suppression are not recommended.

Extinguishing Incipient-Stage Fires

Fire extinguishers should be placed according to NFPA 10, *Standard for Portable Fire Extinguishers, 2022*

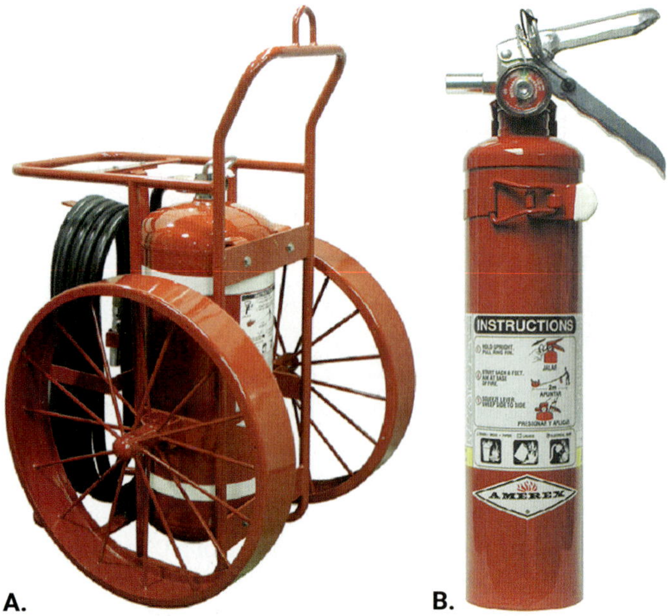

FIGURE 6-1 Portable fire extinguishers can be large or small. **A.** A wheeled fire extinguisher. **B.** A hand-held fire extinguisher.

Courtesy of Amerex Corporation.

Edition, so that they will be available for immediate use on small, incipient-stage fires, such as a fire in a wastebasket. A person familiar with fire extinguishers and with access to a suitable fire extinguisher can usually control this type of fire (**FIGURE 6-2**). If flames spread beyond the area of origin to other contents of the room, the fire may become increasingly difficult and dangerous to control with only a portable fire extinguisher.

One advantage of fire extinguishers is their portability and ease of deployment. At times, a firefighter may use a fire extinguisher from the fire-site premises to control an incipient-stage fire. NFPA 1900, *Standard for Aircraft Rescue and Firefighting Vehicles, Automotive Fire Apparatus, Wildland Fire Apparatus, and Automotive Ambulances, 2024 Edition*, recommends that fire apparatus be equipped with a minimum of one dry-chemical fire extinguisher and one pressurized water extinguisher with a minimum capacity of 2.5 gal (9.5 L). Fire department vehicles that are not equipped with water or fire hose usually carry at least one multipurpose fire extinguisher.

A disadvantage of fire extinguishers is that they are "one-shot" devices. In other words, once the contents of a fire extinguisher have been discharged, the device cannot be used to fight fires until it is recharged or replaced. If the fire extinguisher does not control the fire before it is completely discharged, some other device or method must be employed. This is a serious limitation when compared to a fire hose with a continuous water supply. Therefore, when you use a portable fire extinguisher to control an incipient-stage fire, make sure you do not place yourself in a dangerous situation.

FIGURE 6-2 A trained individual with a suitable fire extinguisher can usually control an incipient-stage fire.

© Jones & Bartlett Learning. Photographed by Glen E. Ellman.

SAFETY TIP

Do not place yourself in a dangerous situation by trying to fight a large fire with a small fire extinguisher. You cannot fight a fire or protect yourself with an empty extinguisher, an undersized extinguisher, or the inappropriate type of extinguisher.

Extinguishing Fires with the Appropriate Extinguishing Agent

Remember that fires are categorized into five classes (**TABLE 6-1**). The right fire extinguisher with the right extinguishing agent needs to be used. For example, using water on Class C fires that involve energized electrical equipment increases the risk of electrocution to firefighters and can cause extensive damage to the electrical equipment. Water should also not be used on fires that involve flammable liquids or gases such as propane (Class B), cooking oils (Class K), and combustible metals (Class D).

As a support person, you must know which fires require which extinguishing agents, which type of fire extinguisher should be used, and how to operate the different types of fire extinguishers. Using an improper type of fire extinguisher or extinguishing agent can spread burning material, cause unnecessary damage, and pose a danger to the fire extinguisher operator.

TABLE 6-1 Types of Fires

Class A	Ordinary combustibles
Class B	Flammable or combustible liquids or gases
Class C	Energized electrical equipment
Class D	Combustible metals
Class K	Kitchen fires involving oils and fats

Courtesy of Guy Peifer.

Methods of Fire Extinguishment

Understanding the nature of fire is key to understanding how fire extinguishers work and how they differ from one another. All fires require four elements: fuel, heat, oxygen, and a sustained chemical reaction (fire tetrahedron). The fire tetrahedron is discussed in greater detail Chapter 5, *Fire Behavior*. If you remove any of these four elements, a fire will not ignite, or the fire will be extinguished. The combustion process begins when the fuel is heated to its ignition temperature, which is the temperature at which it begins to burn. The energy that initiates this process can come from a spark or flame, friction, electrical energy, or a chemical reaction. Once a substance begins to burn, it will generally continue burning as long as an adequate supply of oxygen and fuel to sustain the chemical reaction are present. Removing the heat, oxygen, or fuel interrupts the process.

The extinguishing agents in portable fire extinguishers disrupt the fire tetrahedron by a variety of methods. Some fire extinguishers stop the combustion process by cooling the fuel to below its ignition temperature, some work by blocking the supply of oxygen, and some work by performing both actions. Others release extinguishing agents that interrupt the complex chemical reaction that occurs between the heated fuel and the oxygen.

Water extinguishes fires by cooling the fuel. If the temperature of the fuel falls below its ignition temperature, the combustion process stops.

Creating a barrier that interrupts the flow of oxygen to the flames also extinguishes a fire. Putting a lid on a pan of burning food is an example of this technique (**FIGURE 6-3A**). Applying a blanket of foam to the surface of a burning liquid or gas is another example (**FIGURE 6-3B**). Similarly, surrounding the fuel with a layer of **carbon dioxide (CO_2)**—a colorless, odorless, nontoxic gas that is 1.5 times heavier than air—can cut off the supply of oxygen necessary to sustain the burning process. When the combustion process is stopped by only interrupting the flow of oxygen, be aware that the fuel has not been cooled. This means that if oxygen is reintroduced, the fuel is likely to ignite again.

Some extinguishing agents work by interrupting the molecular chain reactions required to sustain combustion. In some cases, a very small quantity of the agent accomplishes this objective.

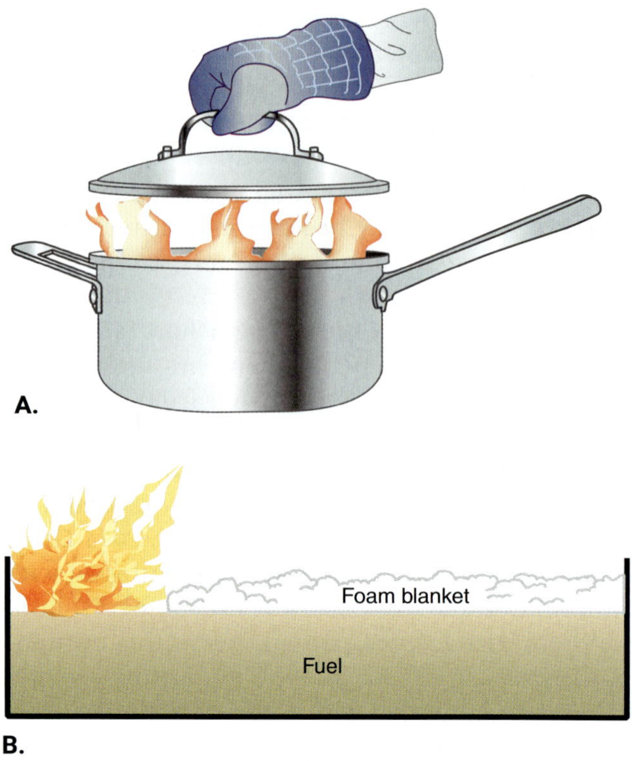

FIGURE 6-3 **A.** Covering a pan of burning food with a lid will extinguish a fire by cutting off the supply of oxygen. **B.** One of the ways a blanket of foam extinguishes fire is by separating the fuel from oxygen.

© Jones & Bartlett Learning

Classification and Rating of Fire Extinguishers

Portable fire extinguishers are classified and rated based on their extinguishing properties and capabilities. In the United States, Underwriters Laboratories Standards & Engagement is the organization that developed the standards, classification, and rating system for portable fire extinguishers. Each fire extinguisher has a specific rating that identifies the classes of fires for which it is both safe and effective. This information is important for selecting the proper fire extinguisher to fight a particular fire. It is also used to determine which types of fire extinguishers should be placed in each location so that incipient-stage fires can be controlled quickly. For example, a commercial kitchen would be an appropriate location for a Class K portable extinguisher.

All portable fire extinguishers are rated with letters that indicate the classes of fire for which the fire extinguisher can be used. Fire extinguishers that are safe and effective for more than one class will be rated with multiple letters. For example, a fire extinguisher that is safe and effective for Class A fires is rated with an *A*, an extinguisher that is safe and effective for Class B fires is rated with a *B*, and one that is safe and effective for both Class A and B fires is rated with both an *A* and a *B*.

The rating on Class A and B fire extinguishers also includes a number. On Class A fire extinguishers, this number indicates the equivalent amount of water it contains after multiplying by 1.25. So a Class A fire extinguisher that is rated 1-A contains the equivalent of 1.25 gal (4.7 L) of water, and a Class A fire extinguisher that is rated 5-A contains the equivalent of 6.25 gal (23.7 L) of water. The higher the number, the greater the extinguishing capability of the fire extinguisher. A typical Class A fire extinguisher carried on a fire apparatus contains 2.5 gal (9.5 L) of water and has a 2-A rating.

On Class B fire extinguishers, the number in the rating indicates the approximate area in square feet (ft^2) of burning fuel that these devices are capable of extinguishing. Certification testing of these fire extinguishers is performed by trained experts who can control a larger fire than a nonexpert user; therefore, the numerical rating is about 40 percent of the area of burning fuel that an expert can consistently extinguish (Underwriters Laboratories 2017). For example, a 10-B rating indicates that a nonexpert user should be able to extinguish a fire in a pan of flammable liquid that is 10 ft^2 (0.9 square meters [m^2]) in surface area, whereas an expert user should be able to extinguish a fire that is 25 ft^2 (2.3 m^2) in surface area. A nonexpert user should be able to use a fire extinguisher rated 40-B to control a flammable liquid pan fire with a surface area of 40 ft^2 (4.7 m^2), whereas an expert user should be able to extinguish a fire with a surface area of 100 ft^2 (9 m^2).

Numbers are used to rate a fire extinguisher's effectiveness only for Class A and B fires; Class C, D, and K fire extinguishers are rated only with the letter indicating the class. For example, if the fire extinguisher can also be used for Class C fires, it contains an agent proven to be nonconductive to electricity and safe for use on energized electrical equipment. So a fire extinguisher that carries a 2-A:10-B:C rating can be used on Class A, B, and C fires. It has the extinguishing capabilities of a 2-A fire extinguisher when applied to Class A fires, has the capabilities of a 10-B fire extinguisher when applied to Class B fires, and can be used safely on energized electrical equipment.

Use fire extinguishers labeled as appropriate for Class B and C fires with caution on Class A fires. They are less effective in extinguishing a common combustible fire than a comparable Class A fire extinguisher would be.

Independent laboratories use standard test fires to rate the effectiveness of fire extinguishers. This testing may involve different agents, amounts, application rates, and application methods. Fire extinguishers are rated not only for their ability to control a specific type of fire but also for the extinguishing agent's ability to prevent rekindling. Some agents successfully suppress a fire by breaking the chemical chain reaction but are unable to prevent the fuel from reigniting because they do not cool the fuel. A rating is given only if the fire extinguisher completely extinguishes the standard test fire and prevents rekindling.

Fire extinguishers that have been tested and approved by an independent laboratory are labeled to clearly designate the classes of fire the unit is capable of extinguishing safely. The traditional lettering system has been used for many years and is still found on many fire extinguishers. The labels consist of the letters corresponding to the class the extinguisher is rated for on colored backgrounds (**FIGURE 6-4**). The shape of the colored backgrounds for Class A, B, and C mirrors the shape of the letter—*A* is in a green triangle, *B* is in a red square, and *C* is in a blue circle. The letter *D* is in a yellow star, and the letter *K* is in

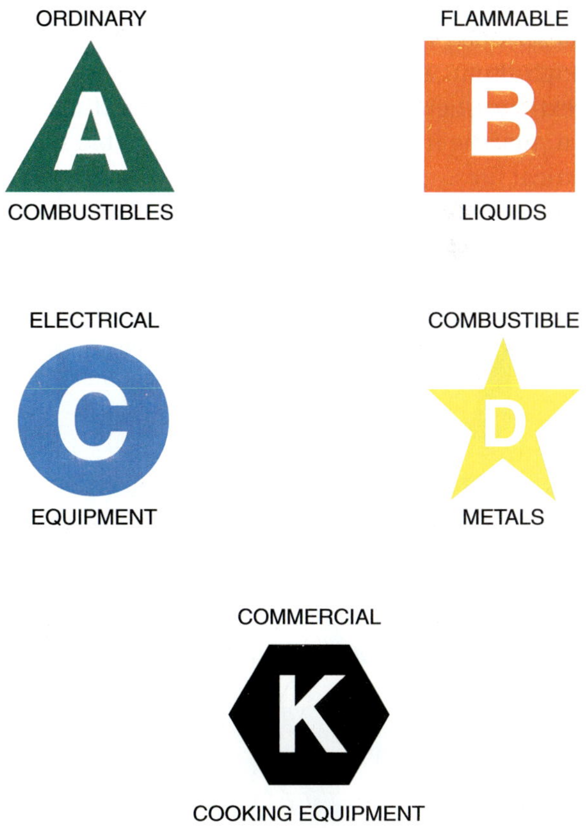

FIGURE 6-4 Traditional letter labels on fire extinguishers often incorporate a shape as well as a letter.

© Jones & Bartlett Learning

a black hexagon. If a letter appears on an extinguisher, it can safely be used on that class of fire. If an extinguisher has more than one letter label, it can safely be used on all the classes of fire identified by those labels.

A universal pictograph system was developed that does not require the user to be familiar with the alphabetic codes for the different classes of fires. This system indicates whether a fire extinguisher is appropriate for use on a particular class of fire. The pictographs are all square icons, each of which is designed to represent a certain class of fire (**FIGURE 6-5**):

- Class A: Burning trash can beside a wood fire
- Class B: A flame and a gasoline can
- Class C: A flame and an electrical plug and socket
- Class D: A flame and a metal gear
- Class K: Flames in a frying pan

Like the lettering system, the presence of an icon indicates that the fire extinguisher has been rated for that class of fire. In addition, if a pictograph label has a red slash across the icon, it means that the fire extinguisher must *not* be used on that type of fire because doing so would create additional risk. A fire extinguisher rated for Class A fires only would show all three icons, but the icons for Class B and C would have red diagonal lines through them. This three-icon array signifies that the fire extinguisher contains a water-based extinguishing agent, making it unsafe to use on flammable liquid or electrical fires.

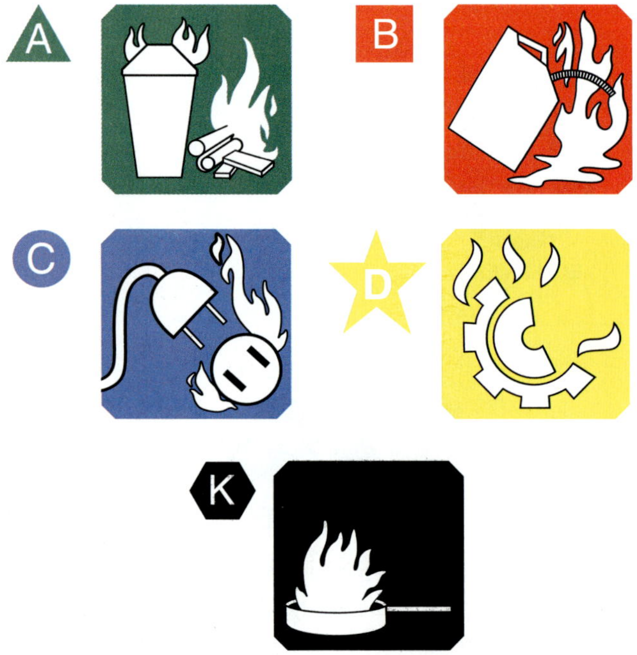

FIGURE 6-5 Icons used to label fire extinguishers.

© Jones & Bartlett Learning

SAFETY TIP

The safest and surest way to extinguish a Class C fire is to turn off the power and treat it like a Class A or B fire. If you are unable to turn off the power, you should be prepared for reignition because the electricity could reignite the fire after it has been extinguished.

Fire Extinguisher Placement

Fire and building codes and regulations require the installation of fire extinguishers in multifamily, public, and commercial buildings so that they will be available to fight incipient-stage fires. NFPA 10, *Standard for Portable Fire Extinguishers, 2022 Edition*, lists the requirements for placing and mounting portable fire extinguishers as well as the appropriate mounting heights in different types of occupancies. The regulations for each type of occupancy specify the maximum floor area that can be protected by each fire extinguisher, the maximum travel distance from the closest fire extinguisher to a potential fire, and the types of fire extinguishers that should be provided. Two key factors must be considered when determining which type of fire extinguisher should be placed in each area: the class of fire that is likely to occur and the potential size of an incipient-stage fire.

Fire extinguishers should be mounted so that they are readily visible and easily accessible (**FIGURE 6-6**). Heavy fire extinguishers should not be mounted high on a wall. If the fire extinguisher is mounted too high, the user might be unable to lift it off its hook or could be injured in the attempt. NFPA 10 requires the following mounting heights for fire extinguishers:

- Fire extinguishers weighing up to 40 pounds (lb; 18 kilograms [kg]) must be mounted so that the top of the extinguisher is not more than 5 feet (ft; 2 meters [m]) above the floor.
- Fire extinguishers weighing more than 40 lb (18 kg) must be mounted so that the top of the extinguisher is not more than 3.5 ft (1 m) above the floor.
- The bottom of an extinguisher must be at least 4 inches (in.; 10 centimeters [cm]) above the floor.

Classifying Area Hazards

Several factors must be considered when determining the number and types of fire extinguishers that should be placed in each area of an occupancy. Among these

FIGURE 6-6 Fire extinguishers should be mounted in locations with unobstructed access and visibility.

© Jones & Bartlett Learning

FIGURE 6-7 Light-hazard areas include offices, churches, and classrooms.

Courtesy of Bill Larkin.

factors are the types of fuels found in the area and the quantities of those materials. Areas where portable fire extinguishers are required are categorized into three risk classifications—light or low hazard, ordinary or moderate hazard, and extra or high hazard—based on the amount and type of combustibles that are present, including building materials, decorations, furniture, and all other contents, including materials in storage. The total quantity of combustible materials present is the building's **fire load** and is measured as the average weight of combustible materials per square foot (or per square meter). The larger the fire load, the larger the potential fire.

Note that the hazard classification for each area is based on the actual amount and type of combustibles that are present, not the occupancy type. Although there are recommended hazard classifications for different types of occupancies, these are simply guidelines based on typical situations.

Light (Low) Hazard

A **light hazard area**, or **low hazard area**, is an area where the quantity, combustibility, and heat release rate of the materials are low, and most materials are arranged so that a fire is not likely to spread. Light hazard areas usually contain limited amounts of Class A combustibles. The area might also contain some Class B combustibles, such as copy machine chemicals or modest quantities of paints and solvents, but all Class B materials in a light hazard area must be kept in closed containers and stored safely. Common light hazard areas are offices, classrooms, churches, assembly halls, and hotel guest rooms (**FIGURE 6-7**).

Ordinary (Moderate) Hazard

An **ordinary hazard area**, or **moderate hazard area**, contains more Class A and B materials than light-hazard locations, and the combustibility and heat release rate of the materials are moderate. Typical examples of ordinary hazard areas are retail stores with onsite storage areas, light manufacturing facilities, auto showrooms, parking garages, research facilities, hotel laundry rooms, and restaurant kitchens (**FIGURE 6-8**). Ordinary hazard areas also include warehouses that contain Class I and II commodities. Class I commodities are noncombustible products stored on wooden pallets or in corrugated cartons that are shrink-wrapped or wrapped in paper. Class II commodities are noncombustible products stored in wooden crates or multilayered corrugated cartons.

Extra (High) Hazard

An **extra hazard area**, or **high hazard area**, contains more Class A and B materials than ordinary-hazard locations, and the combustibility and heat release rate of the materials are high. Examples of extra hazard areas are woodworking shops; service and repair facilities for cars, aircraft, or boats; and kitchens and other cooking areas that have deep fryers, flammable

FIGURE 6-8 Auto showrooms, hotel laundry rooms, and parking garages are classified as ordinary-hazard areas.

© Jones & Bartlett Learning

liquids, or gases under pressure (**FIGURE 6-9**). In addition, areas used for manufacturing processes such as painting, dipping, or coating, as well as facilities used for storing or handling flammable liquids, are classified as extra hazard environments. Warehouses containing products that meet the definitions of Class III, IV, and V materials are also considered extra-hazard locations. These commodities are made of natural fibers, paper products, and products containing plastic.

Determining the Most Appropriate Placement

Most buildings require fire extinguishers that are suitable for fighting Class A fires because ordinary combustible materials, such as furniture, partitions, interior finish materials, paper, and packaging products, are so common. Even where other classes of products are used or stored, there is still a need to defend the facility from a fire involving common combustibles.

In some facilities, a variety of conditions are present. In these occupancies, each area must be individually evaluated, and the extinguisher installation must be tailored to its circumstances. A restaurant is a good example of this situation. The dining areas contain common combustibles, such as furniture, tablecloths, and paper products, that would require a fire extinguisher rated for Class A fires. In the restaurant's kitchen, where the risk of fire involves cooking oils, a Class K fire extinguisher would provide the best defense.

Similarly, within a hospital, fire extinguishers for Class A fires would be appropriate in hallways, offices, lobbies, and patient rooms. Class B fire extinguishers should be mounted in laboratories and areas where flammable anesthetics are stored or handled. Electrical

FIGURE 6-9 Kitchens, woodworking shops, and auto repair shops are considered possible extra-hazard locations.

© Jones & Bartlett Learning

rooms should have fire extinguishers that are approved for use on Class C fires, whereas kitchens need Class K fire extinguishers.

Some areas may need more than one type of fire extinguisher or fire extinguishers with more than one rating. For example, areas that contain both Class A and B combustibles require either a fire extinguisher that is rated for both types of fires or a separate fire extinguisher for each class of fire. A single multipurpose fire extinguisher is generally less expensive than two individual fire extinguishers and eliminates the problem of selecting the proper fire extinguisher for a particular fire. However, it is sometimes more appropriate to install Class A fire extinguishers in general-use areas and place fire extinguishers that are especially effective in fighting Class B or C fires near those specific hazards.

Fire Extinguisher Design

Portable fire extinguishers contain a variety of extinguishing agents—the substances that put out a fire. These extinguishing agents are expelled from the fire extinguisher using pressure, and many fire extinguishers rely on pressurized gas, such as nitrogen, to eject the extinguishing agent. This gas is stored either with the extinguishing agent in the body of the fire extinguisher or externally in a separate cartridge or cylinder. When the gas is stored externally, the extinguishing agent is put under pressure only when it is used. A **stored-pressure fire extinguisher** holds both the extinguishing agent, in wet or dry form, and the expeller gas under pressure in the body of the extinguisher. A **cartridge-/cylinder-operated fire extinguisher** relies on an external cartridge of pressurized gas, which is released only when the fire extinguisher is to be used by pushing down on a lever that punctures the cartridge and pressurizes the cylinder.

Some extinguishing agents, such as CO_2, are **self-expelling agents**, or extinguishing agents that have sufficient vapor pressure at normal operating temperatures to expel themselves from a fire extinguisher. Most self-expelling agents are normally gases but are stored as liquids under pressure in the fire extinguisher. When the confining pressure is released, the agent rapidly expands, causing it to self-discharge. Hand-operated pumps are used to expel the agent when water or water with additives is the extinguishing agent.

Portable Fire Extinguisher Components

Most hand-held portable fire extinguishers have six basic parts (**FIGURE 6-10**):

- A cylinder that holds the extinguishing agent
- A carrying handle
- A nozzle or horn
- A trigger and discharge valve assembly
- A locking mechanism to prevent accidental discharge
- A pressure indicator

Cylinder

The body of the fire extinguisher, known as the **cylinder** or **container**, holds the extinguishing agent. Nitrogen, compressed air, or CO_2 can be used to pressurize the cylinder to expel the agent.

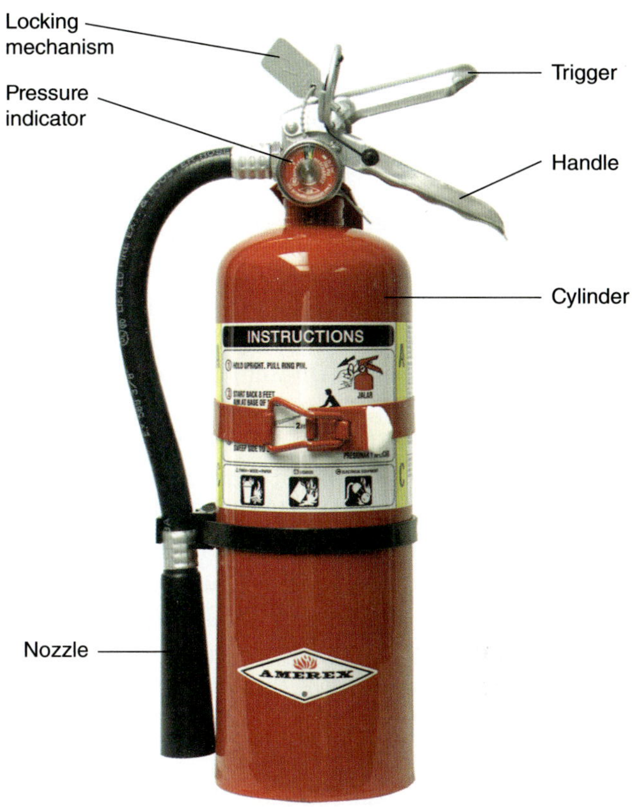

FIGURE 6-10 Portable fire extinguishers have six basic parts.

Courtesy of Amerex Corporation.

Handle

The **handle** is used to carry the portable fire extinguisher and, in many cases, to hold it during use. The actual design of the handle varies from model to model, but all fire extinguishers that weigh more than 3 lb (1.4 kg) have handles. In many cases, the handle is located just below the trigger mechanism.

Nozzle or Horn

The extinguishing agent is expelled through a **nozzle** or **horn**. In some fire extinguishers, the nozzle is attached directly to the valve assembly at the top of the extinguisher. In other models, the nozzle is at the end of a short hose.

Trigger

The **trigger** is the mechanism that is squeezed or depressed to discharge the extinguishing agent. On most portable fire extinguishers, the trigger is a lever located above the handle. On some models, the trigger is a button positioned just above the handle. To release the extinguishing agent, the operator lifts the fire extinguisher by the handle and simultaneously squeezes down on the trigger.

Cartridge/cylinder-operated fire extinguishers usually have a two-step operating sequence. First, a handle or lever is pushed to pressurize the stored agent; then, a trigger-type mechanism incorporated in the nozzle assembly is used to control the discharge.

Locking Mechanism

The **locking mechanism** is a simple, quick-release device that prevents accidental discharge of the extinguishing agent. The simplest form of locking mechanism is a stiff pin, which is inserted through a hole in the trigger to prevent it from being depressed. The pin usually has a ring at the end so that it can be removed quickly.

A special plastic tie, called a **tamper seal**, is used to secure the pin. This seal is designed to break easily when the pin is pulled. (Removing the pin and tamper seal is best accomplished with a twisting motion.) The tamper seal makes it easy to see whether the fire extinguisher has been used and not recharged. It also discourages people from playing or tinkering with the fire extinguisher.

Pressure Indicator

The **pressure indicator** is a gauge that shows whether a fire extinguisher has sufficient pressure to operate properly. Over time, the pressure in a fire extinguisher may dissipate. Checking the gauge first will tell you whether the fire extinguisher is ready for use. Not all fire extinguishers have pressure indicators, so other measures, such as weighing the extinguisher, may be required to determine if the extinguisher is ready to operate. For example, CO_2 extinguishers do not have gauges and must be weighed to determine if they are fully charged.

Pressure indicators vary in terms of both design and sophistication. Most fire extinguishers have a needle gauge. Pressure may be shown in pounds per square inch (psi) or on a three-step scale that indicates if the pressure is too low, in the proper range, or too high. Pressure gauges are usually color-coded, with a green area indicating the proper pressure zone.

Some disposable fire extinguishers intended for home use have an even simpler pressure indicator—a plastic pin built into the cylinder. Pressing on the pin tests the pressure within the fire extinguisher. If the pin pops back up, the fire extinguisher has enough pressure to operate; if it remains depressed, the pressure has dropped below the acceptable level.

Wheeled Fire Extinguishers

A **wheeled fire extinguisher** is a large unit mounted on a wheeled carriage that typically contains between 150 and 350 lb (68 and 158 kg) of extinguishing agent. The wheeled design allows one person to transport the fire extinguisher to the fire. If a wheeled fire extinguisher is intended for indoor use, doorways and aisles must be wide enough to allow for its passage to every area where it could be needed.

Wheeled fire extinguishers usually have a long delivery hose so that the unit can stay in one spot as the operator moves around to attack the fire from more than one side. Usually, a separate cylinder containing nitrogen or some other compressed gas provides the pressure necessary to operate the fire extinguisher.

Wheeled fire extinguishers are most often installed in special hazard areas, such as at military bases, airports, and industrial settings. Models with rubber tires or wide-rimmed wheels are available for outdoor installations.

Types of Fire Extinguishers

Portable fire extinguishers vary according to their extinguishing agent, capacity, effective range, and the time it takes to completely discharge the extinguishing agent. They also have different mechanical designs. This section describes the basic types of extinguishing agents used in portable fire extinguishers, as well as the basic characteristics of the different types of extinguishers. The seven types of fire extinguishers discussed in this section are organized by type of extinguishing agent:

- Water-type fire extinguishers
- Dry-chemical fire extinguishers
- CO_2 fire extinguishers
- Class B foam fire extinguishers
- Wet-chemical fire extinguishers
- Halogenated-agent fire extinguishers
- Dry-powder fire extinguishers

Water-Type Fire Extinguishers

Water is an efficient, plentiful, and inexpensive extinguishing agent that extinguishes fire by cooling the fuel. When it is applied to a fire, it is quickly converted from a liquid into steam, expanding 1600 times, absorbing great quantities of heat in the process. As the heat is removed from the combustion process, the fuel cools below its ignition temperature, and the fire stops burning.

Water-type fire extinguishers are intended for use primarily on Class A fires. Many Class A fuels absorb water, which further lowers the temperature of the fuel and prevents rekindling. Water is a much less effective extinguishing agent for other classes of fires. For example, using a water-type fire extinguisher on hot cooking oil can cause explosive splattering, which can spread the fire and endanger the operator of the fire extinguisher. Many burning flammable liquids simply float on top of water. Because streams of water conduct electricity, it is dangerous to apply a stream of water to any fire that involves energized electrical equipment. If a water-type fire extinguisher is used on a burning combustible metal, a violent reaction can occur. Because of these limitations, plain water is used only in Class A fire extinguishers. Class B foam fire extinguishers, which are a specific type of water extinguisher, are intended for fires involving flammable liquids.

Wetting-Agent and Class A Foam Fire Extinguishers

Plain water beads up and rolls off the surface of the material it is applied to. In 1984, a **Class A foam concentrate**, sometimes called **wet water**, was developed. This concentrate adds surfactants to water. A **surfactant** is a substance that reduces the surface tension of water (the physical property that causes water to bead or form a puddle on a flat surface). This allows the water to better penetrate and soak into Class A materials. The National Institute of Standards and Technology (NIST) estimates that treated water can "wet" Class A materials 20 times more effectively than plain water. A **Class A foam fire extinguisher** contains a solution of water and Class A foam concentrate. This extinguishing agent has foaming properties that allow it to cling to the material with minimal runoff, as well as the ability to reduce surface tension. Class A foam has greater cooling abilities and requires less water to extinguish a fire. A **wetting-agent fire extinguisher** expels water that contains a solution intended to reduce surface tension. Reducing the surface tension allows water to spread over the fire and penetrate more efficiently into Class A fuels.

Both wetting-agent and Class A foam fire extinguishers are available in the same configurations as other water-type fire extinguishers, including handheld stored-pressure models and wheeled units. These fire extinguishers should not be exposed to temperatures below 40°F (4°C).

FIGURE 6-11 Most fire apparatus carry stored-pressure water-type fire extinguishers.

© Daniel Heighton/iStock/Getty Images Plus/Getty Images.

Stored-Pressure Water-Type Fire Extinguishers

A **stored-pressure water-type fire extinguisher** contains water or a water-based extinguishing agent stored under pressure. The most popular kind of stored-pressure water-type fire extinguisher is the 2.5-gal (9.5-L) model with a 2-A rating (**FIGURE 6-11**). Many fire department vehicles carry this type of fire extinguisher for use on incipient stage Class A fires. This type of fire extinguisher expels water in a solid stream with a range of 35 to 40 ft (9 to 12 m) through a nozzle at the end of a short hose. The discharge time is approximately 55 seconds if the fire extinguisher is used continuously. A full fire extinguisher weighs about 30 lb (14 kg).

The recommended procedure for operating a stored-pressure water-type fire extinguisher is to set it on the ground, grasp the handle with one hand, and pull out the ring pin or release the locking latch with the other hand. At this point, the fire extinguisher can be lifted and used to douse the fire. Use one hand to aim the stream at the fire, and squeeze the trigger with the other hand. The stream of water can be turned into a spray by putting a thumb at the end of the nozzle. This technique is often used after the flames have been extinguished as part of the effort to thoroughly soak the fuel.

Stored-pressure water-type fire extinguishers can be recharged at any location that provides water and a source of compressed air. It is recommended that recharging be performed by a trained individual or a company that specializes in fire extinguishers. If that is not possible, however, be sure to follow the manufacturer's instructions to ensure proper and safe recharging.

Water Mist Fire Extinguishers

A **water mist fire extinguisher** is constructed in a manner similar to the stored-pressure water-type extinguisher. They are typically white in color (**FIGURE 6-12**). Instead of the discharge hose and nozzle assembly of a stored-pressure water-type extinguisher, they have a discharge hose that is connected to an applicator wand and a misting nozzle. They are commonly available in 1.75-gal (6.6-L) and 2.5-gal (9.5-L) sizes.

Water mist fire extinguishers contain distilled or de-ionized water, which conducts less electricity than tap water. The water is discharged from the misting nozzle as a fine spray that provides safety from electrical shock. For this reason, water mist fire extinguishers are safe to use on Class A and C fires and are rated at 2-A:C. The operator must be within 5 to 12 ft (2 to 4 m) of the fire in order for this type of fire extinguisher to be effective.

Water mist fire extinguishers are used where reduced water application is required and regular fire extinguishers might cause excessive damage. Typical uses include museums, rare book collections, hospital environments, telecommunication facilities, and "clean room" manufacturing facilities.

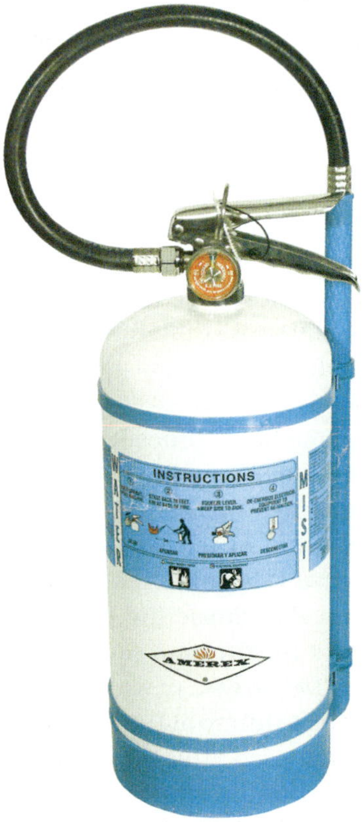

FIGURE 6-12 Water mist fire extinguishers are easily identifiable because they are typically white in color.

Courtesy of Amerex Corporation.

Loaded-Stream Fire Extinguishers

One notable disadvantage of water as an extinguishing agent is that it freezes at 32°F (0°C). In areas that are subject to below-freezing temperatures, a **loaded-stream fire extinguisher** can be used to counteract this limitation. Loaded-stream fire extinguishers discharge a solution of water containing an alkali metal salt that prevents freezing at temperatures as low as −40°F (−40°C). Pressure for these fire extinguishers is supplied by a separate cylinder of CO_2.

The most common model is the 2.5-gal (9-L) unit, which is identical to a typical stored-pressure water-type extinguisher. Hand-held models are available with capacities of 1 to 2.5 gal (4 to 9.5 L) of water and are rated from 1-A to 3-A. Larger units, including a 17-gal (64-L) unit rated 10-A and a 33-gal (124-L) unit rated 20-A, are also available.

Pump Tank Fire Extinguishers

A **pump tank fire extinguisher** uses water as an extinguishing agent, but the water in these devices is not stored under pressure. Instead, the pressure needed to expel the water is provided by a hand-operated, double-acting, vertical piston pump, which moves water out through a short hose on both the up- and the downstrokes. The manually operated pump may be mounted directly on the cylinder of the fire extinguisher, or it may be part of the nozzle assembly. This type of extinguisher is available in sizes ranging from 1-A-rated, 1.5-gal (6-L) units to 4-A-rated, 5-gal (19-L) units. This type of fire extinguisher sits upright on the ground during use. A small bracket at the bottom allows the operator to steady the fire extinguisher with one foot while pumping.

Pump tank fire extinguishers can be used with antifreeze agents. The manufacturer should be consulted for details because some antifreeze agents (such as common salt) can corrode the fire extinguisher or damage the pump. Fire extinguishers with steel cylinders corrode more easily than those with copper or nonmetallic cylinders.

Backpack Fire Extinguishers

A **backpack fire extinguisher** is used primarily outdoors for fighting brush and grass fires. Most of these units have a tank capacity of 5 gal (19 L) and weigh approximately 50 lb (23 kg) when full. Backpack fire extinguishers are rated for Class A fires but do not carry numeric ratings.

The water tank in a backpack fire extinguisher is made of fiberglass, stainless steel, galvanized steel, nylon, canvas, or brass. Backpack fire extinguishers are designed to be refilled easily in the field, such as from a lake or a stream, through a wide-mouth opening at the top. A filter keeps dirt, stones, and other contaminants from entering the tank. Antifreeze agents, wetting agents, or other special water-based extinguishing agents can be used with backpack fire extinguishers.

Most backpack fire extinguishers are operated via hand pumps. The most common design has a trombone-type, double-acting piston pump located at the nozzle, which is attached to the tank by a short rubber hose. To discharge the fire extinguisher, the operator holds the pump in both hands and moves the piston back and forth.

Some models have a compression pump built into the side of the tank. On these devices, it takes about 10 strokes of the pump handle to build up the initial pressure, which is maintained through continuous slow strokes. The operator uses the other hand to control the discharge. A lever-operated shut-off nozzle is provided at the end of a short hose.

Dry-Chemical Fire Extinguishers

A **dry chemical** is an extinguishing agent composed of fine particles that extinguish fire in three ways. First, the finer particles of the chemical vaporize when they reach the high temperature of the flame, and this vapor interrupts the chemistry of the flame. Second, the particles of the dry chemical shield the surface of the fuel from the heat radiation from the flame, thereby reducing the rate at which the burning fuel is being vaporized. And third, when a dry chemical is extensively applied and reaches the surface of the fuel, it smothers the fire by forming an insulating blanket. A **dry-chemical fire extinguisher** delivers a stream of dry chemicals onto a fire. This type of extinguisher is available as both stored-pressure and cartridge-/cylinder-operated fire extinguishers.

The following five compounds are used as dry-chemical extinguishing agents:

- Sodium bicarbonate (rated for Class B and C fires only)
- Potassium bicarbonate (rated for Class B and C fires only)
- Urea-based potassium bicarbonate (rated for Class B and C fires only)
- Potassium chloride (rated for Class B and C fires only)
- Monoammonium phosphate (rated for Class A, B, and C fires)

Sodium bicarbonate (baking soda), the original dry-chemical agent, is often used in small household fire extinguishers. Potassium bicarbonate, potassium chloride, and urea-based potassium bicarbonate all have greater fire-extinguishing capabilities per unit volume for Class B fires than sodium bicarbonate. Potassium chloride is more corrosive than the other dry-chemical extinguishing agents. **Monoammonium phosphate**, also known as **ammonium phosphate**, is a finely ground substance that looks like yellow talcum powder. It is the only dry-chemical extinguishing agent that is rated as suitable for use on Class A fires. Although the other types of dry-chemical fire extinguishers can be used to extinguish Class A fires, the other chemicals do not cool the fuel, so a water dousing would be needed to completely extinguish any smoldering embers and prevent rekindling.

Dry-chemical fire extinguishers offer several advantages over water-type fire extinguishers:

- They are effective on Class B fires.
- They can be used on Class C fires because the chemicals are nonconductive.
- They can be stored and used in areas where temperatures fall below the freezing point.

Dry-chemical fire extinguishers can be used on Class C fires that involve energized electrical equipment, but the chemicals can damage computers, electronic devices, and electrical equipment. The fine particles are carried in the air and settle like a fine dust inside the equipment. Over a period of months, this residue can corrode metal parts, causing considerable damage. Another disadvantage of dry-chemical fire extinguishers is that although the chemicals are not toxic, the cloud of fine dust created when they are discharged in an enclosed environment can make breathing more difficult and impair vision.

The trigger on a dry-chemical fire extinguisher allows it to be discharged intermittently; releasing the trigger stops the flow of the agent. This does not mean, however, that the fire extinguisher can be put aside and used again later. Dry-chemical fire extinguishers usually continue to lose pressure after a partial discharge. Pressure loss occurs even when only a small amount of agent has been discharged and the pressure gauge still indicates that the fire extinguisher is properly charged. This loss of pressure occurs because the agent leaves residue in the valve assembly that allows the stored pressure to leak out slowly.

According to NFPA 10, depending on the fire extinguisher's size, the horizontal range of the discharge stream can be from 5 ft to 30 ft (1.5 m to 9.2 m). Some models have special nozzles that allow for a longer range. The long-range nozzles are useful when the fire involves burning gas or a flammable liquid under pressure or when the operator is working in a strong wind. It is best to have the wind at your back when using a hand-held dry-chemical fire extinguisher.

Most small, hand-held, dry-chemical fire extinguishers are available with capacities ranging from 1 to 30 lb (0.45 to 14 kg) of agent and are designed to discharge their contents completely in as little as 8 to 20 seconds. Larger units may discharge for as long as 30 seconds. Wheeled fire extinguishers are available with capacities of up to 350 lb (159 kg) of agent. Large dry-chemical fire extinguishers also may be mounted on fire apparatus to deal with special risks.

The selection of which dry-chemical fire extinguisher to use depends on the compatibility of different agents with one another and with any products that they might contact. Some dry-chemical extinguishing agents cannot be used in combination with particular types of foam.

SAFETY TIP

The different types of dry-chemical extinguishers look similar. Make sure you check the label before using one to ensure you have the desired extinguishing agent.

Ordinary Dry-Chemical Fire Extinguishers

The first dry-chemical fire extinguishers were introduced in the 1950s and were rated for Class B and C fires only. The industry term for this type of B:C-rated unit that uses sodium bicarbonate as the extinguishing agent is **ordinary dry-chemical extinguisher**. Ordinary dry-chemical fire extinguishers are available in hand-held models with ratings of up to 160-B:C. Larger, wheeled units carry ratings of up to 640-B:C.

In 1959, the U.S. Navy Research Laboratory developed a dry chemical that is twice as effective in extinguishing Class B fires as sodium bicarbonate and five times more effective than CO_2. This new extinguishing agent was named Purple K because of the purple color of the substance and the characteristic lavender tint it gives to flames. The main component is potassium bicarbonate, and it also contains sodium bicarbonate, mica, Fuller's earth, and amorphous silica. It is made **hydrophobic** (repels or does not mix with water) by adding methyl hydrogen polysiloxane. Because they are nonconductive, Purple K extinguishers are rated for Class B fires and Class C fires that involve a flammable liquid.

SAFETY TIP

Purple K should not be confused with Class K extinguishers because they are not effective in class K fires.

Multipurpose Dry-Chemical Fire Extinguishers

During the 1960s, the **multipurpose dry-chemical fire extinguisher** was introduced. These fire extinguishers contain monoammonium phosphate and are rated for Class A, B, and C fires. The chemicals in these fire extinguishers take the form of fine particles that are treated with other chemicals to prevent the particles from absorbing moisture, which could cause packing or caking and interfere with the extinguisher's discharge and flow. When discharged, these chemicals form a crust over Class A combustible fuels, thereby preventing rekindling (**FIGURE 6-13**).

Multipurpose dry-chemical fire extinguishers should never be used on Class K fires, such as those involving deep-fat fryers located in commercial kitchens. The monoammonium phosphate–based extinguishing agent is acidic and does not react with cooking oils to produce the smothering foam needed to extinguish this type of fire. Even worse, the acid will counteract the foam-forming properties of any alkaline extinguishing agent that is applied to the same fire.

Multipurpose dry-chemical fire extinguishers are available as stored-pressure, cartridge, or nitrogen cylinder–type hand-held models with ratings ranging from 1-A to 20-A and from 10-B:C to 120-B:C. Larger wheeled models have ratings ranging from 20-A to 40-A and from 60-B:C to 320-B:C.

Carbon Dioxide Fire Extinguishers

When CO_2 is discharged on a fire, it forms a dense cloud that displaces the air surrounding the fuel because it is

FIGURE 6-13 Multipurpose dry-chemical fire extinguishers can be used for Class A, B, and C fires.

© Jones & Bartlett Learning

heavier than air. This effect interrupts the combustion process by reducing the amount of oxygen that can reach the fuel. The placement of a blanket of CO_2 over the surface of a liquid fuel also disrupts the fuel's ability to vaporize.

CO_2 is both a self-expelling agent and an extinguishing agent. In a portable **carbon dioxide (CO_2) fire extinguisher**, CO_2 is stored under a pressure of 800 psi (5500 kPa), which keeps the CO_2 in liquid form at room temperature. When the pressure is released, the liquid CO_2 rapidly converts to a gas, and the expansion of the gas forces the agent out of the container.

The CO_2 is discharged through a siphon tube that reaches to the bottom part of the storage cylinder; it is then forced through a hose and expelled through a horn or cone-shaped applicator that directs the flow of the agent on the fire. The rapidly moving product through the applicator sometimes creates static electricity and can release a static charge. This is not dangerous and will not ignite any unburned fuel.

When discharged, the CO_2 is very cold and contains a mixture of CO_2 gas and dry ice, which is quickly converted to a gas. The gas forms a visible cloud of dry ice that freezes the moisture in the air when the moisture comes in contact with the CO_2. This helps cool the burning materials and surrounding areas. After the fire is extinguished, the CO_2 should continue to be applied to promote cooling and reduce the chance of reignition.

CO_2 fire extinguishers are rated for Class B and C fires only. This extinguishing agent does not conduct electricity and has two significant advantages over dry-chemical agents: It is not corrosive, and it does not leave any residue because it dissipates into the air.

Gaseous extinguishing agents are a type of **clean agent** that extinguish fires using extinguishing agents that do not leave a residue. Because of this, gaseous agents such as CO_2 can be effective in suppressing fires in areas that contain computers and other sensitive electronic equipment because they can extinguish a fire without causing significant damage to the room or contents. CO_2 fire extinguishers are also used in food preparation areas and in laboratories.

CO_2 fire extinguishers have several limitations and disadvantages:

- CO_2 fire extinguishers are heavier than similarly rated extinguishers that use other extinguishing agents (**FIGURE 6-14**).
- CO_2 fire extinguishers have a short discharge range (3 to 8 ft [1 to 2.5 m]), requiring the operator to be close to the fire, which increases the risk of personal injury. Using a CO_2 extinguisher from an upwind (with the wind at your back) location will give you greater reach.
- CO_2 fire extinguishers do not perform well at temperatures below 0°F (–18°C) or in windy or drafty conditions because the extinguishing agent dissipates before it reaches the fire.
- When CO_2 fire extinguishers are used in confined spaces, the extinguishing agent dilutes the oxygen in the air. If the air is diluted enough, people in the area begin to suffocate.
- CO_2 fire extinguishers are not suitable for use on fires involving pressurized fuel or on Class K fires.

SAFETY TIP

CO_2 discharged into a confined space will reduce the oxygen level in that space.

FIGURE 6-14 Carbon dioxide fire extinguishers are heavy because of the weight of the container and the large quantity of agent needed to extinguish a fire. They also have a large discharge nozzle, making them easily identifiable.

© Jones & Bartlett Learning. Photographed by Glen E. Ellman.

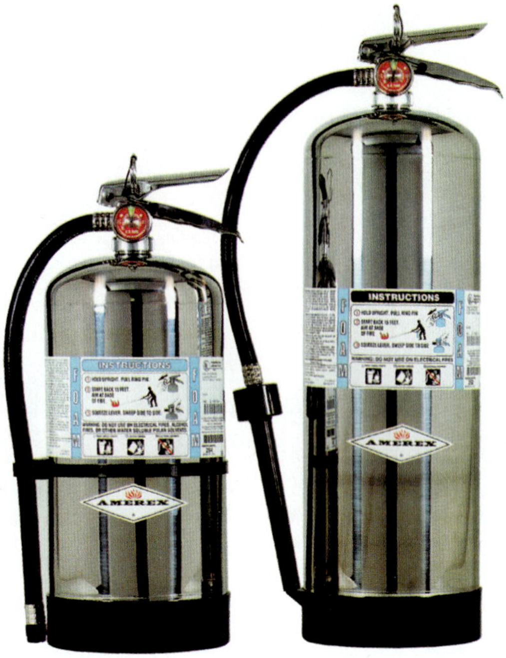

FIGURE 6-15 An AFFF fire extinguisher produces an effective foam for use on Class B fires.

Courtesy of Amerex Corporation.

Smaller CO_2 fire extinguishers contain from 2.5 to 5 lb (1 to 2 kg) of agent and are designed to be operated with one hand. The horn is attached directly to the discharge valve on the top of the fire extinguisher by a hinged metal tube. In larger models, the horn is attached at the end of a short hose and requires two-handed operation. Depending on their size, CO_2 fire extinguishers can discharge completely in 8 to 30 seconds.

CO_2 extinguishers are not single use; they can be recharged after they are used. The trigger mechanism on CO_2 fire extinguishers can be operated intermittently to preserve any remaining agent. The pressurized CO_2 remains in the fire extinguisher, but the extinguisher must still be recharged after use. The fire extinguisher can be weighed to determine how much agent is left in the storage cylinder.

SAFETY TIP

Do not aim the CO_2 fire extinguisher discharge at anyone or allow it to come in contact with exposed skin; frostbite could result.

Class B Foam Fire Extinguishers

Class B foam fire extinguishers are similar in appearance and operation to water-type, wetting agent, and Class A foam fire extinguishers. They discharge a solution of water and concentrates of either **aqueous film-forming foam (AFFF)**—a foam designed to form a blanket over spilled flammable liquids to suppress vapors or on actively burning pools of flammable liquids—or **film-forming fluoroprotein (FFFP) foam**, a foam that contains surfactants that produce a fluid film for suppressing hydrocarbon fuel vapors (**FIGURE 6-15**). The agent is discharged through an **air-aspirating nozzle**, which draws air into the water stream. The result is a foam solution that floats over the surface of a burning liquid, creating a blanket that separates the fuel from oxygen. This blanket prevents the fuel from vaporizing and forms a barrier between the fuel and the oxygen, extinguishing the flames and preventing reignition.

Class B foam fire extinguishers are effective in fighting Class A fires, but they are not suitable for Class C fires or for fires involving flammable liquids or gases under pressure. They are also not intended for use on Class K fires. In addition, although fires involving a **polar solvent**—a water-soluble, flammable liquid, such as alcohol, acetone, ester, and ketone—are considered to be Class B fires, only specifically labeled foam fire extinguishers can be used on these fires. Class B foam fire extinguishers also are not effective at freezing temperatures. Consult the fire extinguisher manufacturer for information on using foam agents effectively at low temperatures.

AFFF and FFFP hand-held stored-pressure fire extinguishers are available in two sizes: 1.6 gal (6 L), rated 2-A:10-B, and 2.5 gal (9 L), rated 3-A:20-B. A wheeled model with a 33-gal (125-L) capacity and a rating of 20-A:160-B is also available.

Both AFFF and FFFP concentrates produce highly effective foams. The decision of which one should be used depends on the foam's compatibility with the flammable liquid involved and other extinguishing agents that could be used on the same fire. Detailed information on the use of AFFF and FFFP is available in NFPA 11, *Standard for Low-, Medium-, and High-Expansion Foam, 2021 Edition.*

Wet-Chemical Fire Extinguishers

A **wet-chemical extinguishing agent** converts the fatty acids in cooking oils or fats to a soap or foam, a process known as **saponification**. They include aqueous solutions of potassium acetate, potassium carbonate, and potassium citrate, either singly or in various combinations. Wet-chemical extinguishing agents are specifically formulated for use in commercial kitchens and food-product manufacturing facilities, especially where food is cooked in a deep fryer. A **wet-chemical fire extinguisher** uses wet-chemical agents to extinguish Class K fires.

Portable Class K wet-chemical fire extinguishers are available in three sizes: 0.8 gal (3 L), 1.5 gal (6 L), and 2.5 gal (9 L). There are no numerical ratings for these fire extinguishers. Built-in, automatic extinguishing systems in commercial kitchens that are equipped with deep-fat fryers, cooking oils, and grills use wet-chemical extinguishing agents. The fixed, automatic fire-extinguishing systems discharge the agent directly over the cooking surfaces. Wet-chemical extinguishing agents are discharged as a fine spray, which reduces the risk of splattering. Cleanup afterward is much easier than after other types of extinguishing agents are used, allowing a business to reopen sooner.

Halogenated-Agent Fire Extinguishers

Almost all fuels consist of hydrogen (H) and carbon (C) atoms, so they are called *hydrocarbons*. Hydrocarbons are discussed further in Chapter 5, *Fire Behavior*. **Halogen** is the family of elements (chemicals listed in the periodic table) that includes fluorine (F), bromine (Br), iodine (I), and chlorine (Cl), which are chemically related. A **halogenated hydrocarbon**, also known as a **halocarbon**, is a hydrocarbon in which at least one hydrogen atom of the hydrocarbon is replaced by a halogen. A common type of halocarbon used in fire suppression is **hydrofluorocarbon (HFC)**. A **halogenated extinguishing agent** is an extinguishing agent made from halogenated hydrocarbons (halocarbons), and a **halogenated-agent fire extinguisher**—also called **clean-agent fire extinguisher**—uses halogenated agents.

There are two broad categories of halogenated agents: Halons and newer, less ozone-depleting Halon alternatives. Halons and Halon alternatives are used primarily for Class B and C fires because they are clean agents that extinguish fire by interrupting the chemical reaction of the fire tetrahedron. Remember that clean agents are nonconductive, noncorrosive, and leave no residue, so they will not harm sensitive equipment or people. Per pound, they are approximately twice as effective at extinguishing fires as CO_2, which means less extinguishing agent is required to extinguish a fire.

Commonly used Halon for fire suppression exists in two forms: as Halon 1211, which is Bromochlorodifluoromethane ($CBrClF_2$), also known as BCF, and as Halon 1301, Bromotrifluoromethane ($CBrF_3$). Halon 1211 is used only in portable extinguishers and is a streaming agent, whereas Halon 1301 is used only in fixed extinguisher installations, typically in cargo holds or engines, and it is a total flooding agent. Both types of Halon are rated for Class B and C fires. Larger-capacity extinguishers are also rated for use on Class A fires.

Halon 1211 is available in hand-held, stored-pressure fire extinguishers with capacities that range from 2 lb (0.9 kg), rated 2-B:C, to 22 lb (10 kg), rated 4-A:80-B:C. These fire extinguishers are not rated for use on Class B fires involving pressurized fuels or on Class K fires. Wheeled Halon 1211 models are available with capacities of up to 150 lb (68 kg) with a rating of 30-A:160-B:C. The wheeled fire extinguishers use a nitrogen booster charge from an auxiliary cylinder to expel the agent. Portable fire extinguishers that use Halon 1211 as the extinguishing agent store the agent as a liquid and discharge it under relatively high pressure.

A 1987 international agreement known as the *Montreal Protocol* limited Halon production and importation because these agents severely damage the earth's ozone layer. As a result of this agreement, Halons can no longer be manufactured or imported. Since then, Halons have been replaced by more environmentally friendly halocarbons that are not subject to the same environmental restrictions. However, Halons are the extinguishing agent in many existing systems and portable fire extinguishers still in use today. Halons can be reclaimed—that is, removed from existing systems or extinguishers that are being taken out of service and processed to remove contaminants—and reclaimed Halons can be used to recharge a system or a portable fire extinguisher that has discharged. Eventually, this supply will be depleted. If no reclaimed Halon is available, that Halon system or extinguisher must be replaced with a new, different type of system or extinguisher. Halon systems that are no longer used must be decommissioned by a certified company that reclaims the existing Halon agent according to strict environmental policies.

A fire extinguisher containing a halogenated extinguishing agent releases a mist of vapor and liquid droplets that disrupts the molecular chain reactions within the combustion process, thereby extinguishing the fire. Halogenated-agent fire extinguishers have a horizontal discharge stream range of 6 to 35 ft (2 to 11 m). These agents dissipate rapidly in windy conditions, as does CO_2, so their effectiveness is limited outdoors. Because halogenated agents also displace oxygen, they should be used with care in confined areas.

Two common halogenated agents, 3M Novec 1230 Fire Protection Fluid and FM-200, have an ozone-depletion potential of zero, which means they are safe for the earth's ozone layer. Both of these halogenated agents extinguish fire by rapidly removing heat. 3M Novec 1230 Fire Protection Fluid evaporates 50 times faster than water. Although these particular Halon alternatives do not contain Halon gas, they still use HFC to suppress fires, which is not as eco-friendly.

Dry-Powder Fire Extinguishers and Extinguishing Agents

Not to be confused with dry-chemical extinguishers, which are rated for Class A, B, and C fires, a **dry-powder fire extinguisher** uses **dry powder**, a chemical compound that is stored in fine granular or powdered form. Dry powder is used to extinguish Class D fires involving combustible metals. Flammable metal fires are not easily extinguished with water, and some metals react violently when water is applied, so a dry-powder extinguishing agent is applied to smother the fire. The dry-powder agent forms a solid crust over the burning metal, which blocks out oxygen and absorbs heat. The extinguishing agents and application methods required to put out Class D fires vary greatly, depending on the specific metal, the quantity of metal involved, and the physical form of the fuel (whether the metal is a solid object or is in the form of grindings or shavings).

SAFETY TIP

Do not confuse dry-chemical and dry-powder agents. Dry-chemical fire extinguishers are rated for Class B and C fires or for Class A, B, and C fires. Dry-powder fire extinguishers are designed for use in suppressing Class D fires.

The most commonly used dry-powder extinguishing agent is formulated from finely ground sodium chloride (table salt) plus additives to help it flow freely over a fire. A thermoplastic material mixed with the agent binds the sodium chloride particles into a solid mass when they come into contact with a burning metal. Dry-powder fire extinguishers using sodium chloride–based agents are available with a 30-lb (14-kg) capacity in either stored-pressure or cylinder/cartridge models. Wheeled models are available with 150- and 350-lb (68- to 160-kg) capacities.

Dry-powder fire extinguishers are often carried on specialty apparatus such as hazardous materials units, but they might be carried on frontline apparatus if there is a known hazard within the primary response area. They also are used in businesses or manufacturing plants where specific hazards are present. It is a good idea to identify the location of these extinguishers during building inspections and in the **preincident plan**—data about a location in your community that describe the layout and hazards at the location and identify response essentials.

Dry-powder extinguishers are easily identified by their yellow cylinder (**FIGURE 6-16**). They have adjustable nozzles that allow the operator to vary the flow of the extinguishing agent. When the nozzle is fully opened, the hand-held models have a range of 6 to 8 ft (2 to 2.5 m). Extension wand applicators are available to direct the discharge from a more distant position.

FIGURE 6-16 Dry-powder fire extinguishers are easily identified by their yellow cylinder.

© Jones & Bartlett Learning. Photographed by Glen E. Ellman.

Class D agents must be applied carefully so that the molten metal does not splatter. Care must be taken to ensure that no water comes in contact with the burning metal because even a trace quantity of moisture can cause a violent reaction. Flammable metals can burn at more than 5000 degrees, hot enough to separate water into its component parts of hydrogen and oxygen, and hydrogen is flammable and has explosive properties.

The same sodium chloride–based dry-powder agent that is used in portable fire extinguishers can be stored in bulk form and applied by hand. Another dry-powder extinguishing agent for Class D fires, graded granular graphite mixed with phosphorus-containing compounds, cannot be expelled from a portable fire extinguisher. Instead, this agent is produced in bulk and must be applied manually from a pail or other container using a shovel or scoop. When applied to a metal fire, the phosphorus compounds release gases that blanket the fire and cut off its supply of oxygen, and the graphite absorbs heat from the fire, allowing the metal to cool to below its ignition temperature. Bulk dry-powder agents are available in 40- and 50-lb (18- and 23-kg) pails and 350-lb (159-kg) drums.

More information about Class D extinguishing agents and fire extinguishers can be found in the NFPA's *Fire Protection Handbook*.

Use of Fire Extinguishers

Fire extinguishers are designed to be simple to operate. Every portable fire extinguisher is labeled with printed operating instructions. An individual with only basic training should be able to use most fire extinguishers safely and effectively.

There are six basic steps in extinguishing a fire with a portable fire extinguisher:

1. Locate the fire extinguisher.
2. Select the proper classification of fire extinguisher.
3. Ensure your personal safety by having an exit route.
4. Transport the fire extinguisher to the location of the fire.
5. Activate the fire extinguisher to release the extinguishing agent.
6. Apply the extinguishing agent to the fire for maximum effect.

SAFETY TIP

Proper technique in using a portable fire extinguisher is important for both safety and effectiveness. Practice using a portable fire extinguisher under the careful supervision of a trained instructor.

Although these steps are not complicated, practice and training are essential for effective fire suppression with a fire extinguisher. Research has shown that the effective use of Class B portable fire extinguishers depends heavily on user training and expertise. A trained expert can extinguish a larger fire than a nonexpert can, using the same extinguisher. As a support person, you should be able to operate any fire extinguisher that you might be required to use, whether it is carried on your fire apparatus, hanging on the wall of your firehouse, or placed in some other location in your community.

SAFETY TIP

Always test a fire extinguisher before entering the area involved. Activate the trigger for long enough to ensure that the agent discharges properly.

Locating Fire Extinguishers

As a support person, you should know which types of fire extinguishers are carried on your department's apparatus and where each one is located. You should also know where fire extinguishers are located in and around the fire station and other workplaces in your department. You should have at least one fire extinguisher in your home and another in your personal vehicle, and you should know exactly where they are located. Knowing the exact locations of fire extinguishers can save valuable time in an emergency.

Selecting the Proper Fire Extinguisher

Selecting the proper fire extinguisher requires an understanding of the classification and rating system for fire extinguishers. Knowing which types of agents are available, how they work, which ratings the fire extinguishers carried on your fire apparatus have, and which extinguisher is appropriate for a particular fire situation is also important. Understanding the fire extinguisher rating system and the different types of agents will enable fire service personnel to determine whether it is suitable for a particular fire situation. A quick look at the label should be all that is needed.

Support personnel should be able to assess a fire quickly, determine whether the fire can be controlled by a fire extinguisher, and identify the appropriate extinguisher to use. Using a fire extinguisher with an insufficient rating may not completely extinguish the fire, which can place the operator in danger of being burned or otherwise injured. If the fire is too large for the fire extinguisher, consider other options, such as obtaining additional extinguishers or making sure that a charged hose line is ready to provide backup.

SAFETY TIP

Before deciding to use a fire extinguisher, size up the fire to ensure that the extinguisher has an adequate capacity and holds the proper extinguishing agent.

Fire service personnel should be able to determine the most appropriate type of fire extinguisher to place in a given area based on the hazards that are present and the types of fires that could occur. In some cases, one type of fire extinguisher might be preferred over another. For example, an extinguishing agent such as CO_2 or a halogenated agent is a better choice than other agents for a fire involving sensitive electronic equipment because it leaves no residue. A dry-chemical fire extinguisher is generally more appropriate than a CO_2 fire extinguisher to fight an outdoor fire because wind will quickly dissipate CO_2. A dry-chemical fire extinguisher would also be the best choice for a fire involving a flammable liquid leaking under pressure from a pipe, whereas foam would be a better choice for a fire involving a liquid spill on the ground.

Ensuring Your Personal Safety

Before using a fire extinguisher, be sure to approach the fire with an exit plan. If the fire suddenly expands or the fire extinguisher fails to control it, you must have a planned escape route. Never let the fire get between you and a safe exit. After the fire has been suppressed, do not turn your back on it. Always watch it and be prepared for it to rekindle until the fire has undergone a complete **overhaul**—the process of finding, exposing, and suppressing any smoldering or hidden pockets of fire in an area that has been burned.

When ordinary civilians use fire extinguishers on incipient stage fires, they are probably wearing their normal clothing. As a member of the fire service, however, you should wear your personal protective clothing and use appropriate personal protective equipment (PPE). Take advantage of the protection they provide.

If you must enter an enclosed area where a fire extinguisher has been discharged, wear full PPE. The atmosphere within the enclosed area will contain a mixture of combustion products and extinguishing agents, and the oxygen content within the space may be dangerously low.

Transporting a Fire Extinguisher

The best method of transporting a hand-held portable fire extinguisher depends on the extinguisher's size, weight, and design. Hand-held portable models can weigh as little as 1 lb (0.45 kg) or as much as 50 lb (23 kg). The ability to handle the heavier fire extinguishers depends on an individual operator's personal strength.

Fire extinguishers with a fixed nozzle should be carried in the favored or stronger hand. This approach enables the operator to depress the trigger and direct the discharge easily. Fire extinguishers that have a hose between the trigger and the nozzle should be carried in the weaker or less-favored hand so that the favored hand can grip and aim the nozzle.

Heavier fire extinguishers may have to be carried as close as possible to the fire and placed upright on the ground. The operator can then depress the trigger with one hand while holding the nozzle and directing the stream with the other hand.

To transport a fire extinguisher, follow the steps in **SKILL DRILL 6-1**.

SKILL DRILL 6-1

Transporting a Fire Extinguisher Support Person, NFPA 1010: 5.3.5

1. Locate the closest fire extinguisher.

2. Assess that the fire extinguisher is safe and effective for the type of fire being attacked. Release the mounting bracket straps.

Continues.

SKILL DRILL 6-1 CONTINUED

Transporting a Fire Extinguisher Support Person, NFPA 1010: 5.3.5

3. Lift the fire extinguisher using good body mechanics. Lift small fire extinguishers with one hand and large extinguishers with two hands.

4. Walk briskly—do not run—toward the fire. If the fire extinguisher has a hose and nozzle, carry the extinguisher with one hand, and grasp the nozzle with the other hand.

© Jones & Bartlett Learning. Photographed by Glen E. Ellman.

Operating a Fire Extinguisher

Practice discharging different types of fire extinguishers in training situations to build confidence in your ability to use them properly and effectively. Most fire extinguishers have simple operation systems. Activating a fire extinguisher to apply the extinguishing agent is a simple operation that involves four steps. The **PASS** acronym is a helpful way to remember these steps:

1. Pull the safety pin.
2. Aim the nozzle at the base of the flames.
3. Squeeze the trigger to discharge the agent.
4. Sweep the nozzle across the base of the flames.

To extinguish a Class A fire with a stored-pressure water-type fire extinguisher, follow the steps in **SKILL DRILL 6-2**.

SKILL DRILL 6-2

Extinguishing a Class A Fire with a Stored-Pressure Water-Type Fire Extinguisher Support Person, NFPA 1010: 5.3.5

1. Size up the fire to determine whether a stored-pressure water-type fire extinguisher is safe and effective for the fire. Ensure the fire extinguisher is large enough to be safe and effective. Ensure your safety. Make sure you have an exit route from the fire. Do not turn your back on the fire.

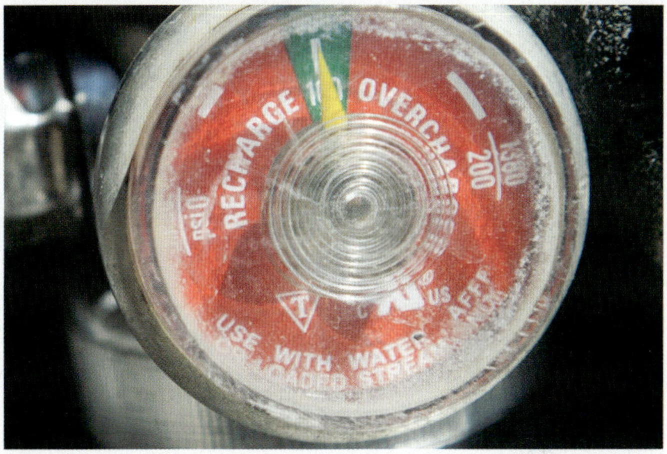

2. Remove the hose and nozzle from its holder if present. Quickly check the pressure gauge to verify that the fire extinguisher is adequately charged.

3. Pull the pin to release the fire extinguisher control valve.

4. Aim the nozzle at the base of the flames, squeeze the trigger, and then sweep the water stream at the base of the flames. You must be within 35 to 40 ft (11 to 12 m) of the fire to be effective.

Continues.

SKILL DRILL 6-2 CONTINUED

Extinguishing a Class A Fire with a Stored-Pressure Water-Type Fire Extinguisher Support Person, NFPA 1010: 5.3.5

5. Overhaul the fire by breaking apart tightly packed fuel to prevent rekindling. Summon additional help if needed.

© Jones & Bartlett Learning. Photographed by Glen E. Ellman.

SKILL DRILL 6-3

Extinguishing a Class A Fire with a Multipurpose Dry-Chemical Fire Extinguisher Support Person, NFPA 1010: 5.3.5

1. Size up the fire to determine whether a multipurpose dry-chemical fire extinguisher is safe and effective for this fire. Ensure the fire extinguisher is large enough to be safe and effective. Ensure your safety. Make sure you have an exit route from the fire. Do not turn your back on the fire. Remove the hose and nozzle.
2. Quickly check the pressure gauge to verify that the fire extinguisher is adequately charged.
3. Remove the hose and nozzle. Pull the pin to release the fire extinguisher control valve.
4. Aim the nozzle at the base of the flames, squeeze the trigger, and then sweep the dry-chemical discharge at the base of the flames. Depending on the size of the fire and fire extinguisher, you must be within 5 to 45 ft (2 to 14 m) of the fire to be effective. Coat the burning fuel with dry chemical.
5. Overhaul the fire by breaking apart tightly packed fuel to prevent rekindling. Summon additional help if needed.

To extinguish a Class A fire with a multipurpose dry-chemical fire extinguisher, follow the steps in **SKILL DRILL 6-3**. *Note:* Multipurpose dry-chemical fire extinguishers are safe and effective on Class A, B, and C fires.

To extinguish a Class B flammable liquid spill or pool fire with a dry-chemical extinguisher, follow the steps in **SKILL DRILL 6-4**. *Note:* Dry-chemical fire extinguishers are safe to use on Class B and C fires. They are effective on Class B fires, but they are not rated for Class A fires.

To operate a CO_2 fire extinguisher, follow the steps in **SKILL DRILL 6-5**. *Note:* CO_2 fire extinguishers are most effective when used in areas without wind. They are designed for Class B and C fires only.

SKILL DRILL 6-4

Extinguishing a Class B Flammable Liquid Fire with a Dry-Chemical Fire Extinguisher Support Person, NFPA 1010: 5.3.5

1. Size up the fire to determine whether a dry-chemical fire extinguisher is safe and effective for the fire. Quickly check the pressure gauge to verify that the fire extinguisher is adequately charged. Ensure the fire extinguisher is large enough to be safe and effective. Ensure your safety. Make sure you have an exit route from the fire. Do not turn your back on the fire.

2. Remove the hose and nozzle. Pull the pin to release the fire extinguisher valve.

3. Aim the nozzle at the near edge of the surface of the burning liquid, squeeze the trigger, and then sweep the dry-chemical discharge across the surface of the burning liquid. Depending on the size of the fire and fire extinguisher, you must be within 5 to 45 ft (2 to 14 m) of the fire to be effective. Start at the near edge of the fire, and work toward the back.

4. Protect from rekindle by keeping a blanket of dry chemical over the fuel. Summon additional help if needed.

© Jones & Bartlett Learning. Photographed by Glen E. Ellman.

SKILL DRILL 6-5

Operating a Carbon Dioxide Fire Extinguisher Support Person, NFPA 1010: 5.3.5

1. Size up the fire to determine whether CO_2 is a safe and effective agent for this fire. Ensure the fire extinguisher is large enough to be safe and effective. Ensure your safety. Make sure you have an exit route from the fire. Do not turn your back on a fire. Take hold of the hose and the horn or nozzle. Pull the pin to release the fire extinguisher control valve.

2. Quickly squeeze the trigger to verify that the fire extinguisher is charged; CO_2 fire extinguishers do not have pressure gauges.

3. Aim the horn or nozzle at the base of the flames, squeeze the trigger, and then sweep at the base of the flames. You must be within 3 to 8 ft (1 to 2.5 m) of the fire to be effective.

4. Overhaul the fire by taking steps to prevent rekindling depending on what type of fuel was involved. You may have to reapply to keep material cooled. Summon additional help if needed.

© Jones & Bartlett Learning. Photographed by Glen E. Ellman.

The Care of Fire Extinguishers

Fire extinguishers must be regularly inspected and properly maintained to ensure that they will be available for use in an emergency. Records must be kept confirming that the required inspections and maintenance have been performed on schedule. The individuals assigned to perform these functions must be properly trained and must always follow the manufacturer's recommendations for inspecting, maintaining, recharging, and testing the equipment. Persons performing maintenance or recharging of portable fire extinguishers should be qualified according to the requirements in NFPA 10.

Inspection

According to NFPA 10, an inspection is a quick check to verify that a fire extinguisher is available and ready for immediate use. Fire extinguishers on fire apparatus should be inspected as part of the regular equipment checks mandated by your department once per month. The personnel charged with inspecting the fire extinguishers should perform the following tasks:

- Ensure that all tamper seals are intact.
- Examine all parts for signs of physical damage, corrosion, or leakage.
- Check the pressure gauge to confirm that it is in the operable range. Some stored-pressure fire extinguishers use compressed air, whereas others use compressed nitrogen. The weight of the fire extinguisher and the presence of an intact tamper seal should indicate that the unit is full of extinguishing agent.
- If the extinguisher is a cartridge-type fire extinguisher, check that the cartridge has not been punctured.
- If the extinguisher is a CO_2 fire extinguisher, determine fullness by weighing the fire extinguisher. The proper weight is labeled on the outside of the fire extinguisher.
- Ensure that the fire extinguisher is properly identified by type and rating.
- Check the hose and nozzle for damage or obstruction by foreign objects.
- Check the **hydrostatic test** date of the fire extinguisher.

If an inspection reveals any problems, the fire extinguisher should be removed from service until the required maintenance procedures are performed. Spare fire extinguishers should be used until the problem is corrected.

Maintenance

The maintenance requirements and intervals for various types of fire extinguishers are outlined in NFPA 10. Maintenance includes an internal inspection as well as any repairs that may be required. These procedures must be performed periodically, depending on the type of fire extinguisher.

Maintenance procedures must always be performed by a qualified person. Some procedures can be performed only at a properly licensed facility. The specific qualifications and training requirements are determined by the manufacturer and the jurisdictional authority. Untrained personnel should never be allowed to perform fire extinguisher maintenance.

Common indications that a fire extinguisher needs maintenance include the following:

- The reading on the pressure gauge is outside the normal range.
- The inspection tag is out of date.
- The tamper seal is broken, especially in fire extinguishers with no pressure gauge.
- The fire extinguisher does not appear to be full of extinguishing agent.
- The hose or nozzle assembly is obstructed.
- There are signs of physical damage, corrosion, or rust.
- There are signs of leakage around the discharge valve or nozzle assembly.

CASE STUDY

You Are the Support Person CONCLUSION

You are the designated on-duty cook for the day, and the kitchen is bare. You go into the grocery store while the rest of your crew waits in the apparatus. As you make your way through the produce section, an employee runs up to you and says that there is a fire in the back room. You quickly radio your officer to inform them of the situation and then immediately proceed to the back room. You locate a small fire in a trash can that is just starting to spread. You quickly scan the room, looking for a fire extinguisher. Not seeing one, you ask the employee where the closest extinguisher is located.

1. **What type of fire extinguisher would be most effective for this fire? Why?**

   Answer: An extinguisher rated for Class A fires would be most effective. Class A fires involve ordinary solid combustible materials such as wood, paper, cloth, rubber, and household rubbish. A water-type fire extinguisher would be ideal because the water would cool the fuel, removing heat from the fire tetrahedron. If a water-type extinguisher is not available, then a multipurpose dry-chemical extinguisher would be the next best choice.

2. **As the fire grows, at what point will using a fire extinguisher become ineffective?**

   Answer: Fire extinguishers work best when the fire is in its incipient stage. If the fire were to spread beyond this stage, attempting to control it with only a fire extinguisher would most likely not be successful and would prove to be dangerous to the user.

3. **How do fire inspectors ensure that a fire extinguisher will work when needed?**

   Answer: Fire extinguishers should be inspected monthly to ensure that all tamper seals are intact; the fire extinguisher is fully charged; there are no signs of physical damage, corrosion, or leakage; the hose and nozzle are not damaged or obstructed by foreign objects; and the hydrostatic test date of the fire extinguisher is current.

WRAP-UP

SUMMARY

- Fire extinguishers are designed to put out small, incipient-stage fires. You need to be able to select the most appropriate extinguisher to use for different types of fires, know which extinguishers must **not** be used for certain fires, be able to operate the most common types of portable fire extinguishers correctly, and know how to inspect and maintain them.
- Fire extinguishers range in size from models that can be operated with one hand to large, wheeled models that contain several hundred pounds of **extinguishing agent**. Extinguishing agents include water, water with additives, dry chemicals, wet chemicals, dry powders, and gaseous agents. Each agent is suitable for a specific type of incipient-stage fire.
- Portable fire extinguishers are classified and rated based on their extinguishing properties and capabilities. Each fire extinguisher has a specific rating that identifies the classes of fires for which it is both safe and effective.
- When determining which type of fire extinguisher should be placed in an area, one must consider the class of fire that is likely to occur and the potential size of an incipient-stage fire.
- A **light hazard area**, or **low hazard area**, is an area where the quantity, combustibility, and heat release rate of the materials are low, and most materials are arranged so that a fire is not likely to spread.
- An **ordinary hazard area**, or **moderate hazard area**, contains more Class A and B materials than

light-hazard locations, and the combustibility and heat release rate of the materials are moderate.

- An **extra hazard area**, or **high hazard area**, contains more Class A and B materials than ordinary-hazard locations, and the combustibility and heat release rate of the materials are high.
- Portable fire extinguishers vary according to their extinguishing agent, capacity, effective range, and the time it takes to completely discharge the extinguishing agent.

KEY TERMS

air-aspirating nozzle A nozzle that draws air into the water stream, creating an aerated or foamy spray that increases the surface area of water droplets, allowing for better heat absorption and faster cooling of a fire, and when used with firefighting foam solutions, aerates the foam mixture.

ammonium phosphate See *monoammonium phosphate.*

aqueous film-forming foam (AFFF) A concentrate based on fluorinated surfactants plus foam stabilizers to produce a fluid aqueous film for suppressing hydrocarbon fuel vapors and usually diluted with water to a 1 percent, 3 percent, or 6 percent solution. (NFPA 11)

backpack fire extinguisher A portable fire extinguisher usually consisting of a 5-gal (19-L) water tank that is worn on the user's back and features a hand-powered piston pump for discharging the water and that is primarily used to fight brush and grass fires.

carbon dioxide (CO_2) A colorless, odorless, electrically nonconductive inert gas that is a suitable medium for extinguishing Class B and C fires. (NFPA 10)

carbon dioxide (CO_2) fire extinguisher A fire extinguisher that uses carbon dioxide gas as the extinguishing agent. It is rated for use on Class B and C fires.

cartridge-/cylinder-operated fire extinguisher A fire extinguisher in which the expellant gas is in a separate container from the agent storage container. (NFPA 10)

Class A foam concentrate A concentrate that, when combined with water, reduces the surface tension of the water and creates a foam. Also called *wet water.*

Class A foam fire extinguisher A fire extinguisher that contains a solution of water and Class A foam concentrate.

clean agent Electrically nonconducting, volatile, or gaseous fire extinguishant that does not leave a residue upon evaporation. (NFPA 10)

clean-agent fire extinguisher A fire extinguisher that uses a halogenated extinguishing agent. Also called *halogenated-agent fire extinguisher.*

container See *cylinder.*

cylinder The body of the fire extinguisher where the extinguishing agent is stored. Also called *container.*

dry chemical A powder composed of very small particles, usually sodium bicarbonate, potassium bicarbonate, or ammonium phosphate based, with added particulate material supplemented by special treatment to provide resistance to packing, resistance to moisture absorption (caking), and the proper flow capabilities. (NFPA 10)

dry-chemical fire extinguisher A fire extinguisher that uses a dry-chemical extinguishing agent and is usually rated for use on Class B and C fires and sometimes on Class A fires.

dry powder Solid materials in powder or granular form designed to extinguish Class D combustible metal fires by crusting, smothering, or heat-transferring means. (NFPA 10)

dry-powder fire extinguisher A fire extinguisher that uses solid materials in powder or granular form to extinguish Class D combustible metal fires by crusting, smothering, or heat-transferring means.

extinguishing agent A material used to stop the combustion process. Extinguishing agents may include liquids, gases, dry-chemical compounds, and dry-powder compounds.

Extra hazard area An occupancy where the total amount of Class A combustibles and Class B flammables is greater than expected in occupancies classed as ordinary (moderate) hazards and the combustibility and heat release rate of the materials are high. Also called *high hazard area.*

film-forming fluoroprotein (FFFP) foam A protein-foam solution that uses fluorinated surfactants to produce a fluid aqueous film for suppressing liquid fuel vapors. (NFPA 10)

fire load The total energy content of combustible materials in a building, space, or area, including

KEY TERMS CONTINUED

furnishings and contents and combustible building elements; expressed in megajoules (MJ). (NFPA 557)

halocarbon See *halogenated hydrocarbon.*

halogen A family of elements (chemicals listed in the periodic table), including fluorine (F), bromine (Br), iodine (I), and chlorine (Cl), that are chemically related.

halogenated-agent fire extinguisher A fire extinguisher that uses a halogenated extinguishing agent. Also called *clean-agent fire extinguisher.*

halogenated extinguishing agent A liquefied gas extinguishing agent that extinguishes fire by chemically interrupting the combustion reaction between fuel and oxygen. Halogenated agents leave no residue. (NFPA 402)

halogenated hydrocarbon A hydrocarbon in which at least one hydrogen atom of the hydrocarbon is replaced by a halogen. Also called *halocarbon.*

handle The grip used for holding and carrying a portable fire extinguisher.

High hazard area See *extra hazard area.*

horn The tapered discharge nozzle of a carbon dioxide fire extinguisher.

hydrofluorocarbon (HFC) A common type of halocarbon used in fire suppression.

hydrophobic The quality of repelling or being unable to mix with water.

hydrostatic testing Pressure testing of a fire extinguisher to verify its strength against unwanted rupture. (NFPA 10)

light hazard area An occupancy where the quantity, combustibility, and heat release of the materials are low and the majority of materials are arranged so that a fire is not likely to spread. Also called *low hazard area.*

loaded-stream fire extinguisher A stored-pressure water-type fire extinguisher that uses an alkali metal salt as a freezing-point depressant.

locking mechanism A device that locks a fire extinguisher's trigger to prevent its accidental discharge.

Low hazard area See *light hazard area.*

moderate-hazard area See *ordinary-hazard area.*

monoammonium phosphate A finely ground substance that looks like yellow talcum powder that is used as an extinguishing agent in dry-chemical fire extinguishers that are rated for Class A, B, and C fires. Also called *ammonium phosphate.*

multipurpose dry-chemical fire extinguisher A fire extinguisher that uses a monoammonium phosphate–based extinguishing agent that is effective on fires involving ordinary combustibles, such as wood or paper, and fires involving flammable liquids and that is rated to fight Class A, B, and C fires.

nozzle A device for use in applications requiring special water discharge patterns, directional spray, or other unusual discharge characteristics. (NFPA 13)

ordinary dry-chemical fire extinguisher A dry-chemical fire extinguisher rated for only Class B and C fires.

Ordinary hazard area An area that contains more Class A and B materials than a light hazard area and where the combustibility and heat release rate of the materials are moderate. Also called *moderate hazard area.*

overhaul A firefighting term involving the process of final extinguishment after the main body of the fire has been knocked down. All traces of fire must be extinguished at this time. (NFPA 1700)

PASS Acronym for the steps involved in operating a portable fire extinguisher: **P**ull pin, **A**im nozzle, **S**queeze trigger, **S**weep across burning fuel.

polar solvent A water-soluble, flammable liquid, such as alcohol, acetone, ester, and ketone.

preincident plan A document developed by gathering general and detailed data that is used by responding personnel in effectively managing emergencies for the protection of occupants, responding personnel, property, and the environment. (NFPA 1660)

pressure indicator A gauge on a pressurized portable fire extinguisher that indicates the internal pressure of the expellant.

pump tank fire extinguisher A nonpressurized, manually operated water-type fire extinguisher that is rated for use on Class A fires. Discharge pressure is provided by a hand-operated, double-acting piston pump.

saponification The process of converting the fatty acids in cooking oils or fats to soap or foam; the action caused by a Class K fire extinguisher.

self-expelling agent An agent that has sufficient vapor pressure at normal operating temperatures to expel itself from a fire extinguisher.

stored-pressure fire extinguisher A fire extinguisher in which both the extinguishing agent and expellant gas are kept in a single container and that includes a pressure indicator or gauge. (NFPA 10)

stored-pressure water-type fire extinguisher A fire extinguisher in which water or a water-based extinguishing agent is stored under pressure.

surfactant A compound that lowers the surface tension (or interfacial tension) between two liquids, between a gas and a liquid, or between a liquid and a solid and that can act as a detergent, wetting agent, emulsifier, foaming agent, and dispersant. (NFPA 1700)

tamper seal A retaining device that breaks when the locking mechanism is released.

trigger The button or lever used to discharge the agent from a portable fire extinguisher.

water mist fire extinguisher A fire extinguisher containing distilled or de-ionized water and employing a nozzle that discharges the agent in a fine spray. (NFPA 10)

wet-chemical extinguishing agent Normally, an aqueous solution of organic or inorganic salts or a combination thereof that forms an extinguishing agent. (NFPA 10)

wet-chemical fire extinguisher A fire extinguisher containing a wet-chemical extinguishing agent for use on Class K fires.

wetting-agent fire extinguisher A fire extinguisher that expels water combined with a concentrate to reduce the surface tension and increase its ability to penetrate and spread.

wet water See *Class A foam concentrate.*

wheeled fire extinguisher A portable fire extinguisher equipped with a carriage and wheels intended to be transported to the fire by one person. (NFPA 10)

REVIEW QUESTIONS

1. What is an extinguishing agent?
2. Which part of the fire tetrahedron should you remove to stop the combustion process?
3. What does a fire extinguisher rating of 2-A:10-B:C tell you?
4. What is the minimum distance above the floor that the bottom of a fire extinguisher should be in multifamily, public, and commercial buildings?
5. What is the term for the retaining device that breaks when the locking mechanism is released?
6. Which type of fire extinguisher can typically be used to extinguish Class A, B, or C fires?
7. What is a polar solvent?
8. Which extinguishing agent should be used on Class K fires?
9. What does the acronym PASS stand for?
10. How do you determine if a carbon dioxide fire extinguisher is completely filled?

DISCUSSION QUESTIONS

1. What type of extinguishing agent should be used on a Class C fire? Why? Describe which agent is best for areas that contain sensitive electronic equipment and explain why.
2. Why is it essential for all fire service personnel to understand the characteristics and operation of the various types of fire extinguishers?
3. While assisting with an apparatus check at the start of your shift, you notice that the tamper seal is missing from the locking pin on the CO_2 extinguisher. What do you do?
4. You are mopping the floors at the firehouse when someone yells, "There's a fire on the stove!" You are told to grab a fire extinguisher to bring into the kitchen. What type of extinguisher do you choose?

REFERENCES

National Fire Protection Association. 2018. *NFPA 402, Guide for Aircraft Rescue and Fire-Fighting Operations.* 2019 Edition. Quincy, MA: National Fire Protection Association.

National Fire Protection Association. 2020. *NFPA 11, Standard for Low-, Medium-, and High-Expansion Foam.* 2021 Edition. Quincy, MA: National Fire Protection Association.

National Fire Protection Association. 2020. *NFPA 1700, Guide for Structural Fire Fighting.* 2021 Edition. Quincy, MA: National Fire Protection Association.

National Fire Protection Association. 2021. *NFPA 10, Standard for Portable Fire Extinguishers.* 2022 Edition. Quincy, MA: National Fire Protection Association.

National Fire Protection Association. 2021. *NFPA 13, Standard for the Installation of Sprinkler Systems.* 2022 Edition. Quincy, MA: National Fire Protection Association.

National Fire Protection Association. 2021. *NFPA 408, Standard for Aircraft Hand Portable Fire Extinguishers.* 2022 Edition. Quincy, MA: National Fire Protection Association.

National Fire Protection Association. 2022. *NFPA 557, Standard for Determination of Fire Loads for Use in Structural Fire Protection Design.* 2023 Edition. Quincy, MA: National Fire Protection Association.

National Fire Protection Association. 2023. *Fire Protection Handbook.* 21st Edition. Quincy, MA: National Fire Protection Association.

National Fire Protection Association. 2023. *NFPA 440, Guide for Aircraft Rescue and Firefighting Operations and Airport/Community Emergency Planning.* 2024 Edition. Quincy, MA: National Fire Protection Association.

National Fire Protection Association. 2023. *NFPA 1660, Standard for Emergency, Continuity, and Crisis Management: Preparedness, Response, and Recovery.* 2024 Edition. Quincy, MA: National Fire Protection Association.

National Fire Protection Association. 2023. *NFPA 1900, Standard for Aircraft Rescue and Firefighting Vehicles, Automotive Fire Apparatus, Wildland Fire Apparatus, and Automotive Ambulances.* 2024 Edition. Quincy, MA: National Fire Protection Association.

National Fire Protection Association. 2023. *NFPA 1, Fire Code.* 2024 Edition. Quincy, MA: National Fire Protection Association.

Underwriters Laboratories. 2017. "Extinguishers and Extinguishing System Units, FWFZ." Accessed June 18, 2018. http://productspec.ul.com/document.php?id=FWFZ.GuideInfo.

Chapter Opener: © Eric Scruggs

CHAPTER

7

Support Person

Firefighter Tools and Equipment

KNOWLEDGE OBJECTIVES

After studying this chapter, you will be able to:

- Describe why it is important for you to know where tools are stored.
- Identify procedures for documenting maintenance performed on tools and equipment.
- Identify procedures, including reporting requirements, for removing a damaged tool from service.
- Explain the importance of returning tools to their assigned locations.
- Describe the supplies needed to clean and inspect tools and equipment.
- Describe how to clean and maintain tools and equipment.
- Describe the equipment used to illuminate an emergency scene.
- Describe how to operate lighting equipment to light exterior and interior scenes.
- Describe the types of lights used to illuminate exterior and interior scenes.
- Describe the safety precautions to take when working with lighting equipment.

SKILL OBJECTIVES

After studying this chapter, you will be able to perform the following skills:

- Demonstrate how to clean and inspect hand tools.
- Demonstrate how to clean and inspect ladders.
- Properly document equipment maintenance following established guidelines.
- Demonstrate proper operation of power and lighting equipment, cords, connectors, and ground-fault interrupter (GFCI) devices.
- Demonstrate where to properly locate lights to illuminate an emergency scene.

ADDITIONAL NFPA STANDARDS

- **NFPA 440**, *Guide for Aircraft Rescue and Firefighting Operations and Airport/Community Emergency Planning, 2024 Edition*
- **NFPA 1410**, *Standard on Training for Emergency Scene Operations, 2020 Edition*
- **NFPA 1550**, *Standard for Emergency Responder Health and Safety, 2024 Edition*
- **NFPA 1660**, *Standard for Emergency, Continuity, and Crisis Management: Preparedness, Response, and Recovery, 2024 Edition*
- **NFPA 1700**, *Guide for Structural Fire Fighting, 2021 Edition*
- **NFPA 1960**, *Standard for Fire Hose Connections, Spray Nozzles, Manufacturer's Design of Fire Department Ground Ladders, Fire Hose, and Powered Rescue Tools, 2024 Edition*

CASE STUDY

You Are the Support Person

Upon reporting for duty, you are notified that you have been detailed to another station to work with the engine company for the tour. This will be your first time working with an engine company, and you are excited about the opportunity. It isn't long before an alarm sounds. "Person trapped, motor vehicle collision, Fourth and Elm," directs the dispatcher. You slip into your gear, take your seat, and put on your seat belt. As the engine pulls out from the station, you think about how this changes your plan for what you thought you would be doing today.

1. How will your role be different for this call?
2. What kinds of tools might be needed for this job?
3. What other kind of knowledge will you need?

Introduction

Fire service personnel use tools and equipment to perform a wide range of activities. The same tools may be used in different ways for fire suppression, rescue operations, and **overhaul**. All crew members must know how to use these tools effectively, efficiently, and safely, even when it is dark or when visibility is limited. They also need to know how to maintain the tools and equipment so that they will always be ready for use.

General Considerations

Hand tools are used to extend or multiply the actions of your body and to increase your effectiveness in performing specific functions. Most of these tools operate using simple mechanical principles. For example, a pike pole extends a firefighter's reach, allowing them to penetrate through a ceiling, and enables them to apply force to pull down ceiling material; an axe multiplies the cutting force one can exert on a given area. In contrast, power tools and equipment have a mechanically driven source of power, such as an electric motor or an internal combustion engine. In certain cases, they are faster and more efficient than hand tools. A **hydraulic tool** uses pressurized fluid to exert force. They can be manually or mechanically powered. If they are mechanically powered, the same hydraulic power source can be used to power multiple hydraulic tools.

General Tool Safety

Safety is a high priority at any incident scene. Before using any kind of tool or equipment, you must be fully aware of how it works and know both the tool's and your individual limitations. You need to operate the equipment so that you, your fellow crew members, victims, and bystanders are not unintentionally injured. Using any tool without the proper training or using the wrong tool for the application could result in serious injury or death. Personal safety also requires the use of proper personal protective equipment (PPE).

The best way to learn how to use tools and equipment properly is under optimal conditions of visibility and safety and with adequate supervision. Your first exposure to and initial use of any tool should be done in an environment where you are able to see what you are doing and practice without endangering yourself or others. As you become more proficient, you should practice using tools and equipment under more difficult working conditions. Eventually, you should be able to use tools and equipment safely and effectively, even in low-light conditions. You must be able to work safely in any environment. You also need to be able to use tools while wearing your protective clothing.

Effective, efficient use of tools and equipment requires that you understand how the tool or device works and the best way to use it to its full potential while using the least amount of energy to accomplish the task. Being effective means that you achieve the desired goal; being efficient means that you produce the desired effect without wasting time or energy. When you are assigned a task on the fireground, your objective is to complete that task safely and quickly. If you waste energy by working inefficiently, you will not be able to perform additional tasks. Knowing which tool is the best one for the task at hand will help you achieve the desired objective. Knowing how to use the tool effectively, following the manufacturer's instructions, and using good technique will ensure you have the energy needed to complete other tasks. Let the tool do the work, don't force it, and pace yourself; exceeding your personal limits will prevent you from completing the required tasks. Many fire departments have standard operating procedures (SOPs) or standard operating guidelines (SOGs) that suggest which tools and equipment could be used in various situations. However, even if these SOPs or SOGs exist, you should understand which tool will allow you to accomplish the task in the most expeditious and efficient manner possible.

New personnel are often surprised by the strength and energy required to perform many tasks. An aggressive, continuous program of physical fitness and training in full gear will enable you to maintain your body in the optimal state of readiness. It does little good to practice using a tool for hours if you are not in good physical condition.

As a support person, you must know where every tool and piece of equipment is carried on your apparatus. Knowing how to use a piece of equipment does you no good if you cannot find it quickly. Your riding position (the seat you are sitting in) will often dictate your responsibilities and what tools you are expected to carry. Some personnel carry a selection of small tools and equipment in the pockets of their coats or pants. Check whether your department requires you to always carry certain tools and equipment, and ask senior firefighters for recommendations about which tools and equipment to carry.

General Carrying Tips

Tools can cause injuries even when they are not in use. Carrying a tool improperly can result in muscle strains, abrasions, or lacerations. Keep the following general carrying tips in mind when you need to carry tools:

- Do not try to carry a tool or equipment that is too heavy or designed to be used by more than one person. Request assistance from another crew member to help you carry it.
- Use your legs, not your back, when lifting heavy tools or equipment.
- Keep sharp edges and points away from your body at all times. Cover or shield these edges with a gloved hand to protect those around you.
- Carry long tools with the head of the tool down toward the ground. Be aware of overhead obstructions and wires, especially when using pike poles and ladders.
- When on the scene, do not leave tools lying on the ground or floor. Tools that are left lying around are tripping hazards. Return unused tools to the tool staging area or to their proper place on the apparatus.

Functions

Some tools are carried on all fire apparatus, whereas other tools are carried only by specific types of companies. A good way to remember the function of each tool is to categorize them by their function. Most of the tools carried by fire departments fit into the functional categories described in **TABLE 7-1**.

Rotating Tools

Tools that rotate things apply a rotational force to make something turn. Many rotating tools are basic

TABLE 7-1 Common Firefighter Tools Categorized by Function

Function	Tool
Rotate	Box-end wrench Gripping pliers Hydrant wrench Open-end wrench Pipe wrench Screwdriver Socket wrench Spanner wrench
Push/pull	Ceiling hook Clemens hook Drywall hook K tool Multipurpose hook Pike pole Plaster hook Roofman's hook Shove knife San Francisco hook
Pry/spread	Claw bar Crowbar Flat bar Halligan tool Hux bar Hydraulic spreader Kelly tool Pry bar Rabbit tool

Continues.

TABLE 7-1 Common Firefighter Tools Categorized by Function CONTINUED

Function	Tool
Strike	Battering ram Chisel Flat-head axe Hammer Mallet Maul Pick-head axe Sledgehammer Spring-loaded center punch
Cut	Axe Bolt cutter Chainsaw Circular saw Cutting torch Hacksaw Handsaw Hydraulic shears Reciprocating saw Rotary saw Seat-belt cutter

Courtesy of Guy Peifer.

hand tools. The rotating tools used most often in the fire service are screwdrivers, wrenches, and pliers, which are used to assemble or disassemble parts that are connected with threaded fasteners. Rotating tools are described in **TABLE 7-2** and shown in **FIGURE 7-1**.

Assembling and disassembling are basic mechanical skills that are routinely employed by firefighters to solve problems. Most fire apparatus carry a tool kit with a selection of open-end wrenches, box-end wrenches, socket wrenches, adjustable wrenches, pipe wrenches, and pliers in a variety of shapes and sizes for different applications. These tool kits also usually include a variety of sizes and types of screwdrivers, including slotted head, Phillips head, Roberts head, Torx, and others. A screwdriver with interchangeable heads is sometimes more useful than a selection of different screwdrivers. This type of screwdriver is carried by many in a pocket of their turnout pants.

Cutting Tools

Cutting tools have sharp edges that sever objects. They come in several forms and are used to cut a wide variety of materials. Firefighters use a wide variety of cutting tools, many of which are described in **TABLE 7-3**. Each of these tools is designed to work on certain types of materials. Remember, people can be injured and tools can be damaged if the tools are used incorrectly.

TABLE 7-2 Tools That Rotate Fasteners for Assembly and Disassembly

Tool	Description
Gripping pliers	A hand tool with a pincer-like working end. In addition to rotating things, gripping pliers can also be used to bend wire or hold smaller objects.
Screwdriver	A tool that is used to turn screws. Two common types of screwdrivers are flat-head screwdrivers, which insert into a screw that has a slot across its head, and Phillips-head screwdrivers, which insert into a screw that has a cross-shaped indentation on its head.
Wrench	A hand tool used to tighten or loosen bolts or turn pipes. Most wrenches come in several sizes.
Box-end wrench	A wrench with a closed end. Some box-end wrenches have a ratchet—a mechanism that rotates in only one direction—on the inside of the closed box end that makes it easier to use.
Open-end wrench	A wrench with an open end.
Combination wrench (**FIGURE 7-2**)	A wrench with a box end on one side and an open end on the other.
Adjustable wrench	An open-ended wrench with one fixed grip and one movable grip so that the opening can be adjusted to accommodate bolts of different sizes.

Tool	Description
Pipe wrench	An adjustable wrench that can be adjusted to fit securely around pipes and other tubular objects.
Hydrant wrench (**FIGURE 7-3**)	A wrench used to open or close a hydrant by rotating the **stem nut** on the top of the hydrant to start or stop the flow of water into the hydrant and to remove the caps from the hydrant outlets. Some hydrant wrenches have a ratchet.
Socket wrench	A wrench that has a socket that fits over a nut or bolt and has a ratcheting handle that you use to tighten or loosen the nut or bolt. You can swap out the socket with other sockets of different sizes.
Speed socket wrench	A long, curved handle with a socket wrench that fits over a nut or bolt at the end. To tighten or loosen the nut or bolt, the handle is rotated.
Spanner wrench	A wrench used to tighten or loosen a fire hose **coupling**, a connection device that you use to connect—or couple—individual lengths of fire hose together and to connect a hose to a fire hydrant or a pump or to nozzles and hose appliances.

Courtesy of Guy Peifer.

FIGURE 7-1 Hydrant wrench, two spanner wrenches, pipe wrench, combination wrench, gripping pliers, Phillips-head screwdriver, and speed socket wrench.

© Jones & Bartlett Learning

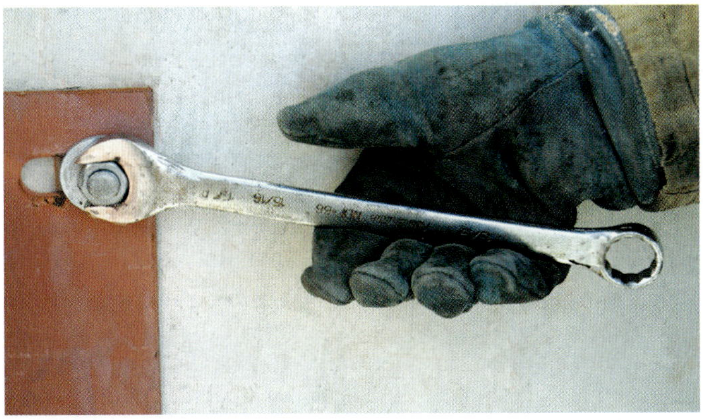

FIGURE 7-2 Using a combination wrench to rotate a bolt.

© Jones & Bartlett Learning

FIGURE 7-3 Using a hydrant wrench to rotate the stem nut on a hydrant to open or close the hydrant.

© Jones & Bartlett Learning. Photographed by Glen E. Ellman.

TABLE 7-3 Cutting Tools

Tool	Description
Axes	Axes are one of the most basic tools in the fire service. The axe head has a wide cutting blade that can be used to chop into a wall, roof, or door. Axe heads vary in weight from 4 to 8 pounds (lb; 1.8 to 3.6 kilograms [kg]), and handles range from 28 to 36 inches (in.; 71 to 94 centimeters [cm]). The weight, size, and type of axe you use depends on the job to be accomplished and your individual ability. Axes are often used to cut through chains or padlocks to open doors or gates. By concentrating the cutting force on a small area, it is possible to break through many chains in just a few seconds.
Flat-head axe (**FIGURE 7-4**)	This axe has a flat head opposite the blade that can be used for striking objects. It can also be used as a striking tool for **forcible entry**, usually in combination with a prying tool, such as a Halligan.
Pick-head axe (**FIGURE 7-5**)	This axe has a pick or a point opposite the blade that can be used for puncturing, pulling, and prying. It can also be used to establish a foothold while working on a sloped roof; for this reason, they are often carried by firefighters assigned to ladder trucks.
Bolt cutter (**FIGURE 7-6**)	This scissors-like hand tool is used to cut through items such as chains or padlocks or to cut through obstacles such as metal fences.
Cutting torch (**FIGURE 7-7**)	A tool that produces an extremely high-temperature flame capable of heating steel until it melts, effectively cutting it. Cutting torches are sometimes used for rescue situations and for cutting through heavy steel objects. These torches produce flames at extremely high temperatures—5700°F (3148°C)—so operators must be specially trained before using them. One drawback of this equipment is that it cannot be used in situations where flammable fuels are present.
Handsaw	A saw that is manually powered.
Carpenter's handsaw (**FIGURE 7-8**)	This saw is designed to cut wood. Saws with large teeth are effective in cutting large timbers or tree branches. Saws with finer teeth are designed for cutting finished lumber.
Coping saw	This saw is used to cut curves in wood. It consists of a handsaw with a narrow blade set between the ends of a U-shaped frame.
Hacksaw (**FIGURE 7-9**)	This saw is designed to cut metal. Different blades can be used, depending on the type of metal being cut. Hacksaws are useful when metal needs to be cut under closely controlled conditions.
Keyhole saw	This specialty saw is narrow and slender and is used to cut keyholes in wood and drywall.
Hydraulic shears (**hydraulic cutters**)	This tool cuts quickly through metal posts and bars. It is powered by a mechanical power source and is used along with hydraulic spreaders and rams to extricate victims from motor vehicles. The same hydraulic power source can be used with all three types of tools.
Power saws	Power saws are mechanically powered by electric motors or gasoline-powered engines, although battery-powered, cordless models of some types of power saws are available. Power saws can accomplish more work than a handsaw in a shorter period of time, allowing firefighters to conserve energy, resulting in less fatigue. Because power saws are dangerous, only trained operators should use them. Some disadvantages to power saws are that they are heavy to carry, can be difficult to start, and may require an electrical connection.
Band saw (**FIGURE 7-10**)	This electrically powered saw has a toothed metal blade stretched over two pulleys in a loop. Some people call it an *endless blade*. The blade cuts in only one direction. The benefit of using a band saw, unlike a reciprocating saw, is that there is no pulling or pushing of the object being cut and very little vibration. The cutting action is uniform with an even load distribution. This makes it ideal for victim extrication in special rescue situations, such as man-versus-machine or impalement injuries.

Tool	Description
Chainsaw	This saw has a gasoline-powered engine or electrically powered motor and is commonly used to cut wood, particularly trees. Firefighters often use special chainsaws called **ventilation saws** to cut **ventilation** openings in roofs constructed of wood, metal, tar, gravel, or insulating materials (**FIGURE 7-11**). Ventilation saws are specifically designed for roof ventilation and have different cutting chains than chainsaws used to cut wood. They also may have a depth gauge on the bar.
Reciprocating saw (**FIGURE 7-12**)	This saw is powered by either an electric or battery motor that rapidly pulls a saw blade back and forth. Different blades are used to cut different materials. Reciprocating saws can be used to cut metal during extrication of a victim from a motor vehicle.
Rotary saw (**FIGURE 7-13**)	This saw may be powered by either gasoline-powered engines or electric motors. Some rotary saws have a round metal blade with teeth. Different blades are used depending on the type of material being cut. Other rotary saws have a flat, abrasive disk made of composite materials designed to wear down as they are used. It is important to match the appropriate saw blade or disk to the material being cut.
Rescue knife	A spring-assisted folding knife that can be used with one hand. Often has a seat-belt cutter and a window breaker as part of the knife.
Seat-belt cutter (**FIGURE 7-14**)	A specialized tool that quickly cuts through a seat belt in a motor vehicle.
Wire cutter or **diagonal cutter**	A hand tool used to cut wire and small-diameter cable.

Courtesy of Guy Peifer.

FIGURE 7-4 Flat-head axe.

Courtesy of Sean Wilson.

SAFETY TIP

Any saw—or any tool—that could generate a spark should not be used in a flammable environment.

Pushing/Pulling Tools

Tools used for pushing and pulling are metal hooks on a head at the end of a pole (**FIGURE 7-15**). These tools extend the reach of the firefighter and increase the firefighter's mechanical advantage on an object.

The poles can be made of wood, fiberglass, or sometimes aircraft steel and come in various lengths from 2 to 12 feet (ft; 0.6 to 3.7 meters [m]). A 4- to 6-ft (1.2- to 1.8-m) pole enables a firefighter to stand on a floor and pull down a ceiling that is 8 to 10 ft (2.4 to 3 m) high. Poles 12 to 14 ft (1.2 to 4.3 m) long are used in rooms with very high ceilings.

The metal hooks attached to the end of the poles are available in many styles that are suited for different

A.

B.

FIGURE 7-5 **A.** Pick-head axe. **B.** A pick-head axe can be used to pry up boards.

A: Courtesy of Sean Wilson; **B:** © Jones & Bartlett Learning. Photographed by Glen E. Ellman.

applications. The different hook designs are intended for different types of ceilings and come in a variety of configurations. Many fire departments use one type of hook for plaster ceilings and another type for drywall ceilings. The end of the pole handles opposite the metal hook might have a D-shaped handle for better pulling power. **TABLE 7-4** describes several types of hooks used in fire department operations.

FIGURE 7-6 Bolt cutters.

Courtesy of Sean Wilson.

FIGURE 7-7 A cutting torch can be used to cut through a metal door.

© Jones & Bartlett Learning

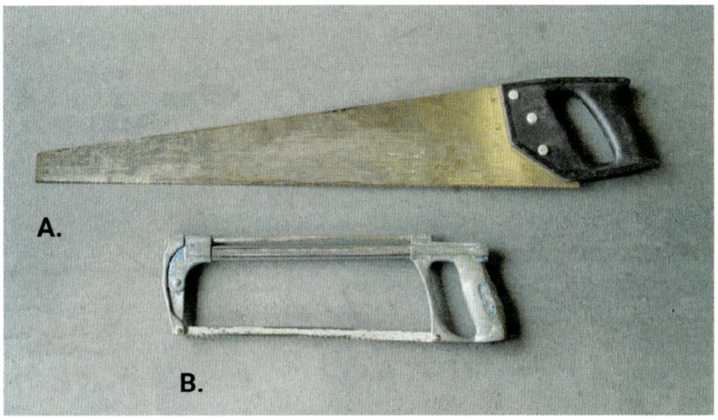

FIGURE 7-8 **A.** Carpenter's handsaw. **B.** Hacksaw.

© Jones & Bartlett Learning

FIGURE 7-9 A firefighter using a hacksaw to cut through metal.

© Jones & Bartlett Learning. Photographed by Glen E. Ellman.

FIGURE 7-10 A band saw cutting a metal pipe.

Courtesy of Stanley Black & Decker, Inc.

FIGURE 7-11 Firefighters using a ventilation saw to cut through a roof.

© Jones & Bartlett Learning. Photographed by Glen E. Ellman.

FIGURE 7-12 The components of a reciprocating saw.

© Jones & Bartlett Learning

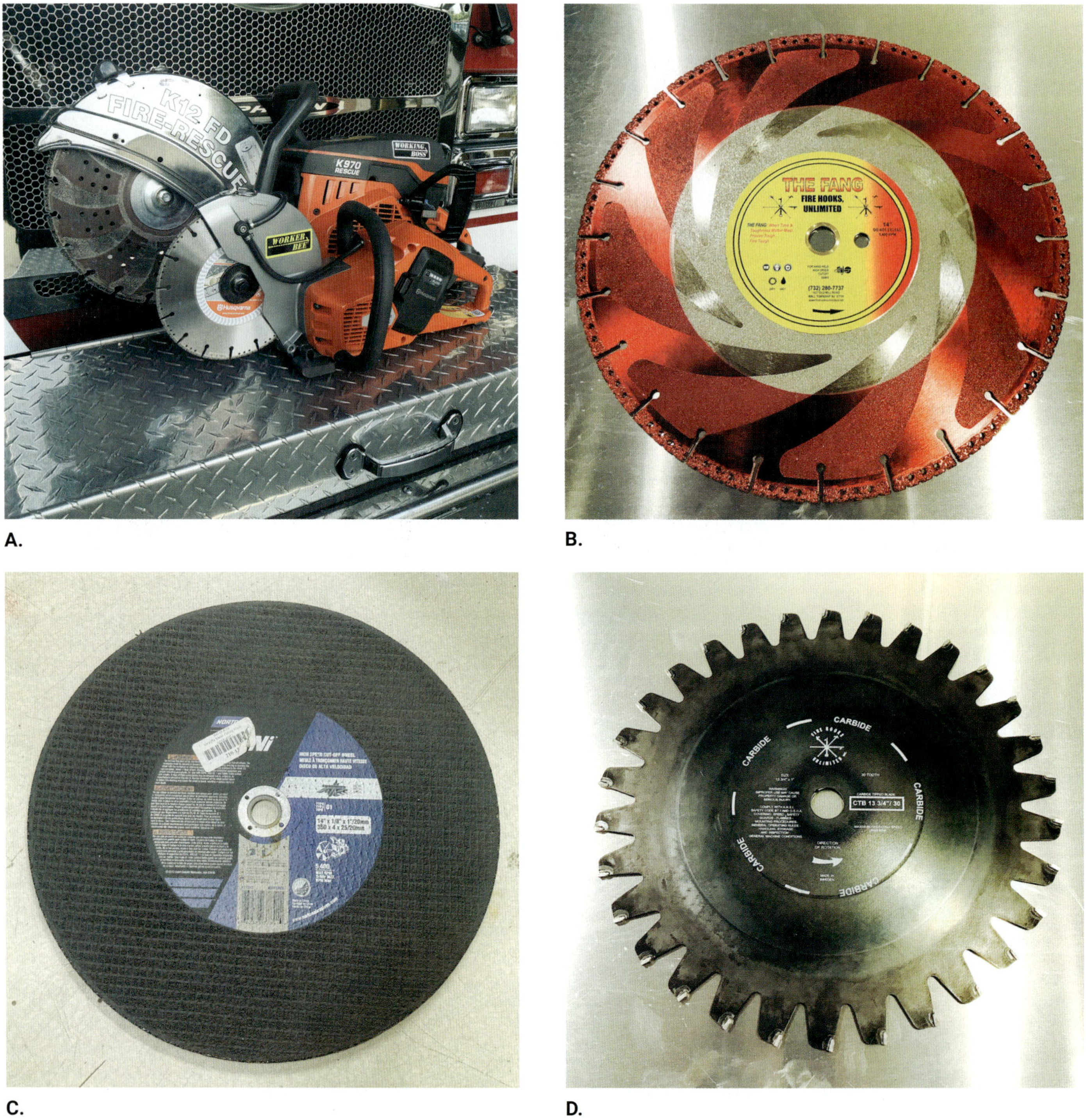

FIGURE 7-13 A rotary saw can cut through metal or wood, depending on the type of blade used. **A.** A rotary saw with a metal cutting blade. **B.** A diamond rescue disk. **C.** A composite abrasive disk. **D.** A carbide-tip disk.

Courtesy of Sean Wilson.

FIGURE 7-14 Seat-belt cutter.

© Jones & Bartlett Learning

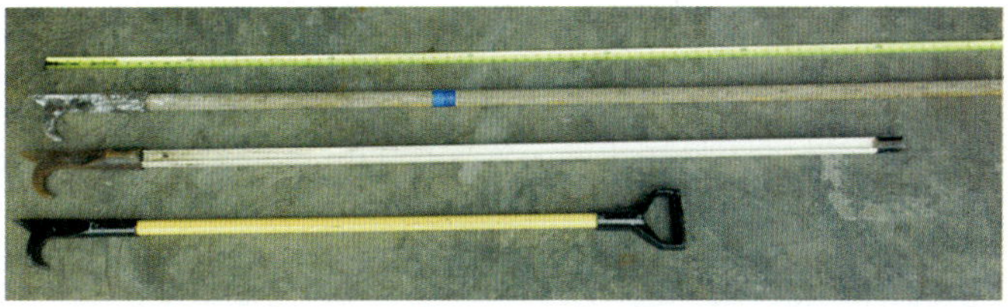

FIGURE 7-15 Hooks come in different sizes and with metal heads that are different shapes.

© Jones & Bartlett Learning. Photographed by Glen E. Ellman.

TABLE 7-4 Tools for Pushing and Pulling (Pike Poles and Hooks)

Tool	Description
Pike pole (**FIGURE 7-16**)	This all-purpose hook is one of the most versatile and common tools used by firefighters. The pole is made of wood or fiberglass, and the metal head has a sharpened point that can be used to punch through a ceiling and a hook that can grab and pull the ceiling down to get to the seat of a fire burning above or to locate an extension of the fire. It can also be used to break windows on a higher floor (**FIGURE 7-17**). Pike poles typically come in lengths of 4 to 12 ft (1.2 to 3.7 m). It is important to **size up** the building upon arrival and bring the right size pole inside. If the pole is too short, the firefighter will not be able to reach the ceiling. If it is too long, they will not be able to use it in a room with a low ceiling.
Ceiling hook	The pole on this tool is wood or fiberglass, and the metal head has a spur at right angles that can be used to probe ceilings and pull the material down to expose the wood behind it.
Clemens hook	A multipurpose tool that can be used for forcible entry and ventilation applications because of its unique head design.
Closet hook	This tool is usually 2 to 4 ft (0.6 to 1 m) long and is used in smaller, tight spaces such as closets and crawl spaces. It is also often used to overhaul upholstery.
Drywall hook	This specialized hook is designed to remove drywall, but it is also used on other materials. It has a honed head to allow for better penetration and a large, toothed head to rip away large areas of material. It is sometimes referred to as the ***universal hook***.
Multipurpose hook	A long pole with a wooden or fiberglass handle and a metal hook.
Plaster hook	A long pole with a pointed head and two retractable cutting blades on the side.
Roofman's hook or **New York hook**	A long pole with a solid metal hook on one end and a chisel on the other end that is often used to force open roof scuttles and hatches as well as pull ceilings and skylights.
San Francisco hook	A multipurpose tool used for forcible entry and ventilation applications. It includes a built-in gas shut-off, allowing firefighters to turn off the gas at the street.

Note: A pick-head axe can also be used to push and pull. Hydraulic spreaders can be used in combination with chains to pull a steering column or a dashboard or even to move a vehicle.

Courtesy of Guy Peifer.

FIGURE 7-16 Using a pike pole to push into a ceiling.

© Jones & Bartlett Learning. Photographed by Glen E. Ellman.

Prying/Spreading Tools

Tools used for prying or spreading may be as simple as a **pry bar** or as mechanically complex as a hydraulic spreader. They come in several sizes and with different features that are designed for different applications, as described in **TABLE 7-5**. Some of these tools are shown in **FIGURE 7-18**.

Striking Tools

Striking tools are used to apply an impact force to an object. They often are employed to gain entrance to a building or a vehicle or to make an opening in a wall or

FIGURE 7-17 Firefighter using a pike pole to break a window.

Courtesy of Steve Redick.

TABLE 7-5 Tools for Prying and Spreading

Tool	Description
Claw bar	A tool with a pointed claw hook on one end and a forked- or flat-chisel pry on the other end that can be used for forcible entry.
Crowbar	A straight bar made of steel or iron with a forked chisel on the working end.
Flat bar	A specialized prying tool made of flat steel with prying ends suitable for performing forced entry.
Halligan tool, **Halligan bar**, or simply **Halligan** (**FIGURE 7-19**)	A tool that incorporates three working ends, making its versatility in prying unmatched. One end of the tool is a bifurcated fork or claw that is rounded on one side; this rounded side is often referred to as the *bevel*. The other end of the tool has two working sides—a pick and an adze. The pick is round and tapers to a sharp point. The **adze** is a curved or straight wedge. Both the pick and the adze are set 90 degrees to the shaft of the tool. Although shorter and longer models are available, a standard Halligan is 30 in. (0.8 m) long. It is used to pry open doors or windows. It was designed in 1948 by New York City firefighter Deputy Chief Hugh Halligan (Long 2007). Many variations of this tool exist, and many names are used to describe it and its parts.

Tool	Description
Hux bar	A multipurpose tool that can be used for forcible entry and ventilation applications because of its unique design. A Hux bar can also be used as a hydrant wrench.
Hydraulic spreader (**FIGURE 7-20**)	A mechanically driven—that is, powered by gasoline, electricity, or batteries—rescue tool that enables you to apply several tons of force on a small area. You must have special training to operate these machines safely. Fire and rescue departments most commonly use this tool for extrication of victims from motor vehicles and machinery. Hydraulic spreaders are available as both full-size corded units used with a gasoline engine intended for vehicle extrication and as small battery-operated units that have changeable tips for extrication or forcible entry. Battery-powered hydraulic tools do not have hoses or generators, so rescue crews can quickly reach the vehicle involved, especially if it is far away from a roadway. Battery-powered tools also do not produce exhaust fumes, making the scene safer.
Kelly tool	A steel bar with two main features—a large pick and a large chisel or fork.
Rabbet tool (also called the **bunny tool**) and **Hydra-Ram** (**FIGURE 7-21**)	Manually powered hydraulic spreading tools used to pry open doors that swing inward by prying the door away from the door jamb at the point where the lock is located.

Courtesy of Guy Peifer.

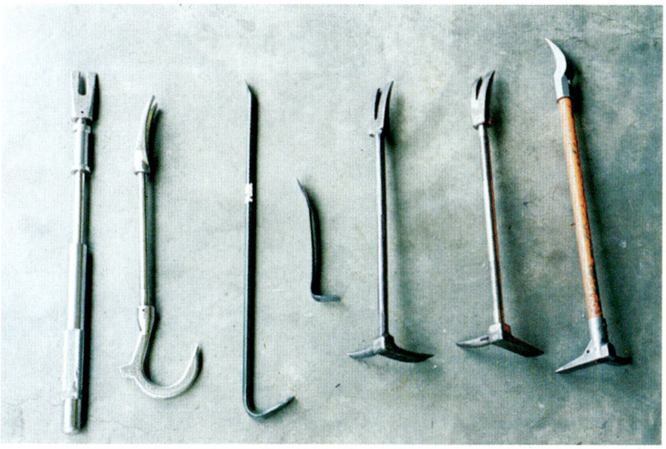

FIGURE 7-18 Tools used for prying and spreading.

© Berta A. Daniels, 2010

roof. This equipment also can be used to force the end of a prying tool into a small opening. Striking tools are described in **TABLE 7-6**, and some of them are shown in **FIGURE 7-22**.

Multiple-Function Tools

Certain tools are designed to perform multiple functions, thereby reducing the number of tools needed to achieve a goal. For example, a flat-head axe can be used as either a cutting tool or a striking tool. Some combination tools can be used to cut, to pry, to strike, and to turn off utilities (**FIGURE 7-23**). The **multi-tool** (**FIGURE 7-24**) is a compact, pocket-size, multiple-function tool that combines several different tools, such as a knife, scissors, wire cutters, pliers, and screwdriver.

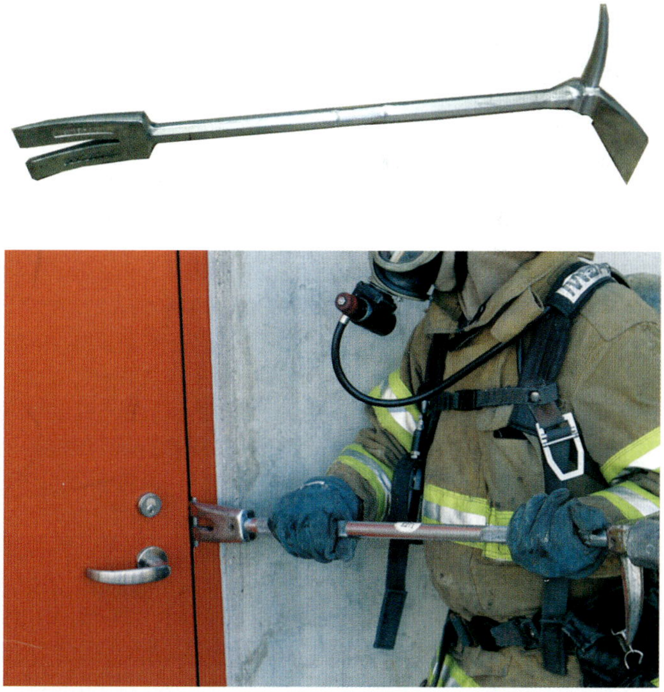

FIGURE 7-19 Using a Halligan tool to pry open a door.

© Jones & Bartlett Learning

A.

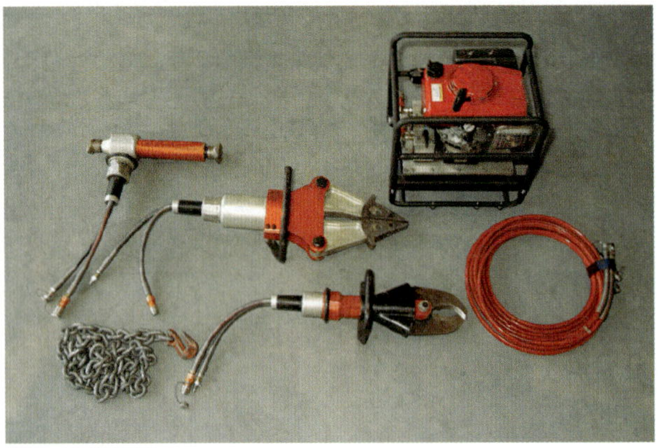

B.

FIGURE 7-20 **A.** Battery-operated hydraulic spreader. **B.** Components of a gasoline-powered hydraulic rescue tool.

A: Courtesy of Sean Wilson; **B:** © Jones & Bartlett Learning

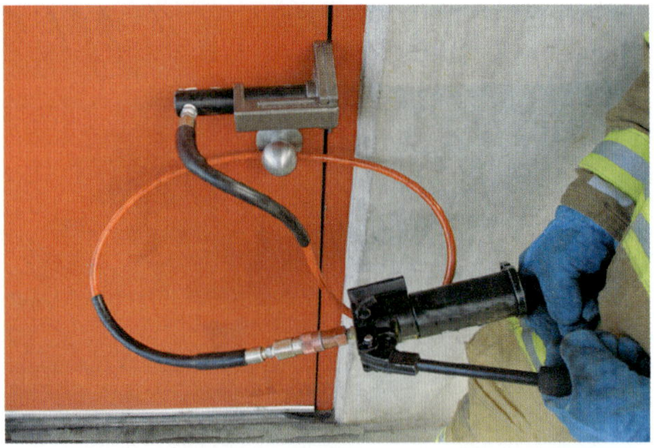

A.

B.

FIGURE 7-21 **A.** Rabbet tool. **B.** Hydra-Ram.

A: © Jones & Bartlett Learning; **B:** Courtesy of Sean Wilson.

TABLE 7-6 Striking Tools

Tool	Description
Battering ram	A heavy metal bar used to break down doors and breach walls. Battering rams come in different sizes and weights. Single-person battering rams are available, but most battering rams are made to be used by two to four people. Battering rams are more commonly used by law enforcement agencies than by fire departments.
Chisel	A metal tool with one sharpened end that can be used to break apart material when used in conjunction with a hammer, mallet, or sledgehammer.
Hammer	A hand tool constructed of solid material with a long handle and a head affixed to the top of the handle, with one side of the head used for striking and the other side used for prying.

Tool	Description
Mallet	A short-handled hammer with a round head.
Maul	A specialized striking tool that weighs 6 lb (3 kg) or more. It has an axe on one side of the head and a sledgehammer on the other side.
Sledgehammer	A hammer that can be one of a variety of weights and sizes. The head of the hammer can weigh from 2 to 20 lb (1 to 9 kg), and the handle may be short, like a carpenter's hammer, or long, like an axe handle.

Note: THE PIG and the flat-head axe can also be used to strike objects or as a striking tool for forcible entry.

Courtesy of Guy Peifer.

FIGURE 7-22 Striking tools (from top): hammer, maul, mallet, sledgehammer.

© Jones & Bartlett Learning. Photographed by Glen E. Ellman.

FIGURE 7-23 Breaching a wall with a multiple-purpose tool.

© Jones & Bartlett Learning. Photographed by Glen E. Ellman.

FIGURE 7-24 A multi-tool.

© Jones & Bartlett Learning

Another example of a multiple-function tool is THE PIG. Designed by a firefighter, it is a combination flat-head and pick-head axe. The pick head can be used for venting a roof (cutting), to pull roof materials, and for overhaul (pushing/pulling), and the flat-head axe blade is used for forcible entry (cutting and striking).

Special-Use Tools

Some fire situations require special-use tools that perform other functions. For example, fire departments located in areas where wildland fires occur frequently may need to carry fire rakes; firefighting brooms; shovels; and combination tools that can be used for raking, chopping, cutting, and leaf blowing.

Most apparatus also carry tools for forcible entry. One example is a **spring-loaded center punch**, which is used to break tempered glass in automobile windows and windows in some buildings (**FIGURE 7-25**). It exerts a large amount of force on a pinpoint-size portion of tempered glass. This disrupts the integrity of the tempered glass and causes the window to shatter into

FIGURE 7-25 A spring-loaded center punch can be used to break a car window safely.

© Jones & Bartlett Learning. Photographed by Glen E. Ellman.

FIGURE 7-26 Shove knife.

© Jones & Bartlett Learning. Photographed by Glen E. Ellman.

FIGURE 7-27 Components of a K tool.

Courtesy of Fire Hooks Unlimited.

FIGURE 7-28 Heavy-duty air bags can be used to lift vehicles in rescue situations.

© Jones & Bartlett Learning. Photographed by Glen E. Ellman.

small, uniform-sized pieces. Another forcible entry tool is a **shove knife**, which is used to release the latch on an outward-swinging door (**FIGURE 7-26**). Pulling the tool down and outward releases the locking mechanism of the door.

There are several types of forcible entry tools that pull lock cylinders so that the lock can be released. This "through the lock" method of forcing a door open minimizes damage to the door because it leaves the door and most of the locking mechanism undamaged. The building owner can have the lock cylinder replaced at a relatively low cost. For example, a **K tool** is a lock-pulling tool that is used with a Halligan and a mallet to pull out a lock cylinder mounted in a wood or heavy metal door (**FIGURE 7-27**).

Fire companies may also carry specialized equipment such as jacks and air bags for lifting heavy objects (**FIGURE 7-28**), a **come along** (a small, hand-operated winch) for lifting or moving heavy objects or bending objects, and tripods. You can learn more about the proper use of this special equipment by taking special rescue courses or during in-service training.

Lighting

Lighting is an important concern at many fires and other emergency incidents. Firefighters performing salvage and overhaul must be able to see where they are going; what they are doing; and which potential hazards, if any, are present. Lights can be portable or mounted on an apparatus. Many emergency incidents occur at night, so exterior lighting is required to illuminate the scene and enable safe, efficient operations. Inside fire buildings, heavy smoke can completely block out natural light. Even if the smoke has cleared, it can still be too dark to safely perform salvage and overhaul operations without additional light. Electrical service is often interrupted or disconnected for safety purposes, so lights mounted on an apparatus and portable lights should be set up to fully illuminate areas during salvage and overhaul. Most fire departments use different types of lighting equipment depending on the particular situation. A **spotlight**, for example, projects a narrow, concentrated beam of light. A **floodlight** projects a more diffuse light over a wide area. Lights can be portable or permanently mounted on fire apparatus.

Although the primary responsibility for fire-scene lighting usually belongs to truck companies or support units, any company can be assigned this duty. Lighting equipment can be mounted on any apparatus, and some fire departments equip special vehicles with high-powered lights to illuminate incident scenes. Departmental SOPs or SOGs, or the incident commander (IC) at the incident scene, will dictate the responsibility for setting up lighting.

Exterior lighting should be provided at all incident scenes during hours of darkness so that firefighters can see what they are doing, recognize hazards, and find victims who need rescuing. Exterior lighting also makes firefighters more visible to drivers who are approaching the scene or maneuvering emergency vehicles. Powerful exterior lights can often be seen through the windows and doors of a dark, smoke-filled building and can therefore guide disoriented firefighters or victims to safety.

Apparatus operators should turn on their apparatus-mounted floodlights and position them for maximum effectiveness so that as much of the area as possible is illuminated. The entire area around a fire building should be illuminated. If some areas cannot be covered with apparatus-mounted lights, use portable lights.

Portable lights can provide interior lighting at a fire scene. Position portable lights inside the building to illuminate interior areas as needed and as time permits. It is important to remember that smoke, in part, consists of particulates. Lighting cannot penetrate or cut through smoke. In many cases, smoke will diffuse light, causing a beam of light to be directed in many directions and angles as the beam bounces off the particulates. This can make it difficult to see the glow of a fire and actually decrease visibility as the room becomes "washed" in light.

When operations reach the salvage and overhaul phase, adequate interior lighting must be provided in all interior areas so that crews can work safely. Proper lighting enables firefighters to see what they are doing and observe any dangerous conditions that need to be addressed. Although exterior lighting is generally needed only at night, interior lighting may be needed even when it is a bright day outside.

Lighting Equipment

Portable lights can be taken into buildings to illuminate the interior. They can also be set up outside the structure to illuminate the fire or emergency incident scene (**FIGURE 7-29**). Portable lights usually range from 300 to 1500 watts and can use several types of bulbs, including quartz bulbs, halogen bulbs, and light-emitting diodes (LEDs).

The electricity for portable lights is supplied by a generator, an inverter, a building's electrical system, or a battery. Portable light fixtures are connected to generators and inverters with electrical cords. These electrical cords should be stored neatly coiled or on permanently mounted reels attached to the fire apparatus. The cord is then pulled from the reel to the place where the power is needed. Portable reels can be taken from the apparatus into the fire scene.

A **junction box** is a device that attaches to electrical cords to provide mobile power outlets. They are placed in convenient locations so that cords for individual lights and electrical equipment can be attached to them. Junction boxes used by fire departments are protected by waterproof covers and are often equipped with small lights so that they can be easily located in the dark.

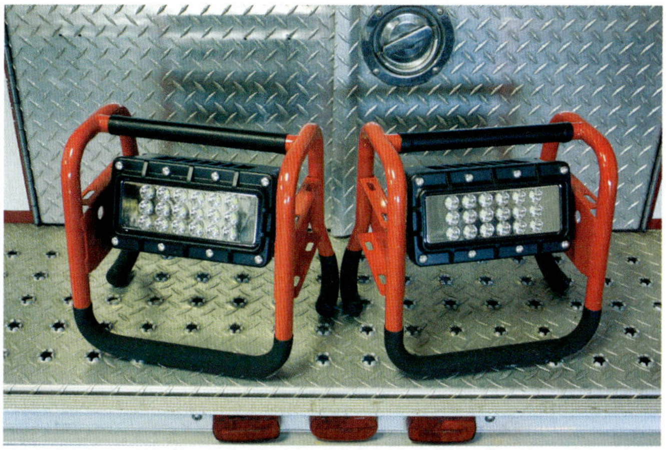

FIGURE 7-29 Portable emergency lights come in various sizes and levels of brightness.

Courtesy of Eric Scruggs.

FIGURE 7-30 A side apparatus-mounted light.

© Jones & Bartlett Learning. Photographed by Bill Larkin.

FIGURE 7-31 Hand lights manufactured for use by firefighters have light outputs ranging from 3500 to 75,000 candlepower.

© Berta A. Daniels, 2010

The connectors and plugs used for fire department lighting have special connectors that attach with a slight clockwise twist. This setup keeps the power cords from becoming unplugged during fire department operations. You should become familiar with the types of electrical connectors in use in your department, and practice making connections until you can connect the pieces of equipment in the dark.

Lights can be permanently mounted on fire apparatus to illuminate incident scenes (**FIGURE 7-30**). In this case, the vehicle operator can immediately illuminate the emergency scene simply by pressing the generator starter button. Some vehicle-mounted lights can be manually raised to illuminate a larger area. Mechanically operated light towers, which can be raised and rotated by remote control, can create near-daylight conditions at a night incident scene.

Battery-powered lights are lightweight, easily transported, do not require power cords, and can be used immediately (**FIGURE 7-31**). Battery-powered lights can provide either personal lighting or scene lighting. Personal lights are carried by the firefighter by hand or attached to their PPE. Scene lights are placed to illuminate a specific area. Battery-powered lights are powered by either disposable or rechargeable batteries, so they have a limited operating time before the batteries need to be recharged or replaced.

Large **hand lights** that project a powerful beam of light are often preferred for interior operations over smaller, personal flashlights. Hand lights offer increased candlepower over smaller, personal flashlights and can be equipped with a shoulder strap or hooked to a belt for easy transport. Every crew member entering a fire building should be equipped with a high-powered hand light.

A personal flashlight is another type of battery-operated light used by firefighters. Always carry a flashlight as part of your PPE. It should be rugged and project a strong light beam. It could be a lifesaver if your primary light source fails. Flashlights are not as powerful as larger hand lights, and they will not operate for as long on one set of batteries.

Safety Principles and Practices

The lighting and power equipment used at a fire scene that is not battery powered generally operates on 110-volt AC (alternating current), which is the same as standard household current. Some systems require higher voltage. All electrical cords, junction boxes, lights, and power tools must be maintained properly and handled carefully to avoid electrical shocks. In addition, all electrical equipment must be properly grounded. Electrical cords must be well insulated, without cuts or defects, and properly sized to handle the required amperage.

Generators should have a **ground-fault interrupter GFCI** to prevent a firefighter from receiving a potentially fatal electric shock. A GFCI senses when there is

SKILL DRILL 7-1

Illuminating an Emergency Scene Support Person, NFPA 1010: 5.3.6

1. Wear PPE. Depending on scene conditions, you may or may not wear self-contained breathing apparatus (SCBA).
2. Inspect all equipment while setting up.
3. Start the portable generator, engage the inverter, or check that there is electrical power in the building.
4. Connect cords, plug adaptors, and lighting equipment.
5. Ensure proper grounding and GFCI use.
6. Position lights so that the scene is adequately and safely illuminated.

Courtesy of Brandon Hausbeck.

a problem with an electrical ground and interrupts the current, shutting down both the power source and the equipment it is feeding.

Some portable generators are equipped with a grounding rod that must be inserted into the ground. Always use a grounding rod if one is provided. Avoid areas of standing or flowing water when placing power cords and junction boxes at a fire scene, and place electrical equipment on higher ground whenever possible.

Follow the steps in **SKILL DRILL 7-1** to illuminate an emergency scene.

Ladders

Ladders are undoubtedly one of the most versatile and valuable tools available to firefighters. Firefighters use ladders for a variety of tasks at an incident scene, including rescuing fellow firefighters, rescuing victims, gaining access to upper floors, and delivering a master stream. Ladders have many additional uses at an incident scene. Every firefighter must be knowledgeable about the many uses of ladders and be proficient in using them.

Uses of Ladders

At a fire scene, ground ladders are used to gain access to upper floors to search for possible victims, to remove victims, or to give firefighters emergency egress. They can provide access to and egress from a window, a cockloft, or an attic. Ladders can also provide a safe pathway between floors, enabling firefighters to avoid a damaged or unsafe stairway. Ladders may be used to gain access to a second-floor window or balcony or to provide stable footing and distribute the weight of firefighters during roof operations. They also can be deployed across a small space to create a bridge or placed to enable a firefighter to climb over a fence or obstruction. Sometimes the best escape route from a fire may be to travel horizontally across a ladder used as a bridge to a nearby building. During salvage operations, a ladder covered with a waterproof tarp can be used as a chute to funnel water out of a building.

Firefighters also use ground ladders in a wide variety of non-firefighting situations. A ground ladder can be used to reach an injured person in a ravine or down a steep highway embankment. When rescuing victims from a collapsed trench, firefighters use ground ladders to access a victim or to aid in bringing them to the surface. Ground ladders are also used to provide access to or escape from other below-grade locations, such as a maintenance hole (the opening to access utility equipment underground). Specialized rescue teams use ladders to provide access to and exit from buses, railcars, or airplanes. Ladders are also used for a variety of functions during a hazardous materials incident. Finally, ladders can be used as ramps to assist in moving equipment during emergencies.

Ladder Construction and Components

Ladders are constructed of metal, fiberglass, or wood. Most fire departments use ladders constructed of metal, usually aluminum, and a few departments use wood ladders. Fiberglass is less frequently used in the fire service. Each material has advantages and disadvantages (**TABLE 7-7**). Following the manufacturer's instructions and cautions for using your ladders is essential.

Ladders must meet the performance requirements described in NFPA 1960, *Standard for Fire Hose Connections, Spray Nozzles, Manufacturer's Design of Fire Department Ground Ladders, Fire Hose, and Powered Rescue Tools, 2024 Edition.* This standard requires manufacturers to adhere to basic requirements such as specifications for the diameter and spacing of rungs; the inclusion of specific labels and markings, including ladder positioning stickers, heat-sensor labels, and electrocution hazard warning labels; and length and width requirements.

All ladders have two beams connected by a series of parallel rungs. The **beam** is the main structural component that runs the length of a ladder. A **rung** is the crosspiece between the two beams, which serves as a step. Per NFPA 1960, the surface area of rungs is required to be skid resistant.

TABLE 7-7 Advantages and Disadvantages of Ladder Materials

Material	Advantages	Disadvantages
Metal	▪ Relatively light in weight ▪ Inexpensive ▪ Easy to repair	▪ Conducts electricity ▪ Subject to failure when exposed to high heat
Wood	▪ Less likely to conduct electricity ▪ Retains more strength when exposed to heat ▪ More flexible than metal ladders	▪ Higher cost ▪ Requires more maintenance ▪ Heavier weight ▪ Combustible
Fiberglass	▪ Less likely to conduct electricity ▪ Increased strength	▪ Increased chance of failure if overloaded ▪ May be combustible

© Jones & Bartlett Learning

There are three main types of ladders, characterized by the type of beam (**FIGURE 7-32**):

- **Trussed beam ladder**: The beams on a trussed beam ladder consist of a top and bottom **rail** connected by a smaller piece called a **truss block**. The rungs are attached to the truss blocks. Trussed beam ladders are usually constructed of aluminum or wood. Longer extension ladders are usually trussed beam ladders because this type of construction results in a lighter-weight ladder.
- **I-beam ladder**: The beams on an I-beam ladder are continuous pieces shaped like an uppercase letter *I*. The top and bottom of the I-beam are a **flange**, and the part between the flanges is the **web**. The rungs are attached to the web. In the fire service, this type of beam is usually made from fiberglass but may also be constructed of metal.
- **Solid beam ladder**: The beams on a solid beam ladder are rectangular-shaped, continuous pieces. Many wood ladders have solid beams. Fiberglass and metal ladders can also have solid beams. Hollow or C-shaped rectangular aluminum beams are also classified as solid beams.

To use and maintain ladders properly, firefighters must be familiar with all the parts of a ladder and the terms used to describe those parts (**FIGURE 7-33**):

- The **tip** is the very top of the ladder.
- The **butt** is the end of the ladder that is placed against the ground when the ladder is raised. It is sometimes called the **heel** or **base**.
- **Butt spurs** are metal spikes attached to the butt of a ladder to prevent the butt from slipping out of position (**FIGURE 7-34**). A **butt plate**, an alternative to butt spurs, is a flat plate attached to the butt of the ladder that swivels to maintain contact with the ground and has spurs on the edges of the plate and cleats on the bottom of the plate.
- The **heat-sensor label** changes color when the ladder is exposed to the amount of heat that could damage the structural integrity of the ladder. They are applied to metal and fiberglass ladders. Heat sensors activate when exposed to a temperature of 300°F (149°C) plus or minus 5 percent. They have an expiration date, at which point the sensor needs to be replaced (**FIGURE 7-35**).
- **Protection plates** are reinforcing pieces on metal and fiberglass ladders that are placed on a ladder at chafing and contact points to prevent damage

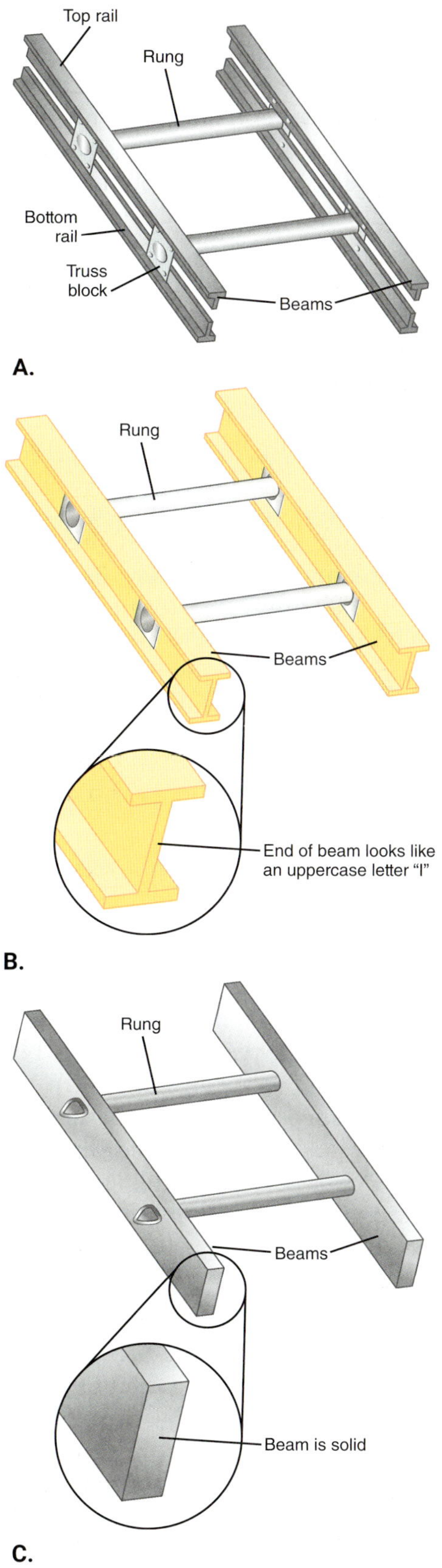

FIGURE 7-32 Three types of beam construction are found in ladders: **A.** Trussed beam. **B.** I-beam. **C.** Solid beam.

© Jones & Bartlett Learning

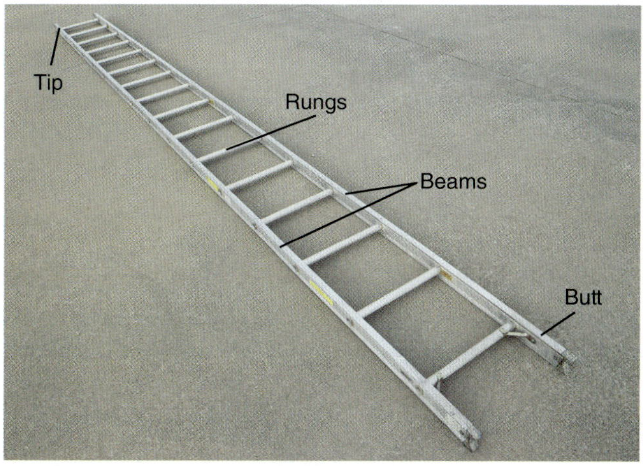

FIGURE 7-33 The basic components of a ladder.

© Jones & Bartlett Learning. Photographed by Glen E. Ellman.

FIGURE 7-34 Butt spurs dig into the ground to prevent the ladder from slipping.

Courtesy of Kevin Lewis.

to the ladder from friction or contact with other surfaces.

- A **tie rod** is a metal bar under each rung of some wood ladders. Tie rods run from one beam to the other and help keep the beams from separating from the rungs.

Ground Ladders

A **fire department ground ladder**, or simply a **ground ladder**, is a portable ladder carried on engines

FIGURE 7-35 A heat-sensor label indicates whether the ladder has been exposed to excessive heat.

Courtesy of Kevin Lewis.

and removed to be used as needed. Most fire apparatus carry an assortment of ground ladders. Types of ground ladders include the following:

- Straight ladder
- Roof ladder
- Extension ladder
- A-frame ladder
- Combination ladders
- Folding ladder

Straight Ladder

A **straight ladder**, also called a **single ladder** or a **wall ladder**, is a single-section, fixed-length ground ladder. These lightweight ladders can be raised quickly, and they can reach the windows and roofs of one- and two-story structures. Straight ladders are commonly 12 to 14 ft (4 to 4.2 m) long but can be as long as 20 ft (6 m).

Roof Ladder

A **roof ladder**, sometimes called a **hook ladder**, is a straight ground ladder that is equipped with roof hooks at one or both ends. **Roof hooks** are spring-loaded, retractable, curved metal pieces attached to the beams at the tip or at both ends of a roof ladder to secure the tip of the ladder to the peak of a pitched roof when the ladder lies flat on the roof. If a straight ladder has roof hooks on both ends, butt spurs are also on both ends.

Roof ladders provide stable footing and distribute the weight of firefighters and their equipment, thereby helping to reduce the risk of structural failure in the roof assembly. They can also be used on flat roofs to distribute weight. In addition to being used on roofs, roof ladders can be used in situations where a straight ladder is needed. Roof ladders can also be used for access to the second floor or a balcony.

Roof ladders in the fire service are usually 12 to 20 ft (4 to 6.6 m) long (**FIGURE 7-36**). Roof ladders are not designed as free-hanging ladders, meaning that they are not specifically designed to support weight when hanging exclusively from the hooks. Nevertheless, they are tested to make sure they can support 1000 pounds (lb; 454 kilograms [kg]) for 1 minute when hanging only by their hooks per NFPA 1932, *Standard on Use, Maintenance, and Service Testing of In-Service Fire Department Ground Ladders, 2020 Edition.*

Extension Ladder

An **extension ladder** is an adjustable-length ground ladder with two or more connected sections that can be extended or retracted to adjust the length of the ladder (**FIGURE 7-37**). An extension ladder is usually heavier than a straight ladder of the same length and normally requires more than one person to set it up. Because the length is adjustable, an extension ladder can replace several straight ladders, and it can be stored in places where a longer straight ladder would not fit.

Extension ladders have additional parts that are not found on straight ladders (**FIGURE 7-38**):

- The **bed section** is the widest section of an extension ladder. It is also called the **base section** because it serves as the base of the ladder—all other sections are raised from the bed section. The butt of an extension ladder is at the bottom of the bed section.
- The bed section supports one or more fly sections. A **fly section** is raised or extended from the bed section or from the previous section if the extension ladder has more than one fly section. When the fly section is not extended—that is, it is all the way down—it is called *fully bedded.* An extension ladder with a bed and one fly section is a two-section ladder, one with a bed and two fly sections is a

FIGURE 7-36 The main use of roof ladders is to distribute weight on a roof. Roof hooks on a straight ground ladder mean that the ladder can be used as a roof ladder.

© Jones & Bartlett Learning. Photographed by Glen E. Ellman.

FIGURE 7-37 An extension ladder can be used at any length, from fully retracted to fully extended.

© Jones & Bartlett Learning. Photographed by Brandon Fryman/JRM Creative.

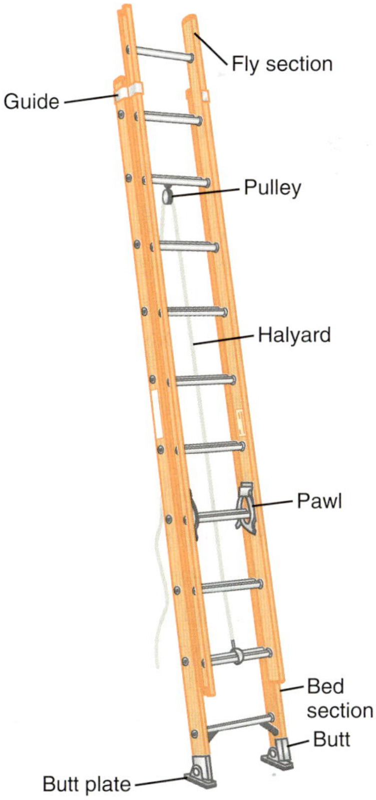

FIGURE 7-38 Extension ladders have additional parts that are not found in straight ladders.

© Jones & Bartlett Learning

three-section ladder, and one with three fly sections is a four-section ladder. (Straight ladders are single-section ladders.)

- The **halyard** is a rope or cable attached to the fly sections that runs through a pulley (or multiple pulleys) near the top of the bed section and down to the ground so that it can be pulled to hoist the fly sections. On extension ladders with three and four sections, wire-rope halyard extensions are attached to the halyard so that the last fly section can be extended fully.

- A **guide** is a strip of metal, wood, or fiberglass that wraps around a fly section and the section below it or slots in the fly section or the section below it to guide the fly section as it is being extended. Some manufacturers recommend that wax be applied between the guides and the beams to ensure smooth operation when extending and collapsing the sections.
- A **pawl** is a mechanical locking device used in pairs to secure and hold extended fly sections in place. They are sometimes called a **ladder lock**, a **dog**, or a **rung lock**.
- A **stop** is a piece of metal, wood, or fiberglass that prevents the fly sections of a ladder from overextending and collapsing the ladder. This is also referred to as a **stop block**.

Fresno Ladder. A **Fresno ladder** is a narrow, two-section extension ladder specifically designed to provide attic access and to be used in any tight space. For example, a Fresno ladder is used as a bridge over a damaged section of an interior stairway. A Fresno ladder is generally short—just 10 to 14 ft (3 to 4 m)—so it has no halyard and is extended manually (**FIGURE 7-39**).

Bangor Ladder. A **Bangor ladder**, also called a **tormentor ladder**, is an extension ladder with a staypole. A **staypole**, also called a **tormentor**, is a long metal pole attached with a swivel joint to the beams at the top of the bed section of the ladder. Each pole has a spur, similar to a butt spur, on the other end. The staypoles are positioned at approximately 45-degree angles from the ladder opposite the climbing side to help stabilize the ladder during raising, lowering, and climbing operations. When the ladder is positioned correctly, the staypoles are planted in the ground on either side for additional stability. The poles help keep these heavy ladders under control while firefighters are maneuvering them into place (**FIGURE 7-40**). Staypoles are required on ladders that are 40 ft (12 m) or longer, but they are sometimes found on some 35-ft (11-m) ladders.

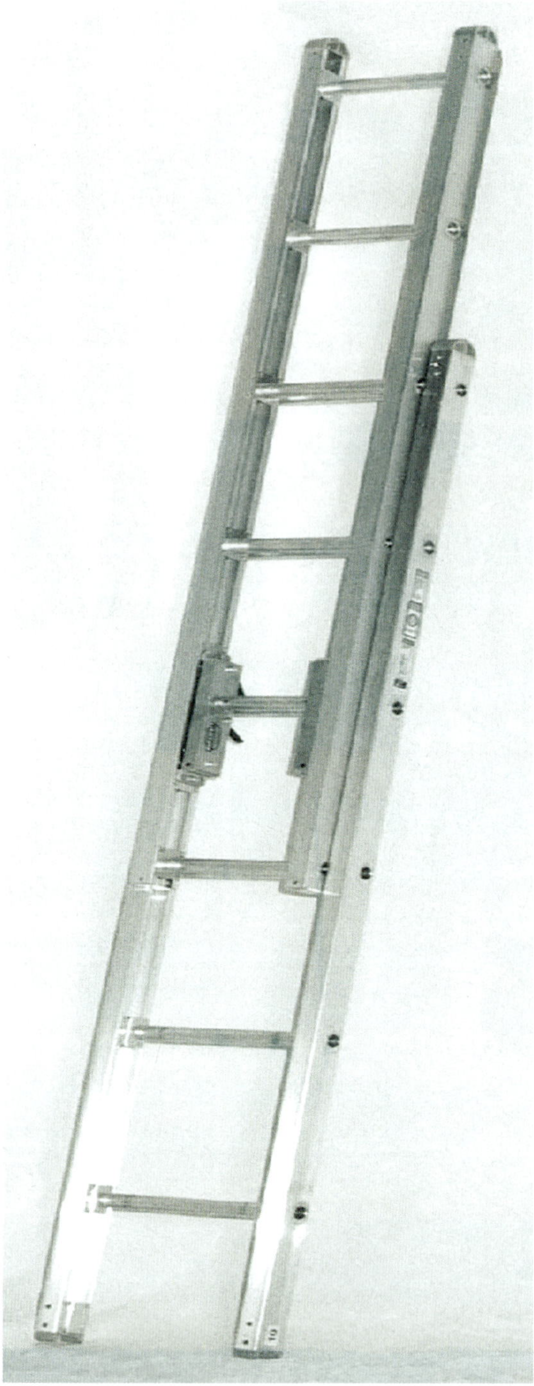

FIGURE 7-39 Fresno ladders are usually short and are extended manually rather than with a halyard.

Courtesy of Duo-Safety Ladder Corporation.

A-Frame Ladder

An **A-frame ladder** is a ladder that has two ladder sections connected with a joint so that it forms an A-shaped structure when it is set up to climb on and closes to fold flat for storage. A common example of an A-frame ladder is a stepladder. It can be used to lower rescuers into a trench or maintenance hole or to raise or lower victims safely. A **ladder A-frame**, sometimes referred to as an **A-frame hoist**, is formed by using rope to attach two straight ladders at the tip in an A shape. A ladder A-frame can be used as a makeshift lift when raising a trapped person.

FIGURE 7-40 Staypoles stabilize a Bangor ladder while it is being raised or lowered.

Reproduced from The Connecticut Fire Academy. 2018. "Practical Skill Training: Skill Sheet 13.4.3: Ground Ladders." Updated March 5, 2018. https://portal.ct.gov/-/media/CFPC/files/NEW-ITEMS-2019/Uploaded-Files/Instructor-Lesson-Plans/Uploaded-Files/Unit-13/SS-1343-Instructor-Reference-Material.pdf

FIGURE 7-41 Combination ladders can be used in several different configurations.

© Jones & Bartlett Learning

Combination Ladder

A **combination ladder**, also called a **multipurpose ladder**, can be used as a stepladder (A-frame ladder) or an extension ladder. Such ladders are convenient for indoor use and for maneuvering in tight spaces. They are generally 6 to 10 ft (2 to 3 m) in length in the A-frame configuration and 10 to 15 ft (3 to 5 m) in length in the extension configuration. A multipurpose ladder can be used as a stepladder or a straight ladder (**FIGURE 7-41**).

Folding Ladder

A **folding ladder**, also called an **attic ladder**, is a narrow, collapsible ladder that is designed to allow access to attic scuttle holes and confined areas (**FIGURE 7-42**). The two beams fold in to enhance the ladder's portability. Folding ladders are commonly available in 8- to 14-ft (2- to 2.4-m) lengths.

Maintenance

All tools and equipment must be properly maintained so that they will be ready for use when they are needed. This means that a tool or other equipment should be in proper working order, in its proper storage location, and ready for immediate use. Every crew member should be able to locate the right tool immediately and be confident that it is ready for use. You must keep equipment clean and free from rust, keep cutting blades sharpened, and keep fuel tanks filled and batteries charged. A tool that is improperly maintained can be dangerous to both the operator and other nearby persons. Every tool and piece of equipment must be ready for use before you respond to an emergency incident. Research and follow the manufacturer's instructions for cleaning, inspecting, and maintaining each piece of equipment. After cleaning and inspecting tools, place them

A. B.

FIGURE 7-42 Folding ladders are designed to be used in narrow spaces or restrictive passages. **A.** A folding ladder in the portable position. **B.** A folding ladder unfolded and ready to use.

A. © Jones & Bartlett Learning. Photographed by Glen E. Ellman; **B.** Courtesy of Kevin Lewis.

back in their proper location, ready for use. If a tool is damaged or inoperative, immediately tag it, take it out of service, document the damage following department procedures, and report it to your supervisor. Generally, tools should be inspected after every use. In addition, thorough, conscientious daily or weekly checks of these items should be performed, particularly with infrequently used tools.

Cleaning and Inspecting Hand Tools

All hand tools should be cleaned completely and inspected, and their conditions should be documented after each use. Remove all dirt, debris, and rust from the tools. If appropriate, use water streams to remove the debris and mild soap to clean the equipment thoroughly. Learn how to safely use the cleaning solutions that your department and the manufacturer specify for cleaning tools and equipment.

Before any tool is placed back into service, it should be inspected for damage. Avoid painting tools because paint may hide defects or visible damage. If painting of a tool is required for identification, do *not* paint axe heads. Painting an axe head hides defects and stress fractures in the metal and may cause the axe head to stick and bind when using it. Keep the number of markings on a tool to a minimum. If an axe needs to be sharpened, it is best to sharpen it by hand with a file. If the axe head edge is sharpened so that the edge is thin, the axe will be prone to chips and cracks.

To clean and inspect hand tools, follow the steps in **SKILL DRILL 7-2**.

Cleaning and Inspecting Power Tools

Just like hand tools, all power tools should be cleaned completely, inspected, and their conditions documented after each use. Remove all dirt and debris from

SKILL DRILL 7-2

Cleaning and Inspecting Hand Tools Support Person, NFPA 1010: 5.5.2

1. Clean and dry all metal parts. Metal tools must be dried completely, either by hand or by air, before being returned to the apparatus. Remove rust with steel wool. Coat unpainted metal surfaces with a light film of lubricant to help prevent rusting. Do not oil the striking surface of metal tools because this treatment may cause them to slip.
2. Inspect wood handles for damage, such as cracks and splinters. Repair or replace any damaged handles. Sand the handle if necessary. Do not paint or varnish a wood handle; instead, apply a coat of boiled linseed oil. Check that the tool head is tightly fixed to the handle.
3. Clean fiberglass handles with soap and water. Inspect for damage. Repair or replace any damaged handles. Check that the tool head is tightly fixed to the handle.
4. Inspect cutting edges for nicks or other damage. Cutting tools should be sharpened after each use. File and sharpen as needed. Power grinding may weaken some tools, so hand sharpening may be required.

© Jones & Bartlett Learning. Photographed by Glen E. Ellman.

the tools. You will likely have to disassemble the tool somewhat to clean it correctly. This is especially true of power saws. If appropriate, use an approved solvent and parts washer to remove the debris and clean the equipment thoroughly. Learn how to safely use the cleaning solutions that your department and the manufacturer specify for power tools and equipment. While cleaning cutting tools, inspect the blade and make sure it is ready for use.

When tools powered by a rechargeable battery are used, make sure the battery is replaced and the used battery is put on a charger. Ensure that tools powered by gasoline engines are refueled with the correct fuel. Some use straight gasoline, whereas others use a 50:1 or 100:1 mixture of gasoline and oil. If a gasoline-powered tool is not likely to be used in the next 12 to 15 days, add a fuel stabilizer. If it is expected to be out of service for more than 15 days, its fuel should be drained and stored in a separate container. To avoid these problems that can occur with ordinary gasoline, many fire agencies use commercial fuel specifically designed for power and emergency equipment. These fuels do not contain ethanol and evaporate cleanly, leaving no residue, which allows a fire agency to keep its tools permanently fueled.

It is poor practice to place a power tool back on the apparatus without cleaning it; ensuring that it is fueled or charged; and ensuring that any parts that need replacement are replaced, such as a chain or blade from a rescue saw.

Cleaning, Inspection, Maintenance, Service Testing, and Storing Ladders

Ground ladders used by the fire service must be able to withstand extreme conditions. They might be dropped, overloaded, or exposed to temperature extremes during use. Ladders must meet the performance requirements described in NFPA 1960. These requirements are based on probable conditions during emergency operations. NFPA 1932 provides general guidance for the use of ground ladders. Ladders must be regularly inspected, maintained, and service tested, following the NFPA standard. In addition, ladders should always be inspected and maintained in accordance with the manufacturer's recommendations.

Cleaning

Ladders must be cleaned regularly to remove any road grime and dirt that have built up on the apparatus during storage. In addition, ladders should be cleaned after each use and before each inspection to ensure that any hidden faults can be observed.

Use a soft-bristle brush and water to clean ladders. A mild, diluted detergent may be used, if allowed by the manufacturer's recommendations. Remove any tar, oil, or grease deposits with a safety solvent as recommended by the manufacturer. Do not get any solvent on the halyard of an extension ladder because contact with solvents can damage halyard ropes. Rinse and dry the cleaned ladder before placing it back on the apparatus.

Inspection and Maintenance

Ladders must be visually inspected at least once per month. Ladders should also be inspected after each use. Ladders are easily damaged while in use. For example, a ladder might be overloaded or used at a low angle during a rescue, or an unexpected shift in fire conditions could bring the ladder in direct contact with flames. Whenever a ladder is used outside of its recommended limits, it should be taken out of service for inspection and testing, even if there is no visible damage. If an inspection reveals any deficiencies, the ladder must be removed from service until repairs are made. All inspections should be recorded according to your department's SOPs.

Maintenance is simply the regular process of keeping the ladder in proper operating condition. All firefighters should be able to perform routine ladder maintenance, such as replacing a halyard or an expired heat-sensor label. But only qualified personnel at a properly equipped repair facility should perform repairs involving the structural or mechanical components of a ladder. As with inspections, all maintenance should be recorded according to your department's SOPs.

TABLE 7-8 describes things to look for during a ladder inspection and the actions you can take to correct deficiencies and maintain the ladders.

TABLE 7-8 Ladder Inspection and Maintenance

Component	What to Look For	Action to Take If a Deficiency Is Found
All ladders		
Beams	Check the beams for cracks, splintering, breaks, gouges, checks, wavy conditions, or deformations.	Immediately place out of service
Rungs	Check all rungs for snugness, tightness, punctures, wavy conditions, worn serrations, splintering, breaks, gouges, checks, or deformations.	Immediately place out of service
Butt spurs and butt plates	Check the butt spurs and butt plates for excessive wear or other defects.	Immediately place out of service
Heat-sensor labels	▪ Check to see if the sensor has expired. ▪ Check to see if the sensors indicate that the ladder has been exposed to excessive heat.	▪ If the heat-sensor labels have reached their expiration date, replace them. ▪ If the heat-sensor label indicates that a ladder has been exposed to excessive heat, remove the ladder from service and conduct a service test.
Bolts and rivets	Check all bolts and rivets for tightness. Bolts on wood ladders should be snug and tight without crushing the wood.	Immediately place out of service

Component	What to Look For	Action to Take If a Deficiency Is Found
Welded joints	Check all welded joints on metal ladders for cracks or defects.	Immediately place out of service
Metal surfaces	Check metal surfaces for signs of surface corrosion.	Immediately place out of service
Fiberglass surfaces	Check fiberglass ladders for loss of gloss on the beams.	Maintain the finish in accordance with the manufacturer's recommendations. This should be done on a regular basis according to the manufacturer's recommendations in addition to when an inspection reveals the need.
Wood surfaces	▪ Check for damage to the varnish finish on wood ground ladders. ▪ Check for signs of wood rot—that is, dark soft spots.	▪ Maintain the finish in accordance with the manufacturer's recommendations. This should be done on a regular basis according to the manufacturer's recommendations in addition to when an inspection reveals the need. ▪ If there are signs of wood rot, take the ladder out of service.
Painted surfaces	Because paint can hide structural defects in a ladder, ensure that ladders are ***not*** painted except for the top and bottom 18 inches (in.; 0.47 m) of each section, which are painted for purposes of identification and visibility.	▪ Remove the paint.
Roof ladders (**FIGURE 7-43**)	▪ Check the roof hooks for sharpness. ▪ Check that the roof hooks operate properly. ▪ Check that the roof hooks do not have rust and other contaminants. ▪ Check for sufficient lubrication.	▪ Take out of service and replace. ▪ Take out of service and replace. ▪ Remove rust and other contaminants. ▪ Lubricate roof hooks.
Extension ladders		
Halyards (**FIGURE 7-44**)	Check the halyard for kinking and fraying and ensure that it moves smoothly through the pulleys.	Replace the halyard if it is kinked, frayed, or otherwise worn or damaged.
Wire-rope halyard extensions	Check the wire-rope halyard extensions on three- and four-section ladders for snugness. This check should be performed when the fly sections are fully bedded to ensure that the upper sections will align properly during operation.	
Guides	▪ Check the guides for chafing. ▪ If the manufacturer requires wax, check for adequate wax between the guides and the beams.	▪ Lubricate the guides in accordance with the manufacturer's recommendations.
Pawls	Check the pawls for proper operation.	Lubricate the pawls following the manufacturer's instructions.

Courtesy of Kevin Lewis.

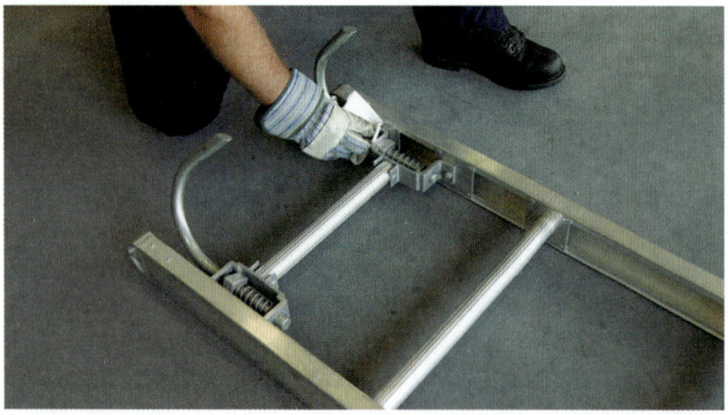

FIGURE 7-43 Roof hooks must operate smoothly.

© Jones & Bartlett Learning

FIGURE 7-44 Extension ladder pawls must operate smoothly.

© Jones & Bartlett Learning. Photographed by Glen E. Ellman.

Storage

When ladders are not in use, they should be placed on racks or in brackets and protected from the weather. Fiberglass ladders can be damaged by prolonged exposure to direct sunlight (**FIGURE 7-45**).

To clean, inspect, maintain, and store ladders, follow the steps in **SKILL DRILL 7-3**.

FIGURE 7-45 Ladders should be stored on racks or in brackets, out of the weather or direct sunlight.

© Jones & Bartlett Learning

SKILL DRILL 7-3

Clean, Inspect, Maintain, and Store a Ladder Support Person, NFPA 1010: 5.5.2

1. Clean all components following national standards and the manufacturer's recommendations. Visually inspect each component of the ladder for wear and damage.

SKILL DRILL 7-3 CONTINUED

Clean, Inspect, Maintain, and Store a Ladder Support Person, NFPA 1010: 5.5.2

2. On extension ladders, lubricate the ladder pawls, guides, and pulleys using the recommended material and according to the manufacturer's recommendations.

3. Perform a functional check of all components.

4. If deficiencies are found and cannot be addressed at the fire station, tag and remove the ladder from service, and if needed, schedule the ladder to go to a repair facility. If the ladder passes inspection or is repaired at the fire station, return the ladder to the apparatus or to the storage area. Complete the maintenance record for the ladder.

© Jones & Bartlett Learning. Photographed by Glen E. Ellman.

CASE STUDY

You Are the Support Person CONCLUSION

Upon reporting for duty, you are notified that you have been detailed to another station to work with the engine company for the tour. This will be your first time working with an engine company, and you are excited about the opportunity. It isn't long before an alarm sounds. "Person trapped, motor vehicle collision, Fourth and Elm," directs the dispatcher. You slip into your gear, take your seat, and put on your seat belt. As the engine pulls out from the station, you think about how this changes your plan for what you thought you would be doing today.

1. **How will your role be different for this call?**

   Answer: Even though you are not responding to a fire, safety should still be a high priority. Traffic is a major safety concern, and you need to ensure that you do not exit the fire engine into oncoming traffic. The perimeter of a safe work zone must be established and maintained throughout the duration of the call.

2. **What kinds of tools might be needed for this job?**

   Answer: Which tools will be needed depends on the severity of the accident, the types of vehicles involved, and many other scene variables. It is important that all tools on the apparatus be well maintained and ready to go because you never know when they will be needed.

3. **What other kind of knowledge will you need?**

   Answer: You will need to know which tools are carried on the apparatus, where the tools and any accessories for them are located on the apparatus, and how each tool works.

WRAP-UP

SUMMARY

- Fire service personnel use tools and equipment to perform a wide range of activities. All crew members must know how to use these tools effectively, efficiently, and safely, as well as how to maintain the tools and equipment so that they will always be ready for use.
- Some tools are carried on all fire apparatus, whereas other tools are carried only by specific types of companies. A good way to remember the function of each tool is to categorize them by their function.
- Rotating tools apply a rotational force to make something turn. Many rotating tools are basic hand tools. The rotating tools used most often by the fire service are screwdrivers, wrenches, and pliers, which are used to assemble or disassemble parts that are connected with threaded fasteners.
- Cutting tools have sharp edges that sever objects. They come in several forms and are used to cut a wide variety of materials.
- Pushing/pulling tools are metal hooks on a head at the end of a pole. These extend the reach of the firefighter and increase the firefighter's mechanical advantage on an object.
- Tools used for prying and spreading may be as simple as a **pry bar** or as mechanically complex as a hydraulic spreader.
- Striking tools are used to apply an impact force to an object. They often are employed to gain entrance to a building or a vehicle or to make an opening in a wall or roof. This equipment also can be used to force the end of a prying tool into a small opening.

- Certain tools are designed to perform multiple functions, thereby reducing the number of tools needed to achieve a goal. Some combination tools can be used to cut, to pry, to strike, and to turn off utilities.
- Some fire situations require special-use tools that perform other functions. For example, fire departments located in areas where wildland fires occur frequently may need to carry fire rakes; firefighting brooms; shovels; and combination tools that can be used for raking, chopping, cutting, and leaf blowing.
- All tools and equipment must be properly maintained so that they will be ready for use when they are needed. This means that a tool or other equipment should be in proper working order, in its proper storage location, and ready for immediate use. Every crew member should be able to locate the right tool immediately and be confident that it is ready for use.
- All hand tools should be cleaned completely and inspected, and their conditions should be documented after each use. Remove all dirt, debris, and rust from the tools. If appropriate, use water streams to remove the debris and mild soap to clean the equipment thoroughly. Learn how to safely use the cleaning solutions that your department and the manufacturer specify for cleaning tools and equipment.
- Just like hand tools, all power tools should be cleaned completely, inspected, and their conditions documented after each use.

KEY TERMS

adjustable wrench An open-ended wrench whose opening can be adjusted to accommodate bolts of different sizes.

adze The curved or straight wedge part of a Halligan tool.

axes Cutting tools that have a wide cutting blade that can be used to chop into a wall, roof, or door.

band saw An electrically powered saw that has a toothed metal blade stretched over two pulleys in a loop.

Bangor ladder An extension ladder with a staypole. Also called a tormentor ladder.

battering ram A tool made of hardened steel with handles on the sides used to force doors and breach walls. Larger versions may be used by as many as four people; smaller versions are made for one or two people.

bolt cutter A cutting tool used to cut through thick metal objects such as bolts, locks, and wire fences.

box-end wrench A hand tool used to tighten or loosen bolts. The end is enclosed, as opposed to an open-end wrench. Each wrench is a specific size, and most have ratchets for easier use.

bunny tool See *rabbet tool.*

carpenter's handsaw A saw designed for cutting wood.

ceiling hook A tool with a long wooden or fiberglass pole that has a metal point with a spur at right angles at one end. It can be used to probe ceilings and pull down plaster lath material.

chainsaw A power saw that uses the rotating movement of a chain equipped with sharpened cutting edges. It is typically used to cut through wood.

chisel A metal tool with one sharpened end that is used to break apart material in conjunction with a hammer, mallet, or sledgehammer.

claw bar A tool with a pointed claw hook on one end and a forked- or flat-chisel pry on the other end. It is often used for forcible entry.

Clemens hook A multipurpose tool that can be used for several forcible entry and ventilation applications because of its unique head design.

closet hook A type of pike pole intended for use in tight spaces, commonly 2 to 4 ft (0.6 to 1.2 m) in length.

combination wrench A hand tool with an open-end wrench on one end and a box-end wrench on the other.

come along A hand-operated tool used for dragging or lifting heavy objects that uses pulleys and cables or chains to multiply a pulling or lifting force.

coping saw A saw designed to cut curves in wood.

coupling One set or pair of connection devices attached to a fire hose that allow the hose to be interconnected to additional lengths of hose or adapters and other firefighting appliances. (NFPA 1960)

crowbar A straight bar made of steel or iron with a forked chisel on the working end that is suitable for performing forcible entry.

cutting torch A torch that produces a high-temperature flame capable of heating metal to its

KEY TERMS CONTINUED

melting point, thereby cutting through an object. Because of the high temperatures (5700°F [3148°C]) that these torches produce, the operator must be specially trained before using this tool.

diagonal cutter See *wire cutter.*

drywall hook A specialized version of a pike pole that can remove drywall more effectively because of its hook design.

flat bar A specialized type of prying tool made of flat steel with prying ends suitable for performing forcible entry.

flat-head axe A tool that has a head with an axe on one side and a flat head on the opposite side.

forcible entry Techniques used by fire personnel to gain entry into buildings, vehicles, aircraft, or other areas of confinement when normal means of entry are locked or blocked. (NFPA 440)

Fresno ladder A narrow, two-section extension ladder specifically designed to provide attic access and to be used in any tight space.

gripping pliers A hand tool with a pincer-like working end that can be used to bend wire or hold smaller objects.

hacksaw A cutting tool designed for use on metal. Different blades can be used for cutting different types of metal.

Halligan See *Halligan tool.*

Halligan bar See *Halligan tool.*

Halligan tool A prying tool that incorporates a sharp tapered pick, a blade (either an adze or wedge), and a fork; it is specifically designed for use in the fire service. Also called a *Halligan* and a *Halligan bar.*

hammer A striking tool.

hand light A small, portable light carried by firefighters to improve visibility at emergency scenes. It is often powered by rechargeable batteries.

handsaw A manually powered saw designed to cut different types of materials. Examples include hacksaws, carpenter's handsaws, keyhole saws, and coping saws.

Hux bar A multipurpose tool that can be used for several forcible entry and ventilation applications because of its unique design. It also may be used as a hydrant wrench.

junction box An electrical enclosure that houses one or more wiring connections. The box protects the connections from environmental conditions and accidental contact.

hydrant wrench A hand tool that is used to operate the valves on a hydrant; it also may be used as a spanner wrench. Some models are plain wrenches, whereas others have a ratchet feature.

Hydra-Ram A one-piece integrated hydraulic forcible entry tool.

hydraulic cutters See *hydraulic shears.*

hydraulic shears A lightweight, hand-operated tool that can produce up to 10,000 lb (4500 kg) of cutting force. Also called *hydraulic cutters.*

hydraulic spreader A lightweight, hand-operated tool that can produce up to 10,000 lb (4500 kg) of prying and spreading force.

hydraulic tool A power tool that uses pressurized fluid to exert force.

Kelly tool A steel bar with two main features: a large pick and a large chisel or fork.

keyhole saw A saw designed to cut keyhole circles in wood and drywall.

K tool A tool that is used to remove lock cylinders from structural doors so that the locking mechanism can be unlocked.

mallet A short-handled hammer.

maul A specialized striking tool, weighing 6 lb (3 kg) or more, with an axe on one side of the head and a sledgehammer on the other side.

multipurpose hook A long pole with a wooden or fiberglass handle and a metal hook on one end used for pulling.

multi-tool A compact, pocket-size, multiple-function tool that combines several different tools, such as a knife, scissors, wire cutters, pliers, and screwdriver.

New York hook See *roofman's hook.*

open-end wrench A hand tool that is used to tighten or loosen bolts. The end is open, as opposed to a box-end wrench. Each wrench is a specific size.

overhaul A firefighting term involving the process of final extinguishment after the main body of the fire has been knocked down. All traces of fire must be extinguished at this time. (NFPA 1700)

pick-head axe A tool that has a head with an axe on one side and a pointed end ("pick") on the opposite side.

pike pole A pole with a sharp point ("pike") on one end coupled with a hook. It is used to make openings in ceilings and walls. Pike poles are manufactured in different lengths for use in rooms of different heights.

pipe wrench A wrench having one fixed grip and one movable grip that can be adjusted to fit securely around pipes and other tubular objects.

plaster hook A long pole with a pointed head and two retractable cutting blades on the side.

power saw A saw that is usually powered by an electric motor or a gasoline engine. The three primary types of mechanical saws are chainsaws, rotary saws, and reciprocating saws.

pry bar A specialized prying tool made of a hardened steel rod with a tapered end that can be inserted into a small area.

rabbet tool A hydraulic spreading tool designed to pry open doors that swing inward. Also called a *bunny tool*.

reciprocating saw A saw that is powered by an electric motor or a battery motor and whose blade moves back and forth.

rescue knife A spring-assisted folding knife that can be used with one hand; often includes a seat belt cutter and a window breaker.

roofman's hook A long pole with a solid metal hook used for pulling. Also called a *New York hook*.

rotary saw A saw that is powered by an electric motor or a gasoline engine and that uses a large rotating blade to cut through material. The blades can be changed depending on the material being cut.

San Francisco hook A multipurpose tool that can be used for several forcible entry and ventilation applications because of its unique design, which includes a built-in gas shut-off and directional slot.

screwdriver A tool used for turning screws.

seat-belt cutter A specialized cutting device that cuts through seat belts.

shove knife A forcible entry tool used to trip the latch on outward-swinging doors.

sledgehammer A hammer that can be one of a variety of weights and sizes.

socket wrench A wrench that fits over a nut or bolt and uses the ratchet action of an attached handle to tighten or loosen the nut or bolt.

spanner wrench A type of tool used to couple or uncouple hose by turning the rocker lugs or pin lugs on the connections.

speed socket wrench A long, curved handle with a socket wrench that fits over a nut or bolt at the end. To tighten or loosen the nut or bolt, the handle is rotated.

spring-loaded center punch A spring-loaded punch used to break automobile glass.

stem nut The large nut at the top of the operating stem in a dry-barrel hydrant that is turned to open the hydrant valve.

ventilation The controlled and coordinated removal of heat and smoke from a structure, replacing the escaping gases with fresh air. (NFPA 1410)

ventilation saws Cutting tools designed for roof ventilation that have a different cutting chain than chainsaws used to cut wood; also may have a depth gauge on the bar.

wire cutter A hand tool used to cut wire and small-diameter cable. Also called a *diagonal cutter*.

wrench A hand tool that comes in several sizes and is used to tighten or loosen bolts.

REVIEW QUESTIONS

1. What is considered a high priority at any incident scene? How does this relate to using tools and equipment?
2. If you are assigned the task of looking for pockets of fire in the ceiling, what functional category would the tool you use fall under? List at least three tools from this category that you could use.
3. List the three working ends on a Halligan.
4. What are the seven phases of incident scene operations?
5. Where can firefighters find instructions for cleaning and maintaining tools and equipment?
6. What needs to happen after a hand tool is used?

DISCUSSION QUESTIONS

1. If the hydrant wrench is missing from the hydrant bag, what can you use to remove the hydrant caps and turn the stem nut?
2. If your company is assigned to go to the roof at a structure fire to assist with ventilation, which cutting hand tool should you bring?
3. As a new firefighter, what could you do if you are unsure of which tool to choose for the task at hand?

REFERENCES

Long, Merritt. 2007. "History of the Halligan Tool." Firefighter Nation, December 24, 2007. Accessed October 4, 2023. http://my.firefighternation.com/group/firefightinghistorymyths/forum/topics/889755:Topic:233830.

National Fire Protection Association. 2019. *NFPA 1407, Standard for Training Fire Service Rapid Intervention Crews.* 2020 Edition. Quincy, MA: National Fire Protection Association.

National Fire Protection Association. 2019. *NFPA 1410, Standard on Training for Emergency Scene Operations.* 2020 Edition. Quincy, MA: National Fire Protection Association.

National Fire Protection Association. 2019. *NFPA 1932, Standard on Use, Maintenance, and Service Testing of In-Service Fire Department Ground Ladders.* 2020 Edition. Quincy, MA: National Fire Protection Association.

National Fire Protection Association. 2020. *NFPA 1700, Guide for Structural Fire Fighting.* 2021 Edition. Quincy, MA: National Fire Protection Association.

National Fire Protection Association. 2023. *NFPA 440, Guide for Aircraft Rescue and Firefighting Operations and Airport/Community Emergency Planning.* 2024 Edition. Quincy, MA: National Fire Protection Association.

National Fire Protection Association. 2023. *NFPA 1550, Standard for Emergency Responder Health and Safety.* 2024 Edition. Quincy, MA: National Fire Protection Association.

National Fire Protection Association. 2023. *NFPA 1660, Standard for Emergency, Continuity, and Crisis Management: Preparedness, Response, and Recovery.* 2024 Edition. Quincy, MA: National Fire Protection Association.

National Fire Protection Association. 2023. *NFPA 1900, Standard for Aircraft Rescue and Firefighting Vehicles, Automotive Fire Apparatus, Wildland Fire Apparatus, and Automotive Ambulances.* 2024 Edition. Quincy, MA: National Fire Protection Association.

National Fire Protection Association. 2023. *NFPA 1960, Standard for Fire Hose Connections, Spray Nozzles, Manufacturer's Design of Fire Department Ground Ladders, Fire Hose, and Powered Rescue Tools.* 2024 Edition. Quincy, MA: National Fire Protection Association.

Chapter Opener: © Eric Scruggs

CHAPTER

8

Support Person

Ropes and Knots

KNOWLEDGE OBJECTIVES

After studying this chapter, you will be able to:

- Describe the four primary types of fire service rope.
- List the two types of life safety rope.
- Describe the characteristics of utility ropes.
- Describe the characteristics of webbing.
- List the four components of the rope maintenance formula.
- Describe the importance of properly maintaining tools and equipment.
- Describe how to preserve rope strength and integrity.
- Describe how to clean rope.
- Describe the reasons for placing rope out of service.
- Describe how to inspect rope.
- Describe how to keep an accurate rope record.
- Describe how to store rope properly.
- List the common types of knots that are used in the fire service.
- List the terminology used to describe the bends in rope that are formed when a knot is tied.
- List the terminology used to describe the parts of a rope when tying knots.
- Describe the characteristics of a safety knot.
- Describe the characteristics of a figure eight knot.
- Describe the characteristics of a half hitch.
- Describe the characteristics of a clove hitch.
- Describe the characteristics of a bowline knot.
- Describe the characteristics of a sheet bend.
- Describe the characteristics of a water knot.
- Describe the methods used to hoist a tool.

SKILLS OBJECTIVES

After studying this chapter, you will be able to perform the following skills:

- Demonstrate how to clean and inspect fire department ropes.
- Properly document equipment maintenance following established guidelines.
- Demonstrate how to place a life safety rope in a rope bag.
- Demonstrate how to properly tie a safety knot.
- Demonstrate how to properly tie a figure eight knot.
- Demonstrate how to properly tie a half hitch.
- Demonstrate how to properly tie a clove hitch in the open.
- Demonstrate how to properly tie a clove hitch around an object.
- Demonstrate how to properly tie a figure eight on a bight.
- Demonstrate how to properly tie a figure eight follow-through.
- Demonstrate how to properly tie a bowline.
- Demonstrate how to properly tie a sheet or Becket bend.
- Demonstrate how to properly tie a figure eight bend.
- Demonstrate how to properly tie a water knot.
- Demonstrate how to hoist an axe.

- Demonstrate how to hoist a pike pole.
- Demonstrate how to hoist a ladder.
- Demonstrate how to hoist a charged hose line.
- Demonstrate how to hoist an uncharged hose line.
- Demonstrate how to hoist an exhaust fan or power tool.

ADDITIONAL NFPA STANDARDS

- **NFPA 2500**, *Standard for Operations and Training for Technical Search and Rescue Incidents and Life Safety Rope and Equipment for Emergency Services, 2022 Edition*

CASE STUDY

You Are the Support Person

You've arrived at a three-story townhouse fire. The engine company has knockdown on the fire on the top floor. The captain of the truck company drops a rope from the third-floor window and calls for the ventilation fan.

1. What type of rope is best used for this activity?
2. What knot will you use to tie the fan with?
3. What other steps will you take before giving the signal for the truck company to hoist the fan?

Introduction

The fire service often utilizes different types of rope to support or achieve tactical objectives. Rope is often used in emergency situations to directly or indirectly save or protect the lives of civilians and firefighters. This chapter will provide the basic information for you to begin developing the skills you'll need to accomplish those essential objectives.

You will learn about the different types of ropes used, the materials they are made of, their construction and characteristics, and how to maintain them. You must be able to choose the appropriate rope for a variety of situations and know how to tie essential knots quickly, accurately, and securely. You should be able to deploy and use ropes quickly to move equipment in a controlled, safe manner.

Understanding Ropes and Webbing

Before we dive into the world of ropes and knots, let's consider some key concepts. A **rope** is an object constructed of fibers that is designed to sustain a weight. In the fire service, rope is used in tasks ranging from lifting and lowering tools to raising victims from a confined space. **Webbing** is a woven material of flat or tubular weave in a long strip that is often used in conjunction with rope. All ropes are able to stretch and elongate, which is one reason why ropes are able to absorb the forces created when objects pull their fibers. There is a limit to how far a rope can stretch, which is measured as the rope's minimum breaking strength. The **minimum breaking strength (MBS)** is the weight limit that a rope can safely support.

SAFETY TIP

Many fire departments use color coding or other visible markings to identify different types of rope. This allows a firefighter to determine quickly if a rope is a life safety rope or a utility rope. The length of each rope should be clearly marked by a tag or a label on the rope bag.

Rope Types and Webbing

Four types of rope are primarily used in the fire service; each of these rope types has a specific function:

- Life safety rope
- Escape rope
- Throwline for water rescues
- Utility rope

Webbing is also used in conjunction with rope in certain conditions. Every member of a fire crew must be able to recognize the type of rope instantly from its appearance and markings.

NFPA 2500, *Standard for Operations and Training for Technical Search and Rescue Incidents and Life Safety Rope and Equipment for Emergency Services, 2022 Edition*, specifies the criteria for the design, construction, and performance of ropes, webbing, and related equipment. NFPA 2500 also requires rope manufacturers to provide detailed instructions for the proper use, maintenance, and inspection of rope and webbing, including the conditions for removing the rope or webbing from service. In addition, NFPA 2500 requires rope manufacturers to supply a list of things to look for when a rope is inspected after it has been used. If the rope does not meet all of the inspection criteria, it must be retired from service.

Life Safety Rope

Only **life safety rope** is used for suspending people. It must be used whenever a rope is needed to support the weight of one or more persons, whether during training or during firefighting, rescue, or other emergency operations. Life safety rope has the most stringent performance requirements for rope in the fire service because when rope is supporting a person, rope failure could result in serious injury or death. Life safety rope is a critical tool and should be used *only* for life-saving purposes.

Life safety rope is identified by the paperwork provided by the manufacturer when it is purchased. There is also a label called a **trailer** along the entire length of the rope inside the outer sheath that indicates that it is life safety rope. The trailer also includes the name of the manufacturer, model number, make number, serial number, and date of manufacture. The paperwork that accompanies the rope upon purchase usually provides a comprehensive description of the classification of use, strength rating, elongation, core and sheath materials, length, diameter, weight, and durability, among other items specific to the type of rope desired.

There are two basic types of life safety rope: general-use life safety rope and technical-use life safety rope. **General use life safety rope** is the most common life safety rope carried by the fire service in the United States. NFPA 2500 lists the requirements for the performance of life safety rope. These requirements are gauged by MBS, elongation, and minimum and maximum diameter of the ropes. The diameter of a general use life safety rope must be at least 7/16 inch (in.; 11 millimeters [mm]) but not larger than 5/8 in. (16 mm). In addition, general-use life safety rope must have an MBS strength of 8992 pound-force (lbf; 40 kilonewtons [kN]). **Technical use life safety rope**, per NFPA 2500, must have a diameter that is at least 3/8 in. (9.5 mm) but not larger than 1/2 in. (12.5 mm). The MBS of technical use safety rope is 4496 lbf (20 kN), which is half of the required breaking strength of general use life safety rope. NFPA 2500 also specifies the limits of elongation of life safety ropes and the testing requirements for life safety ropes with minimum and maximum elongation.

All life safety rope must be resistant to abrasions so that rope does not fray. Life safety rope must also be made using **block creel construction**, which means the rope must be made from fibers that run the entire length of the rope without knots or splices. The continuous filaments produce a rope that is stronger than one constructed of shorter fibers that are twisted or braided together. Chapter 29 in NFPA 2500 provides information on how to select a life safety rope.

Although technical-use life safety rope can be significantly lighter than general-use life safety rope, it is not as strong. Members of highly trained rescue teams who deploy to technical environments, such as mountainous or wilderness terrain, carry this smaller, lighter rope so that they can carry more ropes and longer lengths over rough terrain. Although general use life safety rope allows a much greater margin of safety, the highly trained technical rescue technicians know how to use technical use life safety rope within the limits of its specifications. Rope with a diameter of 7/16 in. (11 mm) falls within the specified range for both general use and technical use life safety rope. But because general use life safety rope has a breaking strength that is double that of technical use life safety rope, technical rescue teams often choose 7/16 in. (11 mm) general use life safety rope instead of technical rescue safety rope of the same diameter.

Escape Rope

When you are responding to an emergency, you should always have a safe way to get out of a situation and reach a safe location. You might be able to go back through the door that you entered, or you might have another exit route, such as through a different door, through a window, or down a ladder. If conditions suddenly change for the worse, having an escape route can save your life. Sometimes, however, you might find yourself in a situation during which conditions deteriorate so rapidly that you cannot use your planned exit route. For example, the stairway you used might collapse behind you, or the room you are in might suddenly flash over, blocking your planned exit route. In such a situation, you might need to take extreme measures to get out of the building.

Two types of ropes were developed for this type of situation. The **escape rope** is intended for emergency, self-rescue situations. The **fire escape rope** is intended specifically for emergency self-rescue from an immediately hazardous environment in which fire or fire products are involved. The performance requirements for fire escape rope state that this type of rope must be able to hold 300 pounds (lb; 136 kilograms [kg]) for 45 seconds at 1112°F (600°C) and for 5 minutes at 750°F (400°C).

Both types of escape rope are designed to carry the weight of only one person and to be used only one time (**FIGURE 8-1**). Both types of escape rope must have a diameter of at least 19/64 in. (7.5 mm) but not larger than 3/8 in. (9.5 mm) and an MBS of 3034 lbf (13.5 kN). Because they are thinner and lighter than life safety ropes, they fit easily into a small packet or pouch and are easy to carry. An escape rope should be replaced with a new rope if it is exposed to an immediately dangerous to life or health (IDLH) environment, even if it was not used. When replacing any rope or equipment, follow the proper procedure for your department, notifying your supervisor or appropriate personnel and using the proper forms to record all of the information.

FIGURE 8-1 An escape rope is designed to be used by only one person.

© Jones & Bartlett Learning. Photographed by Glen E. Ellman.

TIP

Escape ropes are not classified as life safety ropes.

Throwline for Water Rescue

A **throwline** for water rescue is a type of rope that is used during water rescue operations and that floats on top of the water. Throwlines are 50 to 100 feet (ft; 15 to 30 meters [m]) in length and are stored in a **throw bag**, which is attached to one end of the rope and has a layer of sponge foam so that it floats (**FIGURE 8-2A**). The throw

A.

B.

FIGURE 8-2 **A.** Water rescue throwlines are kept in throw bags. **B.** To use, the rescuer holds on to the end of the rope sticking out of the top of the bag and throws the bag, which has the other end of the rope attached to it, to the victim in the water.

© fotomy/Alamy Stock Photo

bag is thrown to a victim in the water while the rescuer holds on to the other end of the rope. As the bag travels through the air, the rope unravels (**FIGURE 8-2B**). The victim can grab the bag or the rope, and then the rescuer can pull the victim toward them. Throwlines also can be used as a tether for rescuers entering the water.

Throwlines must be at least 19/64 in. (7 mm) in diameter but not larger than 3/8 in. (9.5 mm) in diameter, and they must have an MBS of 2923 lbf (13 kN). They are made of material that floats, such as polypropylene, so that the rope does not snag on underwater hazards or get entangled in motorboat propellers. Throwlines do not have the strength or abrasion resistance required to be used as a life safety rope.

Utility Rope

Utility rope is used in most cases when it is not necessary to support the weight of a person, such as when hoisting or lowering tools or equipment (**FIGURE 8-3**).

FIGURE 8-3 Utility ropes are used for hoisting and lowering tools.

Courtesy of Chris Rimm.

Utility ropes can also be used to mark off areas or stabilize objects. NFPA 2500 does not specify performance requirements for utility ropes, and they must never be used in situations where life safety rope is required.

Webbing

As stated in the beginning of the chapter, webbing is a woven material of flat or tubular weave in a long strip (**FIGURE 8-4**). Although it is not actually considered a type of rope, webbing is often used in conjunction with rope. Because of its special characteristics, webbing may even be preferable to rope in certain situations. For example, webbing is better suited to rig anchor points around objects with abrupt bends or corners on them, such as steel I-beams or square columns. In addition, because of its wide, flat surface, webbing may be more abrasion resistant in some rigging applications. **Rigging** is a general term used when building a lifting and lowering system with ropes and rope equipment.

There are hundreds of types and sizes of webbing. Most webbing is made of nylon or polyester and ranges in width. Flat webbing is constructed of a single layer of material; tubular webbing consists of a flattened tube of material. Although flat webbing is less expensive, tubular webbing is more supple and easier to work

FIGURE 8-4 Webbing is available in tubular (left) or flat (right) construction.

Courtesy of Troy Corbin, Country Brook Design, Inc.

with. Escape webbing must have an MBS of 3034 lbf (13.5 kN), in addition to meeting all of the other performance requirements as escape rope in NFPA 2500.

Rope Materials

Ropes are made from many different materials. The earliest ropes were made from naturally occurring vines or natural fibers, such as manila or cotton, that were woven together. The natural fibers were twisted together to form strands. Each strand could contain hundreds of individual fibers of different lengths.

Natural fiber ropes have the following drawbacks:

- Lose their load-carrying ability over time
- Are subject to mildew, which weakens the fibers
- Absorb water when wet, making them susceptible to deterioration
- Once wet, are difficult to dry
- Degrade quickly, even when properly stored

Today, ropes made from natural fibers are no longer acceptable as life safety ropes and are rarely even used as utility ropes. Life safety ropes are always made from synthetic fibers.

Ever since the invention of nylon in 1938, synthetic fibers have been used to make ropes. In addition to nylon, several newer synthetic materials, such as polyester, polypropylene, and polyethylene, are now also used to make ropes. Synthetic fibers have several advantages over natural fibers:

- Synthetic fibers are generally stronger than natural fibers, so it is possible to use a smaller-diameter rope without sacrificing strength.
- Synthetic materials can produce very long fibers that run the full length of a rope to provide greater strength and added safety.
- Synthetic fibers have a longer life span than natural fibers.
- Synthetic ropes are more resistant to rotting and mildew than natural fiber ropes and do not degrade as rapidly.
- Depending on the material, synthetic ropes might also be more resistant to melting and burning than natural fiber ropes.
- Synthetic ropes absorb much less water when they get wet than natural fiber ropes, and they can be washed and dried.
- Some types of synthetic rope can float on water, which is a major advantage in water rescue situations.

FIGURE 8-5 Polypropylene rope is often used in water rescues.

© Jones & Bartlett Learning. Photographed by Glen E. Ellman.

As with all ropes, prolonged exposure to ultraviolet light and exposure to strong acids or alkalis can damage synthetic ropes and decrease their life expectancy. Additionally, all ropes need to be protected from abrasion and sharp objects that can damage or cut the rope fibers.

Nylon is commonly used for ropes because it has a high melting temperature compared to other synthetic materials, it has a high resistance to abrasion, and it is strong but lightweight. Nylon ropes are also resistant to most acids and alkalis. Polyester is another commonly used synthetic fiber for ropes. Some ropes are made of a combination of nylon and polyester or other synthetic fibers. Polypropylene is the lightest of the synthetic fibers used for ropes. Because it does not absorb water and it floats, throwlines are often made of polypropylene (**FIGURE 8-5**). Polypropylene is not, however, a suitable material for life safety or escape rope because it is not as strong as other synthetic materials, it is hard to knot, and it has a low melting point.

Rope Construction

There are several different types of rope construction. The three most common types of rope construction used in the fire service are twisted rope, braided rope, and kernmantle rope.

Twisted Rope

Twisted rope, which is also called **laid rope**, is made of individual fibers twisted into strands. The strands are then twisted together to make the rope (**FIGURE 8-6**). This method of rope construction has been used for hundreds of years. Both natural and synthetic fibers can be used to make twisted rope.

FIGURE 8-6 Twisted rope.

© Jones & Bartlett Learning. Photographed by Glen E. Ellman.

FIGURE 8-7 Braided rope.

© Jones & Bartlett Learning. Photographed by Glen E. Ellman.

There are a few disadvantages to twisted ropes. In a twisted rope, all of the fibers are exposed and subject to abrasion, which can damage the rope fibers and reduce rope strength. Twisted ropes are also more prone to stretching and unraveling when a load is applied. Some fire departments have used twisted ropes for their utility ropes because they are easily distinguishable from life safety ropes. Twisted ropes are suitable for many activities on the fireground, including hoisting tools. They need to be maintained properly per the manufacturers' directions, have sufficient strength for what they will lift or stabilize, have minimal stretch (or elongation), and be replaced when damaged or fail inspection. Finally, twisted ropes are not life safety ropes.

Braided Rope

Braided rope is constructed by weaving or intertwining strands—typically synthetic fibers—together in the same way that hair is braided (**FIGURE 8-7**). Similar to twisted rope, all of the strands in a braided rope are exposed and subject to abrasion. A double-braided rope is a braided rope covered by a protective braided sleeve. This means that only the fibers in the outer sleeve are exposed. The inner core is protected from abrasion (**FIGURE 8-8**).

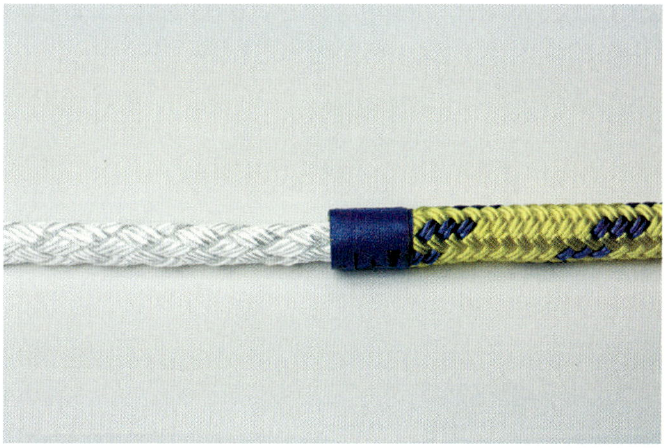

FIGURE 8-8 Double-braided rope.

Courtesy of Steve Hudson.

Like twisted rope, braided rope stretches under a load, but it is not prone to unraveling or twisting. Braided ropes are also often used as utility rope. They serve well as utility rope for moving and hauling equipment and holding items in place. The main disadvantage to using braided rope as utility rope is that it looks very similar to life safety rope, which may cause confusion.

Braided and double-braided ropes are often used in the marine and arborist industries. They are used to lift, lower, and stabilize equipment, vessels, and objects. The fire service typically uses braided rope as utility rope or in the fire/marine divisions.

Kernmantle Rope

Kernmantle rope consists of two distinct parts: the kern and the mantle. The **kern** is the center or core of the rope. The **mantle**, or **sheath**, is a braided covering that protects the kern from dirt and abrasion. Although both parts of a kernmantle rope are made with synthetic fibers, each part may be made from different synthetic fibers. Kernmantle ropes are made with block creel construction, where each individual fiber continues from end to end without splices.

Kernmantle rope is strong and flexible yet relatively thin and lightweight (**FIGURE 8-9**). This construction is well suited for rescue work and is very popular for life safety rope.

FIGURE 8-9 Kernmantle rope with the kern exposed.

Courtesy of Steve Hudson.

Static and Dynamic Kernmantle Rope

All ropes stretch or elongate. A kernmantle rope can be either dynamic or static, depending on how it reacts to an applied load. The differences between dynamic and static ropes result from both the fibers used and the construction method. A **dynamic rope** is designed to stretch more than a static rope when it is loaded. Some dynamic rope can elongate up to 40 percent of the length of the entire rope. Dynamic rope is typically used in safety lines in climbing. The safety rope stretches and cushions the shock when a sudden load is applied—for example, if a climber falls.

A **static rope** has a lower range of elasticity than dynamic rope. A static rope is designed to stretch less when a load is applied and is therefore more suitable for most fire rescue situations in which sudden shock loads are less likely and when limited elongation is desired to have more control over a system, such as when lowering, raising, or rappelling for a rescue. Decreased elongation is also desirable where rope haul systems are used. NFPA 2500 outlines how much life safety ropes can elongate.

A dynamic kernmantle rope is constructed with overlapping or woven fibers in the core (**FIGURE 8-10A**). When the rope is loaded, the core fibers are pulled tighter, which gives the rope its additional elasticity. In the core of a static kernmantle rope, all of the fibers are parallel, which decreases its elongation (**FIGURE 8-10B**).

Most fire departments use life safety ropes that are made with static kernmantle construction. This type of rope is well suited for lowering a person and

A.

B.

FIGURE 8-10 **A.** Dynamic kernmantle rope core with overlapping fibers in the core. Note the trailer label that travels the length of the core under the mantle. **B.** Static kernmantle rope core with parallel fiber in the core. Note the trailer label that travels the length of the core under the mantle.

A. Courtesy of Pigeon Mountain Industries; **B.** © Jones & Bartlett Learning. Courtesy of Steve Hudson.

can be used with a mechanical advantage pulley system for lifting individuals. It can also be used to create a **highline system**, which is a system of ropes or cables that allow rescuers to travel or send an object or a victim in a litter through the air from one building or object to another. Decreased elongation is desirable in these circumstances.

Rope Maintenance

All ropes—especially life safety ropes—need proper care. Proper care ensures a long life for your rope and reduces the chance of equipment failure and accidents. If a rope is not properly maintained, it will degrade much more quickly and have a shorter life span than it should.

People's lives, including the lives of the firefighters in your crew, depend on the proper maintenance of your ropes. All ropes should be maintained and cared for with the same concern as life safety rope because all ropes perform critical tasks. For example, a piece of equipment could be needed on the roof of a building to prevent the spread of fire toward firefighters, and a utility rope could be hoisting that tool. For more information on the care and maintenance of life safety rope, see Chapter 29 in NFPA 2500. There are four parts to the maintenance formula:

- Care
- Clean
- Inspect
- Store

Care for the Rope

You must follow certain principles to preserve the strength and integrity of rope:

- Protect the rope from sharp and abrasive surfaces. Use edge protectors when the rope must pass over a sharp or unpadded surface.
- Protect the rope from rubbing against another rope or webbing. Friction generates heat, which can damage or destroy the rope.
- Protect the rope from heat, chemicals, and flames.
- Protect the rope from prolonged exposure to sunlight. Ultraviolet radiation can damage rope.
- Do not step on the rope! Your footstep could force shards of glass, splinters, or abrasive particles into the core of the rope, damaging the rope fibers.
- Follow the manufacturer's recommendations for rope care.
- Protect the rope from shock loads.

A **shock load** is a load on a rope that suddenly drops or stops. A piece of equipment being raised or lowered that drops suddenly would shock load a utility rope. A rescuer who steps over an edge or off a ledge with slack in the rope would shock load a life safety rope. A shock load places a rope under extra tension. If a rope undergoes shock load, it should be carefully inspected for tears, flat spots, bulges, or any deformations. Although there might not be any visible damage, a severe shock load should be a reason to consider retiring the rope from service. Repeated shock loads can severely weaken a rope so that it can no longer be used safely. Keeping accurate rope records helps identify potentially damaged rope.

Clean the Rope

Dirt can work its way into rope fibers, weakening the rope, so part of maintaining ropes is to wash them after use. There are several ways to clean ropes. Many synthetic ropes can be washed by hand with a scrub brush and mild soap and water. If the manufacturer recommends it, you can attach a rope washer to a garden hose that you can run a rope through. Some manufacturers recommend placing the rope in a mesh bag and washing it in a front-loading washing machine. Do not use washing machines with agitators to wash ropes because they can bind or wind the rope and cause unpredictable or undetectable stresses on the rope. Follow the manufacturer's recommendations for specific care of your rope. When using a washing machine, use the gentle cycle and use a mild detergent that is safe for nylon, such as Woolite or similar type of product sold by rope manufacturers. Do not use bleach because it can damage rope fibers.

Ropes should be dried before they are put away. Do not pack or store wet or damp rope. Most manufacturers recommend air drying ropes by suspending them out of direct sunlight. Do not air dry a rope by laying it on the floor; floors often have had oils or other chemicals on them at one time or another. The use of mechanical drying devices is not recommended because the heat could damage the fibers. Also, rope should never be dried or stored in direct sunlight because the ultraviolet rays will damage the fibers.

Follow the steps in **SKILL DRILL 8-1** to clean fire department ropes.

SKILL DRILL 8-1

Cleaning Fire Department Ropes Support Person, NFPA 1010: 5.5.2

1. Wash the rope by hand with mild soap and water and use a soft brush to clean dirtier areas if needed.

2. Use a rope washer, or put the rope into a mesh bag and then wash it in a washing machine without an agitator if recommended by the rope's manufacturer.

3. Hang the rope out of direct sunlight to air dry it. Inspect the rope and replace it in the rope bag so that it is ready for use.

Steps 1 and 3: © Jones & Bartlett Learning. Photographed by Glen E. Ellman; **Step 2:** Courtesy of Captain David Jackson, Saginaw Township Fire Department.

Inspect the Rope

Life safety ropes must be inspected after each use, whether the rope was used for an emergency incident or in a training exercise. Unused life safety rope should be inspected on a regular schedule. Some departments inspect all rope, including escape rope, throwlines, and utility rope, every 3 months. Each department should obtain the recommended inspection criteria from the manufacturer of each of its ropes.

When you inspect a rope, ask the following questions:

- Has the rope been exposed to heat or flame?
- Has the rope been exposed to abrasion?
- Has the rope been exposed to chemicals?
- Has the rope been exposed to shock loads?
- Are there any depressions, discolorations, or lumps in the rope?

Inspect the rope visually, looking for cuts, frays, discoloration, shiny marks from heat or friction, or other damage as you run it through your fingers. Feel for inconsistencies in the thickness of the rope. Make sure that your grasping hand is gloved to protect yourself from sharp objects that may be embedded in the rope. Because you cannot see the inner core of a kernmantle rope, feel for any flat spots or lumps on the inside. Examine the mantle for discolorations, abrasions, or flat spots, and check for fibers from the kern poking through the mantle. If you have any doubt about whether the rope has been damaged, consult your company officer.

A life safety rope that is no longer usable must be pulled from service and either destroyed or designated as a utility rope. A downgraded rope must be clearly marked so that it cannot be confused with a life safety rope.

Each inspection of a life safety rope or utility rope should be recorded in a **rope record** (**FIGURE 8-11**). NFPA 2500 indicates that at least the following record shall be kept for each life safety rope:

- Individual identification of the rope
- The date the rope was purchased

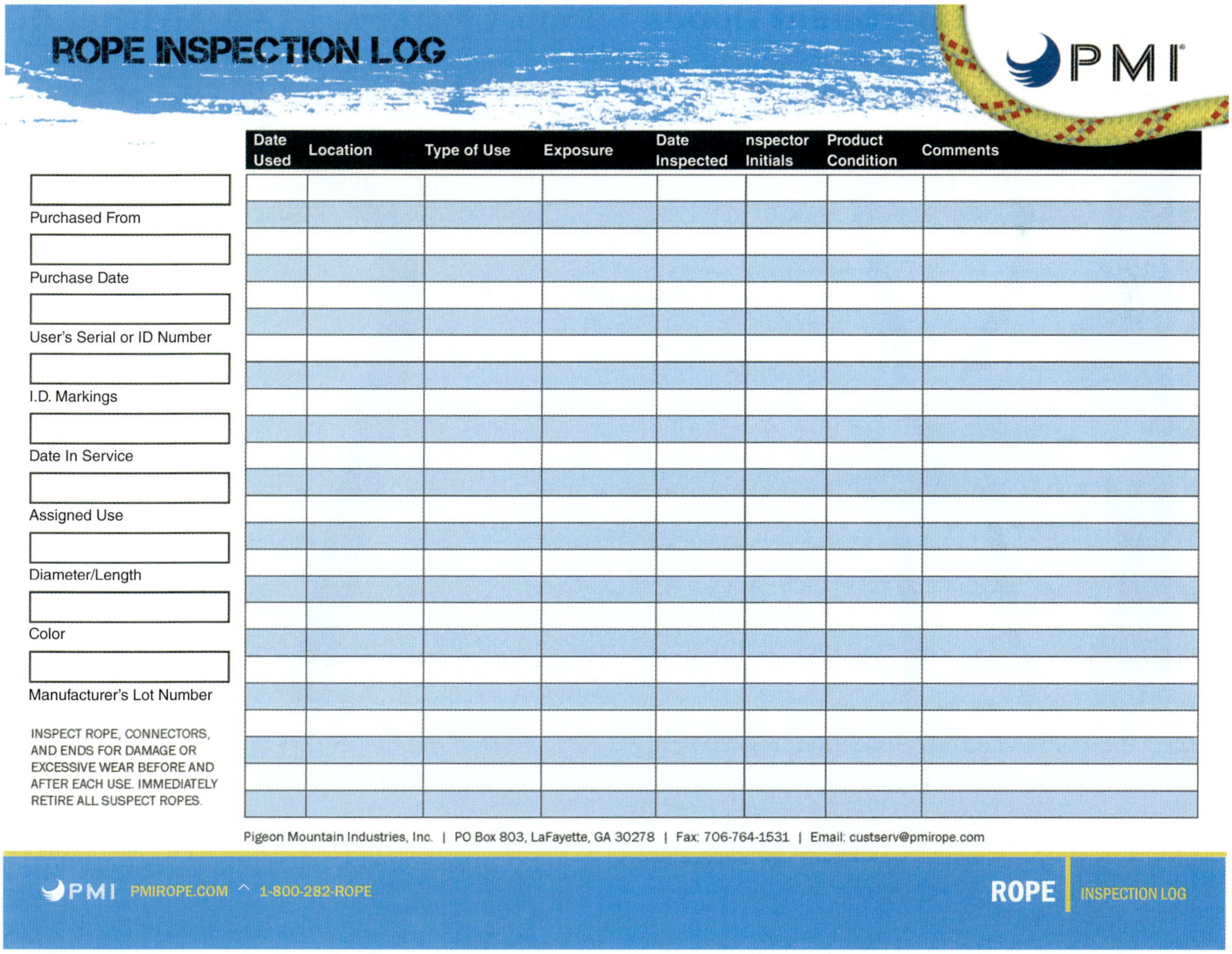
ROPE INSPECTION LOG

PMI

Purchased From

Purchase Date

User's Serial or ID Number

I.D. Markings

Date In Service

Assigned Use

Diameter/Length

Color

Manufacturer's Lot Number

INSPECT ROPE, CONNECTORS, AND ENDS FOR DAMAGE OR EXCESSIVE WEAR BEFORE AND AFTER EACH USE. IMMEDIATELY RETIRE ALL SUSPECT ROPES.

Date Used	Location	Type of Use	Exposure	Date Inspected	nspector Initials	Product Condition	Comments

Pigeon Mountain Industries, Inc. | PO Box 803, LaFayette, GA 30278 | Fax: 706-764-1531 | Email: custserv@pmirope.com

PMI PMIROPE.COM ^ 1-800-282-ROPE ROPE INSPECTION LOG

FIGURE 8-11 Rope inspection should be recorded in a rope record.

Courtesy of Pigeon Mountain Industries.

- Date the rope was placed in service
- The manufacturer's contact information
- The item or model number
- The month and year of manufacture
- The date of use, including how it was used, the weather conditions, any damage, and other conditions related to use
- The date of cleaning and inspection
- Removal from service and date of return
- Other notes deemed relevant by the authority having jurisdiction

Follow the steps in **SKILL DRILL 8-2** to inspect fire department ropes.

Store the Rope

Ropes need to be stored in areas where there is some air circulation, away from temperature extremes, and out of sunlight. Ropes should also not be placed where they are exposed to fumes from gasoline, oils, hydraulic fluids, or other petroleum products because those fumes might damage the ropes. This means that apparatus compartments used to store ropes should be separated from compartments used to store any oil-based products or machinery powered by gasoline or diesel fuel. Finally, do not place any heavy objects on top of the rope to avoid damaging the rope and to ensure that the rope is readily accessible (**FIGURE 8-12**).

SKILL DRILL 8-2

Inspecting Fire Department Ropes Support Person, NFPA 1010: 5.5.2

1. Inspect the rope after each use and at regular intervals.
   - Examine the core by looking and feeling for depressions, flat spots, and lumps.
   - Examine the sheath by looking for discolorations, abrasions, flat spots, and embedded objects.

2. Record the inspection in the rope record. Remove the rope from service if it is damaged.

Courtesy of Steve Hudson.

Some types of rope are stored in a **rope bag** to protect them and to make it easy to deploy them. If a rope bag is used, the entire rope should be stored in the bag. If you leave any section of the rope outside of the bag, that section is exposed to the environment outside the bag and could be damaged by coming into contact with objects. Some rope bags have a hole in the bottom that rope can fit through. Some agencies may advise leaving a small length of rope outside through the hole, but this is not advisable because it exposes that section of rope to the elements.

FIGURE 8-12 Ensure that ropes are stored safely.

© Jones & Bartlett Learning. Photographed by Glen E. Ellman.

Ropes stored in rope bags should be placed in the bag so that the rope is easy to deploy. That is, a firefighter should be able to reach into the rope bag, quickly find the end of the rope, and then pull to get the rope out of the bag without any snags. Avoid storing ropes with pre-made knots in bags unless they are of the immediate pre-rigged deployable types such as throwlines, escape ropes, and fire escape ropes, which need to be ready for use immediately and are stored and rigged for this purpose.

Rope bags can be used to aid in identifying the type of rope stored in the bag. For example, some agencies store all of their utility ropes in black bags; life safety rope in orange, red, or yellow bags; and water rescue rope in blue bags. The color of the bag used can be any variation. The color of the bag does not matter as much as the fact that the agency has a policy so that all members know which rope is in which color bag. If an agency uses colored rope bags to help identify the type of rope, the rope should still have an identifying label on each end to identify it.

Follow the steps in **SKILL DRILL 8-3** to place a life safety rope into a rope bag.

SKILL DRILL 8-3

Placing a Life Safety Rope in a Rope Bag Support Person, NFPA 1010: 5.5.2

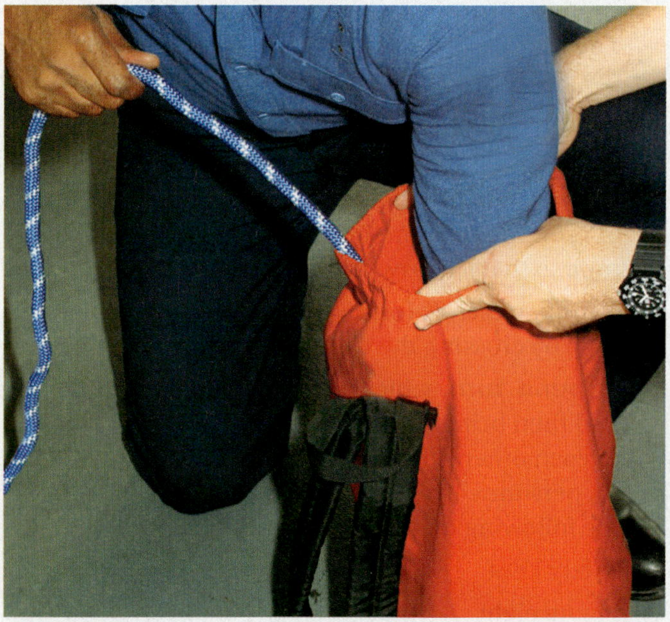

1. Inspect the rope to be certain it is fit for service. Stuff the life safety rope into the rope bag in a random pattern so that it deploys properly. Do not try to coil the rope in the bag because this action will cause the rope to kink and become tangled when it is pulled out.
2. Return the rope and the rope bag to its proper location for storage or service. The ends of the rope should not protrude out of the bag.

© Jones & Bartlett Learning. Photographed by Glen E. Ellman.

Manufacturers typically package rope on spools, or rope is coiled and wrapped in plastic or in a sealed plastic bag to protect it during transport. Ropes are not typically coiled or spooled for storage when they are in service and stored on apparatus. Ropes stored coiled or on spools are typically ropes that an agency has recently purchased, is keeping in storage, or has not been issued for department use yet. Coiled and spooled ropes are exposed to the elements around them, so if these storage methods are used, extra consideration should be given to where they are stored so that they are not exposed to extreme temperatures or stored adjacent to objects that can damage them.

FIGURE 8-13 The sections of a rope used in tying knots.

© Jones & Bartlett Learning. Photographed by Glen E. Ellman.

Knots

A **knot** is a prescribed way of fastening lengths of rope or webbing to objects or to each other. As a support person, you must know how to tie certain knots correctly and when to use each type of knot. Knot tying has often been described as a perishable skill. It cannot be emphasized enough how important it is to practice tying and applying these knots frequently. Practice, practice, practice!

Fire service personnel should be aware that knots decrease the load-carrying capacity of rope by a certain percentage. The reduction varies based on the type of knot, the type of rope, the method by which the rope is loaded, and the test itself. Therefore, providing a percentage reduction would be somewhat arbitrary because it can vary from rope to rope and test to test. Many technical experts consider that the knot will reduce the strength of the rope by up to 50 percent, which still keeps the safety margin within a working range for most situations. With the exception of high-tensioned systems such as horizontal highlines utilized by technical rope rescuers, most fire service personnel do not need to be concerned with knots reducing the load-carrying capacity. Experienced rope technicians select the knot that best fits the situation and understand the safety margin calculated with respect to the load applied and the strength of the rope used. Knowing which knot to use for a particular situation is more important than knowing the percentage of load-bearing reduction the knot will cause.

Instructions for tying knots refer to three parts of the rope (**FIGURE 8-13**):

- The **working end** is the part of the rope you hold as you form the knot.
- The **running end**, or standing end, is the other end of the rope, the part that is not used for forming the knot.

FIGURE 8-14 A bight.

© Jones & Bartlett Learning. Photographed by Glen E. Ellman.

- The **standing part** is the part of the rope between the working end and the running end.

When a knot is tied, a bend is formed in the rope. Three types of bends are as follows:

- A **bight**, which is formed by reversing the direction of the rope to form a U bend without crossing the rope so that it has two parallel ends (**FIGURE 8-14**).
- A **loop**, which is formed by forming a circle in the rope by crossing the rope (**FIGURE 8-15**).
- A **round turn**, which is formed by making a loop and then bringing the two ends of the rope parallel to each other (**FIGURE 8-16**).

All personnel should know how to tie and properly use the following common knots:

- Safety knot
- Figure eight

FIGURE 8-15 A loop.

© Jones & Bartlett Learning. Photographed by Glen E. Ellman.

FIGURE 8-16 A round turn.

© Jones & Bartlett Learning. Photographed by Glen E. Ellman.

- Half hitch
- Clove hitch
- Figure eight on a bight
- Figure eight follow-through
- Bowline
- Sheet bend (Becket bend)
- Figure eight bend
- Water knot (ring bend)

There are many ways to correctly tie each of these knots. Find the method that works for you and use it all the time. Your department might require that you learn how to tie other knots in addition to those listed here.

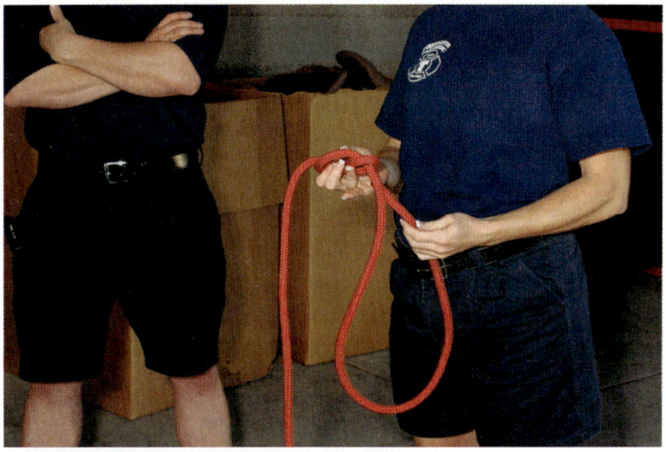

FIGURE 8-17 To maintain your knot-tying skills, practice tying different knots frequently.

© Jones & Bartlett Learning. Photographed by Glen E. Ellman.

After you tie a knot, you should properly **dress** the knot by tightening and removing twists, kinks, and slack from the knot. The configuration of a properly dressed knot should be evident so that it can be easily inspected.

It is important to become proficient in tying knots. Knot-tying skills can be quickly lost without adequate practice (**FIGURE 8-17**). You never know when you will need to use these skills in an emergency situation. For added practice, try tying these knots with your gloves on or in darkness to simulate the conditions you might encounter in an emergency situation.

Safety Knot

The **safety knot** is used to back up or finish knots, such as the clove hitch, bowline, or sheet bend, that have a tendency to become loose when they are not continuously loaded or that can slip when loaded. Safety knots are not needed on many of the knots commonly used in the fire service and by rope rescue operations and technician-level personnel. When properly tied and dressed, these knots are secure on their own as long as you leave a 4- to 6-in. (10- to 15-centimeter [cm]) tail on the leftover working end of the rope. When a safety knot is used to finish or back up another knot, it should be tied close to the primary knot and be properly dressed. Using a safety knot is commonly referred to as "backing up" or "finishing the knot." The overhand knot is often used as a safety knot in the fire service for the knots that need them. Follow the steps in **SKILL DRILL 8-4** to tie an overhand knot as a safety knot.

SKILL DRILL 8-4

Tying an Overhand Knot as a Safety Knot Support Person, NFPA 1010: 5.3.8

1. Take the end of the rope that exits the main knot, usually the shorter end of the rope near the accompanying main knot, and form a loop around the standing part of the rope as close as possible to the accompanying knot.

2. Pass the end of the rope through the loop.

3. Tighten the safety knot by pulling on both ends at the same time.

4. When complete, the safety knot should sit directly next to the accompanying knot.

© Jones & Bartlett Learning. Photographed by Brandon Fryman/JRM Creative.

Figure Eight Knot

A figure eight knot is used to prevent the end of a rope from passing through a device such as a pulley. For this reason, the figure eight knot is sometimes known as a *stopper knot*. When you pull the figure eight knot tight, it has the shape of a figure eight. This knot is the basis of other figure eight knots, including the figure eight on a bight, the figure eight follow-through, and the figure eight bend (all three of these are discussed later in this chapter). When a rope is stored in a rope bag, the figure eight knot is often loosely tied at the end of a rope to make it easy to identify the end of the rope inside the bag.

Follow the steps in **SKILL DRILL 8-5** to tie a figure eight knot.

SKILL DRILL 8-5

Tying a Figure Eight Knot Support Person, NFPA 1010: 5.3.8

1. Form a loop in the rope.

2. Pass the end of the rope around the front of the rope.

3. Thread the end or the rope through the loop from the back.

4. Pull both ends of the rope in opposite directions to tighten the knot.

© Jones & Bartlett Learning. Photographed by Glen E. Ellman.

Hitches

A **hitch** is a knot that can be slipped over or tied around an object to secure the rope to the object and is easily untied when the object is removed. They are typically used on round or cylindrically shaped objects. Hitches are ideal for hoisting many fire service tools such as axes, pike poles, and hose lines. Other types of hitches can be used to adjust the tension in a rope that is used to secure or stabilize an object. The advantage of using hitches is that they can be tied and removed quickly. However, hitches can move or turn around the object they are tied to, so you should tie a safety knot to prevent the hitch from moving if that is a concern. There are many types of hitches, but half hitches and clove hitches are the types used most often on the fireground.

Half Hitch

The half hitch is a loop of rope passed around or over an object or around the rope itself. It is not a secure knot, so it is used only in conjunction with other knots. For example, when hoisting an axe or a pike pole, you use the half hitch to keep the rope aligned with the handle so that the object won't swing on the rope and the entire object can be controlled throughout the hoist. On long objects, you might need to use several half hitches.

Follow the steps in **SKILL DRILL 8-6** to tie a half hitch around an object.

SKILL DRILL 8-6

Tying a Half Hitch Around an Object Support Person, NFPA 1010: 5.3.8

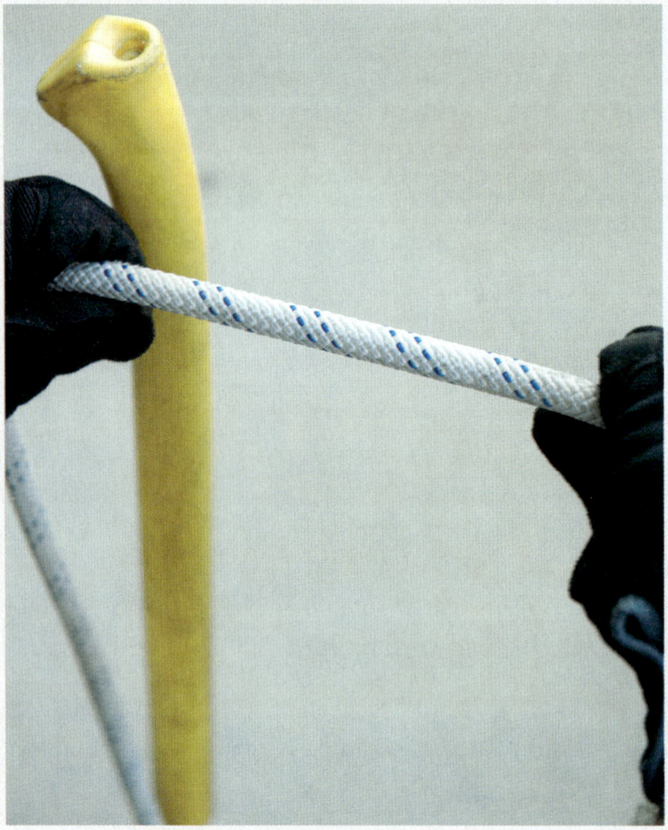

1. Grab the rope with your palms facing away from you.

2. Rotate one hand so that the palm is facing you to form a loop in the rope.

SKILL DRILL 8-6 CONTINUED

Tying a Half Hitch Around an Object Support Person, NFPA 1010: 5.3.8

3. Pass the loop over the end of the object.

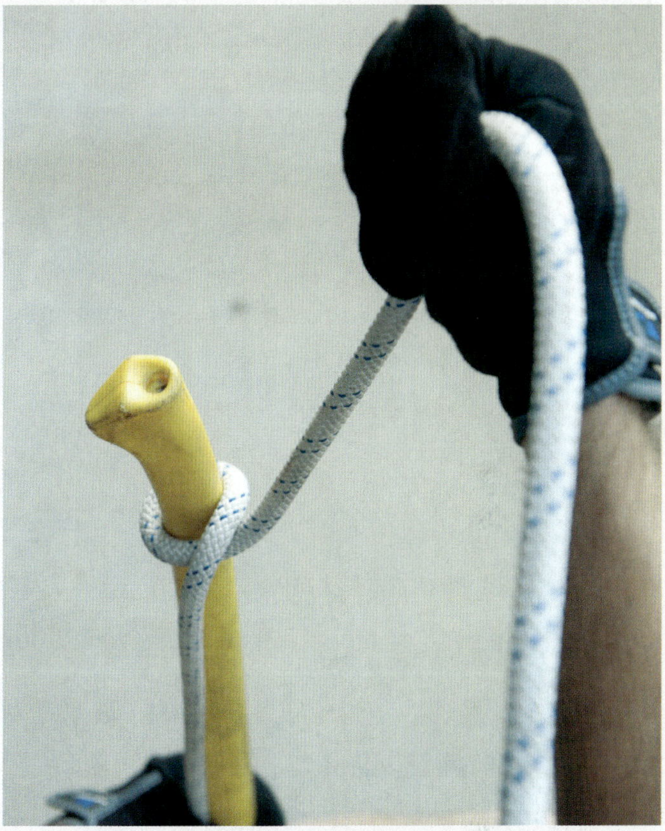

4. Finish the half-hitch knot by positioning it and pulling tight, maintaining tension on the rope.

© Jones & Bartlett Learning. Photographed by Glen E. Ellman.

Clove Hitch

A clove hitch is essentially two half hitches oriented in opposite directions. The clove hitch can be used to tie a hoisting rope around an axe or pike pole. A clove hitch can be tied anywhere in a rope and will hold equally well if tension is applied to either end of the rope or to both ends simultaneously.

There are two methods of tying this knot. A clove hitch tied in the open is used when the knot can be formed and then slipped over the end of an object, such as an axe or a pike pole. It is tied by making two consecutive loops in the rope. Follow the steps in **SKILL DRILL 8-7** to tie a clove hitch in the open.

SKILL DRILL 8-7

Tying a Clove Hitch in the Open Support Person, NFPA 1010: 5.3.8

1. Thread the end of the rope through the loop from the back.

2. Holding on to the rope, uncross your hands to create a loop in each hand with the end of the rope in one hand hanging behind the rope between the two loops and the end of the rope in the other hand hanging in front of the rope between the two loops.

3. Place the loop with the end of the rope hanging in front of the rope between the two loops behind the other loop so that the two loops are on top of each other.

4. Slide the loops over the end of the object. Pull the ends of the rope in opposite directions perpendicular to the object to tighten the knot. If needed, tie a safety knot in the working end of the rope.

© Jones & Bartlett Learning. Photographed by Glen E. Ellman.

A clove hitch may or may not require a safety knot to finish it. The type of object and where on the rope the knot is tied will dictate the need for a safety knot. For a round object such as a large tree or fence post, a clove hitch is tied around the object by passing the end of the rope around the object twice to form the clove hitch. A safety knot is used to finish the knot because the clove hitch will only have tension on one side of the rope and needs the additional stability of a safety knot. For objects with handles such as pike poles and axes, a clove hitch can be formed separately or in the open near the end of the rope and then slipped over the object. A safety knot is then used to finish the knot, and a tag line is attached to guide the object during hoisting. When a clove hitch can be tied on an object using the middle of the rope, then a safety knot typically is not used. One end of the rope is used for hoisting, and the other end is used to guide the object.

If the object is configured so that the clove hitch cannot be slipped over one end, the same knot can be tied around the object by tying two half hitches. For example, if you need to tie a clove hitch around a tall tree, utility pole, or rung of a ladder or to the rail of a litter basket, you can't create the loops and then slide them over the end of those objects. Follow the steps in **SKILL DRILL 8-8** to tie a clove hitch around an object.

SKILL DRILL 8-8

Tying a Clove Hitch Around an Object Support Person, NFPA 1010: 5.3.8

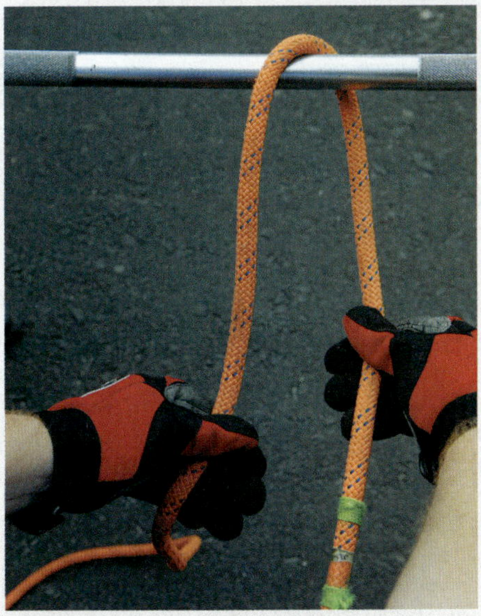

1. Wrap the working end of the rope around the object.

2. Pass the working end of the rope in front of the standing part of the rope to create a loop, and then pass it over the object again, a short distance above or next to the first loop.

3. Pass the working end of the rope under the second loop so that it goes in front of the object between the two loops.

Continues.

SKILL DRILL 8-8 CONTINUED

Tying a Clove Hitch Around an Object Support Person, NFPA 1010: 5.3.8

4. Pull the ends of the rope in opposite directions perpendicular to the object to tighten the knot.

5. If needed, pass the working end of the rope over the object and then tie a safety knot close to the clove hitch around the standing part of the rope.

© Jones & Bartlett Learning. Photographed by Glen E. Ellman.

Loop Knots

A **loop knot** is used to form a loop in the end of a rope. Loops can be used to hoist tools, to secure a person during a rescue, or to secure a rope to a fixed object. When tied properly, these loop knots will not slip yet are easy to untie. Like hitches, there are many types of loop knots. Three common loop knots that fire service personnel need to know how to tie are the figure eight on a bight, the figure eight follow-through, and the bowline.

Figure Eight on a Bight

The figure eight on a bight knot creates a secure loop at the working end of a rope. This loop can be used to attach the end of the rope to a fixed object or a piece of equipment. Follow the steps in **SKILL DRILL 8-9** to tie a figure eight on a bight.

Figure Eight Follow-Through

A figure eight follow-through knot is used to create a secure loop at the end of the rope when the rope must be wrapped around an object or passed through an opening before the loop is formed. It is very useful for attaching a rope to a fixed ring or object that is closed at both ends, such as the handle to a power saw. Follow the steps in **SKILL DRILL 8-10** to tie a figure eight follow-through.

SKILL DRILL 8-9

Tying a Figure Eight on a Bight Support Person, NFPA 1010: 5.3.8

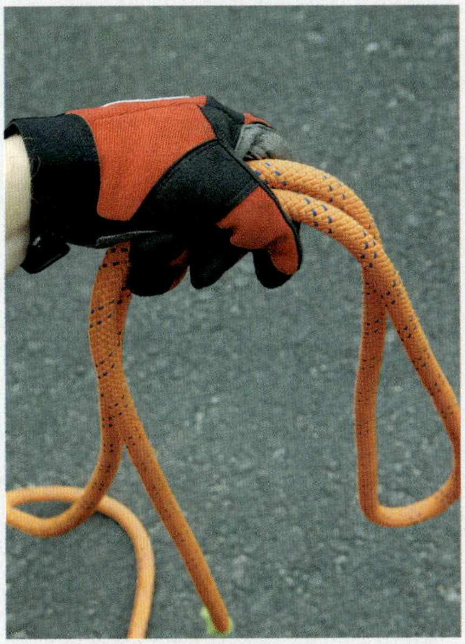

1. Double over a section of the rope to form a bight. Hold the two sides of the bight together as if they were one rope. The closed end of the bight becomes the working end of the rope.

2. Form a loop in the doubled section of the rope by passing the working end of the rope behind the standing part of the rope.

3. Pass the working end of the rope around the standing part of the rope and feed it through the bight from the back.

4. While holding on to the bight part of the working end, pull the ends of the rope that exit the knot, making sure you pull both pieces of rope evenly and maintain the figure eight shape to form a secure loop.

© Jones & Bartlett Learning. Photographed by Glen E. Ellman.

SKILL DRILL 8-10

Tying a Figure Eight Follow-Through Support Person, NFPA 1010: 5.3.8

1. Tie a simple figure eight knot in the rope, leaving enough rope between the knot and the working end to wrap around the object or pass through the opening and still have enough left to tic the follow-through. Do not tighten the knot; leave it loose. Then thread the working end around the object or through the opening.

2. Pass the working end back through the original figure eight knot, following the path of the rope in the original knot in the opposite direction.

3. Once the end of the rope has been threaded through the entire knot, pull the knot tight and evenly in the same manner as to dress the figure eight on a bight.

© Jones & Bartlett Learning. Photographed by Glen E. Ellman.

Bowline

A bowline is also used to form a loop in the end of a rope. The bowline can be used as an alternate knot for a figure eight on a bight or figure eight. This type of knot is frequently used to secure the end of a rope to an object or an anchor point. It also can be used to hoist equipment. A bowline can be tied in the open or after you wrap the rope around something. One advantage to using a bowline is that it can be tied and untied quickly, which is obviously very useful at an emergency scene. The bowline can become loose when it is not under load. For this reason, a bowline needs to be finished or backed up with a safety knot. Follow the steps in **SKILL DRILL 8-11** to tie a bowline.

SKILL DRILL 8-11

Tying a Bowline Support Person, NFPA 1010: 5.3.8

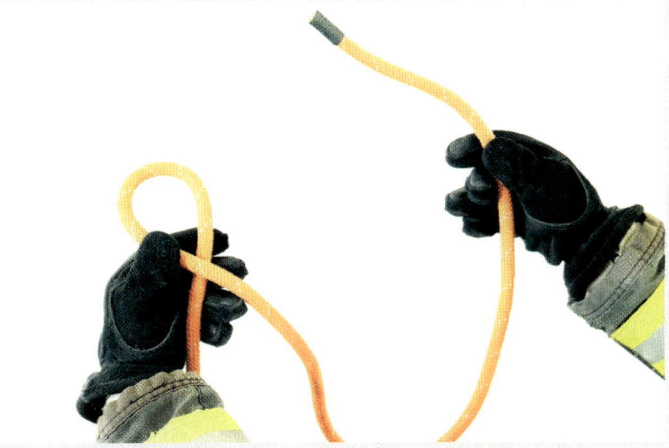

1. If you are tying a bowline in the open, form a small loop near one end of the rope, with the part of the rope leading to the working end on the top of the loop. The point at which you make this small loop determines how big the bowline loop will be. If you are tying the bowline around an object, such as a tree, wrap the rope around the object. You should have a part of the rope in each hand, one from the working end and the other going to the standing end. Form a loop in the rope going toward the standing end, with the part of the rope leading to the working end on top of the loop.

2. Thread the working end up through the loop from the bottom.

3. Pass the working end under the standing part.

Continues.

SKILL DRILL 8-11 CONTINUED

Tying a Bowline Support Person, NFPA 1010: 5.3.8

4. Pass the working end over the standing part and then back down through the same small loop.

5. Tighten the knot by pulling the working end and the standing or running end in opposite directions.

6. Tie a safety knot in the working end of the rope. The overhand safety knot gets tied around the rope that forms the loop.

© Jones & Bartlett Learning. Photographed by Brandon Fryman/JRM Creative.

Bend Knots

A **bend knot** is used to join two ropes or pieces of webbing together. These knots also can be used to join a rope to a chain or a cable. The three types of bend knots that fire service personnel will most often use are the sheet bend, the figure eight bend, and the water knot.

Sheet Bend

The sheet bend (or Becket bend) is used to join two ropes of unequal diameters. A sheet bend is also used to join a rope to a chain or to hoist a rope of a different diameter. It should not be relied upon to support a human load. Follow the steps in **SKILL DRILL 8-12** to tie a sheet bend.

SKILL DRILL 8-12

Tying a Sheet Bend Support Person, NFPA 1010: 5.3.8

1. Form a bight at the working end of one rope (the blue rope) and position your finger along the side of the bight. If the ropes are of unequal diameters, form the bight in the larger rope.

2. Thread the working end of the second rope (the red rope) up through the bight from the back side.

Continues.

SKILL DRILL 8-12 CONTINUED

Tying a Sheet Bend Support Person, NFPA 1010: 5.3.8

3. Wrap the second rope (the red rope) completely around your finger and the top side of the bight (the blue rope) and continue wrapping all the way around the bight. Thread the working end of the second rope (the red rope) between the original bight and under the second rope.

4. Hold the bight on the first rope (the blue rope) in one hand and the running end of the second rope (the red rope) in the other hand, then pull in opposite directions to tighten the knot.

5. Tie a safety knot in the working end of each rope.

© Jones & Bartlett Learning. Photographed by Glen E. Ellman.

Figure Eight Bend

The figure eight bend (or tracer eight) is used to join two ropes that have a similar diameter. When used with life safety rope, this knot can support a human load. Follow the steps in **SKILL DRILL 8-13** to tie a figure eight bend.

Water Knot

The water knot (or ring bend) is used to join webbing of the same or different widths together. When a single piece of webbing is used and the opposite ends are tied to each other, a loop or sling is created. The water knot could also be used to make to several shorter lengths into a single longer length. Follow the steps in **SKILL DRILL 8-14** to tie a water knot.

SKILL DRILL 8-13

Tying a Figure Eight Bend Support Person, NFPA 1010: 5.3.8

1. Tie a figure eight near the end of one rope.

2. Thread the working end of the second rope through the knot from the opposite end and follow the path of the rope in the original knot in the opposite direction.

Continues.

SKILL DRILL 8-13 CONTINUED

Tying a Figure Eight Bend Support Person, NFPA 1010: 5.3.8

3. Pull the knot tight.

© Jones & Bartlett Learning. Photographed by Glen E. Ellman.

SKILL DRILL 8-14

Tying a Water Knot Support Person, NFPA 1010: 5.3.8

1. At one end of the webbing, approximately 6 in. (152 mm) from the end, tie an overhand knot. Do not tighten this knot; keep it loose.

SKILL DRILL 8-14 CONTINUED

Tying a Water Knot Support Person, NFPA 1010: 5.3.8

2. Thread the other end of the webbing through the knot and follow the path of the webbing in the original knot in the opposite direction. Thread it through the knot until approximately 6 in. (152 mm) is poking out from the other side of the knot.

© Jones & Bartlett Learning. Photographed by Glen E. Ellman.

Hoisting

Tying knots and using rope to hoist equipment in emergency situations are skills in which you need to become proficient. In emergency situations, you may need to raise or lower a tool or equipment to firefighters. In all of these situations, it is essential that the correct knots are used, tied correctly, and secured and that the movement of the objects is controlled throughout the hoist.

It is important for you to learn how to raise and lower an axe, a pike pole, a ladder, a charged hose line, an uncharged hose line, an exhaust fan, power tools, and any other item your department uses or carries. When you are hoisting or lowering a tool or any object, make sure no one is standing under the object in case there is a failure and it falls. Also keep the scene clear of people to avoid any chance of an accident.

When hoisting a large object, such as an exhaust fan, attach a **tag line**—a separate rope that personnel on the ground use to guide an object that is being hoisted or lowered—to the object at the opposite end or corner from where the hoisting rope is secured to the object. This is used to help control the path of the object and prevent it from spinning or getting caught on an obstruction, such as an electrical line, a utility pole, or a structural feature. The tag line should be attached with a knot that is easily untied, such as a bowline, so that it can be removed easily to allow the equipment to be placed into service quickly.

To hoist an object, a firefighter carries a utility rope up to the roof, an upper floor, or higher elevation; secures one end of the rope; and then tosses the other end of the rope down to a crew member on the ground. If possible, the firefighter at the higher elevation should secure the end they are holding to an anchor point, but if there is nothing stable to use as an anchor point, that firefighter will need to hold tightly to their end of the rope. The crew member on the ground then ties the tool or piece of equipment to the utility rope using the appropriate knot. If needed, the crew member on the ground also ties a tag line to the piece of equipment. Then the crew member on the ground signals the firefighter at the higher elevation, and the firefighter begins pulling on the rope to hoist the tool or piece of equipment. If there is a tag line attached, the crew member on the ground holds the end of the tag line to help control the object on its way up.

To lower a tool or piece of equipment, essentially the reverse happens. The firefighter at the higher elevation ties the utility rope to the object and secures that end of the rope. If a tag line is needed, the firefighter at the higher elevation ties a tag line to the object. The firefighter tosses the other ends of the utility rope and the tag line down to the crew member on the ground. The firefighter at the higher elevation feeds the rope out to lower the object, ensuring that a reserve of rope remains on the roof. For example, if the utility rope is 150 ft (46 m) and the distance to lower the tool is 40 ft (12 m), then the firefighter would ensure that at least 100 ft (30 m) of rope remains on the roof. If a tag line is used, the crew member on the ground helps control the descent.

Hoisting an Axe

An axe should be hoisted in a vertical position with the head of the axe lower than the handle. After receiving the end of the utility rope from the firefighter above, follow the steps in **SKILL DRILL 8-15** to hoist an axe.

SKILL DRILL 8-15

Hoisting an Axe Support Person, NFPA 1010: 5.3.8

1. Tie the end of the hoisting rope around the handle of the axe near the head using a figure eight on a bight. Slip the knot down the handle from the end to the head.

2. Pass the loop under the head.

3. Place the standing part of the rope parallel to the axe handle.

4. Tie a half hitch along the axe handle to keep the handle parallel to the rope.

SKILL DRILL 8-15 CONTINUED

Hoisting an Axe Support Person, NFPA 1010: 5.3.8

5. Add a second half hitch along the axe handle if necessary.

6. Communicate with the firefighter above that the axe is ready to raise. To release the axe, hold the middle of the handle, release the half hitches, and slip the knot up and off.

© Jones & Bartlett Learning. Photographed by Brandon Fryman/JRM Creative.

Hoisting a Pike Pole

A pike pole is hoisted in a vertical position with the head at the top. After receiving the end of the utility rope from the firefighter above, follow the steps in **SKILL DRILL 8-16** to hoist a pike pole.

Hoisting a Ladder

A ladder is hoisted in a vertical position. A tag line attached to the bottom helps keep the ladder under control as it is hoisted. If it is a roof ladder, the hooks on the ladder should be in the retracted position. After receiving the end of the utility rope from the firefighter above, follow the steps in **SKILL DRILL 8-17** to hoist a ladder.

SKILL DRILL 8-16

Hoisting a Pike Pole Support Person, NFPA 1010: 5.3.8

1. Place a clove hitch over the bottom of the handle and secure it close to the bottom of the handle, finishing it with a safety knot or leaving enough length of rope below the clove hitch to be used as a tag line while raising the pike pole.

2. Place a half hitch around the middle of the handle about halfway up the handle to keep the rope parallel to the handle.

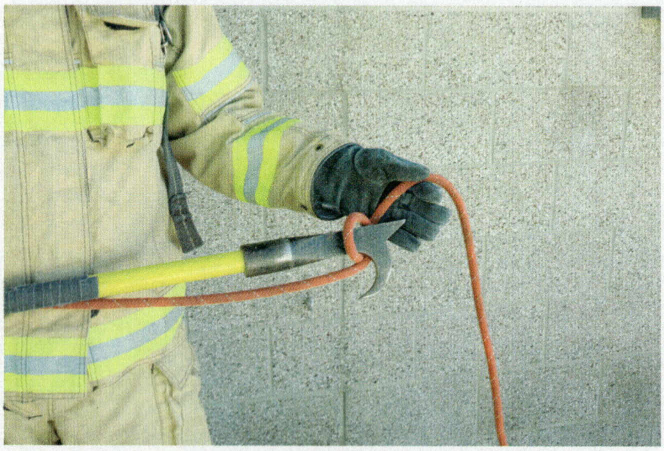

3. Place a second half hitch over the handle, and then secure it under the head of the pike pole.

4. Communicate with the firefighter above that the pike pole is ready to raise.

© Jones & Bartlett Learning. Photographed by Brandon Fryman/JRM Creative.

SKILL DRILL 8-17

Hoisting a Ladder Support Person, NFPA 1010: 5.3.8

1. Tie a figure eight on a bight to create a loop approximately 3 to 4 ft (1 to 1.2 m) in diameter and large enough to fit around both ladder beams (the long vertical sides of the ladder).

2. Starting at the top of the ladder, pass the loop under the rungs, between the beams, until the loop is three or four rungs from the top of the ladder. Pull the end of the loop from under the rungs and toward the top of the ladder.

3. Place the loop over the top tip of the ladder, over the outside of both beams.

Continues.

SKILL DRILL 8-17 CONTINUED

Hoisting a Ladder Support Person, NFPA 1010: 5.3.8

4. Pull the running end of the rope to remove the slack from the rope.

5. Attach a tag line to the bottom rung of the ladder to stabilize it as it is being hoisted.

6. Communicate with the firefighter above that the ladder is ready to be raised. Hold on to the tag line to stabilize the bottom of the ladder as it is being hoisted.

© Jones & Bartlett Learning. Photographed by Brandon Fryman/JRM Creative.

Hoisting a Charged Hose Line

A **charged hose line** is a hose line filled with water and under pressure from the pump. It is almost always preferable to hoist a dry or uncharged hose line because water adds considerable weight to a hose line. Water weighs 8.33 lb per gallon (3.79 kg per liter), which makes hoisting a charged line much more difficult. For example, a 50-ft (15-m) dry hose that is 2½ in. (64 mm) in diameter weighs about 30 lb (14 kg). When that hose is charged and filled with water, it weighs about 140 lb (64 kg). However, there will be occasions when it is necessary to hoist a charged hose line. After receiving the end of the utility rope from the firefighter above, follow the steps in **SKILL DRILL 8-18** to hoist a charged hose line.

SKILL DRILL 8-18

Hoisting a Charged Hose Line Support Person, NFPA 1010: 5.3.8

1. Make sure that the nozzle shut-off handle is in the forward position so that it is completely closed.

2. Tie a clove hitch 1 to 2 ft (30 to 60 cm) behind the nozzle. Add a safety knot to secure the loose end of the rope on the side of the clove hitch away from the nozzle.

3. At the end of the rope closest to the nozzle, make a bight in the rope, even with the nozzle shut-off handle. Insert the bight through the handle opening.

4. Slip the bight loop over the end of the nozzle.

Continues.

SKILL DRILL 8-18 CONTINUED

Hoisting a Charged Hose Line Support Person, NFPA 1010: 5.3.8

5. Pull the working end of the rope to pull the bight tight. This creates a half hitch and secures the handle in the off position while the charged hose line is hoisted.

6. Communicate with the firefighter above that the hose line is ready to hoist. The knot can be released after the line is hoisted by removing the tension from the rope and slipping the bight back over the end of the nozzle.

© Jones & Bartlett Learning. Photographed by Brandon Fryman/JRM Creative.

Hoisting an Uncharged Hose Line

An **uncharged hose line** is a hose line that is not filled with water and not under pressure from a pump. Before hoisting an uncharged hose line, fold the hose back on itself and place the nozzle on top of the hose about 3 ft (1 m) from the bend in the hose that you created by folding the hose. This action eliminates unnecessary stress on the couplings by ensuring that the rope pulls on the hose and not directly on the nozzle. After receiving the end of the utility rope from the firefighter above, follow the steps in **SKILL DRILL 8-19** to hoist an uncharged hose line.

Hoisting an Exhaust Fan or Power Tool

Several types of tools and equipment, such as exhaust fans, chainsaws, circular saws, and any other object that has a strong, closed handle, are hoisted using the same technique. The hoisting rope is secured to the object by passing the rope through the opening in the handle. A figure eight follow-through knot or bowline with a safety knot is then used to close the loop.

You should practice hoisting the actual tools and equipment used in your department. You should be able to perform this task automatically and in adverse conditions.

Remember that you always use utility rope for hoisting tools. You do not want to get oil or grease on designated life safety ropes. If a life safety rope becomes oily or greasy, it should be taken out of service and destroyed so that it will not be used again mistakenly as a life safety rope. The damaged rope can be cut into short lengths and designated and downgraded to be used as utility rope.

After receiving the end of the utility rope from the firefighter above, follow the steps in **SKILL DRILL 8-20** to hoist an exhaust fan.

SKILL DRILL 8-19

Hoisting an Uncharged Hose Line Support Person, NFPA 1010: 5.3.8

1. Fold about 3 ft (1 m) of hose back on itself and place the nozzle on top of the hose.

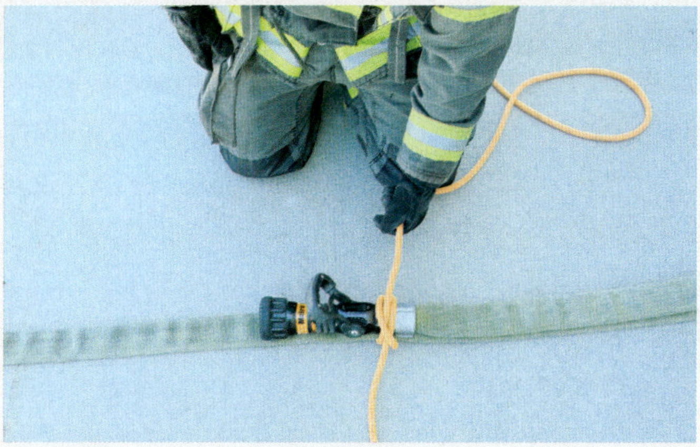

2. Tie a clove hitch around both the nozzle and the hose it is lying on top of, and then tighten the knot securely.

3. Tie a safety knot around the nozzle and the hose below the clove hitch.

4. Tie two half hitches around the folded-over hose, evenly spaced between the nozzle and the bend in the hose created when you folded the hose.

Continues.

SKILL DRILL 8-19 CONTINUED

Hoisting an Uncharged Hose Line Support Person, NFPA 1010: 5.3.8

5. Communicate with the firefighter above that the hose line is ready to hoist. Hoist the hose with the fold at the top and the nozzle pointing down. Before releasing the rope, the firefighters at the top must pull up enough hose so that the weight of the hanging hose does not drag down the hose.

© Jones & Bartlett Learning. Photographed by Brandon Fryman/JRM Creative.

SKILL DRILL 8-20

Hoisting an Exhaust Fan or Power Tool Support Person, NFPA 1010: 5.3.8

1. Tie a figure eight knot in the rope about 3 ft (1 m) from the working end of the rope. Loop the working end of the rope around the fan handle and back to the figure eight knot.

2. Secure the rope by tying a figure eight follow-through.

SKILL DRILL 8-20 CONTINUED

Hoisting an Exhaust Fan or Power Tool Support Person, NFPA 1010: 5.3.8

3. Attach a tag line to the opposite corner on the same side of the fan for better control.

4. Communicate with the firefighter above that the exhaust fan is ready to hoist.

© Jones & Bartlett Learning. Photographed by Brandon Fryman/JRM Creative.

Choosing the Appropriate Knot

Now that we have discussed the basic knots and hitches used in the fire service, you may be asking, with so many options, how do I choose which knot to use? Remember, task objectives drive tactics. For example, the task objective is to get a particular tool to a location on the fireground as fast and safely as possible so that it can be deployed as quickly as possible. With the length of rope available and the type of tool, you must decide which knot or hitch will accomplish that task.

How do you choose? You draw upon your knowledge to match the right solution to the issue at hand. For example, you need to hoist a pike pole to the second floor of a structure so that the salvage and overhaul team can carry out their operations. You know that clove hitches are ideal for tools that are cylindrical, such as an axe or a pike pole, and you apply that solution. If we changed the tool to a chainsaw, then you would use a bowline or figure eight follow-through because there is a handle to tie the knot around. Keep the basics in mind and use that knowledge to make swift and accurate decisions.

TIP

The fire and rescue business is dynamic; this is why fire service personnel need to constantly train, read, study, and stay current with new techniques and standards. Learning about ropes and knots will not end with this chapter.

CASE STUDY

You Are the Support Person CONCLUSION

You've arrived at a three-story townhouse fire. The engine company has knockdown on the fire on the top floor. The captain of the truck company drops a rope from the third-floor window and calls for the ventilation fan.

1. What type of rope is best used for this activity?

   Answer: Utility rope.

2. What knot will you use to tie the fan with?

   Answer: Figure eight follow-through with a loop.

3. What other steps will you take before giving the signal for the truck company to hoist the fan?

   Answer: Attach a tag line; guide the fan up so it doesn't hit anything; communicate.

WRAP-UP

SUMMARY

- The four types of rope primarily used in the fire service are life safety rope, escape rope, throwlines for water rescues, and utility rope. Each of these has a specific function.
- Life safety rope must be used to support the weight of one or more persons, whether during training or during firefighting, rescue, or other emergency operations.
- Escape rope is intended for emergency, self-rescue situations. Fire escape rope is intended specifically for emergency self-rescue from an immediately hazardous environment in which fire or fire products are involved.
- A throwline is used during water rescue operations.
- Utility rope is used in most cases when it is not necessary to support the weight of a person, such as when hoisting or lowering tools or equipment.
- Webbing is a woven material of flat or tubular weave in a long strip. Although it is not actually considered a type of rope, webbing is often used in conjunction with rope and may be preferable to rope in certain situations.
- The three most common types of rope construction used in the fire service are twisted rope, braided rope, and kernmantle rope.
- Twisted rope, which is also called laid rope, is made of individual fibers twisted into strands and then twisted together. Twisted ropes are suitable for many activities on the fireground, including hoisting tools.
- Braided rope is constructed by weaving or intertwining strands together in the same way that hair is braided. Braided ropes serve well as utility rope for moving and hauling equipment and holding items in place.
- Kernmantle rope consists of two distinct parts: the kern and the mantle. Kernmantle rope is well suited for rescue work and is very popular for life safety rope.
- A dynamic rope is designed to stretch more than a static rope when it is loaded. Dynamic rope is typically used in safety lines in climbing.
- A static rope has a lower range of elasticity than dynamic rope and is more suitable for most fire rescue situations in which sudden shock loads are less likely and when limited elongation is desired, such as when lowering, raising, or rappelling for a rescue.
- All ropes—especially life safety ropes—need proper care. Proper care ensures a long life and reduces the chance of equipment failure and accidents.

- Life safety ropes must be inspected after each use, whether the rope was used for an emergency incident or in a training exercise. Unused life safety rope should be inspected on a regular schedule.
- Ropes need to be stored in areas where there is some air circulation, away from temperature extremes, and out of sunlight.
- The safety knot is used to back up or finish knots, such as the clove hitch, bowline, or sheet bend.
- A figure eight knot is used to prevent the end of a rope from passing through a device such as a pulley. This knot is the basis of other figure eight knots, including the figure eight on a bight, the figure eight follow-through, and the figure eight bend.
- A hitch is a knot that can be slipped over or tied around an object to secure the rope to the object and is easily untied when the object is removed. Hitches are ideal for hoisting many fire service tools such as axes, pike poles, and hose lines.
- The half hitch is a loop of rope passed around or over an object or around the rope itself. It is not a secure knot, so it is used only in conjunction with other knots.
- A clove hitch is essentially two half hitches oriented in opposite directions. The clove hitch can be used to tie a hoisting rope around an axe or pike pole.
- A loop knot is used to form a loop in the end of a rope. Loops can be used to hoist tools, secure a person during a rescue, or secure a rope to a fixed object.
- The figure eight on a bight knot creates a secure loop at the working end of a rope. This loop can be used to attach the end of the rope to a fixed object or a piece of equipment.
- A figure eight follow-through knot is used to create a secure loop at the end of the rope when the rope must be wrapped around an object or passed through an opening before the loop is formed.
- A bowline is also used to form a loop in the end of a rope. The bowline can be used as an alternate knot for a figure eight on a bight or figure eight. This type of knot is frequently used to secure the end of a rope to an object or an anchor point. It also can be used to hoist equipment.
- A bend knot is used to join two ropes or pieces of webbing together. These knots can also be used to join a rope to a chain or a cable. The three types of bend knots a firefighter will most often use are the sheet bend, the figure eight bend, and the water knot.
- The sheet bend (or Becket bend) is used to join two ropes of unequal diameters. A sheet bend is also used to join a rope to a chain or to hoist a rope of a different diameter. It should not be relied upon to support a human load.
- The figure eight bend (or tracer eight) is used to join two ropes that have a similar diameter. When used with life safety rope, this knot can support a human load.
- The water knot (or ring bend) is used to join webbing of the same or different widths together. When a single piece of webbing is used and the opposite ends are tied to each other, a loop or sling is created.
- In emergency situations, you may need to help raise or lower a tool or equipment to firefighters. It is essential that the correct knots are used, tied correctly, and secured and that the movement of the objects is controlled throughout the hoist.
- An axe should be hoisted in a vertical position with the head of the axe lower than the handle.
- A pike pole is hoisted in a vertical position with the head at the top.
- A ladder is hoisted in a vertical position. A tag line attached to the bottom helps keep the ladder under control as it is hoisted. If it is a roof ladder, the hooks on the ladder should be in the retracted position.

KEY TERMS

bend knot A knot that joins two ropes or webbing pieces together. (NFPA 2500)

bight The open loop in a rope or piece of webbing formed when it is doubled back on itself. (NFPA 1006)

block creel construction Rope constructed without knots or splices in the yarns, ply yarns, strands or braids, or rope. (NFPA 2500)

braided rope Rope constructed by intertwining strands in the same way that hair is braided.

charged hose line A hose line filled with water and under pressure from the pump.

dress To tighten and remove twists, kinks, and slack from the rope after tying a knot.

dynamic rope A rope typically used for climbing that is designed to be elastic and stretch when loaded.

escape rope Rope dedicated solely for the purpose of supporting people during emergency self-escape (self-rescue); not intended for use in a hazardous

KEY TERMS CONTINUED

environment involving fire or fire products; not classified as a life safety rope. (NFPA 2500)

fire escape rope Rope dedicated solely for the purpose of supporting people during emergency self-escape (self-rescue) from an immediately hazardous environment involving fire or fire products; not classified as a life safety rope. (NFPA 2500)

general use life safety rope A life safety rope with a diameter that is at least 7/8 in. (11 mm) but not larger than 5/8 in. (16 mm), with a minimum breaking strength of 8992 lbf (40 kN).

highline system A system of using rope or cable suspended between two points for movement of persons or equipment over an area that is a barrier to the rescue operation, including systems capable of movement between points of equal or unequal height. (NFPA 1006)

hitch A knot that attaches to or wraps around an object so that when the object is removed, the knot will fall apart. (NFPA 2500)

kern In a kernmantle rope, the center or core of the rope.

kernmantle rope Rope made of two parts, the kern and the mantle.

knot A fastening made by tying rope or webbing in a prescribed way. (NFPA 2500)

laid rope See *twisted rope.*

life safety rope Rope dedicated solely for the purpose of supporting people during rescue, firefighting, other emergency operations, or training evolutions. (NFPA 2500)

loop A piece of rope formed into a circle by crossing the rope.

loop knot A knot that forms a secure loop in the end of a rope.

mantle In a kernmantle rope, the braided covering that protects the kern from dirt and abrasion. Also called *sheath.*

minimum breaking strength (MBS) The result of subtracting 3 standard deviations from the mean result of the lot being tested using the formula in 8.2.5.2. (NFPA 2500)

rigging The process of building a system to move or stabilize a load. (NFPA 1006)

rope A compact but flexible, torsionally balanced, continuous structure of fibers produced from strands that are twisted, plaited, or braided together and that serve primarily to support a load or transmit a force from the point of origin to the point of application. (See also 3.3.153.2, Life Safety Rope.) (NFPA 1006)

rope bag A bag used to protect and store rope so that the rope can be easily and rapidly deployed without kinking.

rope record A record for each piece of rope that includes a history of when the rope was placed in service, when it was inspected, when and how it was used, and which types of loads were placed on it.

round turn A piece of rope looped to form a complete circle with the two ends parallel.

running end The part of a rope that is not used to form a knot.

safety knot A knot used to back up another knot that has a tendency to become loose when not continuously loaded or that can slip when loaded by securing the leftover working end of the rope.

sheath See *mantle.*

shock load An instantaneous load that places a rope under extreme tension, such as when a falling load is suddenly stopped as the rope becomes taut.

standing part The part of a rope between the working end and the running end.

static rope A rope that stretches very little under load.

tag line A rope that personnel on the ground can use to guide an object that is being hoisted or lowered.

technical use life safety rope A life safety rope with a diameter that is at least 3/8 in. (9.5 mm) but not larger than 1/2 in. (12.5 mm), with a minimum breaking strength of 4496 lbf (20 kN) and that is used by members of highly trained rescue teams who deploy to technical environments such as mountainous or wilderness terrain.

throw bag A water rescue system that includes 50 ft to 75 ft (15.24 m to 22.86 m) of water rescue rope, an appropriately sized bag, and a closed-cell foam float. (NFPA 1006)

throwline A floating rope that is intended to be thrown to a person during water rescues or as a tether for rescuers entering the water. (NFPA 2500)

trailer A label that travels the entire length of a life safety rope under the outer sheath that identifies the rope as a life safety rope.

twisted rope Rope constructed of fibers twisted into strands, which are then twisted together. Also called *laid rope.*

uncharged hose line A hose line that is not filled with water and not under pressure from a pump.

utility rope Rope used for securing objects, hoisting equipment, or securing a scene to prevent bystanders from being injured; utility rope must never be used in life safety operations.

webbing Woven material of flat or tubular weave in the form of a long strip. (NFPA 2500)

working end The part of the rope used for forming a knot.

REVIEW QUESTIONS

1. What are the four primary types of ropes used in the fire service?
2. What type of fiber is commonly used for fire service rope?
3. What are the three main rope constructions?
4. What type of construction is required for all life safety rope?
5. What are the four parts of the rope maintenance formula?
6. What does a knot do to the load-carrying capacity of the rope?
7. When hoisting equipment, what should you use to guide it and prevent it from striking obstructions?

DISCUSSION QUESTIONS

1. Describe the characteristics of each of the four types of rope.
2. Describe the four parts of rope maintenance, and explain why each part is necessary.
3. Describe some key factors to consider when using rope to hoist tools or equipment.

REFERENCES

National Fire Protection Association. 2020. *NFPA 1006, Standard for Technical Rescue Personnel Professional Qualifications.* 2021 Edition. Quincy, MA: National Fire Protection Association.

National Fire Protection Association. 2021. *NFPA 2500, Standard for Operations and Training for Technical Search and Rescue Incidents and Life Safety Rope and Equipment for Emergency Services.* 2022 Edition. Quincy, MA: National Fire Protection Association.

Chapter Opener: © Eric Scruggs

CHAPTER

9

Support Person

Water Supply Systems

KNOWLEDGE OBJECTIVES

After studying this chapter, you will be able to:

- Describe the characteristics of dry-barrel hydrants.
- Describe the characteristics of wet-barrel hydrants.
- List examples of suitable static water supply sources.
- Describe the equipment and procedures that are used to access static sources of water.
- Describe the advantages of a portable tank system.
- Describe the characteristics of a mobile water supply apparatus.

SKILLS OBJECTIVES

After studying this chapter, you should be able to perform the following skills:

- Demonstrate the safe operation of a fire hydrant.
- Demonstrate how to safely shut down a fire hydrant.
- Assist the pump driver/operator with drafting.
- Perform the connection and proper placement of an intake hose for drafting operations.
- Set up a portable tank and the equipment necessary to use it.

ADDITIONAL NFPA STANDARDS

- **NFPA 24**, *Standard for the Installation of Private Fire Service Mains and Their Appurtenances, 2022 Edition*
- **NFPA 291**, *Recommended Practice for Water Flow Testing and Marking of Hydrants, 2022 Edition*
- **NFPA 1140**, *Standard for Wildland Fire Protection, 2022 Edition*
- **NFPA 1142**, *Standard on Water Supplies for Suburban and Rural Fire Fighting, 2022 Edition*
- **NFPA 1900**, Standard for Aircraft Rescue and Firefighting Vehicles, Automotive Fire Apparatus, Wildland Fire Apparatus, and Automotive Ambulances, *2024 Edition*
- **NFPA 1910**, *Standard for the Inspection, Maintenance, Refurbishment, Testing, and Retirement of In-Service Emergency Vehicles and Marine Firefighting Vessels, 2024 Edition*
- **NFPA 1960**, *Standard for Fire Hose Connections, Spray Nozzles, Manufacturer's Design of Fire Department Ground Ladders, Fire Hose, and Powered Rescue Tools, 2024 Edition*
- **NFPA 1962**, *Standard for the Care, Use, Inspection, Service Testing, and Replacement of Fire Hose, Couplings, Nozzles, and Fire Hose Appliances, 2018 Edition*

CASE STUDY

You Are the Support Person

Your department has been dispatched to a reported building fire. Upon arrival, you find a barn that is fully involved. The fire attack is clearly going to be a defensive operation, but aggressive actions will be required to protect a house and other buildings that are located approximately 100 yards (91 meters) away from the barn, which is located on a narrow, tree-lined lane. The incident commander says he needs a flow of 500 gallons per minute (gpm; 1893 liters per minute [L/min]) to be established.

1. What water sources might be available?
2. Which type of water supply would be most dependable?
3. Which type of water source would require the most resources to implement?

Introduction

After life safety, one of the primary objectives at a fire is to secure an adequate water supply to get water on the fire and protect exposures. Water cools the fuel, allowing the fire to be extinguished. **Water supply**, the source of water utilized by firefighters; hydraulics, the behavior of flowing water; rural and municipal water supplies; and types of fire hydrants and their operation and maintenance are all critical concepts that you need to understand. This chapter covers the water flow during the first half of its journey for the fire service—from the water source to the fire pumper or fire engine.

When determining a water supply, you need to make sure it is reliable and has sufficient volume and pressure to meet the needs of the incident. The importance of a dependable and adequate water supply for fire-suppression operations is self-evident. The charged hose line (one or more lengths of hose coupled together) is not only the primary weapon for fighting fire but also the firefighter's primary defense against being burned or driven out of a burning building. The basic plan for fighting most fires depends on having an adequate supply of water to safely confine, control, and extinguish the fire.

Water supply cannot be interrupted while crews are working inside a building; firefighters can be injured or killed without an adequate water supply. Firefighters entering a burning building need to be confident that their water supply is both reliable and adequate—for their protection and to extinguish the fire.

The source of water varies in different communities. Some communities have a **municipal water system**, a public water supply network owned and maintained by a government or municipality. Other communities have a **private water system**, a system owned and maintained by a private business entity. These water systems provide water under pressure through fire hydrants (**FIGURE 9-1**). Fire hydrants are usually a reliable and practical source of water, but fire hydrants have limited capacity and will not provide an

FIGURE 9-1 The water that comes from a hydrant is provided by a municipal or private water system.

© Jones & Bartlett Learning. Photographed by Glen E. Ellman.

endless supply of water. Automatic sprinkler systems and standpipe systems may be connected directly to a municipal water source, affecting the available water from the hydrants. Rural or remote areas may depend on a **static water source**, a water source that is not under pressure, such as a pond, lake, or stream, or even a swimming pool, private cistern, or other body of water. Some areas depend on large water tenders (often called *tankers*) that carry up to 5000 gallons (gal; 18,927 liters [L]) of water as their water supply. These sources serve as drafting sites for fire department apparatus to obtain and deliver water to the fire scene. To **draft** is the process of drawing water up through a hose from a static water source and into the pump on an apparatus.

Often, the first available water source is the water carried in a tank on the first-arriving fire apparatus. Fire engines carry a minimum of 300 gal (1100 L) of water and may carry up to 1000 gal (3785 L) of water. This tank water can be used in the initial fire attack while the establishment of an adequate, continuous water supply is secured to support continued fire attack and to ensure the safety of the firefighters on the scene. Departments that do not rely on a public water supply and fire hydrants often have a mobile water supply apparatus that can carry 1000 to 5000 gal (3785 to 18,927 L) of water. The operational plan must ensure that an adequate and reliable water supply is available before the tank becomes empty.

Municipal Water Systems

As the name suggests, most municipal water systems are owned and operated by a local government agency, such as a city, county, or special water district, although some municipal water systems are privately owned. Municipal water is supplied to homes, commercial establishments, and industries, as well as to hydrants and fire protection systems in buildings.

A municipal water system has three major components: a water source, a water treatment facility, and a distribution system.

Water Sources

The water source for a municipal water system can be wells, rivers, streams, lakes, or human-made water storage facilities. The source depends on the geographic and hydrologic features of the area. Underground pipelines or open canals supply some cities with water from sources that are many miles away. The water source needs to be large enough to meet the total demands of the service area. Many municipal water systems draw water from multiple sources to ensure a sufficient supply. In addition, most municipal water systems include large storage facilities to ensure that they will be able to meet the community's water supply demands if access to the primary source or courses is interrupted. The backup supply for some systems can provide water for several months or years. In other systems, however, the supply may last only a few days.

FIGURE 9-2 Impurities are removed at the water treatment facility.

© Hundley Photography/Shutterstock

Water Treatment Facilities

Municipal water systems include a water treatment facility, where impurities are removed from the water (**FIGURE 9-2**). The nature of the treatment system depends on the quality of the untreated source water. Source water that is clean and clear requires little treatment, but some systems must use extensive filtration to remove impurities and foreign substances. Some treatment facilities use chemicals to remove impurities, kill bacteria and harmful organisms, and improve the water's taste. Ultraviolet (UV) light sources are also used to kill bacteria and harmful organisms and to keep the water pure as it moves through the distribution system to individual homes or businesses. All the water in the system must be suitable for drinking.

Water Distribution System

After the water has been treated, it enters the distribution system. The distribution system delivers water from the treatment facility to the end users and fire hydrants through a complex network of underground pipes. This type of pipe is called a **water main**. In most cases, the distribution system also includes pumps, storage tanks, control valves, reservoirs, and other necessary components to ensure that the required volume of water can be delivered where and when it is needed at the required pressure.

Water pressure requirements differ, depending on how the water will be used. Generally, water pressure ranges from 20 to 80 pounds per square inch (psi; 138 to 551 kilopascals [kPa]) at the delivery point. The recommended minimum pressure for water coming from a fire hydrant is 20 psi (138 kPa), but it is possible to operate with lower hydrant pressures under some circumstances. In some locations with high fire risk, higher-pressure systems may be present.

Most water distribution systems rely on an arrangement of pumps to provide the required water pressure. If the pumps stop operating, the pressure is lost, and the system is unable to deliver adequate water pressure or volume to the end users or to hydrants. Most systems that use pumps have multiple pumps and backup power supplies to reduce the risk of a service interruption due to a pump failure. The extra pumps can be used to boost the flow when the pressures in the system drop below a predetermined setting, such as for a major fire or during other high-demand periods.

SAFETY TIP

During operations at a major fire, the water department should be contacted to boost the water supply and pressure to the area of the alarm.

In a **gravity-feed system**, the water source, water treatment facility, and water storage facilities are located on high ground, whereas the end users live in lower-lying areas, such as a community in a valley (**FIGURE 9-3**). This type of system may not require any pumps because gravity provides the necessary pressure to deliver the water downhill. In some systems, the difference between the elevation of the water and the end users is so great that the water pressure created by gravity is so high that pressure-control devices are needed to keep the system from being subjected to excessive pressure.

Many municipal water supply systems use a combination pump and gravity-feed system to deliver water. Pumps can be used to deliver water from the water treatment facility to an **elevated water storage tower** or to storage tanks located on hills or high ground. The elevated water storage facilities maintain the desired water pressure in the distribution system, ensuring that water can be delivered under pressure even if the pumps are not operating (**FIGURE 9-4**). When the elevated water storage facilities need refilling, large supply pumps are used. Additional pumps may be installed to increase the pressure in particular areas, such as for a neighborhood on a hilltop.

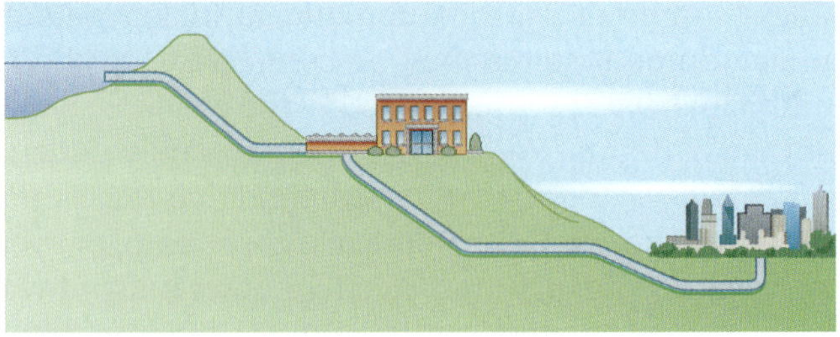

FIGURE 9-3 A gravity-feed system can deliver water to a low-lying community without the need for pumps.

© Jones & Bartlett Learning

FIGURE 9-4 Water stored in an elevated water storage tower can be delivered to end users under pressure.

© Jones & Bartlett Learning. Photographed by Glen E. Ellman.

Combination pump and gravity-feed systems must maintain enough water in elevated storage tanks and in reservoirs to meet anticipated demands. When more water is used than the pumps can supply or if the pumps are out of service, some systems can operate for several days using their elevated storage reserves, but others may be able to function for only a few hours before the system becomes empty.

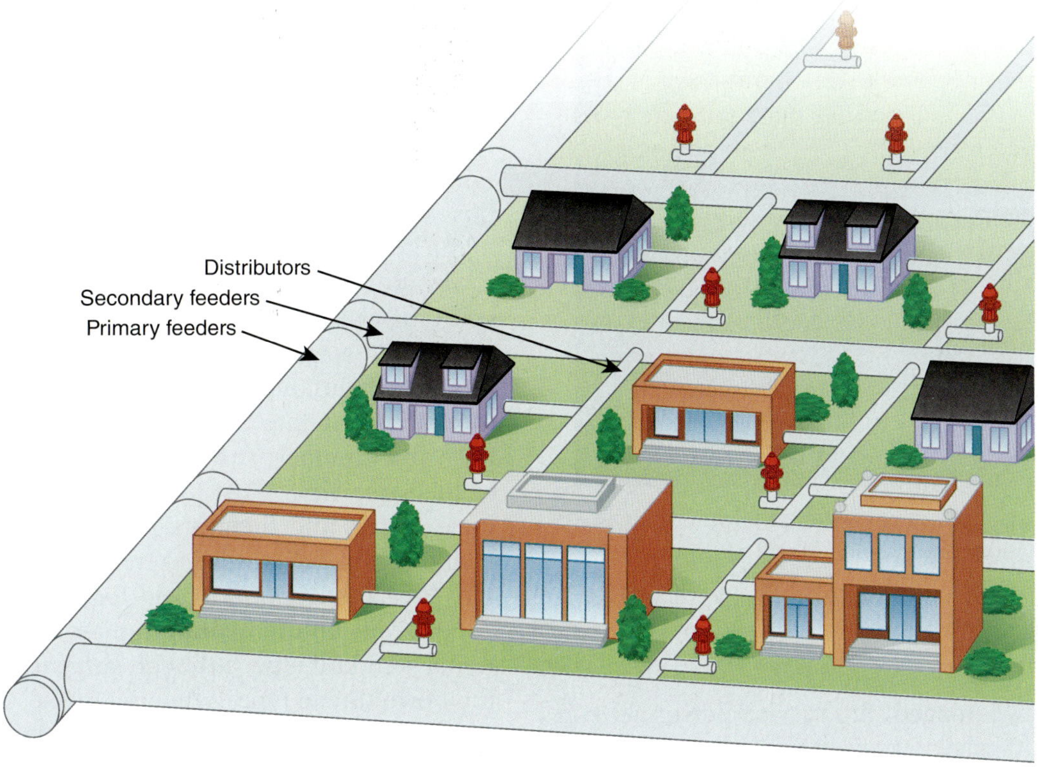

FIGURE 9-5 The underground distribution system includes three different sizes of water main: primary feeders, secondary feeders, and distributors.

© Jones & Bartlett Learning

The underground distribution system that delivers water to end users is made up of three sizes of water main (**FIGURE 9-5**). A large main is called a **primary feeder** or **trunk line** and carries large quantities of water to a section of the municipality. A smaller main is called a **secondary feeder** or **branch line** and is connected to the primary feeder and distributes water to a smaller area such as a neighborhood. The smallest type of main, called a **distributor pipe**, or simply **distributor**, carries water to the users and to hydrants along individual streets.

The size of water mains varies depending on the amount of water needed both for normal consumption and for fire protection in each location. Most jurisdictions specify the minimum size main that can be installed to ensure an adequate flow. Older systems, however, may have undersized water mains. In addition, the volume of water delivered through a water main may decrease if the pipe becomes corroded or partly filled with sediment.

Water mains in a well-designed distribution system are laid out in a grid pattern. A grid arrangement provides **water flow** to a fire hydrant from two or more directions and establishes multiple paths from the water source to each area. This helps ensure an adequate flow of water for firefighting. The grid design also helps minimize downtime for the other portions of the system if a water main breaks or needs maintenance work because the water flow can be diverted around the affected section.

Older water distribution systems may have a **dead-end water main**, which is a main that is supplied from only one direction. Fire hydrants on a dead-end main have a limited water supply. If two or more hydrants on the same dead-end main are used to fight a fire, the hydrant closest to the water source may have more water and greater water pressure than the hydrants farther from the water supply.

Control valves installed at intervals throughout a water distribution system allow different sections of the water supply to be turned off or isolated. These valves are used when a water main breaks or when work must be performed on a section of the system.

A **shut-off valve** is located at each connection point where the underground mains meet the distributor pipes. These valves control the flow of water to individual customers or to individual fire hydrants (**FIGURE 9-6**). If the water system in a building or to

FIGURE 9-6 A shut-off valve controls the water supply to an individual user or fire hydrant. The shut-off valve is located under the cover.

© Jones & Bartlett Learning. Photographed by Glen E. Ellman.

a fire hydrant is damaged, fire service personnel can close the shut-off valve to prevent further water flow.

The fire department should notify the water department when fire operations will require prolonged use of large quantities of water. The water department may be able to increase the normal volume and pressure by starting additional pumps. In some systems, the water department can open valves to increase the flow to a certain area in response to fire department operations at major fires.

Fire Hydrants

Fire hydrants provide water for firefighting purposes. Fire hydrants are part of the municipal water distribution system and connect directly to public water mains. Fire hydrants are also installed on private water systems supplied by the municipal water system or by a separate source. The water source, as well as the adequacy and reliability of the supply to private hydrants, must be identified to ensure that a sufficient water supply will be available when fighting fires.

Most fire hydrants consist of an upright steel casing—the **barrel**—attached to the underground water distribution system. An **outlet** or outlets are provided to connect fire department hose to the hydrant. The outlets may be of various sizes depending on the local jurisdiction, although 2½-inch (in.; 64-millimeter [mm]) connections and one larger connection (4½, 5, or 6 in. [114, 127, or 154 mm])—called a **steamer port**—are common. Fire hydrants are equipped with one or more valves to control the flow of water through the hydrant.

Fire hose needs to be connected to the fire hydrant. The connection on the end of the hose that you use to connect hoses to outlets on fire hydrants and fire pumps is called a **coupling**. Couplings are either threaded or nonthreaded. A **threaded hose coupling** has a male hose coupling with threads on the outside on one end of the hose and a female hose coupling with threads on the inside on the other end of the hose. The female hose coupling also has a **swivel**, which is a ring around the coupling that you turn to secure the female coupling around a male coupling without twisting the hose. Nonthreaded couplings are the same on both ends of the hose and interlock with another nonthreaded coupling of the same type. Most hydrants and fire pumps have male outlets and therefore require a female coupling. (Couplings are discussed in more detail in Chapter 11, *Fire Hose, Fire Appliances and Tools, and Nozzles.*)

Threads on fire hydrant outlets are usually the national standard type, although some jurisdictions have their own thread type. When not in use, outlets are covered by a **hydrant cap** (also called a **discharge cap**).

The two most common types of fire hydrants are the dry-barrel hydrant and the wet-barrel hydrant, both of which are connected to a pressurized water source. A third type of hydrant is the dry hydrant, which drafts water from a static water source.

Dry-Barrel Hydrants

A **dry-barrel hydrant**, also called **frost-proof hydrant**, is used in climates where temperatures fall below freezing (**FIGURE 9-7**). These fire hydrants are connected to a pressurized municipal water system. The valve that controls the flow of water into the barrel of the hydrant is located at the base of the barrel below the frost line to keep the hydrant from freezing (**FIGURE 9-8**). The length of the barrel depends on the climate and the depth of the water main below ground. Water enters the barrel of a dry-barrel hydrant only when it is used. Turning the **stem nut** on the top of the hydrant—called the **bonnet**—rotates the **operating stem** that opens the valve below ground so that water flows up into the barrel of the hydrant. Taking this action will **charge** the hydrant.

Whenever this type of hydrant is not in use, the barrel must remain dry. If the barrel contains standing water, it will freeze in cold weather and render the hydrant inoperable. An opening at the bottom of the barrel allows the water to drain out after each use. When the hydrant valve is fully closed, the drain is fully open. When

FIGURE 9-7 A dry-barrel hydrant.

Courtesy of American AVK Company.

FIGURE 9-8 A dry-barrel hydrant is controlled by an underground valve.

© Jones & Bartlett Learning

FIGURE 9-9 Dry-barrel hydrant with a gate valve attached.

© Jerry Bergquist/Shutterstock

the hydrant valve is opened, the drain closes, which prevents water from being forced out of the drain when the hydrant is under pressure. If the drain becomes clogged, the hydrant may not drain. In this case, it may be necessary to pump water out of the hydrant before it freezes.

Hydrant valves should always be fully opened or fully closed. If the valve is partially opened, the drain is also partially open, so pressurized water can flow out. This leakage can erode the soil around the base of the hydrant and may damage the hydrant or cause the water main to break. A fully opened hydrant valve also allows the maximum flow of water available to fight a fire.

Most dry-barrel hydrants contain only one large valve that controls the flow of water to all outlets. Before that valve is opened, each outlet must be connected to a hose or a gate valve, or it must have a hydrant cap firmly in place before the valve is turned on. A **gate valve** is a valve attached to a hydrant outlet that can be used to turn off the flow of water at that outlet. This allows you to connect hose to that outlet after the hydrant is charged (**FIGURE 9-9**). Although it is not required to use a gate valve, it is the best practice to use one when connecting to a dry-barrel hydrant. If a gate valve is not attached before the hydrant is charged, the hydrant would need to be shut off before additional lines could be added.

Wet-Barrel Hydrants

A **wet-barrel hydrant** is used in locations where temperatures do not drop below freezing. Like dry-barrel hydrants, wet-barrel hydrants are connected to a pressurized water source. These fire hydrants always have water in the barrel and do not have to be drained after

FIGURE 9-10 Each outlet on a wet-barrel hydrant has its own valve that controls the flow of water to that outlet.

Courtesy of American AVK Company.

FIGURE 9-11 A wet-barrel fire hydrant after being struck by a car.

© Don Bartell/Alamy Stock Photo

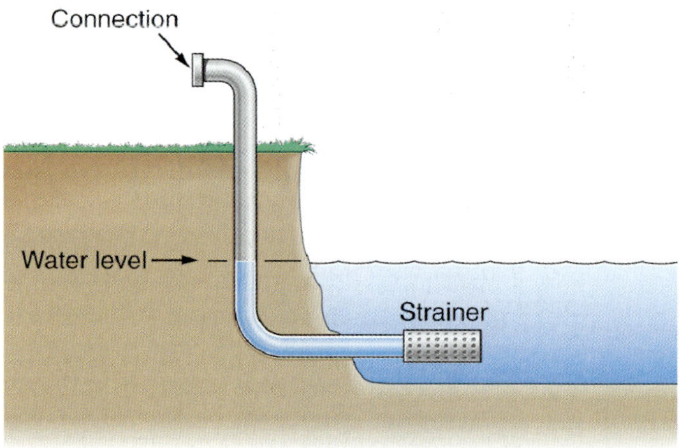

FIGURE 9-12 A dry hydrant, or drafting hydrant, can be placed at an accessible location near a static water source.

© Jones & Bartlett Learning

each use. Each outlet on a wet-barrel hydrant has its own valve that controls the flow of water to that outlet (**FIGURE 9-10**). The operating nut for each individual valve is on the opposite side of each outlet. You can hook up one hose line and begin flowing water, and then later attach a second hose line and open the valve for that outlet without first shutting down the hydrant.

Because wet-barrel hydrants are always full of water and the control valve is aboveground, there is a chance that water will flow uncontrolled if the hydrant is damaged (**FIGURE 9-11**). Newer wet-barrel hydrants have a flapper valve to prevent water flow if the hydrant is damaged or struck by a vehicle.

Dry Hydrants

A **dry hydrant**, also called a **drafting hydrant**, is a fixed piping system with one end in a static water source and the other end accessible to attach a hose. Do not confuse the terms *dry-barrel hydrant* and *dry hydrant*—they are very different. Dry-barrel hydrants are connected to a pressurized municipal water system in areas where the temperature drops below freezing. A dry hydrant is a permanent piping system connected to a static water source such as a stream, pond, or lake. The end of the pipe in the water has a strainer on it. The strainer end should be located below the water's surface and away from any silt or potential obstructions. The other end has a connection for a hard suction hose (**FIGURE 9-12**). The connection should be accessible to fire apparatus, and the connection should be at a convenient height for hook-up (**FIGURE 9-13**). To get water from a dry or drafting hydrant, you must draft the water using a fire pump.

Dry hydrants are often installed in lakes and rivers and close to clusters of buildings where there is a recognized need for fire protection. They may also be installed in farm cisterns or connected to swimming pools on private property to make water available for

FIGURE 9-13 The height of the dry hydrant or drafting hydrant connection is convenient for hook-up to an engine or a portable pump.

Courtesy of Ryan Van Buskirk.

the local fire department. In some areas, dry hydrants provide access to static water sources in areas that are inaccessible to fire apparatus. They allow fire service personnel to reach water under the frozen surface of a lake or river. NFPA 1142, *Standard on Water Supplies for Suburban and Rural Fire Fighting, 2022 Edition,* has more information about dry hydrants.

Fire Hydrant Operation

Fire service personnel must be proficient in operating fire hydrants. First, always ensure that the hydrant is operational before use. Debris in the water distribution system may be forced into the barrel of a hydrant, a leaking ground valve may allow water to enter the hydrant barrel and freeze during inclement weather, or vandals may remove caps and place trash or foreign objects into the hydrant. These materials can obstruct the water flow or damage a fire department pumper if they are drawn into the pump.

It is good practice to check the operation of the hydrant and flush out debris before connecting a hose to a hydrant. To do so, first open one of the outlet caps. Then, on a dry-barrel hydrant, open the hydrant valve just enough to ensure that water flows into the hydrant and flushes out any foreign matter. On a wet-barrel hydrant, open the valve to the outlet, and then close the valve, connect the hose, and then reopen the valve all the way. Fire departments should ensure that inspections and tests to keep hydrants operating smoothly are performed regularly.

Operating a Dry-Barrel Fire Hydrant

To open a dry-barrel hydrant, you remove the cap from the outlet you will be using, then slowly open the hydrant valve to flush out any debris. After closing the hydrant valve, you attach a hose or gate valve to the outlet, then slowly and completely reopen the hydrant valve.

Individual fire departments may have their own variations on this procedure. For example, some departments specify that the wrench used to open the hydrant valve be left on the hydrant. Other departments require that the wrench be removed and returned to the apparatus so that an unauthorized person cannot interfere with the operation. Always follow the standard operating procedures (SOPs) for your department.

SKILL DRILL 9-1 outlines the steps for getting water from a dry-barrel fire hydrant efficiently and safely.

Shutting a hydrant down properly is just as important as opening a hydrant properly. If the hydrant is damaged during shutdown, it cannot be used until it has been repaired. Dry-barrel fire hydrants are located in areas where the temperatures dip below freezing. After the fire is out and before you leave the fire scene, you need to make sure that dry-barrel hydrants are completely drained, even if the weather is warm. During winter months, any water left in the hydrant can freeze. Fire crews may lose valuable time connecting a hose to a frozen hydrant, only to discover that it will not operate. If this happens, firefighters will be without water until the crew can locate a working hydrant and can reposition and reconnect the hose lines. A properly draining hydrant will create suction against a hand placed over the outlet opening. After the hydrant is fully drained, replace the cap. To shut down a dry-barrel hydrant efficiently and safely, follow the steps in **SKILL DRILL 9-2**.

Operating a Wet-Barrel Fire Hydrant

Wet-barrel fire hydrants are located in areas where the temperatures do not normally dip below freezing. These hydrants always have water in their barrels and do not require draining. The steps for operating them are different from the steps required to operate a dry-barrel hydrant. **SKILL DRILL 9-3** lists the steps for opening a wet-barrel fire hydrant efficiently and safely. Follow the steps in **SKILL DRILL 9-4** to shut down a wet-barrel fire hydrant.

SKILL DRILL 9-1

Opening a Dry-Barrel Fire Hydrant Support Person, NFPA 1010: 5.3.4

1. Remove the cap from the outlet you will be using.

2. Look inside the hydrant opening for debris.

3. Check that the remaining caps are tight.

4. Attach the hydrant wrench to the stem nut located on top of the hydrant. Check the top of the hydrant for an arrow indicating the direction to turn to open.

SKILL DRILL 9-1 CONTINUED

Opening a Dry-Barrel Fire Hydrant Support Person, NFPA 1010: 5.3.4

5. Open the hydrant valve enough to verify the flow of water and flush out any debris that may be in the hydrant.

6. Close the hydrant valve to stop the flow of water. Attach the hose or gate valve to the hydrant outlet.

7. When instructed to do so by your officer or the pump driver/operator, start the flow of water by turning the hydrant wrench slowly to avoid a pressure surge. Continue turning the wrench to fully open the valve. This may take 12 or more turns, depending on the type of hydrant.

8. Once the flow of water has begun, you can open the hydrant valve more quickly. Make sure that you open the hydrant valve completely. If the valve is not opened fully, the drain hole will remain open.

© Jones & Bartlett Learning. Photographed by Glen E. Ellman.

SKILL DRILL 9-2

Shutting Down a Dry-Barrel Fire Hydrant Support Person, NFPA 1010: 5.3.4

1. Turn the hydrant wrench until the stem valve is closed.

2. Allow the hose to drain by opening a drain valve or disconnecting a hose connection downstream. Slowly disconnect the hose from the hydrant outlet, allowing any remaining pressure to escape. If a gate valve is attached, slowly open the gate valve to release pressure in the system, disconnect the hose from the gate valve, and then disconnect the gate valve from the hydrant outlet.

3. Leave one hydrant outlet open until the hydrant is fully drained. If you feel suction on your hand when you place it over the opening, the hydrant is still draining. In very cold weather, you may have to use a hydrant pump to remove all of the water and prevent freezing.

4. Replace the hydrant cap. Do not leave or replace the caps on a dry-barrel hydrant until you are sure that the water has completely drained from the barrel.

© Jones & Bartlett Learning. Photographed by Glen E. Ellman.

SKILL DRILL 9-3

Operating a Wet-Barrel Fire Hydrant Support Person, NFPA 1010: 5.3.4

1. Remove the cap from the outlet you will be using.
2. Look inside the hydrant opening for debris.
3. Check that the remaining caps are tight.
4. Attach the hydrant wrench to the stem nut located behind the outlet you will be using. Check the hydrant for an arrow indicating the direction to turn to open.
5. Open the hydrant valve enough to verify the flow of water and flush out any debris in the hydrant.
6. Close the hydrant valve to stop the flow of water.
7. Attach the hose or valve to the hydrant outlet.
8. When instructed to do so by your officer or the pump driver/operator, start the flow of water by turning the hydrant wrench slowly to avoid a pressure surge. Continue turning the wrench to fully open the valve. This may take 12 or more turns, depending on the type of hydrant.
9. Once the flow of water has begun, you can open the hydrant valve more quickly. Make sure that you open the hydrant valve completely.

© Jones & Bartlett Learning. Courtesy of Bill Larkin.

SKILL DRILL 9-4

Shutting Down a Wet-Barrel Fire Hydrant Support Person, NFPA 1010: 5.3.4

1. Turn the hydrant wrench until the valve opposite the outlet you are using is closed.
2. Allow the hose to drain by opening a drain valve or disconnecting a hose connection downstream. Slowly disconnect the hose from the hydrant outlet, allowing any remaining pressure to escape.
3. Replace the hydrant cap.

© Jones & Bartlett Learning. Courtesy of Bill Larkin.

Water Hammer

Water hammer is a surge in pressure caused by suddenly stopping the flow of a stream of water. A fast-moving stream of water has a large amount of **kinetic energy**. If the water suddenly stops moving when a valve is closed abruptly, all the kinetic energy is converted to an instantaneous increase in pressure. Because water cannot be compressed, the additional pressure is transmitted along the hose or pipe as a shock wave. This water hammer can rupture a hose, cause a coupling to separate, or damage the plumbing on a piece of fire apparatus. Severe water hammer can damage an underground piping system. Firefighters have been injured by equipment that was damaged by a water hammer.

A similar situation can occur if a valve is opened too quickly and a surge of pressurized water suddenly fills a hose. The surge in pressure can damage the hose or cause the personnel at the nozzle to lose control of the stream.

To prevent water hammer, always open and close fire hydrant valves slowly. Pump operators need to open and close the valves on fire engines slowly. When operating the nozzle on an attack line, the firefighter must open the nozzle slowly. Just as important, when closing the shut-off valve on an attack line, the firefighter must do it slowly.

Inspecting, Maintaining, and Testing Fire Hydrants

Fire hydrants should be checked on a regular schedule—no less than once per year—to ensure that they are in proper operating condition. During inspections, personnel may encounter some common problems, and they should know how to correct them.

TIP

Depending on the jurisdiction, inspections of fire hydrants may fall to the fire department that utilizes the system or the water department that maintains it. Regardless, you should understand the inspection process and what is required in your jurisdiction.

Inspecting Fire Hydrants

When inspecting a fire hydrant, the first thing to check is its visibility and accessibility. Fire hydrants should be visible from every direction so that they can be easily spotted. They should not be hidden by tall grass, brush, fences, debris, dumpsters, or any other obstructions (**FIGURE 9-14**). In many communities, fire hydrants are painted in bright, reflective colors for increased visibility. Colored reflectors are sometimes mounted next to fire hydrants or placed in the pavement in front of them to make them more visible at night. In winter, fire hydrants must be clear of snow. In areas that experience large amounts of snow, fire hydrants are sometimes marked with a small flag mounted on top of a pole that is attached to the hydrant. In addition, all jurisdictions have laws prohibiting parking within a specified distance from a fire hydrant, but sometimes the public takes a chance. During an inspection, look for evidence of someone habitually parking in front of the hydrant.

FIGURE 9-14 Fire hydrants should not be hidden or obstructed.

Courtesy of Captain David Jackson, Saginaw Township Fire Department.

Fire hydrants are classified based on their **flow rate**. The bonnet and hydrant caps may be color coded to indicate the flow rate of the fire hydrant. Although some jurisdictions use their own color-coding system, the National Fire Protection Association (NFPA) specifies colors to indicate the water flow available from each hydrant at 20 psi (140 kPa) **residual pressure** in NFPA 291, *Recommended Practice for Fire Flow Testing and Marking of Hydrants, 2022 Edition* (**TABLE 9-1**).

Fire hydrants should be installed at an appropriate height above the ground. Their outlets should not be so high or so low that fire crews have difficulty removing the hydrant caps with a wrench and connecting hose lines to them. NFPA 24, *Standard for the Installation of Private Fire Service Mains and Their Appurtenances, 2022 Edition*, requires a minimum of 18 in. (450 mm) from the center of a hose outlet to the finished grade.

TABLE 9-1 Fire Hydrant Colors Indicating Available Flow Rates

Class	Flow Available at 20 psi (138 kPa)	Color
Class AA	1500 gpm (5700 L/min) and higher	Light blue
Class A	1000–1499 gpm (3800–5699 L/min)	Green
Class B	500–999 gpm (1900–3799 L/min)	Orange
Class C	Less than 500 gpm (1900 L/min)	Red

Modified from National Fire Protection Association (NFPA). 2021. *NFPA 291, Recommended Practice for Fire Flow Testing and Marking of Hydrants*. 2022 Edition. Quincy, MA: NFPA.

Fire hydrants that experience snow accumulations may require hydrants to be placed farther above the finished grade. Fire hydrants should be positioned so that the connections—especially the steamer port—face the street.

During a fire hydrant inspection, check the exterior of the hydrant for signs of damage. Open the steamer port of dry-barrel hydrants to ensure that the barrel is dry and free of debris (**FIGURE 9-15**). Make sure that all caps are present and that the outlet threads are in good working order and lightly greased. If the threads on the discharge ports need cleaning, use a steel wire brush and a small triangular file to remove any burrs in the threads. Check the gaskets in the caps to make sure they are not cracked, broken, or missing, and replace worn gaskets with new ones. Follow the manufacturer's recommendations for maintaining any parts that require lubrication.

The second part of the inspection ensures that the hydrant works properly. Open the hydrant valve just enough to confirm that water flows out and flushes any debris out of the barrel. After flushing, shut down the hydrant. A properly draining hydrant will create suction against a hand placed over the outlet opening. When the hydrant is fully drained, replace the cap.

FIGURE 9-15 All hydrants should be checked at least annually.

© Jones & Bartlett Learning. Photographed by Glen E. Ellman.

Rural Water Supplies

Many fire departments protect areas that are not serviced by municipal or private water systems. In these areas, residents usually depend on individual wells or cisterns to supply water for their domestic use. Because there are few or no fire hydrants in these areas, fire crews need to know where water sources are located and how to get water from those sources.

Static Sources of Water

Several potential static water sources can be used for fighting fires in rural areas. Both natural and human-made bodies of water, such as rivers, streams, lakes, ponds, oceans, canals, reservoirs, swimming pools, and cisterns, can be used to supply water for fire suppression (**FIGURE 9-16**). Some areas have many different static sources, whereas others have few or none.

Water from a static source can be used to fight a fire directly if it is close enough to the fire scene. Otherwise, it must be transported to the fire using long hose lines, engine relays, or mobile water supply apparatus.

FIGURE 9-16 Any accessible body of water can be used as a static source.

© Jones & Bartlett Learning. Photographed by Glen E. Ellman.

FIGURE 9-17 A portable pump can be used if the water source is inaccessible to a fire department engine.

Courtesy of the National Interagency Fire Center.

FIGURE 9-18 Drafting with a floating strainer.

Courtesy of Fol-Da-Tank.

Static water sources must be accessible to a fire engine or a portable fire pump. If a road or hard surface is located within 20 ft (6 m) of the static water source, a fire engine can drive close enough to draft water directly from the water source into the pump through a hard suction hose. Some fire departments construct special access points so that engines can approach the water source. The portable pump can assist fire crews in accessing water in areas that are inaccessible to fire apparatus (**FIGURE 9-17**). The portable pump can be hand-carried or transported using an off-road vehicle to the water source. Portable pumps can deliver as much as 500 gpm (1893 L/min).

When selecting a drafting site, take into consideration the vertical lift that will be required. Remember that for every 10 ft (3 m) in elevation, there is a loss of 4.34 psi (29.92 kPa). Vertical lifts greater than 10 ft (3 m) reduce pumping capacity—the maximum water flow that the pump discharges. A vertical lift of 20 ft (6 m) can reduce an engine's pumping capacity by 40 percent. Fire departments should identify and preplan static water sources in the area and practice establishing drafting operations at all of these locations.

Once the drafting location is identified, you should inspect the swivel gaskets on the coupling. The **swivel gasket** is an O-shaped piece of rubber that sits inside the swivel section of the female hose coupling. When a male hose coupling is tightened against it, the swivel gasket forms a seal that stops water from leaking. If the swivel gasket is damaged or missing, the hose coupling will leak.

Next, the appropriate length of hose is selected and connected, and a strainer is placed on the end of the hose. The strainer prevents large debris such as trash, rocks, weeds, small twigs, and animals/fish from entering the pump. Some drafting situations require a floating strainer to prevent suctioning debris and bottom silt (**FIGURE 9-18**). Once the strainer is in place, drafting can begin.

Follow the steps in **SKILL DRILL 9-5** to assist the pump operator with assembling the equipment needed to draft water from a static water supply.

SKILL DRILL 9-5

Assisting the Pump Operator with Drafting Support Person, NFPA 1010: 5.3.4

1. After the pump operator has positioned the engine at the draft site, inspect the coupling for damage or debris.

2. Connect each section of suction hose together, and connect the strainer to the end of the hose that will be placed in the water.

3. Connect the other end of the suction hose to the fire pump.

4. Advance the suction hose assembly into position with the strainer in the water.

Continues.

SKILL DRILL 9-5 CONTINUED

Assisting the Pump Operator with Drafting Support Person, NFPA 1010: 5.3.4

5. Ensure that the strainer assembly has at least 24 in. (0.6 m) of water in all directions around the strainer.

© Jones & Bartlett Learning. Photographed by Glen E. Ellman.

Portable Tanks

Portable tanks carried on fire apparatus can be quickly set up at a fire scene. These tanks typically hold between 600 and 5000 gal (2271 and 18,927 L) of water and should be placed so that they can be accessed from multiple directions. There are two types of portable tanks: metal frame and self-expanding. Metal-frame portable tanks need to be set up before water is added to them. Self-expanding portable tanks are laid flat and then expand vertically as water is added. Sometimes you will need to hold the collar of the tank up until the water level is high enough so that the tank can support itself.

With this supply method, a mobile water supply apparatus is used to fill the portable tank while an engine drafts from the portable tank (**FIGURE 9-19**). As soon as the mobile water supply apparatus discharges all of its water into the portable tank, the driver/operator of that apparatus leaves to get another load of water.

Speed is a primary advantage when using a portable tank system because tankers do not have to hook up to the pumper to transfer the water into the portable tank. Instead, a **dump valve** on the mobile water supply apparatus enables the tankers to offload as much as 3000 gal (11,356 L) of water in 1 minute (**FIGURE 9-20**).

FIGURE 9-19 An engine may be set up to draft water from a portable tank.

Courtesy of Captain David Jackson, Saginaw Township Fire Department.

FIGURE 9-20 A dump valve allows a mobile water supply apparatus to discharge water into the portable tank quickly.

© Jones & Bartlett Learning. Photographed by Glen E. Ellman.

Some apparatus use a pump system to offload the water even faster. The faster the mobile water supply apparatus can offload its water, the quicker it can return to the fill site for another load.

Another advantage of the portable tank system is its ability to expand rapidly. Additional portable tanks can be set up and linked together to increase water storage capacity, additional engines can be used to draft water from the portable tanks, and additional tankers can be used to deliver more water at a faster rate. A series of mobile water supply apparatus may be used as shuttles, dumping water either simultaneously or in sequence.

To set up a portable tank, follow the steps in **SKILL DRILL 9-6**.

SAFETY TIP

Never get between two apparatus or tankers!

Water Shuttle Operations

Water shuttle operations are conducted when large volumes of water from a static water source are needed at a fire scene. They are conducted using mobile water supply apparatus. The number of mobile water supply apparatus needed depends on the distance between the fill site and the fire scene, the time it takes the apparatus to dump and refill, and the flow rate required at the fire scene. In rural areas, several mobile water supply apparatus may be dispatched for a structural fire.

TIP

Mobile water supply apparatus are commonly referred to as *tankers* or *water tenders*. In the western part of the United States where aerial water tankers (helicopters and fixed-wing aircraft) are commonly used in fighting wildland fires, the term *water tender* means a truck-mounted mobile water supply apparatus. In parts of the country where aerial water tankers are not commonly used, the term *tanker* refers to a truck-mounted mobile water supply apparatus.

Fire department engines usually carry at least 500 gal (1893 L) of water in the booster tank, whereas fire department mobile water supply apparatus generally carry 1000 to 3500 gal of water (3785 to 13,249 L). Some tankers can transport as much as 5000 gal (18,927 L) of water.

All the components of a water shuttle operation must be set up so that water moves efficiently from the fill site to the fire scene. If a mobile water supply apparatus is the only source of water for fighting a structural fire, the attack must be carefully planned. Enough water must be available on the scene to supply the required hose lines. Some fire departments begin the attack using water from the booster tank of the first-arriving unit. The mobile water supply apparatus then pump additional water directly into the attack pumper to keep it full. They can also pump into a portable tank, sometimes called a *pond*.

At the fill site, the tankers must be refilled without delay. The routes in both directions must be planned so that tankers make efficient round trips, without having to back up or make U-turns. At the fire scene, the tankers should be able to drive up, dump their water into the portable tanks, and immediately return to the fill site. The portable tanks must be large enough to receive the full load of each tanker as it arrives. An effective water shuttle operation can deliver several hundred gallons of water per minute without interruption. If your department uses mobile water supply apparatus, you need to learn the specific system used by your area.

Departments must practice water shuttle operations to ensure proper coordination and effective water delivery. An effective operation requires preplanning and excellent leadership. If the water supply is exhausted before the fire is extinguished, the attack team will be in serious danger, and the building will probably be lost. Conversely, if the use of water is too conservative, fire-suppression efforts will probably be unsuccessful.

SKILL DRILL 9-6

Setting Up a Portable Tank Support Person, NFPA 1010: 5.3.4

1. Lift the portable tank off the apparatus. This tank may be mounted on a side rack or on a hydraulic rack that lowers it to the ground. Place the portable tank on as level ground as possible beside the engine. The pump operator will indicate the best location.

2. Expand the tank (metal-frame type) or lay it flat (self-expanding type).

3. One crew member helps the pump operator place the strainer on the end of the suction hose, put the suction hose into the tank, and connect it to the engine.

4. A second crew member helps the mobile water supply apparatus driver/operator discharge water into the portable tank. If the tank is self-expanding, the crew members may need to hold the collar until the water level is high enough for the tank to support itself.

© Jones & Bartlett Learning. Photographed by Glen E. Ellman.

CASE STUDY

You Are the Support Person CONCLUSION

Your department has been dispatched to a reported building fire. Upon arrival, you find a barn that is fully involved. The fire attack is clearly going to be a defensive operation, but aggressive actions will be required to protect a house and other buildings that are located approximately 100 yards (91 meters) away from the barn, which is located on a narrow, tree-lined lane. The incident commander says he needs a flow of 500 gallons per minute (gpm; 1893 liters per minute [L/min]) to be established.

1. **What water sources might be available?**

   Answer: A working fire hydrant would be the ideal water source, but because this fire involves a barn, it is most likely quite a distance from a fire hydrant. In this case you need to think outside of the box. Is there a pump that feeds the irrigation system for the fields? These pumps usually have a decent water flow and their own well or water source. A pond, lake, or stream can be a good source for drafting. A nearby swimming pool will provide you with a limited amount of water, maybe enough to extinguish the fire. If not, it will provide enough water to help keep the fire controlled until another water source is located.

2. **Which type of water supply would be most dependable?**

   Answer: The most dependable water supply would be the one that is readily available and provides a constant amount of available flow. Fire hydrants supplied by a municipal water source would seem like the most dependable, but you need to consider the size of the water main and the age of the system. For example, a fire hydrant that flows 125 gpm (473 L/min) is not sufficient for a fire that requires 250 gpm (946 L/min). A fire engine connected to a dry hydrant at a good water source, such as a large pond or lake, that is able to supply the attack fire engine using a short stretch of large-diameter hose would be the most dependable in this scenario.

3. **Which type of water source would require the most resources to implement?**

   Answer: Drafting from a water source that is a distance from the fire and requiring the use of mobile water supply apparatus (tenders or tankers) to deliver the water to the incident would require the most resources. A fire engine at the source would be required to draft the water and fill the tenders. Depending on the terrain and location of the water source, and therefore the mobile water supply apparatus, a fire engine pumping from the mobile water supply apparatus to the attack engines would be needed. Finally, a number of water supply apparatus would be required to shuttle water from the drafting site to the portable ponds.

WRAP-UP

SUMMARY

- The basic plan for fighting most fires depends on having an adequate supply of water to safely confine, control, and extinguish the fire. The water supply must be reliable and have sufficient volume and pressure to meet the needs of the incident.
- The source of water varies in different communities. Some communities have municipal water systems, and others have private water systems.
- Often, the first available water source is the water carried in a tank on the first-arriving fire apparatus. This tank water can be used in the initial fire attack while the establishment of an adequate, continuous water supply is secured.
- Most water distribution systems rely on an arrangement of pumps to provide the required water pressure. Some systems use pumps to provide pressure.

SUMMARY CONTINUED

If the pumps stop operating, the pressure is lost, and the system is unable to deliver adequate water pressure or volume to the end users or to hydrants.

- In a gravity-feed system, the water source is located on high ground, whereas the end users live in lower-lying areas so that gravity can supply the necessary pressure without the use of pumps. Many municipal water supply systems use a combination pump and gravity-feed system to deliver water.
- The fire department should notify the water department when fire operations will require prolonged use of large quantities of water. The water department may be able to increase the normal volume and pressure by starting additional pumps. In some systems, the water department can open valves to increase the flow to a certain area in response to fire department operations at major fires.
- Fire hydrants are part of the municipal water distribution system and connect directly to public water mains. Fire hydrants are also installed on private water systems supplied by the municipal water system or by a separate source. The water source, as well as the adequacy and reliability of the supply to private hydrants, must be identified to ensure that a sufficient water supply will be available when fighting fires.
- The two most common types of fire hydrants are the dry-barrel hydrant and the wet-barrel hydrant, both of which are connected to a pressurized water source. A third type of hydrant is the dry hydrant, which drafts water from a static water source.
- A dry-barrel hydrant, also called a *frost-proof hydrant*, is used in climates where temperatures fall below freezing. These fire hydrants are connected to a pressurized municipal water system. The valve that controls the flow of water into the barrel of the hydrant is located at the base of the barrel below the frost line to keep the hydrant from freezing.
- A wet-barrel hydrant is used in locations where temperatures do not drop below freezing. These hydrants always have water in the barrel and do not have to be drained after each use.
- A dry hydrant, also called a *drafting hydrant*, is a fixed piping system with one end in a static water source and the other end accessible to attach a hose. Dry hydrants are often installed in lakes and rivers and close to clusters of buildings where there is a recognized need for fire protection.
- It is good practice to check the operation of the hydrant and flush out debris before connecting a hose to a hydrant.
- To open a dry-barrel hydrant, remove the cap from the outlet you will be using, and then slowly open the hydrant valve to flush out any debris. After closing the hydrant valve, you attach a hose or gate valve to the outlet and then slowly and completely reopen the hydrant valve.
- If the hydrant is damaged during shutdown, it cannot be used until it has been repaired. After the fire is out and before you leave the fire scene, make sure that dry-barrel hydrants are completely drained, even if the weather is warm. Wet-barrel fire hydrants always have water in their barrels and do not require draining.
- Suddenly starting or stopping the flow of a hydrant can cause a water hammer—a surge in pressure that can rupture a hose, cause a coupling to separate, or damage the plumbing on a piece of fire apparatus. To prevent water hammer, always open and close fire hydrant valves slowly.
- Fire hydrants should be checked on a regular schedule—no less than once per year—to ensure that they are in proper operating condition.
- Fire hydrants should be visible from every direction so that they can be easily spotted. They should not be hidden by tall grass, brush, fences, debris, dumpsters, or any other obstructions.
- Fire hydrants should be installed at an appropriate height above the ground. Their outlets should not be so high or so low that fire crews have difficulty removing the hydrant caps with a wrench and connecting hose lines to them.
- During a fire hydrant inspection, check the exterior of the hydrant for signs of damage.
- The second part of the inspection ensures that the hydrant works properly. Open the hydrant valve just enough to confirm that water flows out and flushes any debris out of the barrel.
- Many fire departments protect areas that are not serviced by municipal or private water systems. Because there are few or no fire hydrants in these areas, fire crews need to know where water sources are located and how to get water from those sources.
- Both natural and human-made bodies of water, such as rivers, streams, lakes, ponds, oceans, canals,

reservoirs, swimming pools, and cisterns, can be used for fighting fires in rural areas. Water from a static source can be used to fight a fire directly if it is close enough to the fire scene. Otherwise, it must be transported to the fire using long hose lines, engine relays, or mobile water supply apparatus.

- Portable tanks carried on fire apparatus can be quickly set up at a fire scene. There are two types of portable tanks: metal frame and self-expanding. Metal-frame portable tanks need to be set up before water is added to them. Self-expanding portable tanks are laid flat and then expand vertically as water is added.
- Water shuttle operations are conducted using mobile water supply apparatus when large volumes of water from a static water source are needed at a fire scene. The number of mobile water supply apparatus needed depends on the distance between the fill site and the fire scene, the time it takes the apparatus to dump and refill, and the flow rate required at the fire scene. In rural areas, several mobile water supply apparatus may be dispatched for a structural fire.

KEY TERMS

barrel The upright steel casing that is the main part of a fire hydrant.

bonnet The top of a hydrant.

branch line See *secondary feeder.*

charge To fill with water under pressure.

coupling A connection device that connects (couples) individual lengths of fire hose together or connects a hose to a fire hydrant or a pump or to nozzles and hose appliances.

dead-end water main A water main that is supplied from only one direction.

discharge cap See *hydrant cap.*

distributor See *distributor pipe.*

distributor pipe The smallest-diameter underground water main pipes in a water distribution system that deliver water to local users within a neighborhood. Also called *distributor.*

draft The use of suction to move a liquid (such as water) from a vessel or source that is below the intake of a pump. (NFPA 1910)

drafting hydrant See *dry hydrant.*

dry-barrel hydrant A type of hydrant with the main control valve below the frost line between the footpiece and the barrel. Also called *frost-proof hydrant.* (NFPA 24)

dry hydrant An arrangement of pipe permanently connected to a water source other than a piped, pressurized water supply system that provides a ready means of water supply for firefighting purposes and that utilizes the drafting (suction) capability of a fire department pump. Also called *drafting hydrant.* (NFPA 1142)

dump valve A large opening from the water tank of a mobile water supply apparatus for unloading purposes. (NFPA 1900)

elevated water storage tower An aboveground water storage tank that is designed to maintain pressure on a water distribution system.

flow rate The quantity of water flowing, usually measured in gallons (or liters) per minute.

frost-proof hydrant See *dry-barrel hydrant.*

gate valve A valve firefighters attach to an outlet on a dry hydrant that can turn off the flow of water at that outlet.

gravity-feed system A water distribution system that depends on gravity to provide the required pressure. The system storage is usually located at a higher elevation than the end users of the water.

hydrant cap The cover on a fire hydrant outlet that is in place when the hydrant is not in use. Also called *discharge cap.*

kinetic energy The energy possessed by an object as a result of its motion.

municipal water system A system having water pipes servicing fire hydrants and designed to furnish, over and above domestic consumption, a minimum of 250 gpm (946 L/min) at 20 psi (138 kPa) residual pressure for a 2-hour duration. (NFPA 1140)

operating stem The steel rod extending from the top of a dry-barrel hydrant to the hydrant valve that firefighters can turn using the stem nut at the top to open and close the main valve.

outlet An opening on a fire hydrant through which water is discharged.

portable tanks Folding or collapsible tanks that are used at the fire scene to hold water for drafting.

KEY TERMS CONTINUED

primary feeder The largest-diameter water main pipe in a water distribution system that carries the greatest amounts of water. Also called *trunk line.*

private water system A privately owned water system that operates separately from the municipal water system.

residual pressure The pressure that exists in the distribution system, measured at the residual hydrant at the time the flow readings are taken at the flow hydrants. (NFPA 24)

secondary feeder The smaller-diameter water main pipe in a water distribution system that connects a primary feeder to a distributor. Also called *branch line.*

shut-off valve A valve whose primary function is to operate in either a fully shut-off or a fully open condition. (NFPA 1960)

static water source A water source such as a pond, river, stream, or other body of water that is not under pressure.

steamer port The large-diameter port on a fire hydrant.

stem nut The large nut at the top of the operating stem in a dry-barrel hydrant that is turned to open the hydrant valve.

swivel A ring around female couplings that you turn to secure the female coupling around the male coupling without twisting the hose.

swivel gasket An O-shaped piece of rubber inside the swivel section of a female hose coupling that forms a seal that stops water from leaking when a male hose coupling is tightened against it.

threaded hose coupling A type of coupling that requires a male fitting and a female fitting to be screwed together.

trunk line See *primary feeder.*

water flow The volume of water moving through a pipe, hose, or nozzle over a period of time, usually expressed in gallons (liters) per minute (gpm or L/min).

water hammer The surge of pressure that occurs when a high-velocity flow of water is abruptly shut off. The pressure exerted by the flowing water against the closed system can be seven or more times that of the static pressure. (NFPA 1962)

water main A generic term for any underground water pipe.

water pressure The application of force by one object against another. When water is forced through the distribution system, it creates water pressure.

water shuttle operations A method of transporting water from a source to a fire scene using a number of mobile water supply apparatus.

water supply A source of water for firefighting activities. (NFPA 1140)

wet-barrel hydrant A type of hydrant that is intended for use where there is no danger of freezing weather and where each outlet is provided with a valve and an outlet. (NFPA 24)

REVIEW QUESTIONS

1. What are the three major components of a municipal water system?
2. What is the recommended minimum pressure for water coming from a fire hydrant?
3. What type of fire hydrant is used in areas subjected to below-freezing temperatures?
4. Why is it important, when operating from a dry-barrel fire hydrant, to ensure the hydrant valve is fully opened?
5. How are the locations of fire hydrants in residential areas identified and documented?
6. When hooking up to a fire hydrant, either wet barrel or dry barrel, what is the first thing you should do before hooking the hose line to it?
7. What is a water hammer? Why is this action not recommended?
8. What class and color is the fire hydrant with the greatest amount of flow?
9. When selecting a drafting site at a static water source, why is a vertical lift greater than 10 ft (3 m) a concern?

DISCUSSION QUESTIONS

1. Describe the differences between a dry-barrel hydrant and a wet-barrel hydrant, and explain why one is installed in a jurisdiction over the other.
2. What is a dead-end main, and why is it important for you to know where they are and what fire hydrants are connected to them?

REFERENCES

National Fire Protection Association. 2017. *NFPA 1962, Standard for the Care, Use, Inspection, Service Testing, and Replacement of Fire Hose, Couplings, Nozzles, and Fire Hose Appliances.* 2018 Edition. Quincy, MA: National Fire Protection Association.

National Fire Protection Association. 2021. *NFPA 24, Standard for the Installation of Private Fire Service Mains and Their Appurtenances.* 2022 Edition. Quincy, MA: National Fire Protection Association.

National Fire Protection Association. 2021. *NFPA 291, Recommended Practice for Water Flow Testing and Marking of Hydrants.* 2022 Edition. Quincy, MA: National Fire Protection Association.

National Fire Protection Association. 2021. *NFPA 1142, Standard on Water Supplies for Suburban and Rural Fire Firefighting.* 2022 Edition. Quincy, MA: National Fire Protection Association.

National Fire Protection Association. 2023. *NFPA 1140, Standard for Wildland Fire Protection.* 2024 Edition. Quincy, MA: National Fire Protection Association.

National Fire Protection Association. 2023. *NFPA 1900, Standard for Aircraft Rescue and Firefighting Vehicles, Automotive Fire Apparatus, Wildland Fire Apparatus, and Automotive Ambulances.* 2024 Edition. Quincy, MA: National Fire Protection Association.

National Fire Protection Association. 2023. *NFPA 1910, Standard for the Inspection, Maintenance, Refurbishment, Testing, and Retirement of In-Service Emergency Vehicles and Marine Firefighting Vessels.* 2024 Edition. Quincy, MA: National Fire Protection Association.

National Fire Protection Association. 2023. *NFPA 1960, Standard for Fire Hose Connections, Spray Nozzles, Manufacturer's Design of Fire Department Ground Ladders, Fire Hose, and Powered Rescue Tools.* 2024 Edition. Quincy, MA: National Fire Protection Association.

Chapter Opener: © Eric Scruggs

CHAPTER

10

Support Person

Fire Hose

KNOWLEDGE OBJECTIVES

After studying this chapter, you will be able to:

- Describe the various sizes of fire hose and how they are used.
- Describe the two types of suction hose.
- Describe types of supply hose.
- List the common types of hose rolls used to organize supply hose.
- List the common types of hose rolls used with attack hose.
- List the common types of hose damage.
- List the common methods for cleaning and drying hose.
- Describe the importance of a hose inspection.
- Describe the procedure for documenting a defective hose and removing it from service.

SKILLS OBJECTIVES

After studying this chapter, you will be able to perform the following skills:

- Demonstrate replacing the swivel gasket on a fire hose.
- Properly demonstrate a straight hose roll.
- Properly demonstrate a single-doughnut hose roll.
- Properly demonstrate a twin-doughnut hose roll.
- Properly demonstrate a self-locking twin-doughnut hose roll.
- Demonstrate how to clean and maintain hose.
- Demonstrate the proper techniques for cleaning and drying different types of hose.
- Properly document equipment maintenance following established guidelines.
- Properly mark a defective section of hose and remove it from service.

ADDITIONAL NFPA STANDARDS

- **NFPA 1142**, *Standard on Water Supplies for Suburban and Rural Firefighting, 2022 Edition*
- **NFPA 1410**, *Standard on Training for Emergency Scene Operations, 2020 Edition*
- **NFPA 1900**, *Standard for Aircraft Rescue and Firefighting Vehicles, Automotive Fire Apparatus, Wildland Fire Apparatus, and Automotive Ambulances, 2024 Edition*
- **NFPA 1960**, *Standard for Fire Hose Connections, Spray Nozzles, Manufacture's Design of Fire Department Ground Ladders, Fire Hose, and Power Rescue Tools, 2024 Edition*

CASE STUDY

You Are the Support Person

Shortly after joining a fire crew, you are assigned to ride along with an engine company for a shift. After a relatively uneventful day, your company is dispatched for a working fire at a large warehouse. When you arrive, there are flames coming from the front of the building. You hear the incident commander instruct the crew to lay a 4-inch (in.; 101-millimeter [mm]) supply line and prepare for an interior attack.

1. Why is it important to understand the function and proper uses of each type of hose carried on your apparatus?
2. What are the advantages of using large-diameter hose for supply hose?

Introduction

One of the primary functions of firefighting is to quickly and effectively apply sufficient quantities of water to extinguish the fire. In order to assist in this objective, it is necessary to understand fire hose, fire hose appliances, and hose tools.

Fire Hose

Fire hose is a flexible tube used to convey water or other extinguishing agents. It is used for two main purposes: as supply hose and as attack hose. **Supply hose** (or **supply line**) is used to deliver water to an **attack engine**—an engine used to pump water through attack lines at the fireground—from a pressurized source such as a fire hydrant or from a **supply engine**—an engine used to pump water to an attack engine using supply lines. These topics are discussed further in Chapter 9, *Water Supply Systems*. The supply engine may be connected to a pressurized water source, such as a fire hydrant, or to a static (unpressurized) water source. An **attack hose** (or **attack line**) is used for fire suppression by discharging water under pressure onto a fire. Most attack hoses carry water directly from the attack engine to a nozzle that is used to direct the water onto the fire. Attack hose that is held by firefighters to attack fire is called a **handline**. NFPA 1960, *Standard for Fire Hose Connections, Spray Nozzles, Manufacture's Design of Fire Department Ground Ladders, Fire Hose, and Power Rescue Tools, 2024 Edition*, defines the specifications and the inspection and testing requirements for the design and construction of new fire hose.

Hose Sizes

Hose size is the inside diameter of the hose when it is filled with water and under pressure. Fire hose ranges in size from ¾ to 6 inches (in.; 19 to 152 millimeters [mm]) (**FIGURE 10-1**).

Small-Diameter Hose

Small-diameter hose (SDH) is either 1½-in. (38-mm) or 1¾-in. (44-mm) hose and is most often used as the primary attack hose for most fires. Both sizes use the same 1½-in. (38-mm) couplings. These two sizes of attack hose are the sizes used most often during basic fire training. SDH can usually be operated by one firefighter, although having a second firefighter on the line makes it much easier to advance and control the hose. SDH usually comes in 50-foot (ft; 15-meter [m]) lengths.

The primary difference between 1½-in. (38-mm) and 1¾-in. (44-mm) hose is the amount of water that can flow through the hose. Depending on the pressure in the hose and the type of nozzle used, a 1½-in. (38-mm) hose can generally flow between 100 and 200 gallons per minute (gpm; 379 and 757 liters per minute [L/min]). An equivalent 1¾-in. (44-mm) hose can

FIGURE 10-1 Fire hose comes in a wide range of sizes for different uses and situations.

© Jones & Bartlett Learning. Photographed by Glen E. Ellman.

flow between 150 and 250 gpm (568 and 946 L/min). This difference is important because the amount of fire that can be extinguished is directly related to the amount of water that is applied to it. A 1¾-in. (44-mm) hose can deliver much more water and is only slightly heavier and more difficult to advance than a 1½-in. (38-mm) hose.

Medium-Diameter Hose

Medium-diameter hose (MDH) is 2½ in. (64 mm) and 3 in. (76 mm). The 2½-in. (64-mm) hose can be used as either attack hose or supply hose, but it is most often used as attack hose. A 2½-in. (64-mm) hose is used as an attack hose for fires that are too large to be controlled by a 1½-in. (38-mm) or 1¾-in. (44-mm) hose. A 2½-in. (64-mm) hose delivers a flow of approximately 250 gpm (946 L/min). It takes at least two firefighters to safely control a 2½-in. (64-mm) hose due to the weight of the hose, the water, and the nozzle reaction force. A 50-ft (15-m) length of dry 2½-in. (64-mm) hose weighs about 30 pounds (lb; 14 kilograms [kg]). When the hose is filled with water, however, it can weigh as much as 140 lb (64 kg). A 2½-in. (64-mm) hose is most often used for interior attacks in large buildings and for exterior attacks. However, 2 in. (50-mm) hose line with 1 ½-in. (38-mm) couplings or 2 ½-in. (65-mm) couplings is increasing in use. The 2 ¼-in. (57-mm) hose line is capable of the same performance as the 2 ½-in. (65-mm) hose with some weight advantages, it is lighter and easier to handle.

The 3-in. (76-mm) hose is also used as both attack and supply hose. When used as an attack hose, the 3-in. (76-mm) hose is used to deliver water to master stream appliances, which are devices that discharge between 350 gpm (1325 L/min) and 1500 gpm (5678 L/min) or more. Like 1½-in. (38-mm) or 1¾-in. (44-mm) hose, MDH usually comes in 50-ft (15-m) lengths.

Large-Diameter Hose

Large-diameter hose (LDH) has a diameter of 3½ in. (89 mm) or more. Standard LDH sizes include 4-in. (101-mm) and 5-in. (127-mm) diameters. Most LDH is constructed as supply hose, but some fire departments use special LDH that can withstand higher pressures. The largest LDH size is 6 in. (152 mm) in diameter. LDH is commonly available in standard lengths of 100 ft (30 m) but can range from 25 to 350 ft (7.6 to 106.7 m) depending on a department's needs. These lengths can produce a water flow between 350 gpm (1325 L/min) and 1500 gpm (5678 L/min). Like 3-in. (76-mm) hose, LDH is also used to deliver water to master stream appliances.

Hose Construction

Most fire hose is constructed with an inner waterproof liner surrounded by one or multiple outer layers or reinforcements. The outer layer or layers is commonly referred to as the *jacket*. A **single-jacket hose** is constructed with one layer of woven fiber. A **multiple-jacket hose** is constructed with two or more layers of woven fibers. In a multiple-jacket hose with two layers—**double-jacket hose**—the outer layer is bonded to the inner woven layer. The jacket serves as a protective covering, whereas the inner layer provides most of the strength needed to keep the hose from rupturing under pressure. The woven fibers of the jacket are treated to resist water and provide added protection from many common hazards that are likely to be encountered at the scene of a fire. The jacket also provides the strength needed to withstand the high pressures exerted by the water inside the hose. This strength is provided by a woven mesh made from high-strength synthetic fibers such as nylon that are resistant to high temperatures, mildew, and many chemicals. These fibers can also withstand some mechanical abrasion.

SAFETY TIP

Gloves should always be worn when handling hose. Metal shavings, glass shards, or other sharp objects may potentially become embedded in the fibers of the hose. These objects could easily sever a muscle or tendon in a bare hand—and end a career.

Rubber-covered hose is a type of double-jacket fire hose that features a durable rubber-like compound on the outer layer or jacket. This construction provides additional protection and durability to the hose, making it suitable for various firefighting applications. In rubber-covered hose, the inner and outer layers of the jacket are bonded together with woven fibers contained between these layers (**FIGURE 10-2**). This bonding

FIGURE 10-2 Rubber-covered hose.

© Jones & Bartlett Learning. Photographed by Glen E. Ellman.

process ensures that the layers remain securely attached and prevents separation or delamination during use. The woven fibers provide reinforcement and strength to the hose, allowing it to withstand high pressure and resist abrasion and damage.

Collapsible fire hose is designed to be compact and lay flat, and it is easily stored when not in use. It is typically made of synthetic materials, such as polyester, nylon, or a blend of synthetic fibers, which means the hose is flexible and foldable. The collapsible nature of this hose makes it convenient for transportation and storage because it takes up less space compared to non-collapsible, rigid hose. When pressurized with water, the hose expands and becomes rigid, allowing for efficient water flow and firefighting operations.

Non-collapsible fire hose, also known as **rigid fire hose**, is a type of fire hose designed to maintain its circular shape and cross-sectional integrity even when not filled with water and pressurized. Unlike regular collapsible hose, non-collapsible hose incorporates an inner liner or element that provides rigidity and prevents the hose from flattening out when not charged. It is typically constructed from durable materials such as PVC (polyvinyl chloride) or synthetic rubber. The reinforcement prevents the hose from collapsing or kinking when pressurized, ensuring a consistent water flow.

It is important to note that both collapsible and non-collapsible fire hose serve specific purposes and have advantages and disadvantages, depending on the firefighting scenario. Each fire department and agency needs to select the appropriate type of hose based on its operational needs, available resources, and the specific requirements of the firefighting situation.

The inner waterproof liner of the hose is the **hose liner**. The hose liner is usually made of a synthetic rubber compound or a thin, flexible membrane material that can be flexed and folded without developing leaks (**FIGURE 10-3**). The hose liner prevents water from leaking out of the hose and provides a smooth inside surface for water to move against. Without this smooth surface, excessive friction would arise between the moving water and the inside of the hose, causing friction loss and reducing the water pressure and flow rate that reaches the nozzle.

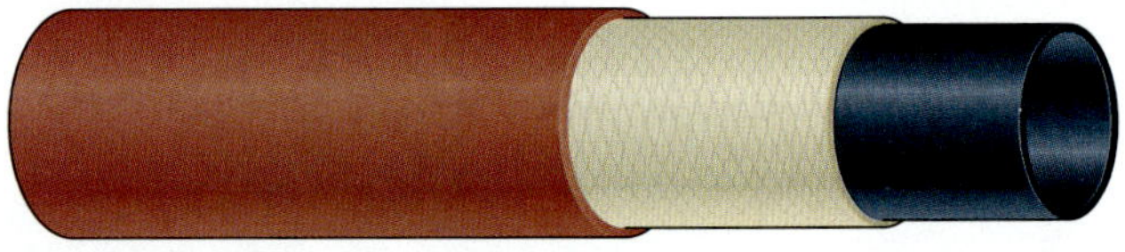

FIGURE 10-3 The hose liner inside a fire hose can be made from synthetic rubber or a variety of other membrane materials.

© Jones & Bartlett Learning

Hose Couplings

Hose usually comes in lengths of 50 to 100 ft (15 to 30 m), although some hose is shorter. At a fire scene, the hose often needs to be much longer to reach the fire. To achieve this, both ends of a length of hose have a coupling permanently attached. A **coupling** is a connection device used to connect—or couple—individual lengths of fire hose together. Couplings are also used to connect a hose to a fire hydrant or a pump or to nozzles and appliances. NFPA 1960 defines the performance requirements for hose couplings, as well as the specifications for the connections of the couplings. The two most common types of supply hose and attack hose couplings are threaded hose couplings and Storz hose couplings.

Threaded Hose Couplings

A **threaded hose coupling** is used on most hose that is 3 in. (76 mm) or less in diameter. A length of fire hose with threaded couplings has a male hose coupling with threads on the outside on one end of the hose and a female hose coupling with threads on the inside on the other end of the hose. The female hose coupling also has a **swivel**, which is a ring around the coupling that must be turned to secure the female coupling around a male coupling without twisting the hose (**FIGURE 10-4**). Most hydrants and fire pumps have male outlets and therefore require a female coupling.

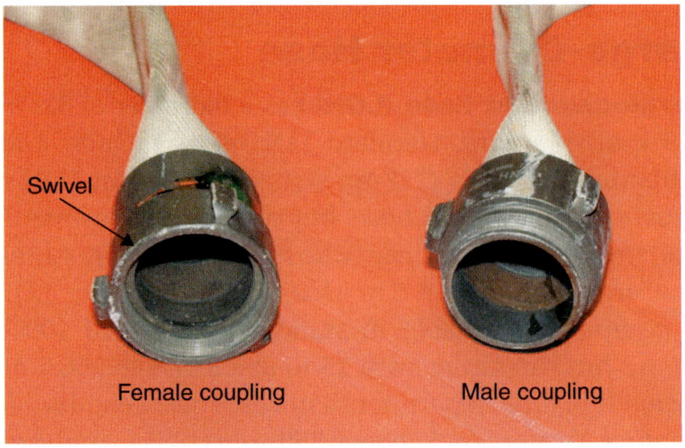

FIGURE 10-4 A set of threaded hose couplings includes one male coupling and one female coupling. The female coupling has threads on the inside of the coupling and a swivel. The male coupling has exposed threads on the outside of the coupling.

© Jones & Bartlett Learning. Photographed by Glen E. Ellman.

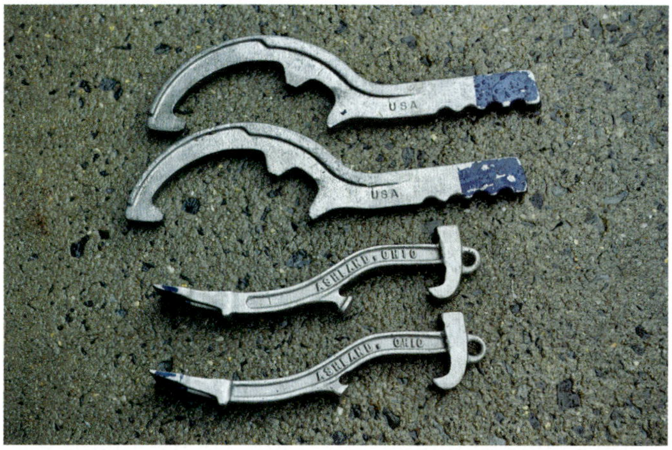

FIGURE 10-5 Spanner wrenches are used to couple or uncouple hose couplings.

© Jones & Bartlett Learning. Photographed by Glen E. Ellman.

FIGURE 10-6 Lugs are extensions or indentations on couplings that provide leverage to aid in the connection and disconnection of hose couplings. **A.** Pin lugs are cylinder-like extensions. **B.** Recessed lugs are circular indentations. **C.** Rocker lugs are rectangular extensions.

© Jones & Bartlett Learning. Photographed by Glen E. Ellman.

When connecting fire hose with threaded couplings, the threads must be properly aligned so that the male and female hose couplings fully engage with minimal resistance. When properly coupled, threaded hose couplings provide a secure connection between two sections of hose and are unlikely to become disconnected during proper use. The swivel on the female hose coupling should be turned until the connection is snug, but only hand tight so that the hose couplings can be easily disconnected.

A disadvantage of threaded hose couplings is that they are prone to cross-threading, meaning the threads of each coupling are not properly aligned with each other. This can result in leakage and possible separation. If leakage does occur after the hose is filled with water, further tightening may be needed, in which case a spanner wrench can be used to gently tighten the couplings until the leakage is stopped (**FIGURE 10-5**). Note that even if you only hand-tighten the connection, you may need to use a spanner wrench to uncouple the hose after it has been pressurized with water. If you do need to use a spanner wrench to couple or uncouple the hose couplings, normally, two spanner wrenches are used together to rotate the two hose couplings in opposing directions.

Threaded hose couplings have a **lug**, which is a protrusion or indentation that provides leverage as you screw couplings together or unscrew them to separate them. On male couplings, the lugs are on the outside of the coupling. On female couplings, the lugs are on the outside of the swivel. Three types of lugs are used: rocker lugs, pin lugs, and recessed lugs (**FIGURE 10-6**). A **rocker lug** (or **rocker pin**) is a rectangular-shaped extension and is the type of lug most commonly used on threaded fire hose couplings. The edges of rocker lugs are beveled to prevent them from catching on objects as the hose is dragged across a surface. A **pin lug** looks like a small cylinder that extends outward from the hose coupling. They are rarely found on fire hose today because the pins tend to snag as the hose is being pulled over rough surfaces. They are, however, found on a **fire department connection (FDC)**—the fire hose connection that fire departments use to pump water into a sprinkler system or a standpipe (a **standpipe** is a vertical pipe that connects a water supply to hose outlets in a multilevel building or sprinkler system). Instead, pin lugs have largely been replaced by rocker lugs. A **recessed lug** is a circular indentation and requires a specially designed spanner wrench called a *booster hose wrench* to engage the lug. Recessed lugs are usually found on the couplings for ¾-in.(19-mm) or 1-in. (25-mm) booster hose.

A **Higbee indicator** is a notch or cut on one of the rocker lugs that indicates the position of the first thread on a coupling. If you align the rocker lugs with the Higbee indicators on a male coupling and a female coupling, the threads will align, and you will be able to screw the couplings together quickly, with a low risk of cross-threading (**FIGURE 10-7**).

To create an airtight seal between couplings, it is important to have a flexible gasket—an O-shaped piece of rubber—to facilitate the seal. On threaded couplings, the gasket sits inside the swivel on the female hose coupling and it is called a **swivel gasket**. When a male hose coupling is tightened against it, the swivel gasket forms a seal that stops water from leaking. If the swivel gasket is damaged or missing, the hose coupling will leak. Although a leaking hose coupling is not a critical problem during most firefighting operations, it can result in unnecessary water damage. During cold weather, a leaking hose coupling can cause ice to form and create a significant safety hazard. If you use a wrench to tighten couplings on an empty hose or if you overtighten couplings on a filled hose, the swivel gaskets can be damaged, causing a leak. Swivel gaskets can also deteriorate with time. The best way to prevent leaks is to make sure the swivel gaskets are in good condition and replace any swivel gaskets that are missing or damaged. To replace the swivel gasket in a female hose coupling, follow the steps in **SKILL DRILL 10-1**.

FIGURE 10-7 Higbee indicators are notches or cuts in rocker lugs that show the position where the threads on a pair of couplings properly align with each other. Higbee indicators are especially helpful when the threads are not visible, such as at night.

© Jones & Bartlett Learning. Photographed by Glen E. Ellman.

Storz Hose Couplings

A **Storz hose coupling** is a nonthreaded coupling that consists of two halves that interlock with each other to form a tight seal (**FIGURE 10-8**). Storz couplings are often referred to as *sexless couplings* because they do not have distinct male or female ends. Instead, they have identical symmetrical lugs and slots that allow them to connect with other Storz couplings of the same size. There are other types of nonthreaded couplings, but Storz couplings are the most common. Storz hose couplings are made for all hose sizes; however, in North America, they are most often used on LDH. Many fire departments use LDH with Storz hose couplings as a supply line between a fire hydrant and an engine.

SKILL DRILL 10-1

Replacing the Swivel Gasket Support Person, NFPA 1010: 5.5.3

1. Feel the swivel gasket with your fingers. If the gasket is dry, brittle, or missing, it must be replaced. If the old gasket is still in place, pull it out with your fingers. Then fold the new swivel gasket in half by bringing the thumb and the forefinger together to create two loops.

2. Place either of the two loops inside the hose coupling, and position it against the gasket seat.

SKILL DRILL 10-1 CONTINUED

Replacing the Swivel Gasket Support Person, NFPA 1010: 5.5.3

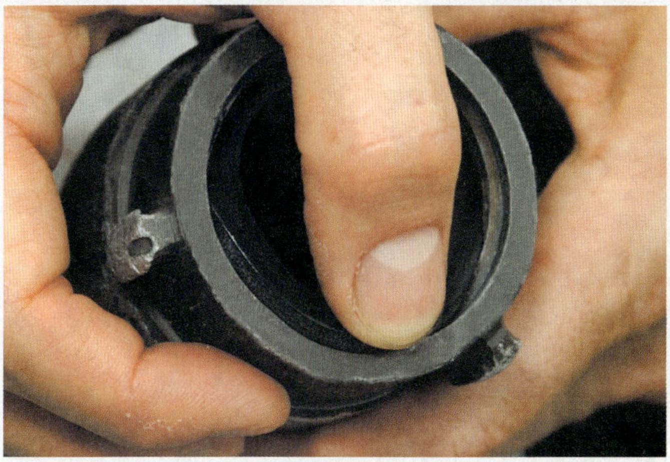

3. Using the thumb, push the remaining unseated portions of the swivel gasket into the hose coupling until the entire swivel gasket is properly positioned against the gasket seat inside the coupling.

© Jones & Bartlett Learning. Photographed by Glen E. Ellman.

FIGURE 10-8 Storz hose couplings on both ends of a length of hose are the same and can be attached to any other Storz coupling of the same diameter.

© Jones & Bartlett Learning. Photographed by Glen E. Ellman.

Because there are no threads to make a connection, Storz hose couplings are connected by mating the two couplings face to face and then turning them clockwise into a locking position. To disconnect a set of Storz-type hose couplings, the two parts are rotated counterclockwise until they release. They may require spanner wrenches for complete coupling and uncoupling.

Storz couplings are more prone to accidental disconnect than threaded couplings if they are not completely coupled. Some Storz hose couplings have a small pin or lever that must be released before uncoupling the hose to prevent an accidental uncoupling of the hose. Storz couplings have gaskets, but they are much more durable and long-lasting compared to gaskets on threaded couplings and rarely require maintenance or replacement.

Storz couplings are also found on some FDCs.

Coupling and Uncoupling Hose

Several techniques are used for connecting and disconnecting hose. Depending on the circumstances, one technique may be more effective than another. Fire personnel should learn how to perform each skill. The following two skill drills describe how to couple hose if you are alone and how to couple hose when working with a second person. To perform the one-firefighter foot-tilt method of coupling fire hose, follow the steps in **SKILL DRILL 10-2**. To perform the two-firefighter method for coupling a fire hose, follow the steps in **SKILL DRILL 10-3**.

SAFETY TIP

Always wear your protective gloves when coupling and uncoupling hose connections.

SKILL DRILL 10-2

Performing the One-Firefighter Foot-Tilt Method of Coupling a Fire Hose with Threaded Couplings Support Person, NFPA 1010: 5.5.3

1. Place one foot on one hose behind the male coupling. Push down with your foot to tilt the male coupling upward.

2. Place one hand behind the female coupling on the other hose and grasp the hose.

3. Place the other hand on the swivel of the female coupling. Bring the two couplings together and align the Higbee indicators. Turn the female coupling counterclockwise until it clicks, which indicates that the threads are aligned. Turn the swivel in a clockwise direction to connect the hose.

© Jones & Bartlett Learning. Photographed by Glen E. Ellman.

SKILL DRILL 10-3

Performing the Two-Firefighter Method of Coupling a Fire Hose with Threaded Couplings Support Person, NFPA 1010: 5.5.3

1. The assisting person picks up one hose, grasping it directly behind the male coupling, and holds it tightly against their body.

2. Hold the female coupling on the other hose firmly with both hands.

3. Bring the female coupling to the male coupling and align the Higbee indicators on the couplings.

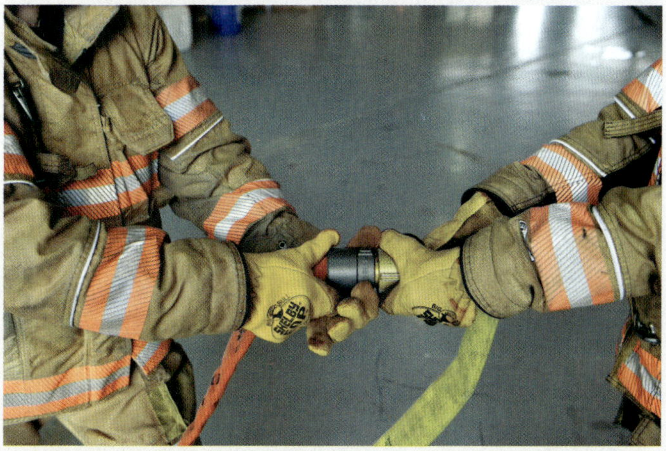

4. Turn the female coupling counterclockwise until it clicks, which indicates that the threads are aligned, and then turn the swivel on the female coupling clockwise to couple the hose.

© Jones & Bartlett Learning. Photographed by Glen E. Ellman.

Hose line sections should never be uncoupled when the hose line is charged—to **charge** a hose line is to fill it with water under pressure. Not only does the water pressure in a charged hose line make it difficult to uncouple the charged line, but the loosened couplings can also become uncoupled and flail around wildly, causing serious injury to firefighters or bystanders in the vicinity. Always shut off the water supply and bleed the pressure from the hose before uncoupling a hose line. If the coupling resists an attempt to uncouple it, check to make sure the pressure is relieved before using spanner wrenches to loosen the coupling. Like coupling hose, you can uncouple hose by yourself or when working with another firefighter. To perform the one-firefighter knee-press method of uncoupling a fire hose, follow the steps in **SKILL DRILL 10-4**. To perform the two-firefighter stiff-arm method of uncoupling a hose, follow the steps in **SKILL DRILL 10-5**. To uncouple a hose using spanner wrenches, follow the steps in **SKILL DRILL 10-6**.

SAFETY TIP

Never attempt to uncouple a charged hose line.

SKILL DRILL 10-4

Performing the One-Firefighter Knee-Press Method of Uncoupling a Fire Hose with Threaded Couplings Support Person, NFPA 1010: 5.5.3

1. Pick up the female coupling.

2. Turn the connection upright, resting the male coupling on a firm surface.

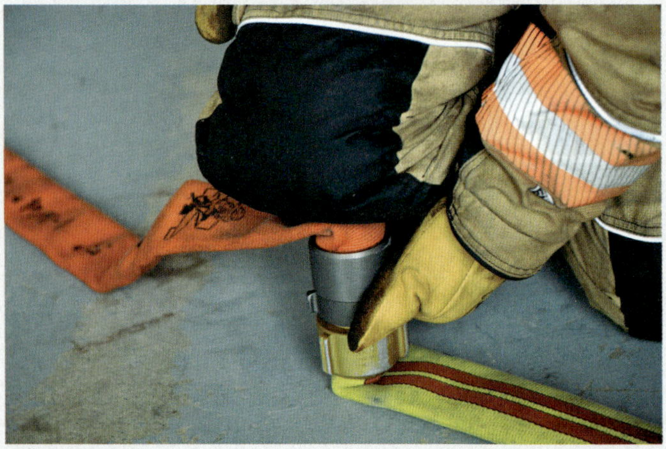

3. Place a knee on the female coupling and press down on it with your body weight to compress the swivel gasket. Turn the swivel on the female coupling counterclockwise to loosen the coupling.

© Jones & Bartlett Learning. Photographed by Glen E. Ellman.

SKILL DRILL 10-5

Performing the Two-Firefighter Stiff-Arm Method of Uncoupling a Fire Hose with Threaded Couplings Support Person, NFPA 1010: 5.5.3

1. Face each other and firmly grasp the hose coupling.
2. With elbows locked straight, push toward each other to compress the swivel gasket.
3. While continuing to push toward each other, one person turns the swivel on the female coupling counterclockwise, loosening the coupling.

© Jones & Bartlett Learning. Photographed by Brandon Fryman/JRM Creative.

SKILL DRILL 10-6

Uncoupling a Hose with Threaded Couplings Using a Spanner Wrench Support Person, NFPA 1010: 5.5.3

1. With the connection on the ground, straddle the hose above the female coupling.

2. Place a spanner wrench on the swivel of the female coupling, with the handle of the wrench to the left.

Continues.

SKILL DRILL 10-6 CONTINUED

Uncoupling a Hose with Threaded Couplings Using a Spanner Wrench
Support Person, NFPA 1010: 5.5.3

3. Place a second spanner wrench on the male coupling, with the handle of the wrench to the right.

4. Push both spanner wrench handles down toward the ground, loosening the connection.

© Jones & Bartlett Learning. Photographed by Glen E. Ellman.

Types of Hose

As mentioned earlier, there are two main types of hose: supply hose and attack hose.

Supply Hose

The **hose bed**—the main storage area on an apparatus for carrying hose—on fire department engines is normally loaded with at least one bed of hose that can be used as supply hose. Sometimes engines are loaded with two beds of supply hose so that they can easily drop a supply hose in two directions.

Supply hose is used to do the following:

- Connect the intake (suction) side of the pump on an attack or supply engine to a fire hydrant
- Connect the discharge side of a pump on a supply engine to the intake side of the pump on an attack engine
- To **draft**—draw water up through a hose—from a **static water source**—a source that is not under pressure, such as a pond, a lake, a stream, or a swimming pool—to the intake side of a pump on an engine or a mobile water supply apparatus or to a portable pump

The size of supply hose varies. It can be 2½ in. (64 mm), 3 in. (76 mm), 4 in. (101 mm), 5 in. (127 mm), or 6 in. (152 mm) in diameter. The choice of diameter is based on the preferences and operating requirements of each fire department. It also depends on the amount of water needed to supply the attack engine, the distance from the source to the attack engine, and the pressure that is available at the source.

A 2½-in. (64-mm) hose may be used as either a supply hose or an attack hose. This size hose has a limited flow capacity as supply hose, but it can be effective at a low to moderate **flow rate**—the number of gallons or liters per minute—and over short distances. When this size hose is used as supply hose, sometimes two parallel lines of 2½-in. (64-mm) hose are used to provide a more effective water supply.

LDH is much more efficient than 2½-in. (64-mm) or 3-in (76-mm) hose for moving large volumes of water over longer distances. Many fire departments use 4-in. (101-mm) or 5-in. (127-mm) hose as their standard supply hose. A single 5-in. (127-mm) supply hose can deliver flow rates exceeding 1500 gpm (5678 L/min) under some conditions. LDH is heavy and difficult to move after it is filled with water, however. NFPA 1900, *Standard for Aircraft Rescue and Firefighting Vehicles, Automotive Fire Apparatus, Wildland Fire Apparatus, and Automotive Ambulances, 2024 Edition,* does not specify the amount of hose that an engine must carry, but a typical engine may carry anywhere from 800 ft (243 m) to 1250 ft (381 m) of supply hose.

Supply hose must be tested annually at a pressure of at least 200 pounds per square inch (psi; 1379 kilopascals [kPa]) or at a pressure not to exceed the service test pressure marked on the hose. It is intended to be used at pressures up to 185 psi (1275 kPa).

Suction Hose

The hose used to connect the intake side of a fire pump to a hydrant or to a static water source is a special type of supply hose called **suction hose**. There are two types of suction hose: soft sleeve hose and hard suction hose.

Soft sleeve hose (known historically as a **soft suction hose**) is a short section (10 to 25 ft [3 to 7.6 m]) of LDH used to transport water from a pressurized source, such as a fire hydrant, to the intake side of the fire pump (**FIGURE 10-9**). The soft sleeve hose allows as much water as possible to flow from the pressurized water source to the intake side of the fire pump through a single line. A soft sleeve hose has threaded female couplings or a Storz or other nonthreaded coupling on both ends. The couplings have rocker lugs to allow for quick tightening by hand. Soft sleeve hose ranges from 2½ in. (64 mm) to 6 in. (152 mm) in diameter and is usually between 10 ft (3 m) and 25 ft (7.6 m) in length.

A **hard suction hose** is a short section of rigid fire hose that is used to draft water from a static water source (**FIGURE 10-10**). It can also be used to carry water from a low-pressure hydrant or draft hydrant to the pumper for drafting operations. The diameter is based on the capacity of the pump but can be as large as 6 in. (152 mm). It normally comes in 10-ft (3-m) or 20-ft (6-m) sections. Hard suction hose can be made from either rubber or plastic. The plastic versions are much lighter and more flexible.

Like soft sleeve hose, hard suction hose has either threaded female couplings or a Storz or other nonthreaded coupling on both ends. Long rocker lugs on the female couplings of hard suction hose assist in tightening the connection. To draft water, it is essential to have an airtight connection at each coupling. Sometimes it may be necessary to gently tap the lugs on the female couplings with a rubber mallet to tighten the hose or to loosen the connection before disconnecting it. Never tap the lugs with anything metal, however, or you could cause damage to the lugs or to the coupling itself.

FIGURE 10-9 Soft sleeve hose.

© Jones & Bartlett Learning. Photographed by Glen E. Ellman.

FIGURE 10-10 Hard suction hose.

© Jones & Bartlett Learning. Photographed by Glen E. Ellman.

Attack Hose

Attack hose is for fire suppression. This hose carries the water from the attack engine to the fire or to an FDC, or from a standpipe system to the fire. The common diameters for attack hose are 1½-in. (38-mm) or 1¾-in. (44-mm), and some fire departments also use 2½-in. (64-mm) attack hose (**FIGURE 10-11**). Each section of attack hose is usually 50 ft (15 m) long. Attack hose is often stored on an engine as a **preconnected attack line**, sometimes called a **preconnect**, in lengths ranging from 150 to 350 ft (45 to 106 m). These travel on the engine already equipped with a nozzle and connected to a pump discharge outlet so that they are ready for immediate use.

FIGURE 10-11 Attack hose used during fire suppression is most commonly 1½-, 1¾-, or 2½-inch (38-, 44-, or 64-mm) lines.

Courtesy of Steve Redick.

Attack hose can be either a multiple-jacket or rubber-covered construction. It must withstand high pressure. Because it is used during fire suppression, it will be subjected to high temperatures, sharp surfaces, abrasion, and other potentially damaging conditions. For this reason, attack hose must be tough. Because firefighters need to be able to maneuver with it, it also needs to be flexible and lightweight.

Attack hose must be tested annually at a pressure of at least 300 psi (2068 kPa) and is intended to be used at pressures of up to 275 psi (1896 kPa).

Booster Hose and Forestry Fire Hose

Booster hose and forestry fire hose are specialized attack hoses. **Booster hose** (or **booster line**) is 1 in. (25 mm) in diameter, noncollapsible, rubber-covered hose usually carried on a hose reel that holds 150 or 200 ft (45 or 61 m) of hose. Booster hose contains a **braided reinforcement**—layers of braided strands of yarn or wire with a layer of rubber between each braid—that gives it a rigid shape. This rigid shape allows the hose to flow water without pulling all of the hose off the reel. Booster hose is lightweight and can be advanced quickly by one person.

Booster hose should not be used for structural or vehicle firefighting because it has limited flow. The normal flow from a 1-in. (38-mm) booster hose is in the range of 40 to 50 gpm (151 to 189 L/min), which is not adequate for extinguishing structure fires. Booster hose use is typically limited to small outdoor fires and dumpster fires.

Forestry fire hose is lightweight and collapsible hose, ¾ in. (19 mm), 1 in. (25 mm), or 1½ in. (38 mm) in diameter, often used to fight wildland and ground cover fires. Large amounts of water are usually not needed for these fires, and the small diameter provides much better maneuverability through brush and trees. This type of hose is sometimes extended for hundreds of feet.

Hose Rolls

An efficient way to transport a single section of fire hose is in the form of a roll. Rolled hose is both compact and easy to manage. A fire hose can be rolled in many different ways, depending on how it will be used. Follow the standard operating procedures (SOPs) of your department when rolling hose.

Straight or Storage Hose Roll

The **straight hose roll**, or **storage hose roll**, is a simple and frequently used hose roll. It is used for general handling and transportation of hose as well as to prepare hose to be stored on a rack (**FIGURE 10-12**). With this roll, the male coupling is at the center of the roll, and the female coupling is on the outside of the roll. To perform a straight hose roll, follow the steps in **SKILL DRILL 10-7**.

Single-Doughnut Hose Roll

The **single-doughnut hose roll** is used when the hose will be put into use directly from its rolled state. With this arrangement, both couplings are on the outside

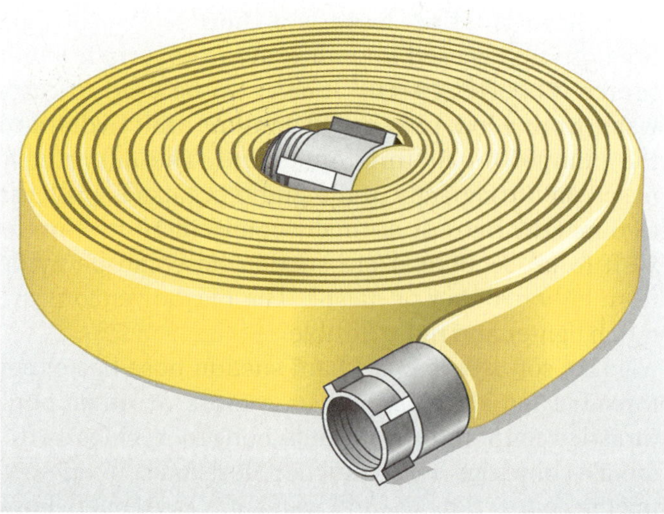

FIGURE 10-12 A straight or storage hose roll is used for transporting and storing hose. The exposed threads of the male coupling stay protected at the center of the roll.

© Jones & Bartlett Learning

SKILL DRILL 10-7

Performing a Straight or Storage Hose Roll Support Person, NFPA 1010: 5.5.3

1. Lay the hose flat and in a straight line. Stand at the end with the male coupling. Fold the male coupling over on top of the hose.

2. Fold the male coupling over on top of the hose again so that the opening is face down.

3. Continue rolling the hose until you reach the female coupling at the other end.

4. Set the hose roll on its side, and then tap any protruding hose flat with your foot.

© Jones & Bartlett Learning. Photographed by Glen E. Ellman.

of the roll, and the rolled hose can be connected and extended by one crew member (**FIGURE 10-13**). With this roll, the threads of the male coupling are protected by the hose rolled on top of it. To perform a single-doughnut roll, follow the steps in **SKILL DRILL 10-8**.

Twin-Doughnut Hose Roll

The twin-doughnut hose roll is used primarily to make a small, compact roll that can be carried easily (**FIGURE 10-14**). This roll can be carried by hand, by a rope, or by webbing. To perform a twin-doughnut hose roll, follow the steps in **SKILL DRILL 10-9**.

Self-Locking Twin-Doughnut Hose Roll

The self-locking twin-doughnut hose roll is similar to the twin-doughnut roll, except that it forms its own carry loop (**FIGURE 10-15**). To perform a self-locking twin-doughnut roll, follow the steps in **SKILL DRILL 10-10**.

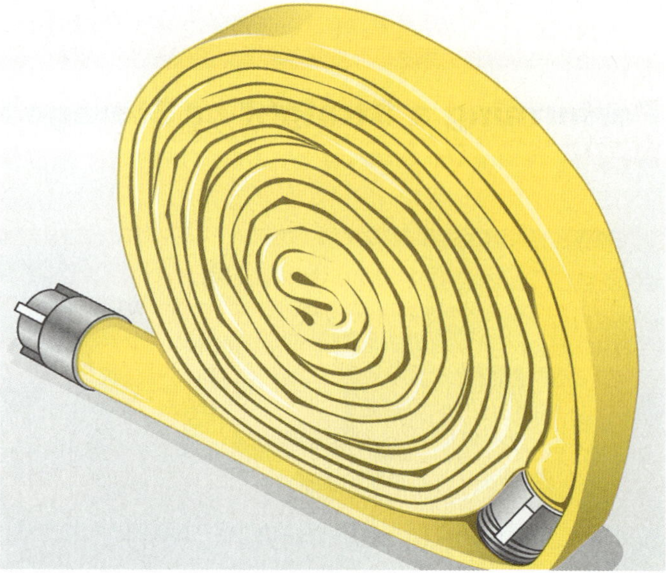

FIGURE 10-13 A single-doughnut hose roll leaves both couplings easily accessible, so it is used when the hose will be put into use directly from its rolled state.

© Jones & Bartlett Learning

SKILL DRILL 10-8

Performing a Single-Doughnut Hose Roll Support Person, NFPA 1010: 5.5.3

1. Lay the hose flat and in a straight line.

2. Locate the midpoint of the hose. From the midpoint, move 5 ft (1.5 m) toward the male coupling end. Pinch the hose at this point.

SKILL DRILL 10-8 CONTINUED

Performing a Single-Doughnut Hose Roll Support Person, NFPA 1010: 5.5.3

3. Start rolling the pinched bit of hose toward the female coupling end.

4. At the end of the roll, wrap the rest of the hose over the male coupling to protect the threads on the male coupling. Set the hose roll on its side, and then tap any protruding hose flat with your foot.

© Jones & Bartlett Learning. Photographed by Glen E. Ellman.

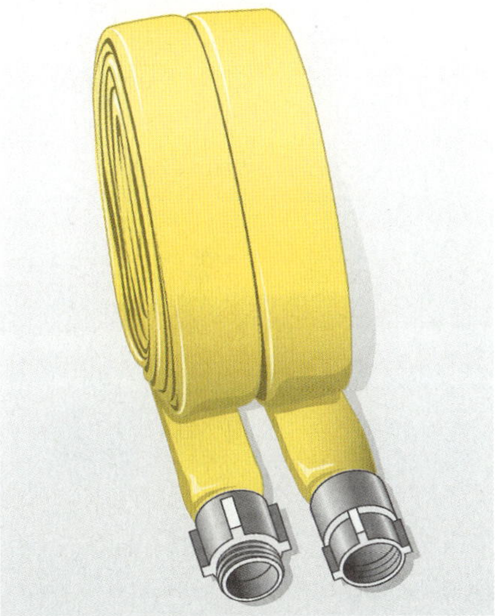

FIGURE 10-14 A twin-doughnut hose roll is used to make a small, compact roll that can be carried easily.

© Jones & Bartlett Learning

Hose Care, Maintenance, and Inspection

Fire hose should be regularly inspected and tested following the procedures in NFPA 1962, *Standard for the Care, Use, Inspection, Service Testing, and Replacement of Fire Hose, Couplings, Nozzles, and Fire Hose Appliances, 2018 Edition.* Hose that is not properly maintained can deteriorate over time and eventually burst. In addition, the swivel gaskets in female couplings need to be checked regularly and replaced when they are worn or damaged.

Causes and Prevention of Hose Damage

Fire hose is a lifeline for firefighters. Every time firefighters respond to a fire, they rely on fire hose to deliver the water needed to attack the fire and protect themselves

SKILL DRILL 10-9

Performing a Twin-Doughnut Hose Roll Support Person, NFPA 1010: 5.5.3

1. Lay the hose flat and in a straight line.

2. Bring the male coupling alongside the female coupling, basically laying each half of the hose next to the other half.

3. With the two halves of the hose laid next to each other, fold over the bend in the hose, and then roll both sections of hose toward the couplings, creating a double roll.

4. Continue rolling until you reach the two ends of the hose. Keep the roll tight and compact as you progress. This will help create a small, easily carried roll.

© Jones & Bartlett Learning. Photographed by Glen E. Ellman.

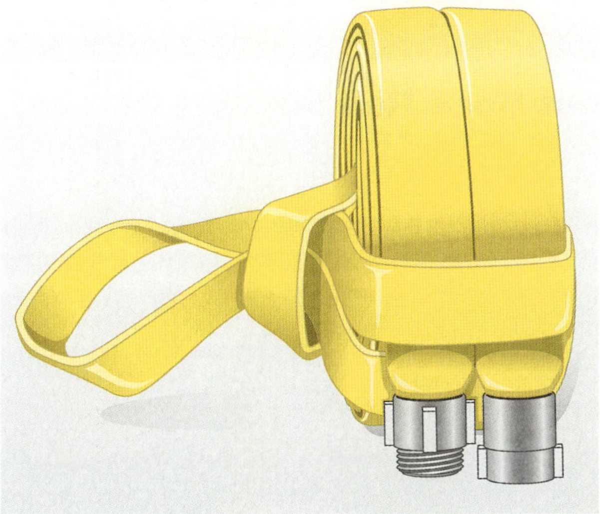

FIGURE 10-15 A self-locking twin-doughnut hose roll forms its own carry loop.

© Jones & Bartlett Learning

from it. Fire hose is a highly engineered product designed to perform well under adverse conditions. Fire service personnel must be careful to prevent damage to the hose that could result in premature or unexpected failure. The factors that most commonly damage fire hose include mechanical causes, heat, cold, ultraviolet (UV) radiation, chemicals, and mildew.

Mechanical Damage

Mechanical damage can occur from many sources. For example, hose that is dragged over rough objects or along a roadway can be damaged by abrasion. Broken glass and sharp objects can cut through the hose. Particles of grit caught in the fibers can damage the jacket or puncture holes in the liner. Reloading dirty hose can cause damage to the fibers in the jacket of the hose. Be especially careful if you need to place a hose through a broken window; if you need to do this, remove any

SKILL DRILL 10-10

Performing a Self-Locking Twin-Doughnut Hose Roll Support Person, NFPA 1010: 5.5.3

1. Lay the hose flat and straighten it out, ensuring there are no knots or tangles. Fold the hose in half, bringing the male coupling and female coupling close to each other. The two couplings should be aligned and facing the same direction.

2. With the hose folded in half, create a small loop or doughnut shape near the folded end of the hose, passing the folded end through the loop. This creates the first doughnut shape.

Continues.

SKILL DRILL 10-10 CONTINUED

Performing a Self-Locking Twin-Doughnut Hose Roll Support Person, NFPA 1010: 5.5.3

3. Bring the top of the loop back toward the couplings to the point where the hose crosses, creating two loops on either side. One loop should be slightly larger than the other.

4. Grasp the inner part of each of the two loops, and then begin rolling the two sides of the hose toward the couplings.

5. When you reach the couplings, pass the larger loop over the rolled hose and through the smaller loop.

6. Hold the roll by the smaller loop, which you can place over your shoulder to carry the hose roll.

© Jones & Bartlett Learning. Photographed by Glen E. Ellman.

protruding sharp edges of glass first. Fire hose is also likely to be damaged if it is run over by a vehicle; if traffic must drive over a hose in a roadway, use a hose bridge.

Hose couplings can be damaged by dropping them on the ground. In particular, the exposed threads on male couplings are easily damaged if they are dropped. Avoid dragging hose couplings because this practice can cause damage to the threads and to the swivels.

Heat, Cold, and UV Radiation

Hose can be damaged by heat and cold as well as by prolonged exposure to UV radiation (sunlight). Heat is an obvious concern when fighting a fire. Hose is not fireproof. A hose that is directly exposed to a fire can burn through and burst quickly. Burning embers and hot coals can cause small leaks or weaken the hose enough that it will burst under pressure. Avoid storing a hose in places where it will come in contact with hot surfaces, such as a heating unit or the exhaust pipe on a vehicle. If the apparatus is parked outside, use a hose cover to protect the hose from sunlight. It is important to keep the hose from becoming brittle. Also, to prevent UV radiation damage, do not dry hose in sunlight.

Cold temperatures can also damage hose. If a hose freezes, the hose liner can rupture, and fibers in the jacket can break. When fire crews are working in below-freezing temperatures, water should be kept flowing through the hose to prevent it from freezing. If a line must be shut down temporarily, the nozzle should be left partly open to keep the water moving. If this is necessary, direct the stream to a safe location away from the fire scene where it will not cause additional water damage or create a slipping hazard when frozen. When a line is no longer needed, the hose should be drained and rolled before it freezes. Taking steps to prevent ice buildup helps maintain hose integrity and firefighter safety in cold-weather operations.

If a hose does freeze, do not attempt to bend a section of frozen hose. Hose that is frozen or encased in ice can be thawed out with a generator that produces steam. You can also carefully use the blunt side of a flathead axe to lightly pound on the ice encasing the hose, being careful not to damage the hose itself. The frozen hose is then transported back to the fire station to thaw. In situations where a hose is frozen solid, it may be necessary to transport the hose back to the fire station on a flatbed truck.

Chemicals

Many chemicals can damage fire hose. You may encounter chemicals at incidents in facilities where chemicals are manufactured, stored, or used, as well as in locations where their presence is not anticipated. Most vehicles contain a wide variety of chemicals that can damage fire hose, including battery acid, gasoline, diesel fuel, antifreeze, motor oil, and transmission fluid. Hose may come in contact with these chemicals at vehicle fires or at the scene of a collision where the chemicals are spilled on the roadway. Supply hose, in particular, often comes in contact with residues from these chemicals when hose lines are laid in the roadway. Remove chemicals from the hose as soon as possible, and then wash the hose with an approved detergent, thoroughly rinse it, and let it dry completely.

Mildew

Mildew is a fungus that can grow on fabrics and materials in warm, moist conditions. A fire hose that has been packed away while it is still wet and dirty is a natural breeding ground for mildew. This fungus feeds on nutrients found in many natural fibers, and this can cause the fibers to rot and deteriorate.

In the days when cotton fibers were used in hose jackets, mildew was a major problem. Hose had to be washed and completely dried after every use before it could be placed back on the apparatus. Modern fire hose is made from synthetic fibers that are resistant to mildew, and most types can be repacked without drying first. The fibers in rubber-covered hose are protected from mildew. Nevertheless, mildew may still grow on exposed fibers if they are soiled with contaminants that will provide mildew with the necessary nutrients.

SAFETY TIP

Whenever a fire hose has suffered possible damage, it should be thoroughly inspected and tested according to Chapter 4 in NFPA 1962 before it is returned to service.

Cleaning and Maintaining Hose

It is important to properly clean fire hose. Because hose can be made from different materials, the exact steps needed to clean and maintain it vary. When hose becomes dirty, it is important to clean it as soon as possible. For a mild cleaning, cool water and a soft brush may be adequate. For dirty hose, it may be necessary to use a mild detergent, especially if the hose has come in contact with petroleum-based products.

TIP

In general, try to prevent hose from coming in contact with petroleum and abrasive substances whenever possible.

Some fire departments use specially manufactured hose washers for washing and drying fire hose. A simple hose washer is a cylindrical device with water jets inside to which you attach a supply line. To use it, first lay the hose out straight and scrub it with a mild detergent to remove debris and dirt. Then insert one end of the scrubbed hose into the cylinder of the hose washer. When the water is turned on to the hose washer, it

FIGURE 10-16 A simple hose washer.

© Jones & Bartlett Learning. Photographed by Glen E. Ellman.

FIGURE 10-17 A mechanical hose washer.

Courtesy of Circul-Air Corp.

FIGURE 10-18 A hose drying cabinet.

Courtesy of Circul-Air Corp.

sprays out of the jets inside the cylinder onto the hose. Slowly pull the hose through the washer, rinsing off the detergent and debris (**FIGURE 10-16**).

More complex hose washers consist of a large cabinet-style mechanical device. These may contain an automatic feed that moves hose through a power washing cycle and then squeegees some of the water off the hose before it leaves the washing machine (**FIGURE 10-17**). This type of machine enables one person to wash a large quantity of hose in a fairly short period of time.

Although most fire hose is rubber-jacketed hose, which is more resistant to water damage than other fabrics, fire hose—even rubber-jacketed hose—should not routinely be stored wet for the following reasons:

- Storing wet hose promotes mold and mildew growth internally, which can weaken the hose liner.
- Water trapped inside the hose can corrode metal couplings over time.
- The additional weight of water makes the hose heavier to handle and deploy.
- Hard water deposits can build up on the exterior jacket.
- Ice that forms in freezing temperatures can cause internal damage.

The best practice is to limit wet storage duration as much as possible and dry hose before reloading by hanging it to drain and air dry internally and externally or using drying cabinets, towers, or racks.

Fire hose can be dried in several different ways. Some fire stations have angled racks with slats on which wet hose can be placed. The angled construction allows water to drain from the inside of the hose, and the slats allow the outside of the hose to dry. Other fire stations are built with hose towers in which one end of the clean and wet hose is hoisted to the top of the tower. This allows water to drain from the inside of the hose and allows the outside of the hose to dry. A third method of drying hose is to place wet hose that has been loosely folded or coiled in a heated hose drying cabinet (**FIGURE 10-18**). Heated air is then circulated through the cabinet. Some heated hose drying cabinets can also be used for drying personal protective equipment.

Always follow the hose manufacturer's guidelines and procedures for washing and drying fire hose. If your department has hose washing and drying equipment, follow the instructions from those manufacturers to be sure the hose is properly cleaned and dried. Some hose washing equipment does not completely clean the hose couplings. Hose couplings may require special attention to be sure they are clean and will operate easily and effectively under emergency conditions.

To manually clean hose that is dirty or contaminated, follow the steps in **SKILL DRILL 10-11**.

Hose Inspections

Visual hose inspections should be performed at least quarterly. A visual inspection should also be performed after each use, either while the hose is being cleaned and dried or when it is reloaded onto the apparatus. Always visually inspect any hose that has come in direct contact with a fire. In addition, hose that has not been used in 30 days should be unpacked, inspected, cleaned, and reloaded. If any defects are found, that length of hose should be immediately removed from service and tagged with a description of the problem. The appropriate notifications must be made to have the hose repaired.

To inspect hose for defects and mark a defective hose, follow the steps in **SKILL DRILL 10-12**.

SKILL DRILL 10-11

Cleaning and Drying Hose Support Person, NFPA 1010: 5.5.3

1. Lay the hose out flat. Rinse the hose with water.

2. Gently scrub the hose with mild detergent, paying attention to soiled areas.

Continues.

SKILL DRILL 10-11 CONTINUED

Cleaning and Drying Hose Support Person, NFPA 1010: 5.5.3

3. Turn over the hose and repeat steps 1 and 2. Give a final rinse to the hose with water and let dry.

© Jones & Bartlett Learning. Photographed by Glen E. Ellman.

SKILL DRILL 10-12

Inspecting a Hose for Defects and Marking a Defective Hose Support Person, NFPA 1010: 5.5.3

1. Thoroughly inspect the fire hose and couplings to identify any defects or damage. Look for common defects in the hose such as holes, cuts, bulges, cracks, or other signs of wear and tear that compromise the hose's integrity. Also inspect the couplings, looking carefully for damage to the threads, lugs, swivels, and gaskets.

2. If a defect is identified, assess the severity and extent of the defect. It is essential to determine whether the hose can still be used or if it needs to be removed from service.

3. Mark the damaged area so that it is easily visible and identifiable to all personnel who handle the hose. Use marking tools approved by your fire department or agency, such as colored tape or paint, or other marking devices that are durable and visible.

4. Document the hose defect, including its location, type, and extent, by recording this information in the hose records.

5. If the defect is severe and compromises the hose's safety and performance, remove the hose from service immediately, place an out-of-service tag (or other method dictated by your department's SOPs) on it, and notify your superiors. Replace the hose with a new one or have it repaired by a qualified professional following the manufacturer's guidelines.

CASE STUDY

You Are the Support Person CONCLUSION

Shortly after joining a fire crew, you are assigned to ride along with an engine company for a shift. After a relatively uneventful day, your company is dispatched for a working fire at a large warehouse. When you arrive, there are flames coming from the front of the building. You hear the incident commander instruct the crew to lay a 4-in. (101-mm) supply line and prepare for an interior attack.

1. **Why is it important to understand the function and proper uses of each type of hose carried on your apparatus?**

   Answer: One of the primary functions of firefighting is to quickly and effectively apply sufficient quantities of water to the fire to extinguish it. In order to achieve this objective, it is necessary to understand the tools that enable you to complete this task.

2. **What are the advantages/disadvantages of using LDH for supply hose?**

   Answer: LDH commonly uses Storz couplings and is much more efficient than SDH for moving large volumes of water over longer distances. However, LDH is heavy and difficult to move once filled with water.

WRAP-UP

SUMMARY

- Fire hose is used for two main purposes: as supply hose and as attack hose. Supply hose (or supply line) is used to deliver water to an attack engine from a pressurized source such as a fire hydrant or from a supply engine.
- Small-diameter hose (SDH) is the primary attack hose for most fires. SDH can usually be operated by one firefighter, although having a second firefighter on the line makes it much easier to advance and control the hose. SDH usually comes in 50-foot (ft; 15-meter [m]) lengths.
- Medium-diameter hose can be used as either attack hose or supply hose, but it is most often used as attack hose. It takes at least two firefighters to safely control due to the weight of the hose, the water, and the nozzle reaction force.
- Large-diameter hose (LDH) is constructed as supply hose, but some fire departments use special LDH that can withstand higher pressures.
- Most fire hose is constructed with an inner waterproof liner surrounded by one or multiple outer layers commonly referred to as jackets. A single-jacket hose is constructed with one layer of woven fiber. A multiple-jacket hose is constructed with two or more layers of woven fibers.
- Rubber-covered hose is a type of double-jacket fire hose that features a durable rubber-like compound on the outer layer or jacket. This construction provides additional protection and durability to the hose, making it suitable for various firefighting applications.
- Collapsible fire hose is flexible, foldable, compact, and it lays flat. It is easily stored when not in use.
- Non-collapsible fire hose, also known as rigid fire hose, is designed to maintain its circular shape and cross-sectional integrity even when not filled with water and pressurized.
- Hose is usually of 50 to 100 ft long. At a fire scene, the hose often needs to be much longer to reach the fire. To achieve this, both ends of a length of hose have a coupling permanently attached. Couplings are also used to connect a hose to a fire hydrant or a pump or to nozzles and appliances.

SUMMARY CONTINUED

- A threaded hose coupling is used on most hose that is 3 in. (76 mm) or less in diameter. A length of fire hose with threaded couplings has a male hose coupling with threads on the outside on one end of the hose and a female hose coupling with threads on the inside on the other end of the hose.
- When connecting fire hose with threaded couplings, the threads must be properly aligned. When properly coupled, threaded hose couplings provide a secure connection between two sections of hose and are unlikely to become disconnected during proper use.
- To create an airtight seal between couplings, you need a flexible gasket to facilitate the seal. On threaded couplings, the gasket sits inside the swivel on the female hose coupling and it is called a swivel gasket.
- A Storz hose coupling is a nonthreaded coupling that consists of two halves that interlock with each other to form a tight seal. Storz couplings are often referred to as sexless couplings because they do not have distinct male or female ends.
- Suction hose is used to connect the intake side of a fire pump to a hydrant or to a static water source. There are two types of suction hose: soft sleeve hose and hard suction hose.
- Soft sleeve hose is a short section of LDH used to transport water from a pressurized source, such as a fire hydrant, to the intake side of the fire pump. The soft sleeve hose allows as much water as possible to flow from the pressurized water source to the intake side of the fire pump through a single line.
- A hard suction hose is a short section of rigid fire hose that is used to draft water from a static water source. It can also be used to carry water from a low-pressure hydrant or draft hydrant to the pumper for drafting operations.
- Attack hose is for fire suppression. It carries the water from the attack engine to the fire or to an FDC, or from a standpipe system to the fire.
- Booster hose and forestry fire hose are specialized attack hose. Booster hose is non-collapsible, rubber-covered hose usually carried on a hose reel. Booster hose contains a braided reinforcement that gives it a rigid shape. Booster hose should not be used for structural or vehicle firefighting because it has limited flow.
- Forestry fire hose is lightweight, collapsible hose often used to fight wildland and ground cover fires. Large amounts of water are usually not needed for these fires, and the small diameter provides much better maneuverability through brush and trees.
- The straight hose roll, or storage hose roll, is a simple and frequently used hose roll used for general handling and transportation of hose and to prepare hose to be stored on a rack. With this roll, the male coupling is at the center of the roll, and the female coupling is on the outside of the roll.
- A fire hose can be rolled in many different ways, depending on how it will be used. Follow the standard operating procedures (SOPs) of your department when rolling hose.
- Hose can be damaged by heat and cold as well as by prolonged exposure to UV radiation (sunlight). If a hose freezes, the hose liner can rupture and fibers in the jacket can break.
- Many chemicals can damage fire hose. Hose may come in contact with these chemicals at vehicle fires or at the scene of a collision.
- A fire hose that has been packed away while it is still wet and dirty is a natural breeding ground for mildew. This fungus feeds on nutrients found in many natural fibers and this can cause the fibers to rot and deteriorate.
- When hose becomes dirty, it is important to clean it as soon as possible. For a mild cleaning, cool water and a soft brush may be adequate. For dirty hose, it may be necessary to use a mild detergent, especially if the hose has come in contact with petroleum-based products.
- Fire hose can be dried in several different ways. Always follow the hose manufacturer's guidelines and procedures for washing and drying fire hose. If your department has hose washing and drying equipment, follow the instructions from those manufacturers to be sure the hose is properly cleaned and dried.
- Visual hose inspections should be performed at least quarterly. A visual inspection should also be performed after each use, either while the hose is being cleaned and dried or when it is reloaded onto the apparatus.

KEY TERMS

adapter Any device that allows fire hose couplings to be safely interconnected with couplings of different sizes, threads, or mating surfaces, or that allows fire hose couplings to be safely connected to other appliances. (NFPA 1960)

attack engine An engine used to pump water through attack lines at the fireground.

attack hose Hose designed to be used by trained firefighters and fire brigade members to combat fires beyond the incipient stage. Also called *attack line.* (NFPA 1962)

attack line See *attack hose.*

booster hose A non-collapsible hose used under positive pressure having an elastomeric or thermoplastic tube, a braided or spiraled braided reinforcement, and an outer protective cover. Also called *booster line.* (NFPA 1962)

booster line See *booster hose.*

braided reinforcement A hose reinforcement consisting of one or more layers of interlaced spiraled strands of yarn or wire, with a layer of rubber between each braid. (NFPA 1962)

chafing block A sturdy rubber, plastic, or wooden block placed under a fire hose where it lays on the ground or rests against hard surfaces to raise the hose off the ground and provide a smooth contact surface and protect the hose from abrasion, friction, and wear against rough surfaces like asphalt, concrete, or debris.

charge To fill with water under pressure.

collapsible fire hose Fire hose typically made from synthetic materials that make the hose flexible and foldable.

coupler A hose appliance that allows two hoses to be connected together and flow into a single hose.

coupling A connection device that connects (couples) individual lengths of fire hose together or connects a hose to a fire hydrant or a pump or to nozzles and hose appliances.

deck gun A device that is permanently mounted on and operated from a vehicle and equipped with a piping system that delivers water to the gun.

double-female adapter A hose adapter that is used to join two male hose couplings.

double-jacket hose A hose constructed with two layers of woven fibers.

double-male adapter A hose adapter that is used to join two female hose couplings.

draft The use of suction to move a liquid (such as water) from a vessel or source that is below the intake of a pump. (NFPA 1910)

fingers Individual, adjustable vanes or protrusions on the face of a fog-stream nozzle that are responsible for shaping the water stream into a specific pattern.

fire department connection (FDC) A connection through which the fire department can pump supplemental water into the sprinkler system, standpipe, or other system furnishing water for fire extinguishment to supplement existing water supplies. (NFPA 13)

fire hose A flexible conduit used to convey water or other extinguishing agents. (NFPA 1960)

fire hose appliance A piece of hardware (excluding nozzles) generally intended for connection to fire hose to control or convey water. Also called *hose appliance.* (NFPA 1962)

fire hose tool A device that assists firefighters with handling, manipulating, connecting, and using a fire hose.

fire stream A stream of water or extinguishing agents.

flow rate The number of gallons or liters of fluid streamed per unit of time.

forestry fire hose A hose designed to meet specialized requirements for fighting wildland fires. (NFPA 1960)

handline A hose and nozzle that can be held and directed by hand. (NFPA 11)

hard suction hose A short section of supply hose that is used to draft water from a static source such as a river, lake, or portable drafting basin to the suction side of the fire pump on a fire department engine or into a portable pump.

Higbee indicator An indicator on both the male and female threaded couplings that indicates where the threads start. These indicators should be aligned before firefighters start to thread the couplings together.

hose appliance See *fire hose appliance.*

hose bed The main storage area on an apparatus for carrying hose.

hose bridge A device that protects a hose when it is necessary for a vehicle to drive over a hose. Also called *hose ramp.*

KEY TERMS CONTINUED

hose clamp A device used to compress a fire hose to stop water flow.

hose hoist See *hose roller.*

hose jacket A device used to stop a leak in a fire hose or to join hose lines that have damaged couplings.

hose liner The inside portion of a hose that is in contact with the flowing water; also called *hose inner jacket.*

hose ramp See *hose bridge.*

hose record A written history of each individual length of fire hose.

hose roller A device that is placed on the edge of a roof and is used to protect hose as it is hoisted up and over the roof edge. Also called *hose hoist.*

hose size An expression of the internal diameter of the hose. (NFPA 1962)

hose tool See *fire hose tool.*

ladder pipe A monitor that attaches to the rungs of a vehicle-mounted aerial ladder. (NFPA 1960)

large-diameter hose (LDH) A hose 3.5 in. (89 mm) or larger that is designed to move large volumes of water to supply master stream appliances, portable hydrants, manifolds, standpipe and sprinkler systems, and fire department pumpers from hydrants and in relay. (NFPA 1410)

lug A protrusion or indentation on a hose coupling that aids in securing and tightening the connection between two couplings.

master stream appliance A device that discharges high-volume water streams, usually between 350 gpm (1325 L/min) and 1500 gpm (5678 L/min), though much larger capacities are available. Also called *master stream device.*

master stream device See *master stream appliance.*

medium-diameter hose (MDH) A hose 2½-in. (64-mm) or 3-in. (76-mm) in diameter most often used as attack hose, but can be used as supply hose.

multiple-jacket A construction consisting of a combination of two separately woven reinforcements (double jacket) or two or more reinforcements interwoven. (NFPA 1962)

non-collapsible fire hose Fire hose typically made from durable materials such as PVC or synthetic rubber so that it maintains its shape and structure. Also called *rigid fire hose.*

nozzle A constricting appliance attached to the end of a fire hose or monitor to increase the water velocity and form a stream. (NFPA 1960)

pin lug A lug that looks like a small cylinder that extends outward from a hose coupling.

portable monitor A monitor that can be lifted from a vehicle-mounted bracket and moved to an operating position on the ground by not more than two people. (NFPA 1960)

preconnect See *preconnected attack line.*

preconnected attack line Attack hose that travels on the engine already equipped with a nozzle and connected to a pump discharge outlet so it is ready for immediate use. Also called *preconnect.*

recessed lug A lug that is circular indentation and requires a specially designed spanner wrench called a booster hose wrench to engage.

reducer A fitting used to connect a small hose line or pipe to a larger hose line or pipe. (NFPA 1142)

rigid fire hose See *non-collapsible fire hose.*

rocker lug A rectangular-shaped, bevel-edged lug that extends from a coupling. Also called *rocker pin.*

rocker pin See *rocker lug.*

rubber-covered hose A hose whose outside covering is made of rubber, which is said to be more resistant to damage.

single-doughnut hose roll A hose roll that has both female couplings on the outside of the roll and the male coupling is protected by the hose rolled on top of it and that is used when the hose will be put into use directly from its rolled state.

single-jacket hose A construction consisting of one woven jacket. (NFPA 1962)

small-diameter hose (SDH) A hose 1½-in. (38-mm) or 1¾-in. (44-mm) hose that is most often used as the primary attack hose for most fires.

soft sleeve hose A short section of large-diameter supply hose that is used to provide water from the large steamer outlet (the large-diameter port) on a fire hydrant or other pressurized water source to the suction side of the fire pump. Also called *soft suction hose.*

soft suction hose See *soft sleeve hose.*

standpipe A pipe and attached hose valves and hose (if provided) used for conveying water to various parts of a building for fire-fighting purposes. (NFPA 1140)

static water source A water source that is not under pressure, such as a pond, a lake, a stream, or a swimming pool.

storage hose roll See *straight hose roll.*

Storz hose coupling A hose coupling that has the property of being both the male and the female coupling. It is connected by engaging the lugs and turning the coupling a one-third turn.

straight hose roll A hose roll that has the male coupling at the center of the roll and the female coupling on the outside of the roll and that is used for general handling and transportation of hose as well as to prepare hose to be stored on a rack. Also called *storage hose roll.*

suction hose A hose that is designed to prevent collapse under vacuum conditions so that it can be used for drafting water from below the pump (lakes, river, wells, etc.). (NFPA 1960)

supply engine An engine used to pump water to an attack engine using supply lines that may be connected to a pressurized water source such as a fire hydrant or to a static (unpressurized) water source.

supply hose Hose designed for the purpose of moving water between a pressurized water source and a pump that is supplying attack lines. Also called *supply line.* (NFPA 1960)

supply line See *supply hose.*

swivel A ring around female couplings that you turn to secure the female coupling around the male coupling without twisting the hose.

swivel gasket An O-shaped piece of rubber inside the swivel section of a female hose coupling that forms a seal that stops water from leaking when a male hose coupling is tightened against it.

threaded hose coupling A type of coupling that requires a male fitting and a female fitting to be screwed together.

REVIEW QUESTIONS

1. Fire hose is used for what two main purposes?
2. What size hose is usually used as the primary attack hose?
3. What is the 2½-in. (64-mm) hose most often used as?
4. What sizes can large-diameter hose (LDH) be?
5. What is the difference between single-jacket and multiple-jacket hose?
6. What is the difference between threaded couplings and Storz couplings?
7. What is supply hose used for?
8. List four common hose rolls.
9. How often should fire hose be visually inspected for serviceability?

DISCUSSION QUESTIONS

1. Why is proper selection and deployment of the right type of fire hose for the specific conditions so critical for effective firefighting operations?
2. How can improper storage and cleaning of fire hose lead to potential damage over time?
3. Why is routine inspection and testing important for maintaining fire hose in good operating condition?

REFERENCES

National Fire Protection Association. 2017. *NFPA 1962, Standard for the Care, Use, Inspection, Service Testing, and Replacement of Fire Hose, Couplings, Nozzles, and Fire Hose Appliances.* 2018 Edition. Quincy, MA: National Fire Protection Association.

National Fire Protection Association. 2019. *NFPA 1410, Standard on Training for Emergency Scene Operations.* 2020 Edition. Quincy, MA: National Fire Protection Association.

REFERENCES CONTINUED

National Fire Protection Association. 2023. *NFPA 11, Standard for Low-, Medium-, and High-Expansion Foam.* 2024 Edition. Quincy, MA: National Fire Protection Association.

National Fire Protection Association. 2020. *NFPA 1700, Guide for Structural Fire Fighting.* 2021 Edition. Quincy, MA: National Fire Protection Association.

National Fire Protection Association. 2021. *NFPA 13, Standard for the Installation of Sprinkler Systems.* 2022 Edition. Quincy, MA: National Fire Protection Association.

National Fire Protection Association. 2021. *NFPA 24, Standard for the Installation of Private Fire Service Mains and Their Appurtenances.* 2022 Edition. Quincy, MA: National Fire Protection Association.

National Fire Protection Association. 2021. *NFPA 1140, Standard for Wildland Fire Protection.* 2022 Edition. Quincy, MA: National Fire Protection Association.

National Fire Protection Association. 2021. *NFPA 1142, Standard on Water Supplies for Suburban and Rural Firefighting* 2022 Edition. Quincy, MA: National Fire Protection Association.

National Fire Protection Association. 2023. *NFPA 1900, Standard for Aircraft Rescue and Firefighting Vehicles, Automotive Fire Apparatus, Wildland Fire Apparatus, and Automotive Ambulances.* 2024 Edition. Quincy, MA: National Fire Protection Association.

National Fire Protection Association. 2023. *NFPA 1910, Standard for the Inspection, Maintenance, Refurbishment, Testing, and Retirement of In-Service Emergency Vehicles and Marine Firefighting Vessels.* 2024 Edition. Quincy, MA: National Fire Protection Association.

National Fire Protection Association. 2023. *NFPA 1960, Standard for Fire Hose Connections, Spray Nozzles, Manufacturer's Design of Fire Department Ground Ladders, Fire Hose, and Powered Rescue Tools.* 2024 Edition. Quincy, MA: National Fire Protection Association.

Chapter Opener: © Eric Scruggs

CHAPTER

11

Support Person

Supply Line and Attack Line Evolutions

KNOWLEDGE OBJECTIVES

After studying this chapter, you will be able to:

- Describe the procedures used to connect supply lines to a fire hydrant.
- Describe the common techniques used to load supply hose.
- Describe the common techniques used to carry and advance supply hose.
- Describe the types of loads used to organize attack hose.

SKILLS OBJECTIVES

After studying this chapter, you will be able to perform the following skills:

- Attach a soft sleeve hose to a fire hydrant.
- Properly demonstrate a forward hose lay.
- Properly demonstrate a reverse hose lay.
- Properly demonstrate a split hose lay.
- Properly demonstrate a flat hose load.
- Properly demonstrate a horseshoe hose load.
- Properly demonstrate a preconnected flat hose load.
- Properly demonstrate an accordion hose load.
- Properly demonstrate a minuteman hose load.
- Properly demonstrate a triple-layer hose load.
- Drain a fire hose.

ADDITIONAL NFPA STANDARDS

- **NFPA 11**, *Standard for Low-, Medium-, and High-Expansion Foam, 2021 Edition*
- **NFPA 13**, *Standard for the Installation of Sprinkler Systems, 2022 Edition*
- **NFPA 25**, *Standard for the Inspection, Testing, and Maintenance of Water-Based Fire Protection Systems, 2021 Edition*
- **NFPA 1140**, *Standard for Wildland Fire Protection, 2022 Edition*
- **NFPA 1410**, *Standard on Training for Emergency Scene Operations, 2020 Edition*
- **NFPA 1962**, *Standard for the Care, Use, Inspection, Service Testing, and Replacement of Fire Hose, Couplings, Nozzles, and Fire Hose Appliances, 2018 Edition*

CASE STUDY

You Are the Support Person

In the early morning hours before sunrise, your engine company is dispatched to a report of a residential structure fire. The driver/operator sees another company approaching an intersection just ahead and close to the dispatched address. The officer on your engine radios the other company and advises them to help perform a split hose lay.

1. Why is it important to understand supply line evolutions?
2. What is the primary purpose of supply line evolutions?

Introduction

Putting water on the fire is an essential part of what we do. Ensuring a dependable water supply is a critical fireground operation that must be accomplished as soon as possible. This topic is discussed further in Chapter 9, *Water Supply Systems.* Ideally, a water supply should be established at the same time as the size-up. The methods of moving water through fire hose must be efficient during an emergency when time is of the essence. Fire hose is used for two main purposes: to supply water from a water source to the pumper (supply hose) and to carry water from the pumper to a nozzle for fire attack (attack hose). These methods are discussed further in Chapter 10, *Fire Hose.* When lengths of hose are connected to each other, the result is referred to as a *hose line.* A hose line is a single appliance in a water delivery system. A supply line or an attack line is the system with all of the included appliances.

To become proficient at moving and placing fire hose, fire service personnel practice fire hose evolutions. An **evolution** is a standard set of steps for working on the fireground and during training. Each fire department must set up equipment and conduct regular training so that crew members are prepared to deploy and use hose lines at a fire scene. Every member of the crew should know how to perform all of the standard evolutions quickly and efficiently. When an officer calls for a particular evolution to be performed, each crew member should know exactly what to do.

Supply Line Evolutions

Supply line evolutions, also called **supply line operations**, are the steps involved in deploying and laying supply hose between a water source and an attack or supply engine, attaching hose to the connections on a pump and a hydrant, carrying and advancing supply hose, and loading the hose onto the apparatus so that it can be deployed quickly and easily. To **deploy hose**—or to **pull hose**—means to remove the hose from the hose bed or other storage location. To **lay** or **flake out** hose means arranging hose so it can easily be pulled toward the fire. An **attack engine** is an engine used to pump water through attack lines at the fireground. A **supply engine** is an engine used to pump water to an attack engine using supply lines. Most hose is carried in the **hose bed** on the apparatus—the main storage area for carrying hose.

Drafting from a static water source is one way to deliver water, but the most common way to deliver water to the fire pump on an engine is from a pressurized water source such as a fire hydrant. Drafting is described in Chapter 9, *Water Supply Systems.* The descriptions in this chapter involve connections between a pump and a hydrant.

Connecting an Engine to a Fire Hydrant

Supply hose is used to deliver the water needed to fight the fire over a short distance. In most cases, a short length of soft sleeve hose is used to transport water from a pressurized source, such as a fire hydrant, to the suction side of the fire pump. A soft sleeve hose is a large-diameter supply hose, usually 10 to 25 feet (ft; 3 to 7.6 meters [m]) long, used to connect the steamer port on the hydrant to the suction side of the fire pump. See Chapter 10, *Fire Hose,* for a detailed description. Although it is uncommon, the connection can also be made with a hard suction hose. The short length of soft sleeve hose is usually not carried on the hose bed. Instead, it is carried in a compartment next to the pump panel or sometimes in a compartment in the front bumper. On some engines, it is preconnected to an intake on the pump. To attach a soft sleeve hose to a fire hydrant, follow the steps in **SKILL DRILL 11-1**.

SKILL DRILL 11-1

Attaching a Soft Sleeve Hose to a Fire Hydrant Support Person, NFPA 1010: 5.3.4

1. After the engine driver/operator positions the engine so that the suction side of the pump is approximately 10 ft (3 m) from the fire hydrant, step off of the apparatus, and get the hydrant wrench and all necessary tools. Remove the hose from the hose bed, along with any needed adaptors.

2. If the soft sleeve hose is not preconnected, attach it to the suction side of the pump on the engine. In some cases, it may be necessary to use an adapter.

3. Unroll the hose toward the hydrant.

4. Remove the cap on the steamer port on the hydrant. Visually inspect the inside of the outlet for debris, and if needed, carefully insert a gloved hand and remove the debris. Attach the hydrant wrench to the stem nut on the fire hydrant, and then check the hydrant for an arrow indicating the direction to turn the stem nut to open the hydrant. Open the hydrant valve enough to verify the flow of water and flush out any debris that may be in the hydrant. Close the valve to stop the flow of water.

Continues.

SKILL DRILL 11-1 CONTINUED

Attaching a Soft Sleeve Hose to a Fire Hydrant Support Person, NFPA 1010: 5.3.4

5. Attach the soft sleeve hose to the fire hydrant.

6. Ensure that there are no kinks or sharp bends in the hose that might restrict the flow of water.

7. When the engine driver/operator signals to charge the hose, open the hydrant valve slowly to avoid a water hammer. Check all connections for leaks and tighten the couplings if necessary.

8. Where required, place chafing blocks under the hose where it contacts the ground to prevent mechanical abrasion.

© Jones & Bartlett Learning. Photographed by Glen E. Ellman.

Laying Supply Hose

The objective of laying supply hose is to remove supply hose from the hose bed and place it between a water supply source and an engine. This can be done using a forward hose lay, a reverse hose lay, or a split hose lay (**FIGURE 11-1**). After the hose is laid, it is connected to the water supply and to the pump. Each fire department will determine its own preferred methods and procedures for supply line operations based on available apparatus, water supply, and regional considerations.

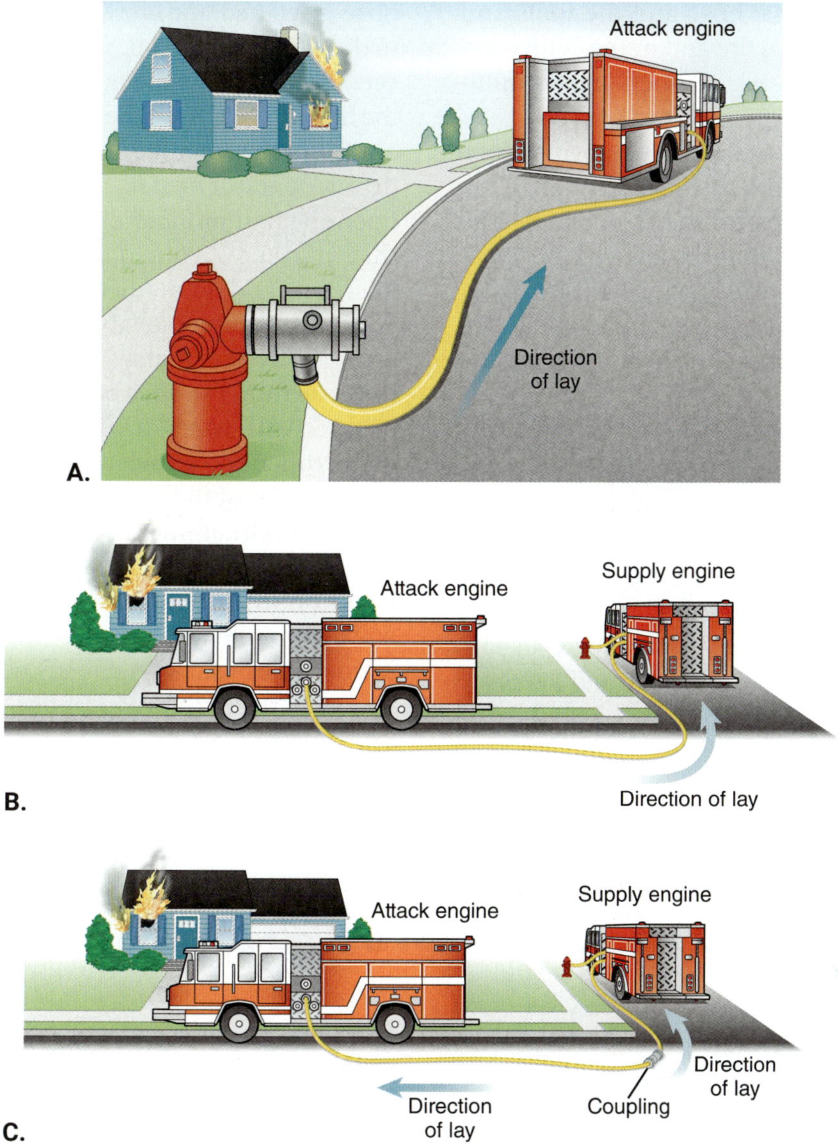

FIGURE 11-1 Laying supply hose can be done using a forward hose lay, a reverse hose lay, or a split hose lay. **A.** A forward hose lay is made from the water source to the attack engine, which is then driven forward toward the fire. **B.** A reverse hose lay is usually done with two engines and is made from the attack engine at the fireground to a water source. **C.** A split hose lay is done when the water source is too far from the fire to use a single supply hose. The attack engine performs a forward hose lay from a specific point, usually an intersection. The supply engine stops at the location where the attack engine's hose was dropped and connects a supply hose to the attack engine's hose. The supply engine then performs a reverse hose lay to the water source.

© Jones & Bartlett Learning

Forward Hose Lay

The most common way of laying hose is the forward hose lay. To execute the **forward hose lay**, also called a **straight hose lay**, the supply hose is pulled from the hose bed on the attack engine and anchored near the water source, usually the hydrant. The engine is slowly driven forward toward the fire, unfolding the hose from the hose bed as it drives—that is, laying out the hose. This method allows the engine company to establish a water supply without assistance from another company and places the attack engine close to the fire. It is most often used by the first-arriving engine company at the scene of a fire. A benefit of the forward hose lay is that the pumper is on the scene, allowing crew members to access additional hose and the tools and equipment that are carried on the apparatus. A forward hose lay can be performed using 2½-inch (in.; 64-millimeter [mm]) hose or larger.

To perform the forward hose lay, the engine stops near the fire hydrant, and a crew member steps off of the apparatus, grasps the coupling on the hose that will be used, and walks toward the hydrant, pulling the hose behind them. When they have walked about 6 ft (1.8 m) past the hydrant, they return to the hydrant and loop the end of supply hose around the base of the hydrant. Once the hose is secured, the crew member at the hydrant signals to the engine driver/operator that the apparatus can proceed to the fire. As the apparatus slowly moves forward, the hose unfolds from the apparatus and onto the ground.

SAFETY TIP

When performing a forward hose lay, the crew member who is connecting the supply hose to the fire hydrant must not stand between the hose and the fire hydrant. When the apparatus starts to move off, the hose could become tangled and suddenly be pulled taut. Anyone standing between the hose and the fire hydrant could be seriously injured.

The crew member at the hydrant waits until the driver/operator signals that they are ready for them to open the hydrant and charge the hose line. Make sure that you know your department's signal to charge a hose line, and do not become so excited or rushed that you mistakenly open the hydrant prematurely. If you open the hydrant before the driver/operator uncouples the hose from the hose bed, the hose bed could become charged with water. If you open the hydrant before the driver/operator either connects the other end to the suction side of the pump or clamps the hose, depending on the situation and department's standard operating procedure (SOP), the loose hose line will discharge water on the ground at the fire scene. Either situation will disrupt the water supply operation and could cause serious injuries.

To perform a forward hose lay, follow the steps in **SKILL DRILL 11-2**.

SKILL DRILL 11-2

Performing a Forward Hose Lay from a Hydrant Support Person, NFPA 1010: 5.3.4

1. The pump driver/operator stops the fire apparatus approximately 10 ft (3 m) past the fire hydrant.

2. Step off of the apparatus and get the hydrant wrench and all necessary tools. Grasp the coupling on the hose in the hose bed, and then walk toward the hydrant, pulling the hose behind you. When you have walked about 6 ft (1.8 m) past the hydrant, walk back to the hydrant, and then wrap the hose around the base of the hydrant to form one full loop, or secure the hose as specified by your department's SOP. Do not stand between the hose and the fire hydrant.

SKILL DRILL 11-2 CONTINUED

Performing a Forward Hose Lay from a Hydrant Support Person, NFPA 1010: 5.3.4

3. Signal the engine driver/operator to proceed to the fire.

4. Once the apparatus has moved off and a length of supply hose has been removed from the apparatus and is lying on the ground, remove the appropriate-size fire hydrant cap from the outlet nearest to the fire, and then check the operating condition of the hydrant.

5. Unwrap the hose from the base of the hydrant, and then attach the hose to the open outlet on the hydrant. An adapter may be needed if a large-diameter hose (LDH) with a Storz-type coupling is used.

6. Attach the hydrant wrench to the stem nut on the fire hydrant and check the hydrant for an arrow indicating the direction to turn the stem nut to open the hydrant. Wait for the attack engine driver/operator to signal that they are ready for the line to be charged.

Continues.

SKILL DRILL 11-2 CONTINUED

Performing a Forward Hose Lay from a Hydrant Support Person, NFPA 1010: 5.3.4

7. When the attack engine driver/operator signals to charge the hose, open the hydrant valve slowly to avoid a water hammer. Make sure you open it fully. Follow the supply hose to the engine and remove any kinks.

© Jones & Bartlett Learning. Photographed by Brandon Fryman/JRM Creative.

Reverse Hose Lay

The **reverse hose lay** is the opposite of the forward hose lay. In the reverse hose lay, the hose is laid out from the fire to the water source, such as a fire hydrant, in the direction opposite to the flow of the water. This can be a useful tactic when multiple pumpers arrive on a fire scene at close to the same time. The attack engine begins fire attack using the booster tank on the engine while the supply engine does a reverse lay to the water source.

TIP

The supply pumper can drop crew members at the scene to help with fire attack before starting the reverse lay.

To perform the reverse hose lay with two engines, the supply engine stops with its hose bed close to an intake on the attack engine. The supply hose is pulled from the hose bed of the supply engine and, if possible, is anchored to a stationary object. The supply engine is then driven to the fire hydrant (or alternative water source), laying out the hose as it goes. At the fire area, the supply hose is connected to an intake on the attack engine. At the water source, the other end is uncoupled from the hose bed on the supply engine and connected to a discharge outlet on the supply engine. The supply engine driver/operator then uses additional sections of supply hose to connect the intake side of the pump on the supply engine to the water source, then pumps water to the attack engine. Like the forward lay, the reverse lay can be performed using a 2½-in. (64-mm) hose or larger.

When performing a reverse lay with threaded couplings, you may find yourself with the wrong end of the hose to make the required connection. Make sure a double-male adaptor and a double-female adaptor are easily accessible so that you can join two couplings of the same sex if needed.

To perform a reverse hose lay, follow the steps in **SKILL DRILL 11-3**.

Split Hose Lay

A **split hose lay** is performed by two engine companies in situations where hoses can or must be laid in two different directions to establish a water supply. This evolution could be used when the attack engine must approach a fire either along a dead-end street with no hydrant or down a long driveway. To perform a split hose lay, the attack engine drops the end of its supply hose where the hose from the two engines will meet, such as at a corner or an intersection. The attack engine then performs a forward lay toward the fire.

SKILL DRILL 11-3

Performing a Reverse Hose Lay with Two Engines Support Person, NFPA 1010: 5.3.4

1. The attack engine driver/operator positions the attack engine at the fireground. The supply engine driver/operator positions the supply engine so that its hose bed is close to an intake on the attack engine. Pull sufficient hose from the hose bed on the supply engine to reach from the supply engine to an intake on the pump on the attack engine. Anchor the hose to a stationary object if possible. Do not stand between the hose and the stationary object.

2. The supply engine driver/operator slowly drives to the fire hydrant, laying out the supply hose.

3. The attack engine driver/operator connects their end of the supply hose to an intake on the pump on the attack engine.

4. At the supply engine, the driver/operator uncouples the supply hose that was laid out from the hose bed and attaches it to a discharge outlet on the supply engine's pump.

Continues.

SKILL DRILL 11-3 CONTINUED

Performing a Reverse Hose Lay with Two Engines Support Person, NFPA 1010: 5.3.4

5. The supply engine driver/operator connects another line to the intake side of the pump on the supply engine. Connect the other end of that line to the hydrant after checking the operating condition of the hydrant (see SKILL DRILL 11-1). Upon the signal from the attack engine driver/operator, the supply engine driver/operator charges the supply line. If the attack engine driver/operator was using the booster tank while waiting for the supply engine to charge the line, they switch over to the supply hose.

© Jones & Bartlett Learning. Photographed by Glen E. Ellman.

The supply engine stops at the same location, pulls off enough hose to connect to the end of the supply line that is already there, and then performs a reverse hose lay to the hydrant. With the two lines connected together, the supply engine can pump water to the attack engine. A split hose lay often requires coordination by two-way radio because the attack engine must advise the supply engine of the plan and indicate where the end of the supply line will be dropped. In many cases, the attack engine is out of sight when the supply engine arrives at the location where the attack engine's line was dropped. And as with a reverse hose lay performed with two engines, when performing a split lay with threaded couplings, make sure a double-male adaptor and a double-female adaptor are easily accessible so that you can join two couplings of the same sex if needed.

To perform a split hose lay, follow the steps in **SKILL DRILL 11-4**.

Loading Supply Hose

Hose can be loaded into the hose bed in several ways, depending on how it will be laid out at a fire scene. Hose must be easily removable from the hose bed, without kinks or twists, and without the possibility of becoming caught or tangled. The ideal hose load is easy to load, avoids wear and tear on the hose, has minimum sharp bends, and allows the hose to lay out of the hose bed smoothly and easily.

SAFETY TIP

Many modern hose beds are several feet above the ground, and a fall could result in significant injuries, so when loading hose on an apparatus, always use caution in climbing up and down on the apparatus. If you are loading hose at a fire scene, watch out for wet, slippery surfaces; ice; or other hazards. Also, wet hose can be heavy, and you may need to reach, stretch, or lift the hose to get it into the hose bed. Use caution! Also, wear appropriate personal protective equipment (PPE) when loading hose. At a minimum, wear your helmet, gloves, and boots.

When loading hose, it is best to observe the following practices:

- Drain all of the water out of the hose before loading it.
- Roll the hose first to remove all the air for a flatter hose load. (Rolling hose is described in Chapter 10, *Fire Hose*.)

SKILL DRILL 11-4

Performing a Split Hose Lay Support Person, NFPA 1010: 5.3.4

1. The attack engine driver/operator stops at the location where the supply line will be dropped. A crew member from the attack engine removes the end of the supply hose from the hose bed and anchors it to a stationary object if possible. This crew member stays at the drop location until either the attack engine has driven off or the supply engine arrives, according to the department's SOPs, and then they walk to the fire.
2. The attack engine driver/operator slowly drives toward the fire, laying out the hose line. When the attack engine arrives at the fireground, the driver/operator can start pumping from the booster tank on the engine.
3. When the supply engine arrives at the location where the supply hose was dropped, it stops. A crew member from the supply engine pulls off enough supply hose to connect to the hose end laid in the street by the attack engine and then connects the two lines (see FIGURE 11-1C). If threaded couplings are used, a double-male or double-female adaptor may be required. Remount the apparatus; the supply engine driver/operator then proceeds slowly to the hydrant.
4. When the supply pumper arrives at the water source, uncouple the supply line from the hose bed. The supply engine driver/operator connects this line to a discharge outlet on the supply engine's pump.
5. The supply engine driver/operator connects another section of supply hose to the intake side of the pump on the supply engine. Connect the other end of this line to the hydrant after checking the operating condition of the hydrant (see SKILL DRILL 11-1).
6. Upon the signal from the attack engine driver/operator, the supply engine driver/operator charges the supply line. If the attack engine driver/operator was using the booster tank while waiting for the supply engine to charge the line, they switch over to the supply hose.

- Do not load hose too tightly. Leave enough room so that you can slide a hand between the folds of hose. If hose is loaded too tightly, it may not lay out properly.
- Load hose so that couplings do not have to turn around as the hose is pulled out of the hose bed. Make a short fold in the hose close to the coupling to keep the hose properly oriented. This short fold is called a *Dutchman*. The Dutchman fold allows the hose to deploy or unload smoothly, and it prevents the coupling from turning and becoming stuck as the hose is laid out (**FIGURE 11-2**).
- Couple sections of hose with the flat sides of the hose oriented in the same direction.
- Check that swivel gaskets are in place before coupling hose.
- Tighten couplings so that they are hand-tight only. With a good gasket, the hose should not leak.
- If your department's SOPs require it, load the appliances for hydrant connections.

Three hose loads are commonly used for loading supply hose: the flat hose load, the horseshoe hose load, and the accordion hose load (**FIGURE 11-3**). Any one of these methods can be used to load hose for either a forward hose lay or a reverse hose lay. You need to learn the specific hose loads used by your fire department.

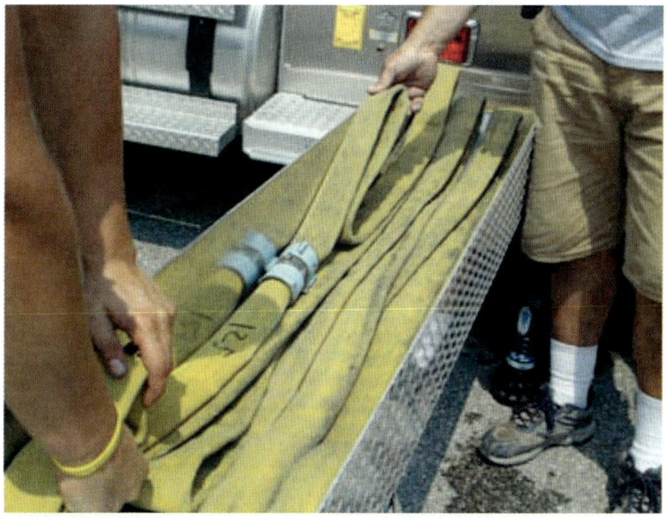

FIGURE 11-2 The Dutchman fold prevents the coupling from turning and becoming stuck as the hose is laid out.

Courtesy of the Westbury Fire Department.

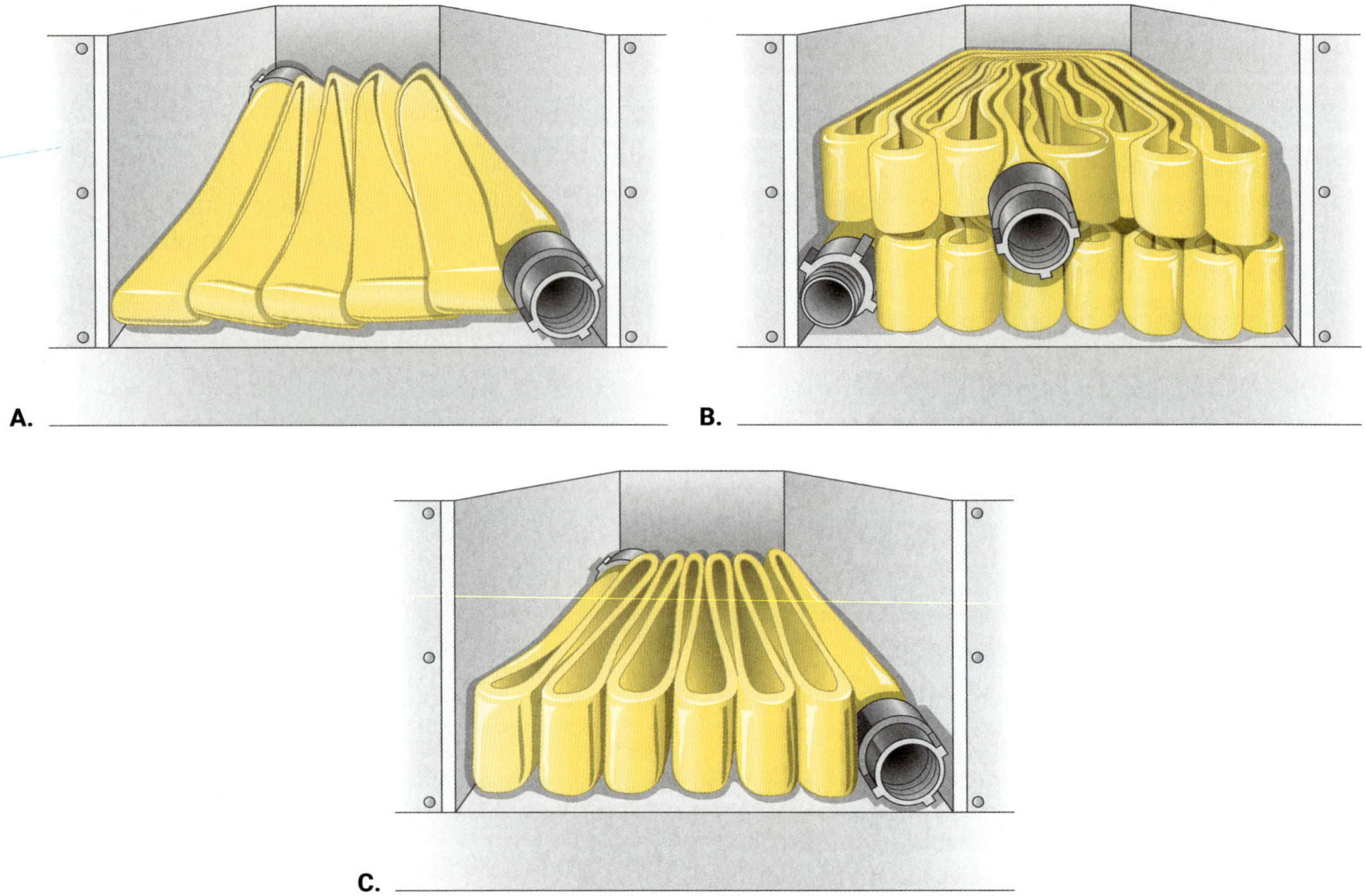

FIGURE 11-3 Three hose loads are commonly used to load supply hose onto the apparatus. **A.** The flat hose load. **B.** The horseshoe hose load. **C.** The accordion hose load.

© Jones & Bartlett Learning

TIP

When loading hose, remember that the time and attention that go into loading the hose properly will be valuable when you need to unload the hose at a fire.

The end of the hose bed closest to the cab (at the front of the vehicle) is the front of the hose bed. The end closest to the tailboard (at the back of the vehicle) is the rear of the hose bed.

If you are loading supply hose with threaded couplings, first determine whether the hose will be used for a forward hose lay or a reverse hose lay. Generally, hydrants require a female coupling. Therefore, to load hose for a forward hose lay, place the male coupling in the hose bed first, and to load hose for a reverse hose lay, place the female coupling in the hose bed first.

Before loading any hose, the amount of hose that will be loaded needs to be determined.

Flat Hose Load

With the **flat hose load**, the hose is laid flat and stacked on top of the previous section. This method is the easiest loading technique to implement, and it can be used for any size hose, including LDH. In fact, the flat hose load is the only load that should be used with LDH. Because the hose is placed flat in the hose bed, it should lay out flat without twists or kinks. The flat hose load minimizes wear and tear on the edges of the hose from the movement and vibration of the vehicle during travel. Many variations of the flat hose load exist, so follow your department's SOPs. To perform a flat hose load, follow the steps in **SKILL DRILL 11-5**.

Horseshoe Hose Load

To load hose with the **horseshoe hose load**, place the hose on its edge and position it around the perimeter of the hose bed in a U-shape. At the completion

SKILL DRILL 11-5

Performing a Flat Hose Load Support Person, NFPA 1010: 5.5.3

1. Start the hose load by placing the coupling at the front of the hose bed and run the hose to the rear of the hose bed, aligned up against one side of the hose bed. If you are loading supply hose with threaded couplings, place the male coupling in the hose bed first to set up for a forward hose lay, and place the female coupling in the hose bed first to set up for a reverse hose lay.

2. When you reach the rear of the hose bed, fold the hose back on itself at the rear of the hose bed.

3. Run the hose back to the front of the hose bed on top of the previous length of hose. While laying the hose to the front of the hose bed, angle the hose to the side of the previous fold.

4. Continue laying the hose in neat folds until the required amount of hose is loaded. To avoid the ends getting too high because the folds are on top of one another, make every other layer of hose slightly shorter, or alternate the folds. Position the last coupling as close as possible to the rear of the bed so that it is within reach from the rear of the apparatus.

© Jones & Bartlett Learning. Photographed by Brandon Fryman/JRM Creative.

of the first U-shape, the hose is folded inward to form another U-shape in the opposite direction. This continues until a complete layer is filled; then another layer is started above the first. When the hose load is complete, the hose in each layer is in the shape of a horseshoe. A major advantage of the horseshoe hose load is that it contains fewer sharp bends than the other hose loads.

There are a few disadvantages to the horseshoe load. A horseshoe hose load cannot be used for LDH because the hose tends to fall over when it stands on its edge. The load also causes more wear on the hose than the flat hose load because the weight of the hose is supported only by the edges, and these edges are subject to the vibration of the hose bed while the motor is running. When laying out a horseshoe hose load, the hose tends to lay out in a wave-like manner from one side of the street to the other.

Many variations of the horseshoe hose load exist, so follow your department's SOPs. To perform a horseshoe hose load, follow the steps in **SKILL DRILL 11-6**.

Accordion Hose Load

The **accordion hose load**, like the horseshoe hose load, requires the hose to be placed on its edge. But instead of forming a U-shape along the edges of the hose bed, the hose is laid side to side in the hose bed. This load is easy to implement and makes the hose easy to deploy and carry. Each crew member can carry multiple folds from the hose bed. When the hose load is complete, the hose in each layer is in the shape of accordion pleats.

Because the hose is stacked on its side in an accordion hose load, this load has the same disadvantages as the horseshoe hose load. That is, this load cannot be used for LDH, and it causes more wear on the hose than the flat hose load.

SKILL DRILL 11-6

Performing a Horseshoe Hose Load Support Person, NFPA 1010: 5.5.3

1. Start the hose load by placing the coupling at the rear of the hose bed. If you are loading supply hose with threaded couplings, place the male coupling in the rear of the hose bed to set up for a forward hose lay, and place the female coupling in the rear of the hose bed to set up for a reverse hose lay. Lay the first length of hose on its edge against one of the walls of the hose bed, and run the hose to the front of the hose bed along the wall.

2. When you reach the front of the hose bed, lay the hose across the width of the bed and continue down the opposite side toward the rear of the hose bed.

SKILL DRILL 11-6 CONTINUED

Performing a Horseshoe Hose Load Support Person, NFPA 1010: 5.5.3

3. When the hose reaches the rear of the hose bed, fold the hose back on itself, then lay it back toward the front of the hose bed along the second side. Keep the hose tight to the previous row of hose around the hose bed until it is back to the rear of the hose bed on the starting side. Fold the hose back on itself again, and then run the hose back to the rear of the hose bed, packing the hose tight to the previous row.

4. Continue to pack the hose in the same manner. Each fold of hose will decrease the amount of space available inside the horseshoe. Once the center of the horseshoe is filled in, begin a second layer by bringing the hose from the rear of the hose bed and laying it around the perimeter of the hose bed on top of the first layer. Complete additional layers using the same pattern you used for the first layer. Position the last coupling as close as possible to the rear of the bed so that it is within reach from the rear of the apparatus.

© Jones & Bartlett Learning. Photographed by Glen E. Ellman.

Many variations of the accordion hose load exist, so follow your department's SOPs. To perform the accordion hose load, follow the steps in **SKILL DRILL 11-7**.

Combination Hose Load in a Split Hose Bed

A **split hose bed** is a hose bed that is divided into two or more sections. Split hose beds have several advantages:

- One compartment in a split hose bed can be loaded for a forward hose lay with the female coupling at the rear of the hose bed, and the other side can be loaded for a reverse hose lay with the male coupling at the rear of the hose bed. This allows a line to be laid in either direction without adaptors.
- When 2½-in. (64-mm) or 3-in. (76-mm) hose is used, two parallel hose lines can be laid at the same time. This is referred to as **dual hose lines**. Dual hose lines are beneficial if the situation requires more water than one hose line can supply.
- The split hose beds can be used to store hose of different size. For example, one side of the hose bed could be loaded with 2½-in. (64-mm) hose that can be used as supply hose or as attack hose. The other side of the hose bed could be loaded with 5-in. (127-mm) hose for use as a supply hose.
- Hoses from all sections of the hose bed can be laid out as a single hose line by coupling the end of the last length of hose in one bed to the beginning of the first length of hose in the next bed. This is called a **combination hose load**. When laying a combination hose load, all of the hose is laid out of one hose bed first, and then the hose continues to lay out from the second hose bed. When the two sides of a split bed are loaded with the hose in opposite directions, either a double-female or double-male adaptor is used to make the connection between the two hose beds.

SKILL DRILL 11-7

Performing an Accordion Hose Load Support Person, NFPA 1010: 5.5.3

1. Start the hose load by placing the coupling at the rear of the hose bed. If you are loading supply hose with threaded couplings, place the male coupling in the rear of the hose bed first to set up for a forward hose lay, and place the female coupling in the rear of the hose bed first to set up for a reverse hose lay. Lay the first length of hose on its edge against one of the walls of the hose bed, and run the hose to the front of the hose bed along the wall.

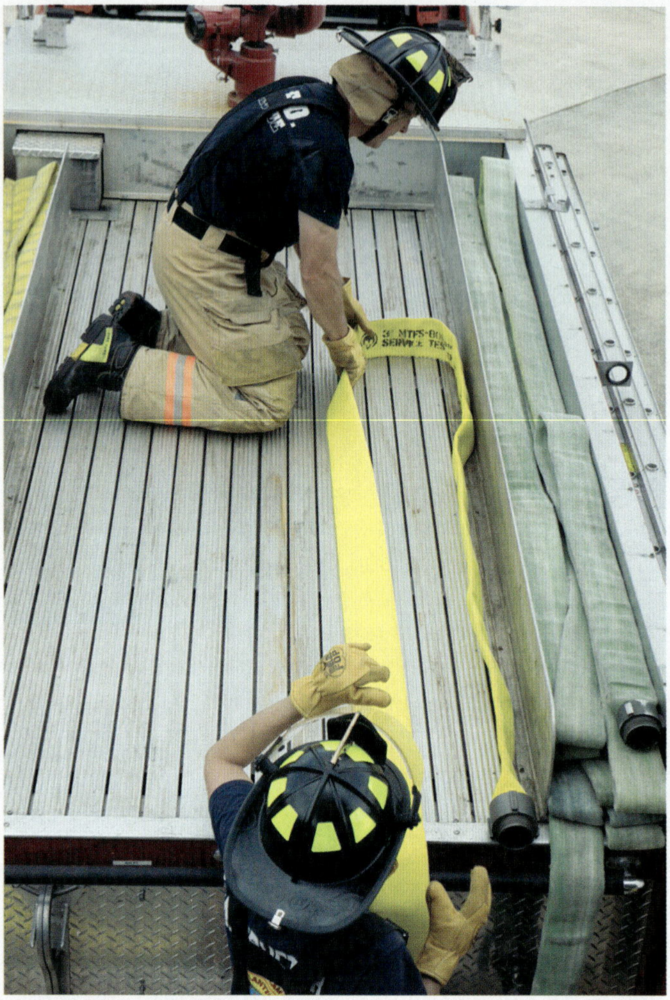

2. When you reach the front of the hose bed, fold the hose back on itself. Run the hose back to the rear of the hose bed, and then fold the hose back on itself so that the bend is even with the edge of the hose bed.

SKILL DRILL 11-7 CONTINUED

Performing an Accordion Hose Load Support Person, NFPA 1010: 5.5.3

3. Continue to lay folds of hose along the hose bed until the required amount of hose is loaded. Alternate the length of the hose folds at each end to allow more room for the folded ends.

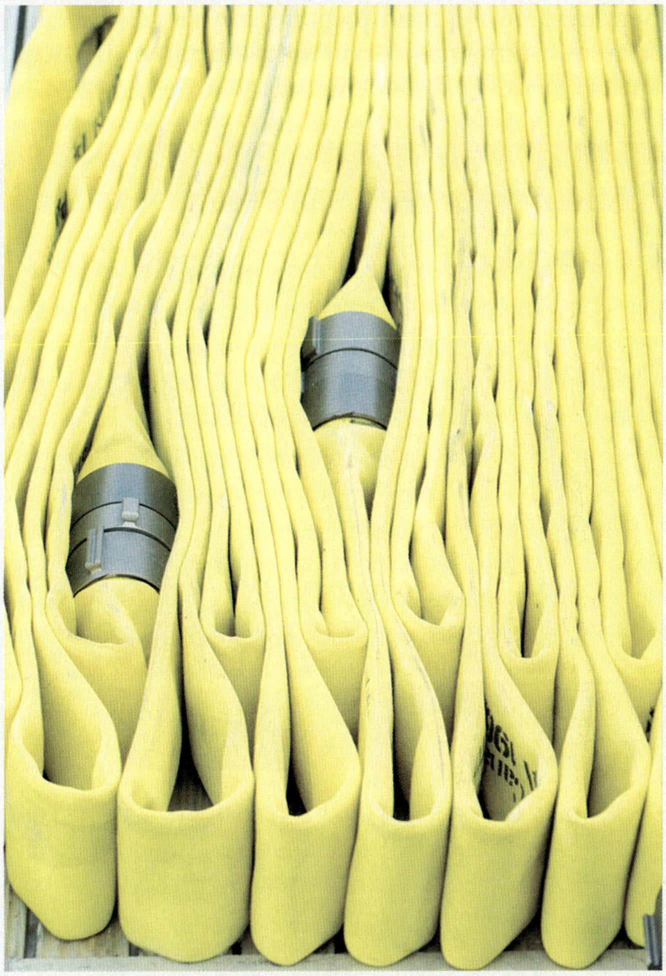

4. Continue to pack the hose in the same manner. Once the bottom layer is completed, begin a second layer by angling the hose upward to begin the second tier. Complete additional layers using the same pattern you used for the first layer.

© Jones & Bartlett Learning. Photographed by Glen E. Ellman.

Supply Hose Carries and Advances

Several different techniques are used to carry and advance supply hose. The best technique for a particular situation will depend on the size of the hose, the distance over which it must be moved, and the number of crew members available to perform the task. Although the same techniques can be used for both supply lines and preconnected attack lines, the working hose drag and the shoulder carry are almost always used with supply lines.

Whenever possible, a hose line should be laid out and positioned as close as possible to where it will be operated before it is charged with water. A charged hose line is much heavier and more difficult to maneuver than a dry hose line. A suitable amount of extra hose should be available to allow for maneuvering after the line is charged.

Working Hose Drag

The working hose drag is used to deploy hose from a hose bed and advance the line over a relatively short distance to the desired location. Depending on the size and length of the hose, several crew members may be required to perform this task. To perform a working hose drag, follow the steps in **SKILL DRILL 11-8**.

Shoulder Carry

The shoulder carry is used to transport multiple sections of connected hose over a longer distance than is practical to drag the hose. It is also useful when a hose line must be advanced around obstructions. For example, this technique can be used to stretch an attack line from the front of a building around to the rear entrance and up to the second floor. The shoulder carry can also

SKILL DRILL 11-8

Performing a Working Hose Drag Support Person, NFPA 1010: 5.3.4

1. Grasp the coupling on the hose in the hose bed and place the end of the hose over one shoulder so that the coupling is at chest height. Hold on to the coupling with the other hand.

2. Walk in the direction you want to advance the hose.

3. As the next hose coupling is ready to come off the hose bed, a second crew member grasps that coupling and places the hose over their shoulder.

4. Continue this process until enough hose has been pulled out of the hose bed, and then another crew member uncouples it from the hose bed.

© Jones & Bartlett Learning. Photographed by Glen E. Ellman.

be used to stretch a supply line from a water source to an attack engine when the hose cannot be laid out by another engine. This technique requires practice and good teamwork to be successful. To perform a shoulder carry, follow the steps in **SKILL DRILL 11-9**.

TIP

By working together to complete tasks efficiently, fire service personnel can achieve their goal of extinguishing the fire in the shortest period of time.

If hose is loaded with the accordion load, one crew member should be able to load hose onto their own shoulder.

Attack Line Evolutions

Attack line evolutions, also called **attack line operations**, are standard methods of working with attack lines to deliver water from an attack engine to a handline, which discharges the water onto the fire.

The attack engine is usually positioned close to the fire, and attack lines are stretched from the attack engine to the fire manually by firefighters. In some situations, an attack engine drops an attack line at the fire and drives from the fire to a fire hydrant or other water source. This procedure is similar to the reverse hose lay for supply lines.

Most engines are equipped with preconnected attack lines, which are lengths of attack hose that travel on the engine equipped with a nozzle and connected to a pump discharge outlet. The connection to the discharge outlet may be in the hose bed or in another location, depending on the configuration of the pumper. Preconnected hose lines are intended for immediate use as attack lines. Additional attack hose that is not preconnected is also carried on the apparatus.

SKILL DRILL 11-9

Performing a Shoulder Carry from a Flat or Horseshoe Load Support Person, NFPA 1010: 5.3.4

1. Stand on the ground at the tailboard of the apparatus. Grasp the end of the hose and place it over one shoulder so that the coupling is at chest height. Hold on to the coupling with the other hand. Have another crew member place additional hose on your shoulder so that the ends of the folds reach about knee level, front and back. Have the other crew member continue to place folds on your shoulder, but only as much as you can safely carry.

2. Hold the hose to prevent it from falling off your shoulder and move forward about 15 ft (4.6 m).

Continues.

SKILL DRILL 11-9 CONTINUED

Performing a Shoulder Carry from a Flat or Horseshoe Load Support Person, NFPA 1010: 5.3.4

3. A second crew member stands on the ground at the tailboard, another crew member places additional hose on their shoulder in the same manner, and then both crew members holding hose move forward about 15 ft (4.6 m). When enough hose has been unloaded onto crew members, the hose is uncoupled from the hose bed and carried to the desired location.

4. All of the crew members start walking in the direction the hose is needed. As they walk, the person on the rear starts offloading hose from their shoulder. Once they have laid out all of their hose, the next person starts laying out hose from their shoulder. Each crew member lays out their supply of hose in turn until the entire length is laid out.

© Jones & Bartlett Learning. Photographed by Glen E. Ellman.

Loading Preconnected Attack Lines

Attack hose is loaded on the apparatus in such a manner that it can be quickly and easily deployed. There are many ways to load attack lines into a hose bed, and this section presents only a few of the most common hose loads. Your department might use a variation of one of these techniques. It is important for you to master the hose loads used by your department.

The most common preconnected attack line is 200 ft (61 m) of 1¾-in. (44-mm) hose. Many engines are also equipped with preconnected 2½-in. (64-mm) hose lines to enable them to make a quick attack on larger fires. Some departments use different lengths, for example, 150 ft (46 m) or 250 ft (76 m), depending on the size of the structures in their area. Fire service personnel should be able to pull an attack line that is long enough to reach the fire but not so long that an excess of hose might slow down the operation and become tangled.

Preconnected attack lines can be placed in several locations on the apparatus. For example, a section of a hose bed at the rear of the apparatus can be loaded with a preconnected attack line. Many engines have a transverse hose bed installed above the pump and loaded so that the hose can be pulled off from either side of the apparatus, sometimes referred to as "cross-lays." Preconnected attack lines can also be loaded into trays mounted on the side of the apparatus. Short preconnected attack lines can be stored in a compartment in the front bumper on many engines. This attack line is often used for vehicle or dumpster fires because the apparatus can drive up close to the incident and a longer hose line is not needed. (This hose line is sometimes referred to as the "trash line.") Booster hose, 1-in. (25-mm) hose stored on reels, is another type of preconnected attack line. Booster hose reels can be mounted in a variety of locations on fire apparatus. Remember that booster hoses should not be used for structural or vehicle firefighting because they cannot flow the amount of water required.

Attack lines are loaded in a manner that ensures that one or two crew members can remove the hose quickly from the hose bed and advance the hose to the fire. Whichever hose load technique is used, the hose must be able to be deployed efficiently and completely—efficiently so that it does not become tangled as it is being removed from the hose bed and advanced, and completely so that it can be charged with water. Laying out attack hose should not require multiple trips between the engine and the point of fire attack.

The hose should also be able to be deployed efficiently around obstacles and corners. If additional crew members are available, they can help move hose line around obstacles. It should also be possible to reload the hose quickly and with minimal personnel.

There is no perfect hose load that works well for every situation. The three most common hose loads for preconnected attack lines are the flat hose load, the minuteman hose load, and the triple-layer hose load. Variations in circumstances may make one type of hose load preferable for your department.

Preconnected Flat Hose Load

The **preconnected flat hose load** is loaded in a similar manner to the flat hose load used for supply lines (**FIGURE 11-4**). The flat hose load is easy to load and easy to deploy. It can be deployed on the shoulder or by grasping the loop of the hose line with your arm. It can be used in jurisdictions that have varying types of structures. To perform a preconnected flat hose load, follow the steps in **SKILL DRILL 11-10**.

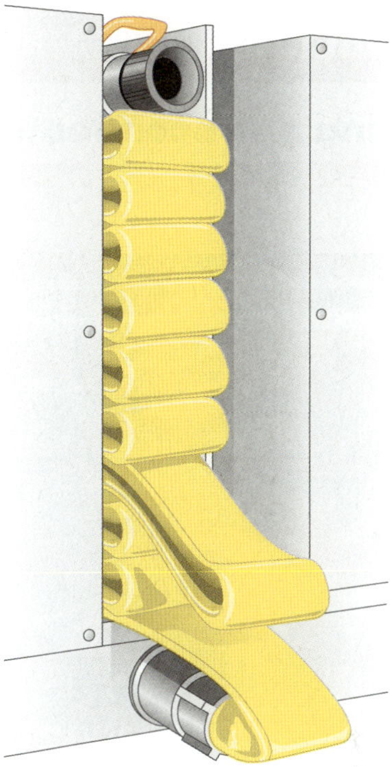

FIGURE 11-4 The preconnected flat hose load is loaded in a similar manner to the flat hose load used for supply lines, except the female end of the hose line is preconnected to a pump discharge outlet, and a nozzle is attached to the male end of the hose line.

© Jones & Bartlett Learning

SKILL DRILL 11-10

Performing a Preconnected Flat Hose Load Support Person, NFPA 1010: 5.5.3

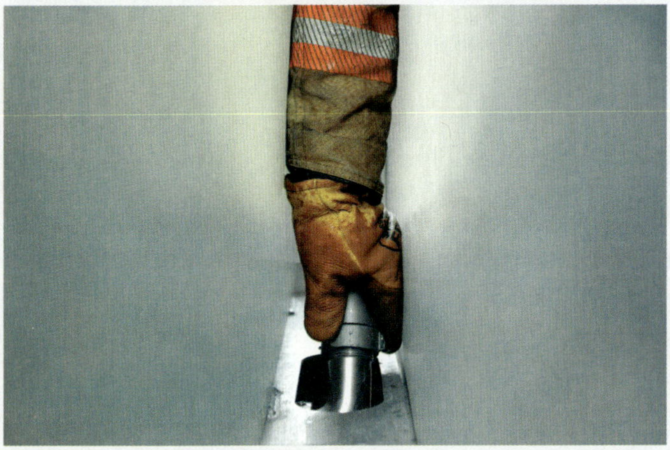

1. Attach the female end of the hose to the preconnect discharge outlet.

2. Begin flat loading the hose in the hose bed.

Continues.

SKILL DRILL 11-10 CONTINUED

Performing a Preconnected Flat Hose Load Support Person, NFPA 1010: 5.5.3

3. Load the first layer of hose and make the first fold even with the edge of the hose bed. Then, make a loop, or ear, on the second or higher fold. This loop will be used as a pulling handle.

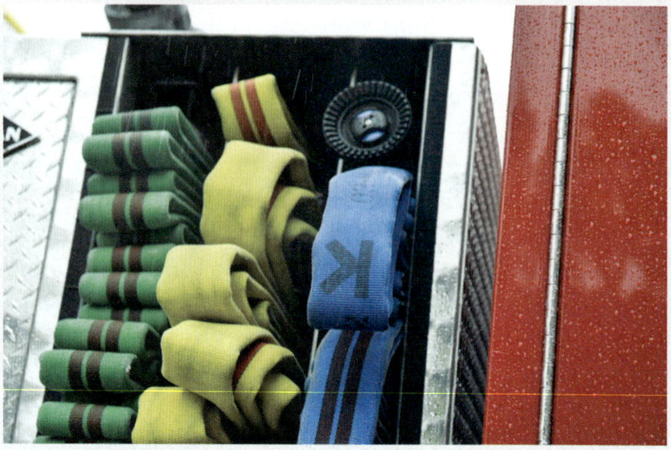

4. Flat load the remainder of the hose in the hose bed. A second loop can be made toward the top of the load and used as an arm loop. Attach the nozzle to the male coupling at the end of the hose.

© Jones & Bartlett Learning. Photographed by Glen E. Ellman.

Minuteman Hose Load

The **minuteman hose load** can be deployed and carried over one shoulder by a single crew member (**FIGURE 11-5**). The crew member drops one fold or loop of hose at a time from the shoulder as they advance toward the fire. This load avoids needing to maneuver the hose line around obstacles, helps to prevent sharp kinks, and avoids dragging the hose on the ground. Because you carry this load on your shoulder, you should have sufficient hose that you can position to pull into the structure if needed. To perform a minuteman hose load, follow the steps in **SKILL DRILL 11-11**.

The **triple-layer hose load** is suited for departments that generally respond to fires in one- or two-story single-family dwellings (**FIGURE 11-6**). It is a hose-loading method in which the hose is folded back onto itself to reduce the overall length to one-third before loading the hose in the bed. This method is used by engine companies with minimal personnel. One person can clear the hose bed and deploy this type of hose load. It can be deployed and charged before entering the building. The triple-layer hose load requires at least two crew members and must be practiced often. It also

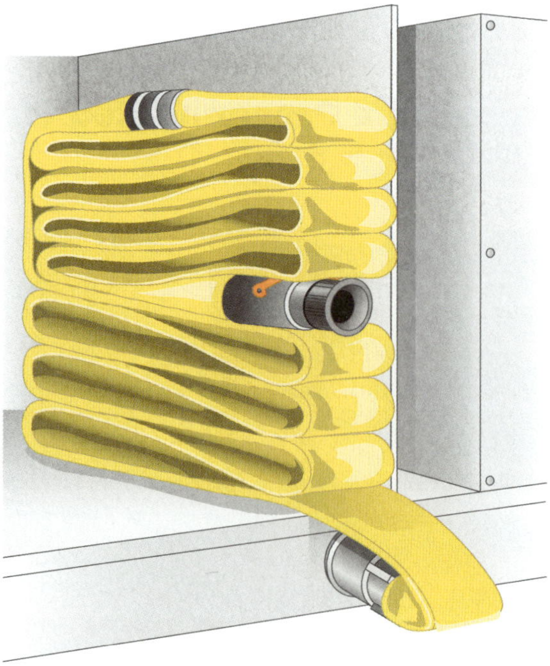

FIGURE 11-5 The minuteman hose load allows a single crew member to drop one fold or loop of hose at a time from the shoulder as they advance toward the fire.

© Jones & Bartlett Learning

SKILL DRILL 11-11

Performing a Minuteman Hose Load Support Person, NFPA 1010: 5.5.3

1. Connect the female end of the first length of hose to the preconnect discharge outlet on the engine.

2. Lay the first section of hose in the hose bed using a flat hose load. At the second fold, make a loop or ear. This loop will be a handle to grab when deploying the hose.

3. Flat load the second section of hose. Place the male coupling end of the second section of hose at the side of the hose bed.

4. Couple the remaining two sections of hose together, and then attach the nozzle to the male coupling on the last section of hose. Place the connected nozzle on top of the second section previously loaded.

Continues.

SKILL DRILL 11-11 CONTINUED

Performing a Minuteman Hose Load Support Person, NFPA 1010: 5.5.3

5. Continue flat loading the remaining sections of hose into the hose bed.

6. When you reach the female coupling of the last section of hose, connect the female coupling from the last section of hose to the male coupling from the second section of hose. When you are finished, the nozzle and the loops should be sticking out of the hose bed. This will make it easy to quickly grab the nozzle and the folds when you need to deploy the hose.

© Jones & Bartlett Learning. Photographed by Brandon Fryman/JRM Creative.

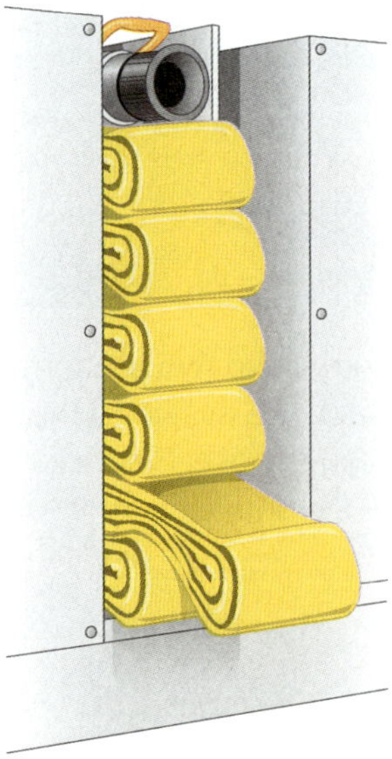

FIGURE 11-6 The triple-layer hose load method is a hose-loading method in which the hose is folded back onto itself to reduce the overall length to one-third before loading the hose in the bed.

© Jones & Bartlett Learning

requires space in which to lay out the hose behind the engine. To perform a triple-layer hose load, follow the steps in **SKILL DRILL 11-12**.

Draining and Picking Up Hose

After hose is used, it must be drained of water before it is put back into service. To do this, lay a section of hose straight on a flat surface, and then lift one end of the hose to shoulder level. Gravity causes the water to flow to the lower portion of the hose. Proceed down the length of hose, letting it feed over the shoulder and back to the ground behind the crew member to be rolled or folding the hose back and forth over your shoulder to be carried to another location. As you do this, all the water will eventually flow out of the hose. When you reach the end of the section, the hose is ready to be rolled or carried. To drain a hose, follow the steps in **SKILL DRILL 11-13**.

SKILL DRILL 11-12

Performing a Triple-Layer Hose Load Support Person, NFPA 1010: 5.5.3

1. Attach the female end of the hose to the preconnect discharge outlet on the engine. Connect the sections of hose together.

2. Extend the hose from the hose bed. Pick up the hose two-thirds of the distance from the preconnect discharge outlet to the hose nozzle.

3. Carry this fold back to the apparatus, and then place it on the ground at the rear of the apparatus, forming three layers of hose.

4. With a second crew member—and possibly additional crew members—pick up the entire length of folded hose.

Continues.

SKILL DRILL 11-12 CONTINUED

Performing a Triple-Layer Hose Load Support Person, NFPA 1010: 5.5.3

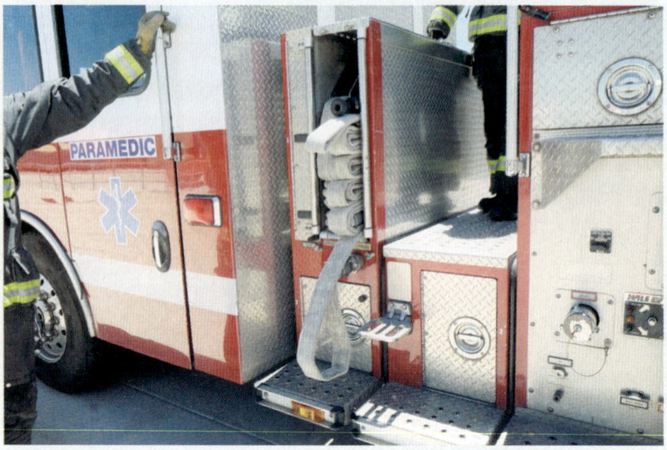

5. Lay the triple-folded hose in the hose bed just like you would a flat load with the nozzle on top and as close as possible to the rear of the bed so that it is within reach from the rear of the apparatus.

© Jones & Bartlett Learning. Photographed by Brandon Fryman/JRM Creative.

SKILL DRILL 11-13

Draining a Hose Support Person, NFPA 1010: 5.5.3

1. Lay the section of hose straight on a flat surface.

SKILL DRILL 11-13 CONTINUED

Draining a Hose Support Person, NFPA 1010: 5.5.3

2. Starting at one end of the hose, lift the hose to shoulder level.

3. Move down the length of hose, laying it on the ground or folding it back and forth over the shoulder.

4. Continue down the length until the entire hose is drained on the ground or on your shoulder.

© Jones & Bartlett Learning. Photographed by Glen E. Ellman.

Unloading the Hose Bed

Hose should be unloaded from the hose bed and then reloaded on a regular basis to place the bends in different portions of the hose. Leaving bends in the same locations for long periods of time weakens the hose at the bends. Hose might also need to be unloaded to transfer it to another apparatus. It could also be necessary to unload the hose for annual testing to be conducted. Before the hose is unloaded, you should take advantage of the fact that the hose bed is empty to clean the hose bed.

CASE STUDY

You Are the Support Person CONCLUSION

In the early morning hours before sunrise, your engine company is dispatched to a report of a residential structure fire. The driver/operator sees another company approaching an intersection just ahead and close to the dispatched address. The officer on your engine radios the other company and advises them to help perform a split hose lay.

1. **Why is it important to understand supply line evolutions and attack line evolutions?**

   Answer: Fire service personnel must be efficient at deploying and placing hose into the needed positions during a fire response. Hose evolutions prepare us to do that.

2. **What is the primary purpose of supply line evolutions?**

   Answer: To become more efficient at deploying and handling fire hose during emergencies.

WRAP-UP

SUMMARY

- The methods of moving water through fire hose must be efficient during an emergency when time is of the essence. Fire hose is used for two main purposes: to supply water from a water source to the pumper (supply hose) and to carry water from the pumper to a nozzle for fire attack (attack hose).
- To become proficient at moving and placing fire hose, fire service personnel practice fire hose evolutions. Each fire department must set up equipment and conduct regular training so that crew members are prepared to deploy and use hose lines at a fire scene.
- Drafting from a static water source is one way to deliver water, but the most common way is from a pressurized water source such as a fire hydrant.
- Supply hose is used to deliver water over a short distance. In most cases, a short length of soft sleeve hose is used to transport water from a pressurized source, such as a fire hydrant, to the suction side of the fire pump.
- Laying supply hose can be done using a forward hose lay, a reverse hose lay, or a split hose lay. After the hose is laid, it is connected to the water supply and to the pump.
- The most common way of laying hose is the forward hose lay, where the supply hose is pulled from the hose bed on the attack engine and anchored near the water source, usually the hydrant. The engine is slowly driven forward toward the fire, unfolding the hose from the hose bed as it goes.
- In the reverse hose lay, the hose is laid out from the fire to the water source, such as a fire hydrant, in the direction opposite to the flow of the water. This can be a useful tactic when multiple pumpers arrive on a fire scene at close to the same time.

- A split hose lay is performed by two engine companies when hoses must be laid in two different directions to establish a water supply. The attack engine drops the end of its supply hose where the hoses from the two engines will meet. The attack engine then performs a forward lay toward the fire. The supply engine stops at the same location, pulls off enough hose to connect to the end of the supply line that is already there, and then performs a reverse hose lay to the hydrant.
- Hose must be easily removable from the hose bed, without kinks or twists, and without the possibility of becoming caught or tangled. The ideal hose load is easy to load, avoids wear and tear on the hose, has minimum sharp bends, and allows the hose to lay out of the hose bed smoothly and easily.
- With a flat hose load, the hose is laid flat and stacked on top of the previous section. This method is the easiest loading technique to implement, and it can be used for any size hose, including LDH.
- For the horseshoe hose load, place the hose on its edge and position it around the perimeter of the hose bed in a U-shape; then fold it inward to form another U-shape in the opposite direction. A major advantage of the horseshoe hose load is that it contains fewer sharp bends than the other hose loads.
- The accordion hose load, like the horseshoe hose load, requires the hose to be placed on its edge. But instead of forming a U-shape along the edges of the hose bed, the hose is laid side to side in the hose bed. This load is easy to implement and makes the hose easy to deploy and carry.
- Several different techniques are used to carry and advance supply hose. The best technique for a particular situation will depend on the size of the hose, the distance over which it must be moved, and the number of crew members available to perform the task.
- The working hose drag is used to deploy hose from a hose bed and advance the line over a relatively short distance to the desired location. Depending on the size and length of the hose, several crew members may be required to perform this task.
- The shoulder carry is used to transport multiple sections of connected hose over a longer distance than is practical to drag the hose. It is also useful when a hose line must be advanced around obstructions. This technique requires practice and good teamwork to be successful.
- Attack line evolutions, also called *attack line operations*, are standard methods of working with attack lines to deliver water from an attack engine to a handline, which discharges the water onto the fire.
- Most engines are equipped with preconnected attack lines, which are lengths of attack hose that travel on the engine equipped with a nozzle and connected to a pump discharge outlet. Preconnected hose lines are intended for immediate use as attack lines.
- The preconnected flat hose load is loaded in a similar manner to the flat hose load used for supply lines. The flat hose load is easy to load and easy to deploy. It can be deployed on the shoulder or by grasping the loop of the hose line with your arm. It can be used in jurisdictions that have varying types of structures.
- The minuteman hose load can be deployed and carried over one shoulder by a single crew member. The crew member drops one fold or loop of hose at a time from the shoulder as they advance toward the fire.
- The triple-layer hose load is suited for departments that generally respond to fires in one- or two-story single-family dwellings. The hose is folded back onto itself to reduce the overall length to one-third before loading the hose into the bed. The triple-layer hose load requires at least two crew members and must be practiced often.
- After hose is used, it must be drained of water before it is put back into service.
- Hose should be unloaded from the hose bed and then reloaded on a regular basis to place the bends in different portions of the hose. Leaving bends in the same locations for long periods of time weakens the hose at the bends.

KEY TERMS

accordion hose load A method of loading hose on a vehicle that results in a hose appearance that resembles accordion sections. This is achieved by standing the hose on its edge and laying it side to side in the hose bed.

attack engine An engine used to pump water through attack lines at the fireground.

attack line evolutions The delivery of water from an attack engine to a handline, which discharges the water onto the fire. Also called *attack line operations.*

KEY TERMS CONTINUED

attack line operations See *attack line evolutions.*

combination hose load A hose-loading method used when one long hose line is needed; the end of the last length of a hose in one bed of a split hose bed is coupled to the beginning of the first length of hose in the opposite bed.

deploy hose To remove hose from the hose bed or other storage location. Also called *pull hose.*

dual hose lines Two parallel hose lines laid at the same time.

evolution A set of prescribed actions that result in an effective fireground activity. (NFPA 1410)

flake out See *lay.*

flat hose load A hose-loading method in which the hose is laid flat and stacked on top of the previous section.

forward hose lay A method of laying a supply line where the line starts at the water source and ends at the attack engine. Also called *straight hose lay.*

horseshoe hose load A hose-loading method in which the hose is laid on its edge around the perimeter of the hose bed so that it resembles a horseshoe.

hose bed The main storage area on an apparatus for carrying hose.

lay To arrange hose on the ground. Also called *flake out.*

minuteman hose load A hose-loading method that allows a single firefighter to deploy and advance the necessary amount of hose from the shoulder; avoids having to maneuver the hose lines around obstacles and helps to prevent sharp kinks.

preconnected flat hose load A hose-loading method in which the hose is laid flat and stacked on top of the previous section, similar to the flat hose load but with one or two loops sticking out of the load to make it easier to deploy.

pull hose See *deploy hose.*

reverse hose lay A method of laying a supply line where the supply line starts at the attack engine and ends at the water source.

split hose bed A hose bed arranged to enable the engine to lay out either a single supply line or two supply lines simultaneously.

split hose lay A scenario in which the attack engine forward lays a supply line from an intersection to the fire, and the supply engine reverse lays a supply line from the hose left by the attack engine to the water source.

straight hose lay See *forward hose lay.*

supply engine An engine used to pump water to an attack engine using supply lines.

supply line evolutions The process of laying supply hose to deliver water from a water supply source to an attack engine. Also called *supply line operations.*

supply line operations See *supply line evolutions.*

triple-layer hose load A hose-loading method in which the hose is folded back onto itself to reduce the overall length to one-third before loading the hose into the hose bed. This load method reduces deployment distances.

REVIEW QUESTIONS

1. Which supply hose load must be used with LDH?
2. Which attack hose load can be quickly loaded onto one firefighter's shoulder?
3. Where is a hose clamp placed to stop the flow of water in a damaged fire hose?
4. What causes the water to flow out of the fire hose when you are picking it up?
5. What happens to fire hose if bends are left in the same place for long periods of time?

DISCUSSION QUESTIONS

1. What are the advantages and disadvantages of the forward lay?
2. What are the advantages of picking up the fire hose as you are draining the hose?

REFERENCES

National Fire Protection Association. 2017. *NFPA 1962, Standard for the Care, Use, Inspection, Service Testing, and Replacement of Fire Hose, Couplings, Nozzles, and Fire Hose Appliances.* 2017 Edition. Quincy, MA: National Fire Protection Association.

National Fire Protection Association. 2019. *NFPA 1410, Standard on Training for Emergency Scene Operations.* 2020 Edition. Quincy, MA: National Fire Protection Association.

National Fire Protection Association. 2021. *NFPA 13, Standard for the Installation of Sprinkler Systems.* 2022 Edition. Quincy, MA: National Fire Protection Association.

National Fire Protection Association. 2021. *NFPA 1140, Standard for Wildland Fire Protection.* 2022 Edition. Quincy, MA: National Fire Protection Association.

National Fire Protection Association. 2022. *NFPA 25, Standard for the Inspection, Testing, and Maintenance of Water-Based Fire Protection Systems.* 2023 Edition. Quincy, MA: National Fire Protection Association.

National Fire Protection Association. 2023. *NFPA 1960, Standard for Fire Hose Connections, Spray Nozzles, Manufacturer's Design of Fire Department Ground Ladders, Fire Hose, and Powered Rescue Tools.* 2024 Edition. Quincy, MA: National Fire Protection Association.

Chapter Opener: © Eric Scruggs

CHAPTER

12

Support Person

Hazardous Materials Regulations, Standards, and Laws

NOTE: Content within this chapter meets the intent of **NFPA 470, Hazardous Materials/Weapons of Mass Destruction (WMD) Standard for Responders, 2022 Edition**, which includes chapter 5, "Professional Qualifications for Hazardous Materials/WMD Awareness Level Personnel (**NFPA 1072**)."

KNOWLEDGE OBJECTIVES

After studying this chapter, you will be able to:

- Identify the difference between hazardous materials/weapons of mass destruction incidents and other emergencies.
- Identify the location of the emergency response plan (ERP) and/or standard operating procedures (SOPs).
- Define the terms *hazardous materials* (or *dangerous goods*, in Canada) and *weapons of mass destruction*.
- Understand the differences between the standards and federal regulations that govern hazardous materials response activities.
- Describe the different levels of hazardous materials training: awareness, operations, technician, specialist, and incident commander.
- Explain the need for a planned response to a hazardous materials incident.

SKILLS OBJECTIVES

There are no skills objectives for awareness-level personnel for this chapter.

CASE STUDY

You Are the Support Person

At 1330 hours, your engine company responds to a report of a suspicious odor at a small plastics manufacturing company. When you arrive, you see a cargo delivery truck parked at the loading dock, with the motor still idling. From your vantage point, a few hundred feet away, you notice a liquid leaking from the back of the truck. On one side of the truck, you can see a black-and-white, diamond-shaped placard that reads "CORROSIVE," with the number 8 at the bottom. The facility security guard reports that the truck has been left unattended for at least an hour.

1. Is this an incident? How would you go about analyzing the scene to better understand the problem? What other pieces of information would you want?
2. You are trained to the awareness level. Would this level of training allow you to put on chemical protective equipment, enter the hazardous area, attempt to determine the nature of the leak, and fix the problem?
3. What reference sources would you want to access for more information on the placard?

Introduction

Fire service personnel, law enforcement personnel, and emergency medical services (EMS) routinely respond to a variety of incidents. These incidents may include structural fires, emergency medical calls, automobile accidents, confined-space rescues, water rescues, and acts of terrorism. All of these, and other situations, may involve hazardous substances that threaten lives, property, and/or the environment. When an incident clearly involves a hazardous material, or when you suspect the presence of a hazardous materials release, the nature of the incident changes, and so must the mentality of the responder (**FIGURE 12-1**).

Hazardous materials incidents are handled in a more deliberate fashion than structural firefighting, and your level of training will dictate the actions you can take to resolve the situation. Unknown substances could be present, or the setting in which the release occurred might be such that it takes time to get a full picture of the incident. For example, train derailments, which have a large incident footprint and typically require the involvement of many response agencies, may also include dangerous chemical substances. These types of factors do not mean that hazardous materials incidents are more or less complicated than fighting fires, but they do mean that responders often take more time to get oriented to the potential hazards and to define a rational approach to solving the situation without becoming exposed to the released substance. If a rescue is required during a hazardous materials incident, or if the situation is imminently dangerous in some other way and requires quick action, events may move quickly.

FIGURE 12-1 The ability to recognize a potential hazardous materials/weapons of mass destruction incident is a critical first step to ensuring your safety. Notice the types of containers adjacent to the spill, the color of the liquid on the floor, and the presence or absence of any reaction between the liquid and the flooring material. Any or all of these observations may be key to forming initial impressions of the scene.

© Jones & Bartlett Learning. Courtesy of Rob Schnepp.

All personnel and responders must understand that actions taken at hazardous materials/weapons of mass destruction (WMD) incidents are largely dictated by the chemicals or hazards involved; environmental influences such as wind, rain, and temperature; and the way the chemicals behave during the release. In short, the nature and circumstances of the response as a whole dictate the decisions made by the personnel and responders at the incident.

TIP

The goal of the responder is to favorably change the outcome of the hazardous materials/WMD incident. This is accomplished through sound planning and by establishing safe and reasonable response objectives based on the level of training. Don't do it if you're not trained to do it!

Additionally, personnel operating at the scene of a hazardous materials/WMD incident must be conscious of the potential or actual law enforcement aspect of the incident. Especially where terrorist or other criminal acts are suspected, responders should be mindful of evidentiary issues associated with the incident. Being mindful of potential evidence at a hazardous materials/WMD event may assist later efforts to identify, capture, and prosecute the person(s) responsible for the act. To that end, all responders on the scene must be cognizant of the impact that their presence and actions will have on potential evidence. Although evidence preservation should not impede the efforts to eliminate the problem or slow life-saving operations, all responders should be diligent in remembering that their actions and observations may play a vital role in the successful prosecution of a criminal suspect.

When responding to hazardous materials/WMD incidents, make a conscious effort to change your perspective. Slow down, think about the problem and available resources, and then take well-considered actions to solve it.

Additionally, initial and ongoing actions may be guided by your authority having jurisdiction (AHJ) and local or emergency response plans and/or standard operating procedures (SOPs). An AHJ is the "organization, office, or individual responsible for enforcing the requirements of a code or standard, or for approving equipment, materials, an installation, or a procedure" (NFPA 470). Basically, the AHJ is the governing body that sets operational policy and procedures for the jurisdiction in which you operate. Every responder should have knowledge of and access to response plans and be trained on how to implement the actions specified in those plans in accordance with their level of training. For example, the AHJ might identify a set of tasks that responders would be expected to perform in their course of duty and match the training competencies to address those tasks.

From a broad perspective, the goals of this section are to help you do the following:

- Recognize the presence of a hazardous materials incident.
- Take initial actions, including establishing scene control zones.
- Implement the incident command system (ICS).
- Use basic reference sources, such as the *Emergency Response Guidebook* (*ERG*).
- Select personal protective clothing based on the anticipated task(s).
- Implement product control measures when needed (for operations and hazardous materials technicians).
- Perform appropriate decontamination.

Ultimately, this section will help you understand where you fit into a full-scale hazardous materials response. The chapters in this section focus on awareness-level personnel. The rest of the chapters focus on operations-level responders, including those responders assigned mission-specific responsibilities. There is some overlap between awareness and operations, but the intent is to make a clear delineation between the two levels.

What Is a Hazardous Material?

The first points to understand, before diving into the standards and regulations that govern hazardous materials response, are the definitions of *hazardous material* and *weapon of mass destruction.*

The U.S. **Department of Transportation (DOT)**—the government agency that publicizes and enforces rules and regulations that relate to the transportation of many hazardous materials—defines a **hazardous material** as any substance or material that is capable of posing an unreasonable risk to human health, safety, or the environment. This definition includes hazardous wastes, marine pollutants, and elevated-temperature materials. It also includes illicit laboratories, environmental crimes, and industrial sabotage. Adding to that,

NFPA 470 defines a hazardous material as "[m]atter (solid, liquid, or gas) or energy that when released is capable of creating harm to people, the environment, and property, including weapons of mass destruction (WMD)." For brevity, this section will refer to hazardous materials and WMD simultaneously. In United Nations model codes and regulations, hazardous materials are called *dangerous goods.*

Around the world, the term **weapon of mass destruction (WMD)** may have different definitions, abbreviations, and acronyms. In general terms, a WMD can be thought of as "any weapon or material that is designed to cause death or serious injury or damage to buildings, structures, or the environment, such as an explosive or incendiary bomb, rocket, or grenade . . . containing or delivering a toxic or dangerous chemical, biological agent, toxin, or vectors; or a weapon designed to release dangerous levels of radiation" (NFPA 470). The most common way of describing WMDs is with the acronym CBRN—that is, as chemical, biological, radiological, or nuclear weapons or materials.

To make things simple, a hazardous material can be almost anything, depending on the situation. Milk, for example, is not routinely regarded as a hazardous substance, but 5000 gallons of milk leaking into a creek does, in fact, pose an unreasonable risk to the environment (**FIGURE 12-2**). A release of ethylene glycol (antifreeze) from a motor vehicle accident may present a low hazard to responders but a high hazard to certain animals. A large chlorine gas release also fits the definition, as would any substance used as a WMD or with criminal intent. Regarding the threat of terrorism, it would be unwise to believe that your jurisdiction is immune to deliberate criminal acts. Such an event could happen anywhere, at any time, with any type of substance.

FIGURE 12-2 A hazardous material can be found anywhere. This photo, for example, was taken at a food packaging plant that has a large ammonia system and numerous chemicals in a storage area. Not a typical backdrop for significant chemical use and storage!

© Jones & Bartlett Learning. Courtesy of Rob Schnepp.

Manufacturing processes sometimes generate hazardous wastes. A **hazardous waste** is what remains after a process or manufacturing activity has used a substance and the material is no longer pure. Hazardous waste can be just as dangerous as pure chemicals. It can also consist of a mixture of several chemicals, which may make it difficult to determine how the substance will react when it is released or if it encounters other chemicals. The illegal production of methamphetamine, for example, may produce a dangerous mixture of chemical wastes.

Regulations and Standards

To understand where you fit in when it comes to hazardous materials/WMD incidents, you must first recognize some of the regulatory drivers that apply to hazardous materials response, beginning with the difference between a regulation and a standard.

A **regulation** is a mandate that is issued and enforced by a governmental body. For example, the **Occupational Safety and Health Administration (OSHA)**, which is part of the U.S. Department of Labor, is the U.S. federal agency that regulates worker safety and, in some cases, responder safety. The **Environmental Protection Agency (EPA)** is the U.S. federal agency that ensures safe manufacturing, use, transportation, and disposal of hazardous substances.

Standards are "documents that contain mandatory . . . requirements . . . in a form . . . suitable for reference by another standard or code or for adoption into law" (NFPA 1). Standards are issued by nongovernmental entities and are generally consensus based. A standard may be voluntary, meaning that an agency such as a fire department is not required to adopt and follow the standard completely, or it could be mandatory, which would carry the weight of law. If an agency does adopt a voluntary standard, then it must meet all the requirements in the standard. For example, organizations such as the National Fire Protection Association (NFPA)—the association that develops and maintains nationally recognized, minimum, consensus standards covering many areas of fire safety and hazardous materials—issue voluntary, consensus-based standards that any member of the public can comment on before committee members agree to adopt them. The technical committee responsible for periodically revising any NFPA standard meets regularly to revise, update, and possibly change the standard, reviewing and acting on public comments during the revision process. Once a standard is finalized, agencies may choose to adopt it.

OSHA regulations require minimum standards for federal and private-sector workers. However, OSHA cannot require states to apply these regulations to state and local government workers. Some states create their own state plan, which must at least meet the OSHA requirements and be approved by OSHA. Some state plans extend the protection to state and local government employees, and some do not. States without a state plan must, in addition to meeting OSHA regulations, meet the regulations for hazardous waste operations described in Title 40, "Protection of the Environment," of the Code of Federal Regulations, Part 311, "Worker Protection," usually referred to as *40 CFR 311*. The **Code of Federal Regulations (CFR)** is a collection of permanent rules published in the *Federal Register* that represent broad areas of interest governed by federal regulation; these rules are organized into 50 titles, which are updated annually. States with a state plan must include regulations for hazardous waste operations that meet the minimum regulations described in 40 CFR 311, "Worker Protection."

NFPA standards governing hazardous materials/WMD response come from the NFPA's Technical Committee on Hazardous Materials Response Personnel. This committee includes more than 30 members from private industry, the fire service, law enforcement, professional organizations, and governmental agencies. In 2021, this group consolidated three published standards, especially important to personnel who may be called upon to respond to hazardous materials/WMD incidents, into one new master standard: NFPA 470, *Hazardous Materials/Weapons of Mass Destruction (WMD) Standard for Responders*. Beginning in 2022, NFPA 470 integrated and replaced NFPA 472, *Standard for Competence of Responders to Hazardous Materials/Weapons of Mass Destruction Incidents*; NFPA 1072, *Standard for Hazardous Materials/Weapons of Mass Destruction Emergency Response Personnel Professional Qualifications*; and NFPA 473, *Standard for Competencies for EMS Personnel Responding to Hazardous Materials/Weapons of Mass Destruction Incidents*.

Responders should understand the relationship between OSHA regulations and NFPA standards when it comes to hazardous materials/WMD response. OSHA regulations are the law that governs hazardous materials/WMD responders; NFPA standards are guidelines that agencies can choose to adopt. NFPA 470 is clear that responders shall receive any additional training to meet applicable DOT; EPA; OSHA; and other state, local, or provincial occupational health and safety regulatory requirements.

Levels of Training

Generally speaking, the former document NFPA 472 outlined training competencies for all hazardous materials responders, whereas the former NFPA 1072 identified their minimum professional qualifications, referred to as *job performance requirements* (JPRs). JPRs are written statements that describe specific job tasks, list the items necessary to complete those tasks, and define measurable or observable outcomes and evaluation areas for the tasks. JPRs are based on the requisite knowledge and requisite skills described in the standard.

The NFPA's latest revisions leave the intent and linkage between competency and JPRs unchanged. That is, they have simply been relocated to the new consolidated document, NFPA 470. In total, NFPA 470 contains 48 chapters with revised and updated content throughout, including the following, which are relevant to this text:

- Competencies for hazardous materials/WMD awareness-level personnel are found in NFPA 470, chapter 4.
- JPRs for hazardous materials/WMD awareness-level personnel (formerly found in NFPA 1072) are now found in NFPA 470, chapter 5.
- Competencies for hazardous materials/WMD operations-level responders are found in NFPA 470, chapter 6.
- JPRs (formerly found in NFPA 1072) for operations-level responders are now found in NFPA 470, chapter 7.
- Competencies for operations-level responders assigned mission-specific responsibilities are found in NFPA 470, chapter 8.
- JPRs for operations-level responders assigned mission-specific responsibilities (formerly NFPA 1072) are now found in NFPA 470, chapter 9.

There are many other chapters in NFPA 470 with content specific to hazardous materials technicians and mission-specific competencies for technicians, incident commanders, safety officers, and other specialist employees. NFPA 475, *Recommended Practice for Organizing, Managing, and Sustaining a Hazardous*

Materials/Weapons of Mass Destruction Response Program, remains a stand-alone document and is not contained in NFPA 470. The intent of NFPA 475 is to establish common criteria for the organization, management, programmatic elements, deployment of personnel, and resources for those entities responsible for the hazardous materials/WMD emergency preparedness function.

In the United States, NFPA standards, as well as EPA and OSHA regulations, are important to those personnel called upon to respond to hazardous materials/WMD incidents. The OSHA regulation containing the hazardous materials response competencies is commonly referred to as **HAZWOPER (HAZardous Waste OPerations and Emergency Response)**. The complete HAZWOPER regulation can be found in CFR, Title 29, 1910.120, and specifics for emergency response are located in subsection *q*. The training levels found in HAZWOPER are similar to the training levels found in the NFPA 470 standard and are identified as awareness, operations, technician, specialist, and incident commander. (*Specialist* is recognized only in the OSHA HAZWOPER regulation.) The content in NFPA 470 is scheduled to be updated every 5 years, the same as NFPA 472. This is a much more frequent revision cycle than that applied to the OSHA HAZWOPER regulation. This rapid revision cycle may be the source of some differences in the definitions of the levels of training presented here.

The following descriptions provide a broad overview, as found in NFPA 470 and the OSHA HAZWOPER regulation, of the different levels of hazardous materials/WMD responders and their training competencies. When reading NFPA 470, it is important to understand that the standard is organized to first spell out the *tasks* that a responder (awareness, operations, technician, incident commander) may be called upon to perform at the scene. The subsequent *training competencies* follow. To be clear, NFPA 470 and the HAZWOPER regulations are not "how to respond" documents. They provide no direction on how to plug leaking containers, use detection and monitoring devices, or decide which level of protection to wear in a certain situation. Instead, they are intended to provide guidance on the competencies associated with the various training levels.

The professional qualifications, as mentioned earlier, are intended to identify the JPRs that personnel should be able to perform to carry out their assigned job duties relative to their level of training. JPRs are to be accomplished in accordance with the requirements of the AHJ. Personnel at the awareness level must meet the JPR requirements defined in NFPA 470, chapter 5. Operations-level responders must also meet the JPR requirements defined in chapter 7 of NFPA 470. The mission-specific competencies for operations-level responders are found in NFPA 470, chapter 9.

Responders employ a **strategy**—that is, a "general course of action, direction, or plan to accomplish incident objectives" (NFPA 470). The two main operational strategies of awareness-level personnel are nonintervention mode or defensive mode. **Nonintervention mode** is a strategic decision whereby responders do not operate near the hazardous materials container. Instead, their efforts focus on public protection actions only, and they allow the container or product to take its natural course. The nonintervention route is taken when the adverse risk to emergency responders is greater than the benefit of taking action. **Defensive mode** is the mode of operation in which direct contact with the material or container is avoided and efforts are focused on controlling or limiting the effects of the release. Responders operating in defensive mode subject themselves to medium risk.

An **operations-level responder**, unlike awareness-level personnel, may choose to operate in an offensive mode. This occurs when responders have direct contact with the material or container and take aggressive action to control the release. **Offensive mode** carries a higher risk to the responders than they typically encounter. *Offensive mode is beyond the scope of awareness-level responders.*

Once a strategy and an operational mode are chosen, responders decide on the **tactics**—"specific actions or tasks taken to achieve strategies"—to use (NFPA 470).

To stay focused on the intent of this text—that is, to provide awareness-level personnel and operations-level responders with appropriate content consistent with the current edition of NFPA 470—we will discuss in detail only the competencies, tasks, and JPRs that awareness-level personnel and operations-level responders are expected to perform. A general overview of the training competencies for the other response levels—technician, specialist, and incident commander—is provided but is not intended to be definitive or comprehensive.

Awareness-Level Personnel

Awareness-level personnel are those persons "who, during the course of their normal duties, could encounter an emergency involving hazardous materials/WMD and who are expected to recognize the presence of the hazardous materials/WMD, protect themselves, call for trained personnel, and secure the scene" (NFPA 470). Per NFPA 470, a person with awareness-level training is not considered to be a *responder*. Instead, these individuals are referred to as awareness-level *personnel*. Persons receiving this level of training are *not* typically called to the scene to respond; rather, awareness-level personnel, such as public works employees or fixed-facility security personnel, function in support roles.

Tasks that awareness-level personnel may be expected to perform on the scene include the following duties:

- Analyze the incident to detect the presence of hazardous materials/WMD.
- Use the *ERG* to identify the name; United Nations/North American Hazardous Materials Code (UN/NA) identification number; type of marking, label, or placard; and container shape of the hazardous material/WMD involved.
- Identify potential hazards from the current edition of the *ERG*, the safety data sheet (SDS), shipping papers, or other approved reference sources.
- Initiate and implement protective actions consistent with the AHJ emergency response plan, SOPs, SDS, and the current edition of the *ERG*.
- Initiate the notification process and communicate as necessary.

The items in this list are all described in Chapter 13, "Recognizing and Identifying the Hazards." As with the awareness level, the operations-level JPRs are based on requisite knowledge and requisite skills.

OSHA's HAZWOPER view of the awareness level is somewhat different than that defined in NFPA 470, mostly because it designates awareness-level personnel as "responders." Per OSHA (1910.120[q] [6][i]), "first responders at the awareness level are individuals who are likely to witness or discover a hazardous substance release and who have been trained to initiate a response sequence by notifying the proper authorities of the release. They would take no further action beyond notifying the authorities of the release." Based on the OSHA HAZWOPER regulation, first responders at the awareness level should have sufficient training or experience to objectively demonstrate competency in the following areas:

- An understanding of what hazardous substances are and the risks associated with them
- An understanding of the potential outcomes of an incident
- The ability to recognize the presence of hazardous substances
- The ability to identify the hazardous substances, if possible
- An understanding of the role of the first responder awareness individual in the response plan
- The ability to determine the need for additional resources and notify the communication center

Operations Level

Per NFPA 470, operations-level responders are those persons who are tasked to respond "to hazardous materials/WMD incidents for the purpose of taking action "to protect nearby persons, the environment, or property from the effects of the release" (NFPA 470). These persons may also have competencies that are specific to their response mission, expected tasks, and equipment and training as determined by the AHJ, that is, the mission-specific competencies listed earlier in the chapter.

The phrase "tasked to respond" may raise a question about your role as a responder to incidents involving a hazardous material/WMD. To clarify this point, consider this scenario: If your local response system—professional or volunteer—is activated (e.g., if a 911 call is made), and your agency responds to the scene to provide on-scene emergency service at a hazardous materials/WMD incident, you are viewed as a responder. Personnel termed *responders* may include fire and rescue assets, law enforcement personnel, EMS workers (both basic life support [BLS] and advanced life support [ALS] providers), experts or other employees from private industry, and other allied professionals.

NFPA 470 continues to expand the scope of an operations-level responder by separating the operations-level suite of competencies into two distinct categories: core competencies and mission-specific competencies. The core competencies are based on those vital tasks that operations-level personnel should perform on the scene of a hazardous materials/WMD incident. Some of those tasks are listed here:

- Analyze the scene of a hazardous materials/WMD incident to determine the scope of the incident.
- Survey the scene to identify containers and materials involved.
- Collect information from available reference sources.
- Predict the likely behavior of a hazardous material.
- Estimate the potential harm the substances might cause.
- Plan a response to the release, including selection of the correct level of personal protective clothing.
- Perform decontamination.

- Preserve evidence.
- Evaluate the status and effectiveness of the response.

This abbreviated list serves as only an illustration of the core competencies. Keep in mind that the core training competencies are designed to describe the skills, knowledge, and abilities to safely accomplish the tasks listed. Core competency training is required of all operations-level responders on the scene, no matter what their function. One of the goals of the NFPA 470 standard is to better match the expected tasks that may be required of the responder with the training that the responder should receive.

In addition to undertaking core competency training, an individual AHJ may find the need to do more training based on an identified or anticipated mission-specific need. To that end, those responders who are expected to perform additional missions, beyond the core competencies, must be trained to carry out those mission-specific responsibilities. These mission-specific competencies are *nonmandatory* and should be viewed as optional. NFPA 470 is designed to allow each AHJ to pick and choose the training program that makes the most sense for its jurisdiction. To that end, operations-level responders may end up performing a limited suite of technician-level skills but do not have the broader knowledge and abilities of a hazardous materials technician.

NFPA 470 provides a mechanism to ensure that operations-level responders, including those with mission-specific training, do not go beyond their level of training and equipment. This is done by using technician-level personnel to provide direct guidance to the operations-level responder operating on a hazardous materials/WMD incident. Operations-level responders are expected to work under the direct control of a **hazardous materials technician** or allied professional who can continuously assess and/or observe the actions of the operations-level responder *and* provide immediate feedback. A hazardous materials technician or an allied professional may provide guidance through direct visual observation or through assessments communicated by the operations-level responder(s).

The goal of mission-specific competencies is to provide operations mission-specific responders with the knowledge and skills needed to perform the assigned responsibilities in a safe and effective manner. Examples of these specialty areas include those listed here (NFPA 470):

- Personal protective equipment (PPE)
- Mass decontamination
- Technical decontamination
- Evidence preservation and public safety sampling
- Product control
- Detection, monitoring, and sampling
- Victim rescue and recovery
- Illicit laboratory incidents
- Radiological hazard–specific tasks
- Disablement/disruption of improvised explosive devices (IEDs), improvised WMD dispersal devices, and operations at improvised explosives laboratories
- Diving in contaminated water environments
- Evidence collection

This text discusses all of these mission-specific competencies except for disablement/disruption of IEDs, improvised WMD dispersal devices, and operations at improvised explosives laboratories and diving in contaminated water environments.

Per the OSHA HAZWOPER regulation, first responders at the operations level are individuals who respond to releases or potential releases of hazardous materials incidents as part of their normal duties for the purpose of protecting nearby persons, property, or the environment from the effects of the release (**FIGURE 12-3**). The OSHA HAZWOPER regulation mandates that the operations-level responders must be trained to respond in a defensive fashion, without trying to stop the release directly, while avoiding contact with the released substance. Their function is to contain the release from a safe distance, keep it from spreading, and prevent or reduce the potential for human exposure. First responders at the operations level are expected to have received at

FIGURE 12-3 Law enforcement and fire department operations-level responders at a joint task force training exercise.

© Jones & Bartlett Learning. Courtesy of Rob Schnepp.

least 8 hours of training or to have sufficient experience to objectively demonstrate competency in the following areas, in addition to those listed for the awareness level:

- Knowledge of the basic hazard and risk assessment techniques
- Knowledge of how to select and use the proper PPE
- An understanding of basic hazardous materials terms
- Knowledge of how to perform basic control, containment, and/or confinement operations within the capabilities of the resources and PPE available with their unit
- Knowledge of how to implement basic decontamination procedures
- An understanding of the relevant SOPs and termination procedures

Technician/Specialist Level

Hazardous materials technicians are those persons who respond "to hazardous materials/WMD incidents using a risk-based response process to analyze a problem involving hazardous materials/WMD, plan a response to the problem, implement the planned response, evaluate progress of the planned response and adjust accordingly, and assist in terminating the incident" (NFPA 470).

These persons may have additional competencies that are specific to their response mission, expected tasks, and equipment and training as determined by the AHJ. A number of very detailed training competencies and JPRs for technician-level training are outlined in chapters 10 and 11 of NFPA 470. Technician-level personnel are integral to the NFPA 470 standard because they are, in many cases, intended to "supervise" the activities of on-scene operations-level responders.

Per the OSHA HAZWOPER regulation, individuals at the technician level respond to hazardous material releases or potential releases for the purpose of stopping the release (**FIGURE 12-4**). Conceptually, this is similar in intent to NFPA 470, in that a technician will likely assume a more proactive role than a responder at the operations level. Technicians typically function at a higher level in terms of their cognitive approach to the response. They should be proficient at implementing a comprehensive, risk-based approach to solving the problem. They will approach the point of release to plug, patch, or otherwise mitigate the problem.

Per the OSHA HAZWOPER regulation, hazardous materials technicians should have received at least 24 hours of training equal to the first responder operations level. In addition, technicians should have the competency, knowledge, and understanding necessary to fulfill the following duties:

FIGURE 12-4 Hazardous materials technician during a product control training exercise.

© Jones & Bartlett Learning

- Implement the employer's emergency response plan.
- Classify, identify, and verify known and unknown materials by using field survey instruments and equipment.
- Function within an assigned role in the ICS.
- Select and use proper specialized chemical PPE.
- Understand hazard and risk assessment techniques.
- Perform advanced control, containment, and/or confinement operations within the capabilities of the resources and PPE available with the unit.
- Understand and implement decontamination procedures.
- Understand termination procedures.
- Understand basic chemical and toxicological terminology and behavior.

The **hazardous materials specialist** is identified only in the OSHA HAZWOPER standard. "Hazardous materials specialists are individuals who respond with and provide support to hazardous materials technicians" (OSHA 1970). They act as the incident-site liaison with federal, state, local, and other government authorities. This level of responder receives more specialized training than a hazardous materials technician. This individual's "duties parallel those of the hazardous materials technician, however, [the specialist's] duties require a more directed or specific knowledge of the various substances they may be called upon to contain"

(OSHA 1970). Practically speaking, however, the two levels are not dramatically different.

Incident Commander

The **incident commander (IC)** is the person "responsible for all incident activities, including the development of strategies and tactics and the ordering and the release of resources" (NFPA 470). The IC must receive any additional training necessary to meet applicable governmental occupational health and safety regulations and the specific needs of the jurisdiction.

The OSHA HAZWOPER regulation requires specific IC hazardous materials–level training for those persons assuming command of a hazardous materials incident requiring action beyond the operations level. Individuals trained as ICs should have at least operations-level training as well as additional training specific to commanding a hazardous materials incident. ICs, who will assume control of the incident scene beyond the first responder awareness level, must receive at least 24 hours of training equal to the first responder operations level and have competency in the following areas:

- Know and implement the employer's ICS.
- Know how to implement the employer's emergency response plan.
- Know and understand the hazards and risks associated with chemical protective clothing.
- Know how to implement the local emergency response plan.
- Know how to initiate the state emergency response plan and the Federal Regional Response Team.
- Know and understand the importance of decontamination procedures.

For more complete information regarding hazardous materials/WMD ICs, refer to chapters 12 and 13 of NFPA 470.

In addition to the initial training requirements for all response levels listed earlier, OSHA regulations require annual refresher training of sufficient content and duration to ensure that responders maintain their competencies or that they demonstrate competency in those areas at least yearly. Consult your local agency for more specific information on refresher training, other hazardous materials laws, regulations, and regulatory agencies.

Other Governmental Agencies

In addition to OSHA and the EPA, several other governmental agencies are concerned with various aspects of hazardous materials/WMD response. The DOT, for example, promulgates and publishes laws and regulations that govern the transportation of goods by highway, rail, pipeline, air, and, in some cases, marine transport.

In addition to creating standardized training for hazardous materials response and hazardous waste site operations, the **Superfund Amendments and Reauthorization Act of 1986 (SARA)** created a method and standard practice for a local community to understand and be aware of the chemical hazards in that community. Under SARA Title III, the **Emergency Planning and Community Right-to-Know Act (EPCRA)** requires a business that handles certain types and amounts of chemicals to report the storage type, quantity, and storage methods to the fire department and the local emergency planning committee. This activity may be quite complex and should be undertaken only by those personnel who fully understand the requirements and the process.

The **local emergency planning committee (LEPC)** gathers and disseminates information about hazardous materials to the public. These voluntary organizations are made up of members of industry, transportation, media, fire and law enforcement agencies, and the public at large; they are established to meet the requirements of EPCRA and SARA. Essentially, LEPCs ensure that local resources are adequate to respond to a chemical event in the community. Responders should be familiar with their local LEPC and know how their department works with this committee.

In addition, each state has a **State Emergency Response Commission (SERC)**. The SERC acts as the liaison between local and state levels of authority. Its membership includes representatives from agencies such as the fire service, law enforcement services, and elected officials. The SERC is charged with the collection and dissemination of information relating to hazardous materials emergencies.

Preplanning

It is a mistake to assume that a response to a hazardous materials/WMD incident begins when the alarm sounds. In reality, the response begins with initial training, continuing education, and preplanning activities at target hazards and other potential problem areas throughout the jurisdiction or response district (**FIGURE 12-5**). A **target hazard** includes any occupancy type or facility that presents a high potential for loss of life or serious impact to the community resulting from a fire, explosion, or chemical release.

Preplanning activities enable agencies to develop logical and appropriate response procedures for

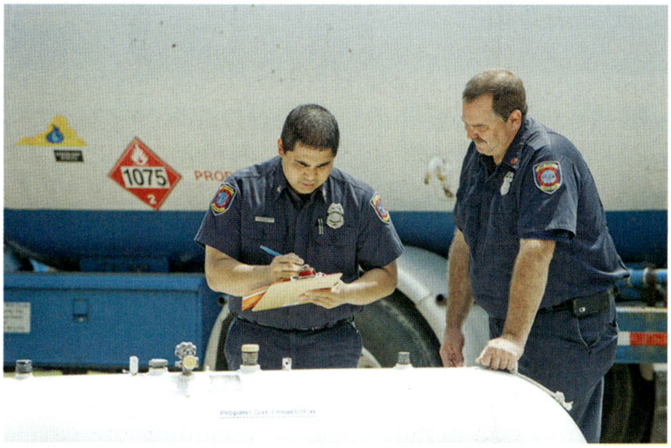

FIGURE 12-5 Conduct preincident planning activities at target hazards throughout the jurisdiction—they could pay off during an actual incident!

© Jones & Bartlett Learning. Photographed by Glen E. Ellman.

anticipated incidents. Planning should focus on the real threats that exist in your community or adjacent communities where you could be assisting. Preplanning at major target hazards should include discussions and information sharing with the LEPC.

Once the threats have been identified, fire departments, law enforcement agencies, public health offices, and other governmental agencies should determine how they will respond and work together in case of a large-scale incident. In many cases, the move toward interoperability before an incident will make the actual response work run more smoothly. Remember this important point: People making good decisions solve problems effectively. When people are acquainted with each other before the event happens, they tend to work together better. Get to know your peers at whatever level you operate.

CASE STUDY

You Are the Support Person CONCLUSION

At 1330 hours, your engine company responds to a report of a suspicious odor at a small plastics manufacturing company. When you arrive, you see a cargo delivery truck parked at the loading dock, with the motor still idling. From your vantage point, a few hundred feet away, you notice a liquid leaking from the back of the truck. On one side of the truck, you can see a black-and-white, diamond-shaped placard that reads "CORROSIVE," with the number 8 at the bottom. The facility security guard reports that the truck has been left unattended for at least an hour.

1. **Is this an incident? How would you go about analyzing the scene to better understand the problem? What other pieces of information would you want?**

   Answer: Based on the initial observations, it is reasonable to classify this as an incident. You could make further attempts to locate the driver and note any phone numbers or business names on the vehicle or trailer. Using the *ERG* to correlate the corrosive placard to an action guide will offer additional tactical guidance.

2. **You are trained to the awareness level. Would this level of training allow you to put on chemical protective equipment, enter the hazardous area, attempt to determine the nature of the leak, and fix the problem?**

   Answer: No, awareness-level personnel are not considered to be responders per NFPA 470. Actions taken at this level are focused on remaining at a safe distance from the release, securing the scene, and notifying other, more highly trained responders per the policies and procedures of the AHJ.

3. **What reference sources would you want to access for more information on the placard?**

   Answer: The *ERG* is a good source of information in the early stages of an incident. The *ERG* can provide more information based on the placard or label; it also provides guidance on identifying the type of container profile and offers some suggested initial actions.

WRAP-UP

SUMMARY

- A hazardous material is any substance or material capable of posing an unreasonable risk to human health, safety, or the environment.
- Hazardous materials incidents are handled more deliberately than structural firefighting, and your level of training will dictate the actions you can take to resolve the situation.
- Awareness-level personnel must be able to recognize hazards, protect themselves, call for trained personnel, and secure the scene (**NFPA 470**).
- Operations-level responders are tasked with protecting nearby people, the environment, or property from the effects of the release (**NFPA 470**).
- Hazardous materials technicians/specialists respond to hazardous materials/WMD incidents using a risk-based response process to analyze a problem involving hazardous materials/WMD, plan a response to the problem, implement the planned response, evaluate the progress of the planned response and adjust accordingly, and assist in terminating the incident (**NFPA 470**).
- The IC is responsible for all incident activities at a hazardous materials incident, including the development of strategies and tactics and the ordering and release of resources (**NFPA 470**).
- Preplanning activities enable agencies to develop logical and appropriate response procedures for anticipated incidents. Planning should focus on the real threats that exist in your community or adjacent communities where you could be assisting.

KEY TERMS

awareness-level personnel Personnel who, during the course of their normal duties, could encounter an emergency involving hazardous materials/weapons of mass destruction (WMD) and who are expected to recognize the presence of the hazardous materials/WMD, protect themselves, call for trained personnel, and secure the scene. (NFPA 470)

Code of Federal Regulations (CFR) A collection of permanent rules published in the *Federal Register* that represent broad areas of interest governed by federal regulation and that are organized into 50 titles, which are updated annually.

defensive mode The mode of operation in which direct contact with the material or container is avoided and responders' efforts focus on controlling or limiting the effects of the release.

Department of Transportation (DOT) The U.S. government agency that publicizes and enforces rules and regulations that relate to the transportation of many hazardous materials.

Emergency Planning and Community Right-to-Know Act (EPCRA) Legislation that requires a business that handles chemicals to report on those chemicals' type, quantity, and storage methods to the fire department and the local emergency planning committee.

Environmental Protection Agency (EPA) The U.S. federal agency that ensures safe manufacturing, use, transportation, and disposal of hazardous substances.

hazardous material Matter (solid, liquid, or gas) or energy that, when released, is capable of creating harm to people, the environment, and property, including weapons of mass destruction (WMD) as defined in 18 U.S. Code, Section 2332a, as well as any other criminal use of hazardous materials, such as illicit labs, environmental crimes, or industrial sabotage. (NFPA 470)

hazardous materials specialist Defined in the OSHA HAZWOPER regulation only, a hazardous materials specialist responds with and provides support to hazardous materials technicians and acts as the incident-site liaison with federal, state, local, and other government authorities regarding site activities.

hazardous materials technician A person who responds to hazardous materials/weapons of mass destruction (WMD) incidents using a risk-based response process to analyze a problem involving hazardous materials/WMD, plan a response to the problem, implement the planned response, evaluate the progress of the planned response and adjust accordingly, and assist in terminating the incident. (NFPA 470)

hazardous waste A substance that remains after a process or manufacturing plant has used some of the material and the substance is no longer pure.

HAZWOPER (HAZardous Waste OPerations and Emergency Response) The federal OSHA regulation that governs hazardous materials waste site and response training (specifics to emergency response can be found in 29 CFR 1910.120[q]).

incident commander (IC) The individual responsible for all incident activities, including the development of strategies and tactics and the ordering and release of resources. (NFPA 470)

local emergency planning committee (LEPC) A committee comprising members of industry, transportation, the public at large, media, and fire and law enforcement agencies that gathers and disseminates information on hazardous materials stored in the community and ensures that there are adequate local resources to respond to a chemical event in the community.

nonintervention mode The operating mode in which responders do not operate near the hazardous materials container and focus their efforts on public protection actions only, allowing the container or product to take its natural course.

Occupational Safety and Health Administration (OSHA) Part of the Department of Labor, the U.S. federal agency that regulates worker safety and, in some cases, responder safety.

offensive mode The mode of operation in which responders have direct contact with the material or container and take aggressive action to control the release.

operations-level responder A person who responds to hazardous materials/weapons of mass destruction (WMD) incidents for the purpose of implementing or supporting actions to protect nearby persons, the environment, or property from the effects of the release. (NFPA 470)

regulation A mandate issued and enforced by a governmental body, such as the Occupational Safety and Health Administration (OSHA) or the Environmental Protection Agency (EPA).

State Emergency Response Commission (SERC) A group that acts as a liaison between the local and state levels by collecting and disseminating information relating to hazardous materials emergencies. SERC includes representatives from agencies such as the fire service, law enforcement services, and elected officials.

strategy The general course of action, direction, or plan to accomplish incident objectives. (NFPA 470)

Superfund Amendments and Reauthorization Act of 1986 (SARA) One of the first U.S. laws to affect how fire departments respond in a hazardous materials emergency.

tactics Specific actions or tasks taken to achieve strategies involving the deployment and directing of resources at an incident. (NFPA 470)

target hazard Any occupancy type or facility that presents a high potential for loss of life or serious impact to the community resulting from a fire, explosion, or chemical release.

weapon of mass destruction (WMD) Any weapon or material that is designed to cause death or serious injury or damage to buildings, structures, or the environment, such as an explosive or incendiary bomb, rocket, or grenade containing or delivering a toxic or dangerous chemical, biological agent, toxin, or vectors or a weapon designed to release dangerous levels of radiation. (NFPA 470)

REVIEW QUESTIONS

1. Why should hazardous materials incidents be handled in a more deliberate fashion than structural firefighting?
2. What is a hazardous material, according to the DOT?
3. What is the most common way of describing WMDs?
4. What is the difference between a regulation and a standard?
5. Do NFPA 470 and OSHA's HAZWOPER describe awareness-level personnel and operations-level responders in the same manner?
6. What are the roles of OSHA, EPA, LEPC, and SERC in hazardous materials response and planning?
7. How do preplanning activities enable agencies to develop logical and appropriate response procedures for anticipated incidents?

DISCUSSION QUESTIONS

1. What is meant by the phrase "tasked to respond," and how is that phrase affected by the differences found in OSHA's definition of awareness-level training as opposed to the NFPA 470 perspective on awareness-level personnel?
2. Is it mandatory that an AHJ provide mission-specific training to all operations-level responders?
3. Why is preincident planning and identifying target hazards especially important in communities where target hazards include a potential for a hazardous waste release?

REFERENCES

National Fire Protection Association. 2023. *NFPA 1, Fire Code*. 2024 Edition. Quincy, MA: National Fire Protection Association.

National Fire Protection Association. 2021. *NFPA 470, Hazardous Materials/Weapons of Mass Destruction (WMD) Standard for Responders*. 2022 Edition. Quincy, MA: National Fire Protection Association.

Occupational Safety and Health Administration. 1970. "1910.120—Hazardous Waste Operations and Emergency Response." Accessed September 15, 2021. www.osha.gov/laws-regs/regulations/standardnumber/1910/1910.120.

"Worker Protection." Code of Federal Regulations, title 40, part 311 (2020): 393–394.

Chapter Opener: © Eric Scruggs

CHAPTER

13

Support Person

Recognizing and Identifying the Hazards

NOTE: Content within this chapter meets the intent of **NFPA 470 *Hazardous Materials/Weapons of Mass Destruction (WMD) Standard for Responders* 2022 Edition**, which includes chapter 5, "Professional Qualifications for Hazardous Materials/WMD Awareness Level Personnel (**NFPA 1072**)."

KNOWLEDGE OBJECTIVES

After studying this chapter, you will be able to:

- Describe how to approach a scene size-up when potentially hazardous materials are involved.
- Identify and describe the types of containers that are often used to contain hazardous materials.
- Describe the purpose and types of various transportation and facility markings for hazardous materials.
- Identify and describe the four routes of entry that harmful substances can take in the human body.

SKILLS OBJECTIVES

After studying this chapter, you will be able to:

- Use the *Emergency Response Guidebook* (*ERG*).

CASE STUDY

You Are the Support Person

During an odor investigation in a light-industrial area of your district, you are directed by a citizen to an abandoned open-head 55-gallon steel drum just off the city street. The person points you to a local shopkeeper, who made the 911 call. The shopkeeper reports that the drum must have been illegally dropped off overnight and that he smelled an unusual odor when he walked by it. You notice a green label on the side of the drum reading "nonhazardous waste." You and the other members of the engine company are trained to the hazardous materials awareness level (based on NFPA 470). Even though you are not tasked to respond to a hazardous materials incident, you now find yourself on the scene of a potential hazardous materials incident.

1. What initial actions would you take based on your level of training?
2. What does the label on the drum signify?
3. How would you go about obtaining more information on the potential contents and potential hazards of the contents of the drum? What notifications would you make?

Introduction

More than 4 billion tons of hazardous materials are shipped annually in the United States by rail, highway, sea, and air. Per the **Chemical Abstracts Service (CAS)**, which produces the largest databases on chemical information, including the CAS Registry (www.cas.org), more than 182 million organic and inorganic substances are registered for use in commerce in the United States, and several thousand new substances are introduced each year. The bulk of the new chemical substances are industrial chemicals, household cleaners, and lawn care products. Given the myriad substances available, responding to almost any situation could end up including a hazardous material or potentially hazardous material. Thus, even though awareness-level personnel are not considered to be responders per NFPA 470, it is possible to find yourself in the midst of a hazardous materials incident. To that end, it is critical to be familiar with the emergency response plans and other standard operating procedures (SOPs) developed by the authority having jurisdiction (AHJ).

Chemicals are used and/or stored in warehouses, hospitals, laboratories, industrial occupancies, residential garages, bowling alleys, home improvement centers, garden supply stores, restaurants, and scores of other facilities or businesses in your response area. So many different chemicals exist in so many different locations that you could encounter almost anything during any type of incident.

Identifying the kinds and quantities of hazardous materials used and stored by local facilities should be an integral part of any comprehensive community response plan. Additionally, the AHJ should develop emergency response plans and other methods to identify high-hazard occupancies and provide personnel with guidance on response procedures and other SOPs to handle hazardous materials incidents at those locations. Also, responding to a false alarm or some other type of incident can be a good opportunity to walk through a location for the purpose of preplanning a future response.

Basic Scene Size-Up

When you develop an emergency response, you should use a **risk-based response process**, which is a "systematic process based on the facts, science, and circumstances of the incident, by which responders analyze a problem involving hazardous materials/weapons of mass destruction (WMD) to assess hazards and consequences, develop an incident action plan (IAP), and evaluate the effectiveness of the plan" (NFPA 470). An **incident action plan (IAP)** is "an oral or written plan approved by the incident commander (IC) containing the incident objectives reflecting the overall strategy for managing the incident" (NFPA 470). This plan must be carried out as safely as possible, should not involve contacting the released substance in any way, and above all must be well thought-out. As awareness-level personnel, the IAP should be in the nonintervention or defensive mode.

At any hazardous materials incident, the first action should always be to approach the scene from a safe location and direction. The traditional rules of staying uphill and upwind are a good place to start. It may

FIGURE 13-1 Notice the details at the scene and think about what they mean.

© Jones & Bartlett Learning. Courtesy of Rob Schnepp.

make sense to use binoculars and view the scene from a safe distance, looking for labels, placards, container shapes, and other clues that could help awareness-level personnel understand the scene. Be sure to question anyone involved in the incident—a wealth of information may be available to you if you simply ask the right person. Take enough time to assess the scene and interpret other clues, such as dead animals near the release, discolored pavement, dead grass, visible vapors or puddles, and any other indicators that may help identify the presence of a hazardous material. A wet area near unidentified containers on an asphalt parking lot may not initially seem significant, but perhaps it is a hot day, and the "wet" area still looks wet an hour after the containers were noticed—well past the point that water would have evaporated. Perhaps the substance is a hydrocarbon, and the wet-looking pavement is actually a solvent that has permeated the asphalt (**FIGURE 13-1**).

TIP

First-responding firefighters, law enforcement personnel, or representatives of other allied agencies should place a high priority on analyzing the scene from a safe distance.

In some cases, it may be possible to detect the presence of a hazardous materials incident based on information relayed in the initial dispatch, from persons on the scene, or based on your own knowledge of the response area. Clues that are seen or heard may also provide valuable information from a distance, enabling you to take precautionary steps. Vapor clouds at the scene, for example, are a signal to move yourself and others away to a place of safety; the sound of an alarm from a toxic gas sensor in a chemical storage room or laboratory may also serve as a warning to retreat. Some highly vaporous and odorous chemicals—chlorine and ammonia, for example—may be detected by smell a long way from the actual point of release. These and other clues may alert awareness-level personnel to the presence of a hazardous atmosphere.

In other cases, you may have to put on your detective hat and search for clues that may indicate whether a hazardous substance is present. At all times, departmental SOPs and your level of training, along with your information-gathering efforts at the scene, should guide your initial and ongoing actions. Use your senses, but do so carefully to avoid becoming contaminated or exposed. The senses that are typically safe to employ on a regular basis are those of sight and sound. Generally, the farther you are from the incident when you notice a problem, the safer you will be. Using any of your senses in a way that will bring you close to the chemical should be done with caution or should be avoided. When it comes to hazardous materials incidents, "leading with your nose" is not a good tactic—but using binoculars from a distance is.

TIP

When you think about all of the places a hazardous materials incident could occur, do not limit your thinking. Explore your response district—you may be surprised by how many kinds of clues you find.

A thorough and thoughtful size-up will help clarify the problem you are facing and help you to decide whether taking an offensive action is within your scope of training. Once you have a basic idea of what happened and have determined whether danger may be present, you can begin to formulate a plan for addressing the incident. That plan begins with taking the basic actions known by the mnemonic SIN:

- **S**afety
- **I**solate
- **N**otify

Safety

Scene size-up, although important in any incident, is especially important during hazardous materials incidents. The ability to read the scene is a critical skill, and

FIGURE 13-2 Personnel and responders make decisions based on the available clues.

© Kali9/E+/Getty Images

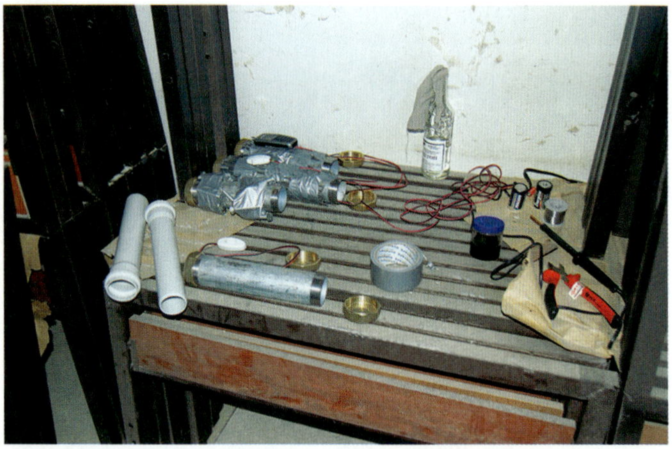

FIGURE 13-3 Pay attention. Situational awareness is key. With awareness-level training, what dangers do you see in the photo, and what would you do if you encountered this situation?

© Jones & Bartlett Learning. Courtesy of Rob Schnepp.

personnel and responders must interpret the available clues and weave them together to make informed decisions and operate safely. It is not enough to simply scan the incident scene—you must train yourself to stop for a moment and pay attention to all of the clues that may be present (**FIGURE 13-2**).

SAFETY TIP

Pay attention to the scene and think before you act: It could save your life.

Looking at something is nothing more than pointing your eyes in the right direction; *seeing*, by contrast, entails taking in the visual clues and piecing them together to form a conclusion. This is the basis for situational awareness (**FIGURE 13-3**). The scenario described in the "You Are the Support Person" feature at the beginning of the chapter, for example, contains a few clues. An **open-head drum** (in which the lid is secured by a bolted clasp-type ring that circles the entire head of the drum) typically contains solid materials or types of substances that cannot be poured or pumped easily. Moreover, steel drums do not usually contain corrosives. The **label** also provides some preliminary information, although it may not be reflective of the substance(s) inside. Meanwhile, the shopkeeper mentioned detecting an odor, which the personnel should follow up and better understand. Each point does not tell the whole story, but collectively, they provide a place from which to start your size-up.

If a chemical incident occurs at a fixed facility, attempt to locate key personnel early in the incident. Generally speaking, most fixed facilities that use and/or store a significant number of chemicals will have an environmental health and safety (EH&S) department. In many cases, EH&S departments employ certified industrial hygienists, chemists, chemical engineers, and/or certified safety professionals to ensure safe work practices are maintained at the site. These industry experts may be a valuable source of information for understanding the chemical inventory of the facility, the ventilation systems, high-hazard chemical storage areas, container profiles, and specialized areas such as clean rooms or refrigeration systems. Additionally, representatives of the EH&S department can connect personnel and responders with **safety data sheets (SDSs)** ("formatted information, provided by chemical manufacturers and distributors of hazardous products, about chemical composition, physical and chemical properties, health and safety hazards, emergency response, and waste disposal of the material" [NFPA 470]), site security, maintenance workers, and other important employees at the facility.

It is important to find the right people—those who have the keys, unique site knowledge, and instructions on how to get around the facility. In many cases, facilities have written response plans with current contact information and/or on-duty representatives who can assist in the event of an emergency. Be prepared, however, to find outdated phone lists, personnel who are no longer in a particular position, inaccurate inventories, and other shortcomings that may hamper your ability to get timely, accurate, and actionable information. Also, you may be speaking with a person who is not familiar with the substance, the process, or the area where a release occurred and who may not have all the answers you are seeking when trying to determine a course of action.

Although it is not possible to know each chemical by name, it is possible to identify many of the more common commodities through available identification systems, such as placards, labels, and other signage. Alternatively, other methods, such as detection devices, eyewitness accounts, visible indicators, and container profiles, may be used to identify the presence of a hazardous material and gain an understanding of the potential hazards. Knowing how to access the proper source of information will help awareness-level personnel interpret visual clues that may signal the possible presence of a hazardous materials incident. You must train yourself to take the time to look at the whole scene so that you can identify the available critical indicators and fit them together with what is known about the problem.

The following steps are suggested initial actions for awareness-level personnel. Your AHJ may also provide more specific and detailed procedures in a variety of response plans.

- Stay upwind, uphill, and out of the problem.
- Obtain a briefing from those involved in the incident prior to acting. These individuals may include bystanders, law enforcement personnel, emergency medical services (EMS) responders, facility representatives, and other responders.
- From a safe distance, attempt to make a positive identification of the released substance. If possible, obtain the correct SDS and shipping papers, and identify markings or labels on the container or transport vehicle. Also consult the ***Emergency Response Guidebook*** **(ERG)** or some other suitable reference source. The *ERG* is a reference book developed by the U.S. Department of Transportation (DOT), Transport Canada, and the Secretariat of Transport and Communications of Mexico. It is "written in plain language, to guide emergency [personnel and] responders in their initial actions at an incident scene" (NFPA 470). It is distributed free of charge to every public response vehicle in the United States. Consulting the *ERG* may allow you to identify the potential hazards of the substance involved.

Isolate

After ensuring your safety, the next step is to isolate and deny entry to the scene. The first operational priority is to separate the people from the problem—life safety is always the first consideration. This is typically accomplished by establishing hazard control zones, areas at hazardous materials/WMD incidents within an established perimeter that are designated based on safety and the degree of hazard. **Hazard control zones** allow on-scene personnel to quickly identify the area of highest contamination and prevent accidental entry by untrained or unprotected public safety personnel or civilians.

If people are exposed to or threatened by the release, do not move past this step until you have addressed the life-safety issues. This could include evacuating affected people from the environment; having potential victims shelter in place until a transient problem, such as a fast-moving vapor cloud or other situation, passes; decontaminating the scene; and rendering medical care.

Isolation and denial of entry to hazard control zones may be accomplished by having law enforcement personnel provide a physical presence, using scene control identifiers such as barrier tape that reads "Danger" or "Caution," or creating physical barriers with fences or other materials.

With hazardous materials incidents, it is important to establish clear and visible command. Establish the command post in an area where you are protected from both the incident and the weather and where you have access to communications and technical reference materials. The command post is typically staffed by responders trained beyond the awareness level.

Next, determine your response objectives and begin to formulate a basic IAP. This plan must be carried out as safely as possible, should not involve contacting the released substance in any way, and—above all—must be well thought-out. Remember, for awareness-level personnel, this plan should focus on a nonintervention or defensive mode.

Another isolation objective may be to identify and remotely secure potential ignition sources when flammable liquids and gases have been released. Common examples of ignition sources include open flames from pilot lights or other sources, arcs that occur when electrical switches are turned on or off, static electricity, and smoking materials.

Notify

Awareness-level personnel will likely need to notify and seek the assistance of other responders—for example, operations-level responders, hazardous materials technicians or other responders with additional training, law enforcement, and other technical experts. Large-scale incidents usually draw upon the resources of many agencies. Awareness-level personnel should know the basic notification procedures for reporting hazardous materials emergencies and requesting assistance from local and regional authorities. For example, you may need to notify regulatory agencies such as the local fish and game agency, the state-level Office of Emergency Services, or county-level agencies such as an air quality control board. The emergency response plan of the AHJ should include a current and comprehensive contact list of local, state, and federal resources available and help you identify the key players in your jurisdiction.

It is not possible in this text to identify all types of communications equipment and procedures necessary for making such notifications. Therefore, awareness-level personnel should be familiar with all communications equipment, radio frequencies, and protocols for using the communications equipment provided by the AHJ. SOPs will help you identify the various notifications that need to be made and should define the points of contact for local, state, and federal resources that might be called upon for assistance during hazardous materials emergencies.

Basic Container Recognition

In basic terms, a **container** is "a receptacle, piping, or **pipeline** used for storing or transporting material of any kind" (NFPA 470). The container type, size, and material of construction often provide important clues about the nature of the substance inside.

For example, a **drum** is a barrel-like storage vessel used to store a wide variety of substances, including food-grade materials, corrosives, flammable liquids, and grease. Drums may be constructed of low-carbon steel, polyethylene, cardboard, stainless steel, nickel, or other materials (**FIGURE 13-4**). Gasoline or waste solvents (from legitimate or illegitimate processes) may be stored in a 55-gallon steel drum with two capped openings (2 inches [51 millimeters (mm)] and ¾ inch [19 mm]) on the top (**FIGURE 13-5**). Sulfuric acid, at 97 percent concentration, could be found in a polyethylene drum that might be black, red, white, or blue. In most cases, there is no correlation between the color of the drum and the possible contents. The same sulfuric acid might also be found in a 1-gallon amber glass container. Hydrofluoric acid, by contrast, is incompatible with silica (glass) and would be stored in a plastic container. Steel or polyethylene drums, bags, high-pressure gas cylinders, railroad tank cars, plastic buckets, aboveground and underground storage tanks, cargo tanks, and pipelines are all representative examples of how hazardous materials are used, stored, and shipped.

Some very recognizable chemical containers, such as 55-gallon drums and compressed gas cylinders, can be found in almost every type of manufacturing facility. Stainless-steel containers may hold particularly dangerous chemicals. Liquids that must be kept cold are often stored in Thermos-like Dewar containers, which are designed to maintain the appropriate temperature (**FIGURE 13-6**).

Personnel and responders should not rely solely on the type of container when making a determination about its contents because there are numerous examples of finding substances in the wrong type of container. The container might also lack any legitimate markings to alert personnel to the possible contents. In any case, it is important to look closely at a container and form an opinion about the material inside.

FIGURE 13-4 Drums may be constructed of many different types of materials, including polyethylene, cardboard, or stainless steel.

© Jones & Bartlett Learning. Courtesy of Rob Schnepp.

FIGURE 13-5 An abandoned steel drum. Notice the configuration of the holes on top.

© Jones & Bartlett Learning. Courtesy of Rob Schnepp.

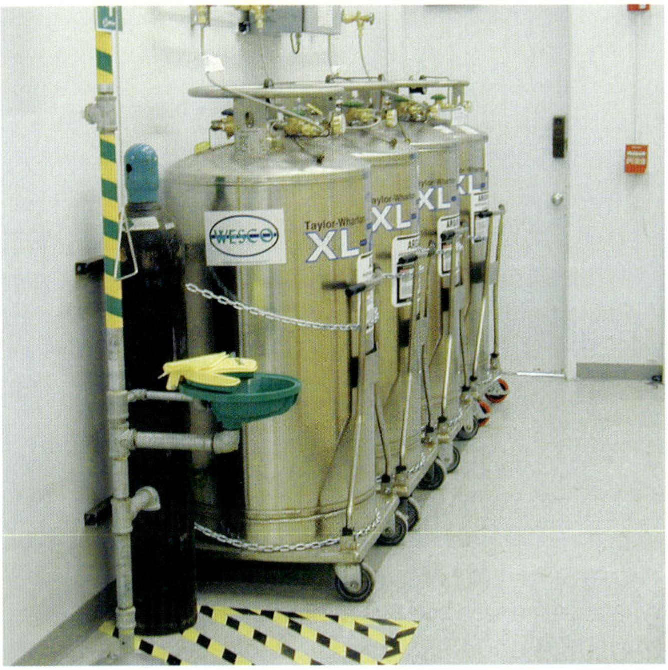

FIGURE 13-6 A series of Dewar containers stored adjacent to a compressed gas cylinder.

© Jones & Bartlett Learning. Courtesy of Rob Schnepp.

The following subsections describe a few types of containers with which awareness-level personnel should be familiar.

Drums

Drums are easily recognizable, barrel-like containers. As mentioned earlier, drums are used to store a wide variety of substances and may be constructed of a variety of materials. Generally, the nature of the chemical dictates the construction of the storage drum. Steel utility drums, for example, hold flammable liquids, cleaning fluids, oil, and other noncorrosive chemicals. Polyethylene drums are used for corrosives such as acids, bases, oxidizers, and other materials that cannot be stored in steel containers. Cardboard drums hold solid materials such as soap flakes, sodium hydroxide pellets, and food-grade materials. Stainless-steel drums and other heavy-duty drums generally hold materials that are too aggressive (i.e., too reactive) for either plain steel or polyethylene.

A **closed-head drum** has a permanently attached lid with one or more small openings. This opening is called a **bung** (see FIGURE 13-5). Typically, these openings are threaded holes sealed by caps that can be removed only by using a special tool called a *bung wrench* (**FIGURE 13-7**). Closed-head drums usually have one 2-inch (51-mm) bung and one ¾-inch (19-mm) bung. The larger bung is used to pump product from the drum; the smaller bung functions as a vent.

FIGURE 13-7 A bung wrench is used to operate the openings on the top of a closed-head drum.

© Jones & Bartlett Learning. Photographed by Glen E. Ellman.

As mentioned earlier, an open-head drum has a removable lid fastened to the drum with a ring (**FIGURE 13-8**). The ring is tightened with a clasp or a threaded nut-and-bolt assembly. These containers typically contain a product in solid form. This is an example of the type of drum described in the opening scenario.

Carboys

Some corrosives and other types of chemicals are transported and stored in a vessel called a *carboy* (**FIGURE 13-9**). A **carboy** is a glass, plastic, or steel container that holds 5 to 15 gallons of product. Glass carboys are often placed in protective wood, foam, fiberglass, or steel boxes to help prevent breakage. For example, nitric acid, sulfuric acid, and other strong acids are often transported and stored in thick glass carboys protected by a wooden or polystyrene crate to shield the glass container from damage during normal shipping.

Cylinders

A **cylinder** is a container that has a circular cross-section and is designed to store liquids or gases under pressure higher than 40 pounds per square inch (psi; 276 kilopascals [kPa]). Cylinders do *not* include portable tanks, multiunit tank car tanks, cargo tanks, or tank cars (NFPA 1).

FIGURE 13-8 An open-head drum has a lid that is fastened with a ring; the ring is tightened with a clasp or a nut-and-bolt assembly.

Courtesy of Globalindustrial.com.

FIGURE 13-9 A carboy is used to transport and store corrosive chemicals.

Courtesy of EMD Chemicals, Inc.

Uninsulated compressed gas cylinders are used to store substances such as nitrogen, argon, helium, and oxygen (**FIGURE 13-10**). They come in a range of sizes and have variable internal pressures. An oxygen cylinder used for medical purposes, for example, has a pressure reading of approximately 2000 psi when full. By comparison, the large compressed gas cylinders found at a fixed facility may have pressure readings of 5000 psi or greater.

FIGURE 13-10 Compressed gas cylinders in storage. What clues do you see that might indicate an unsafe situation?

© Jones & Bartlett Learning. Courtesy of Rob Schnepp.

FIGURE 13-11 Compressed gas cylinder failure due to heat generated by fire.

© Jones & Bartlett Learning. Courtesy of Rob Schnepp.

The high pressures exerted by these cylinders create a potential for danger. If the cylinder is punctured, the valve assembly fails, or the cylinder falls over and damages the valve, a rapid release of compressed gas or liquid may occur that turns the cylinder into an unpredictable missile. Also, if the cylinder is heated rapidly, it could explode with tremendous force, spewing product and metal fragments over long distances. Compressed gas cylinders have pressure-relief valves, but those valves may not be sufficient to relieve the pressure created during a fast-growing fire (**FIGURE 13-11**).

FIGURE 13-12 Small propane cylinder failure due to heat generated by fire.

© Jones & Bartlett Learning. Courtesy of Rob Schnepp.

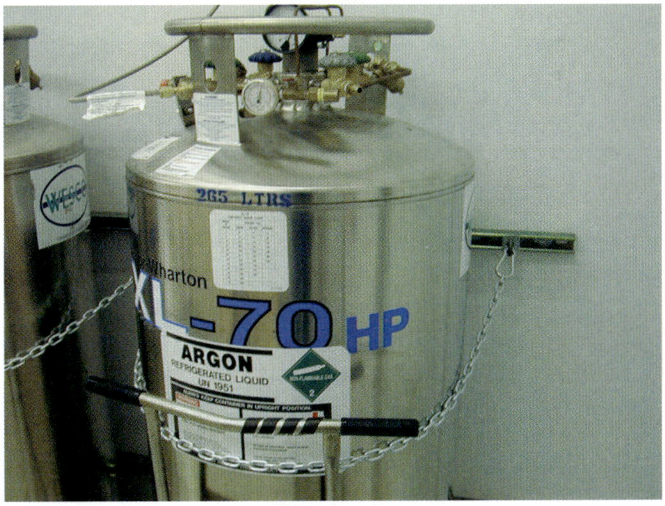

FIGURE 13-13 A small Dewar container.

© Jones & Bartlett Learning. Courtesy of Rob Schnepp.

> **SAFETY TIP**
>
> Cryogens are stored in a liquid state. Beware of skin exposure! Significant injuries, like those associated with thermal burns, can occur when skin meets one of these liquids.

A propane cylinder is a compressed gas cylinder. Propane cylinders have lower pressures (200–300 psi) than compressed gas cylinders and contain a liquefied gas. Liquefied gases such as propane are subject to the phenomenon known as **BLEVE (boiling liquid/expanding vapor explosion)**. BLEVEs occur when pressurized liquefied materials, such as propane or butane inside a closed vessel, are exposed to a source of high heat (**FIGURE 13-12**). Water and other liquids capable of expanding also have the potential to BLEVE.

The low-pressure **Dewar container** is another commonly encountered cylinder type (**FIGURE 13-13**). As mentioned previously, a Dewar container is designed to hold a **cryogenic liquid**, or **cryogen**, which is a liquid that has a boiling point lower than –130°F (–90°C) at an absolute pressure of 14.7 psi (101.3 kPa).

Typical cryogens include oxygen, helium, hydrogen, argon, and nitrogen. Under normal atmospheric conditions, each of these substances is a gas. A complex process turns them into liquids when they reach a specific low temperature, and these liquids can be stored and used for long periods of time. Nitrogen, for example, becomes a liquid at –320°F (–196°C) and must be kept at that temperature if it is to remain in a liquid state.

Cryogens pose a substantial threat if the Dewar container fails to maintain the low temperature of the cryogenic liquid. Cryogens have large expansion ratios—even larger than the expansion ratio of propane (270:1). Cryogenic helium, for example, has an expansion ratio of approximately 750:1. If one volume of liquid helium is warmed to room temperature and vaporized in a totally enclosed container, it can generate a pressure of more than 14,500 psi (99,974 kPa). To counter this possibility, cryogenic containers usually have two pressure-relief devices: a pressure-relief valve and a frangible (easily broken) metal disk.

Transportation and Facility Markings and Information Sheets

The presence of markings on buildings, packages, boxes, and containers often enables personnel to identify a released chemical. When used correctly, marking systems and information sheets indicate the presence of a hazardous material and provide clues about the substance. Marine pollutants and environmentally hazardous substances, for example, pose a risk to aquatic life and the marine ecosystem. Transportation markings denoting those substances may not be seen often, especially in areas without significant bodies of water. You may also find other transportation markings, such as those for elevated-temperature materials (e.g., asphalt and molten sulfur or other liquids transported at temperatures above 212°F [100°C] or 464°F [240°C]), consumer commodities intended for retail sale, and inhalation hazards, in your jurisdiction (**FIGURE 13-14**).

A. B. C. D.

FIGURE 13-14 Examples of markings indicating hazardous chemicals in a container. **A.** Environmental hazard. **B.** Inhalation hazard. **C.** Elevated temperature. **D.** Commodity.

A–D: Courtesy of the U.S. Department of Transportation; **C:** Reproduced from U.S. Department of Transportation. 2020. DOT Chart 17: Hazardous Materials Markings, Labeling and Placarding Guide. www.phmsa.dot.gov/sites/phmsa.dot.gov/files/2021-11/USDOT%20Chart%2017.pdf

Safety Data Sheets

A common source of information about a chemical is the SDS specific to that substance (**FIGURE 13-15**). Essentially, an SDS provides basic information about the chemical make-up of a substance, the potential hazards it presents, appropriate first aid in the event of an exposure, and other pertinent data for safe handling of the material. Although the SDS is not a definitive response tool, it may be a key resource for understanding the chemical and physical properties of a substance. The Occupational Safety and Health Administration (OSHA) requires that an SDS be available to workers who need to be around hazardous materials, whether they are used or stored.

An SDS typically includes the following details (the following list serves as an example and is not a complete list) (NFPA 470, 6.2.2[1], pages 470–425):

- Hazard identification
- Composition/information on ingredients
- First-aid measures
- Firefighting measures
- Accident release measures
- Handling and storage
- Exposure controls/personal protection
- Toxicological information
- Physical and chemical properties

When responding to a hazardous materials incident at a fixed facility, personnel should ask the site representative for an SDS for the spilled material. All facilities that use or store chemicals are required by law to have an SDS on file for each chemical used or stored in the facility. Many sites, but especially those that stock many different chemicals, may keep this information archived in a computer database. An SDS can also be obtained from staffed national resource centers or found on the transporting vehicle.

> **TIP**
>
> The Globally Harmonized System of Classification and Labelling of Chemicals (GHS) uses a standard methodology to define and classify hazards posed by chemical substances, along with a standard method—the SDS—to communicate the hazards. Information about the GHS can be found in the *ERG*.

The National Fire Protection Association 704 Marking System

The National Fire Protection Association (NFPA) has developed its own system for identifying hazardous materials. NFPA 704, *Standard System for the Identification of the Hazards of Materials for Emergency Response*, outlines a marking system characterized by a set of diamonds that are found on the outside of buildings, on doorways to chemical storage areas, and on fixed storage tanks. This marking system is designed for fixed-facility use. Personnel can use the NFPA diamonds to understand the broad hazards posed by chemicals stored in a building or part of a building.

The **NFPA 704 hazard identification system** uses a diamond-shaped symbol of any size, which is itself broken into four smaller diamonds, each representing a property or characteristic of a substance or group of substances (**FIGURE 13-16**). The blue, red, and yellow diamonds each contain a numerical rating in the range of 0–4, with 0 being the least hazardous and 4 being the most hazardous (**TABLE 13-1**).

Tanner Industries, Inc.

SAFETY DATA SHEET

Section 1. Identification	
Product Name:	**Ammonia, Anhydrous**
Synonyms:	Ammonia
CAS REGISTRY NO:	7664-41-7
Supplier:	Tanner Industries, Inc. 735 Davisville Road, Third Floor Southampton, PA 18966
Website:	www.tannerind.com
Telephone (General):	215-322-1238
Corporate Emergency Telephone Number:	**800-643-6226**
Emergency Telephone Number:	**Chemtrec: 800-424-9300**
Recommended Use:	Various Industrial / Agricultural

Section 2. Hazard(s) Identification

Hazard: Acute Toxicity, Corrosive, Gases Under Pressure, Flammable Gas, Acute Aquatic Toxicity

Classification:
Acute Toxicity, Inhalation (Category 4)
Skin Corrosion / Irritation (Category 1B)
Serious Eye Damage / Irritation (Category 1)
Gases Under Pressure (Liquefied gas)
Flammable Gases (Category 2)
Acute Aquatic Toxicity (Category 1)

Note: (1 - Most Severe / 4 - Least Severe)

Pictogram:

Signal word: **Danger**

Hazard statements:
Harmful if inhaled.
Causes severe skin burns and serious eye damage.
Flammable gas.
Contains gas under pressure; may explode if heated.
Very toxic to aquatic life.

Precautionary statements:
Avoid breathing gas/vapors.
Use only outdoors or in well-ventilated area.
Wear protective gloves, protective clothing, eye protection, face protection.
Keep away from heat, sparks, open flames and other ignition sources. No smoking.

FIGURE 13-15 An example of an SDS for anhydrous ammonia (first page only).

Courtesy of Tanner Industries, Inc., Southampton, PA.

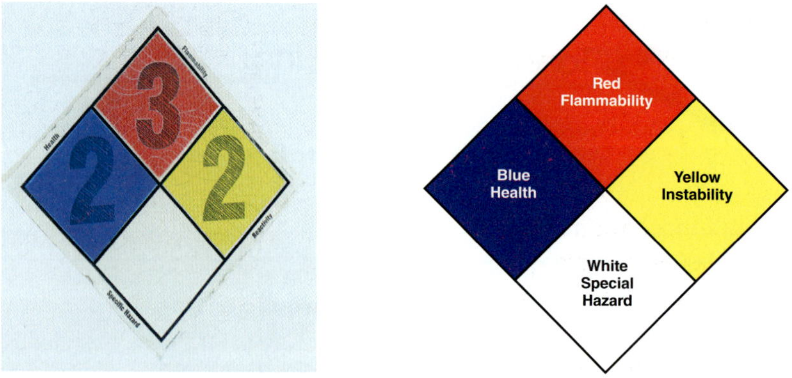

A. **B.**

FIGURE 13-16 **A.** Example of a placard using the NFPA 704 hazard identification system for fixed-facility use. **B.** Each color used in the diamond represents a particular property or characteristic.

A: © Jones & Bartlett Learning. Photographed by Glen E. Ellman; **B:** © Jones & Bartlett Learning

TABLE 13-1 Hazard Levels in the NFPA Hazard Identification System

Flammability Hazards (Red Diamond)		**Instability Hazards (Yellow Diamond)**	
4	Materials that will rapidly or completely vaporize at atmospheric pressure and normal ambient temperature or that are readily dispersed in air and that will burn readily	4	Materials that in themselves are readily capable of detonation or of explosive decomposition or reaction at normal temperatures and pressures
3	Liquids and solids that can be ignited under almost all ambient temperature conditions	3	Materials that in themselves are capable of detonation or explosive decomposition or reaction but require a strong initiating source or that must be heated under confinement before initiation
2	Materials that must be moderately heated or exposed to relatively high ambient temperatures before ignition can occur	2	Materials that readily undergo violent chemical change at elevated temperatures and pressures
1	Materials that must be preheated before ignition can occur	1	Materials that in themselves are normally stable but can become unstable at elevated temperatures and pressures
0	Materials that will not burn under typical fire conditions	0	Materials that in themselves are normally stable, even under fire exposure conditions
Health Hazards (Blue Diamond)		**Special Hazard (White Diamond)**	
4	Materials that, under emergency conditions can be lethal		*SA* indicates simple asphyxiant. *W* indicates water reactivity. *OX* indicates oxidizer.
3	Materials that, under emergency conditions, could cause serious or permanent injury		
2	Materials that, under emergency conditions, could cause incapacitation or possible residual injury		
1			Materials that, under emergency conditions, can cause serious irritation
0			Materials that, under emergency conditions, would offer no hazard beyond that of ordinary combustible material

See the NFPA 704 standard for more detailed information on the criteria that qualify materials as 0-1-2-3-4 in each of the four quadrants.

Data from National Fire Protection Association (NFPA). 2020. *NFPA 704: Standard System for the Identification of the Hazards of Materials for Emergency Response*. 2022 Edition. Quincy, MA: NFPA.

The blue diamond (at the nine o'clock position) indicates the health hazard posed by a material alone or perhaps within a group of other chemicals. When an NFPA diamond represents hazards posed by several different substances, the most severe characteristic of any of the substances may be used as the basis for the hazard level within any of the four colored diamonds. For example, if any one of the substances in a grouping of chemicals could be fatally toxic, that single substance causes a 4 to appear in the blue diamond. All other substances could be much less hazardous, but the one causing the 4 represents the health hazard for the group. The same logic holds true for the red diamond at the 12 o'clock position, which indicates flammability, and the yellow diamond at the three o'clock position, which indicates reactivity.

The six o'clock position on the symbol represents special hazards and has a white background. The special hazards in use include W, OX, and SA. *W* indicates unusual reactivity with water and is a caution about the use of water in either firefighting or spill control response. *OX* indicates that the material is an oxidizer. *SA* indicates that the material is a simple **asphyxiant** gas (nitrogen, helium, neon, argon, krypton, or xenon). For complete information on the NFPA 704 system, consult www.NFPA.org/704.

To accurately compare the DOT marking system and the NFPA 704 system, remember this important difference:

- The DOT hazardous materials marking system is used when materials are being transported from one location to another.
- The NFPA 704 hazard identification system is designed for fixed-facility use.

Hazardous Materials Information System

The **Hazardous Materials Information System (HMIS)** is a color-coded marking system that identifies the hazard level of chemicals in a container so that personnel can work safely around the chemicals. In Canada, a similar system, the Workplace Hazardous Materials Information System (WHMIS), is used. HMIS has helped employers comply with the Hazard Communication standard established by OSHA. The HMIS is similar to the NFPA 704 marking system and uses a numerical hazard rating with colored horizontal columns (**FIGURE 13-17**).

The HMIS is more than just a label; it is a method used by employers to give their personnel necessary information to work safely around chemicals and includes training materials to inform workers of chemical hazards in the workplace. The HMIS is not required by law but rather is a voluntary system that employers choose to use to comply with OSHA's Hazard Communication standard. Although the use of the HMIS is voluntary, informing personnel about hazardous materials is not. If an employer chooses not to use the HMIS, another system that complies with OSHA's requirements must be used. In addition to describing the chemical hazards posed by a substance, the HMIS provides guidance about the personal protective equipment (PPE) that employees need to use to protect themselves from workplace hazards. Letters and icons specify the different levels and combinations of protective equipment required.

FIGURE 13-17 The HMIS helps employers comply with the Hazard Communication standard.

© Jones & Bartlett Learning

Personnel and responders must understand the fundamental difference between the NFPA 704 marking system and HMIS. NFPA 704 is intended to be understood by all levels of hazardous materials personnel and responders; HMIS is intended for the employees of a facility. Although HMIS is not a response information tool, it can give clues about the presence and nature of the hazardous materials found in the facility.

Military Hazardous Materials/ Weapons of Mass Destruction Markings

The U.S. military has developed its own marking system for hazardous materials. The military system serves primarily to identify detonation, fire, and special hazards.

In general, hazardous materials within the military marking system are divided into four categories based on the relative detonation and fire hazards they pose:

- Division 1 materials are considered mass detonation hazards and are identified by a number 1 printed inside an orange octagon (**FIGURE 13-18A**).

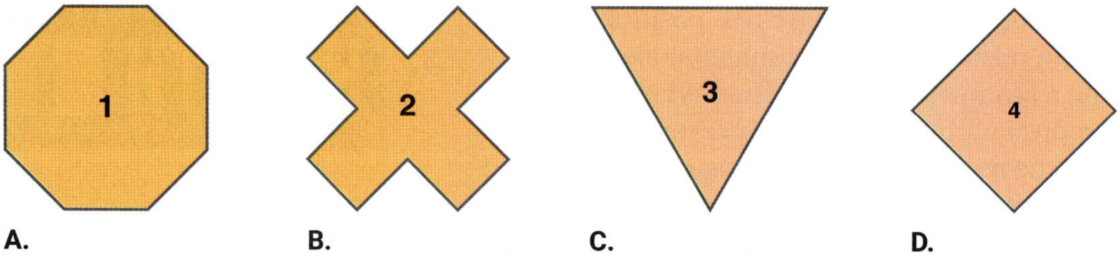

FIGURE 13-18 Military hazardous materials/WMD markings. **A.** Mass detonation hazards. **B.** Explosion-with-fragment hazards. **C.** Mass fire hazards. **D.** Moderate fire hazards.

© Jones & Bartlett Learning

- Division 2 materials have explosion-with-fragment hazards and are identified by a number 2 printed inside an orange *X* (**FIGURE 13-18B**).
- Division 3 materials are mass fire hazards and are identified by a number 3 printed inside an inverted orange triangle (**FIGURE 13-18C**).
- Division 4 materials are moderate fire hazards and are identified by a number 4 printed inside an orange diamond (**FIGURE 13-18D**).

Chemical hazards in the military system are depicted by colors. Toxic agents, such as sarin or mustard, are identified by the color red. Harassing agents, such as tear gas and smoke producers, are identified by yellow. White phosphorus is identified by white. Specific PPE requirements are identified using pictograms. Military shipments containing hazardous materials/WMD are not required, by exception, to be placarded with the DOT marking system.

Shipping Papers

Shipping papers are required by the DOT whenever materials are transported from one place to another (NFPA 498). They identify the names and addresses of the shipper, the receiver, and the material being shipped, and they specify the quantity and weight of each part of the shipment. Additionally, shipping papers allow the reader to match the chemical name found on the shipping papers with the mode of transportation. Shipping papers for road and highway transportation are called a **bill of lading**, or **freight bill**, and are located in the cab of the vehicle (**FIGURE 13-19**). Drivers transporting chemicals are required by law to have a set of shipping papers on their person or within easy reach inside the cab at all times.

A bill of lading and other types of shipping papers may provide additional information about a hazardous substance, such as its packaging group designation. The **packaging group designation** is another system used by shippers to identify special handling requirements or hazards. Some DOT hazard classes require shippers to assign packaging groups based on the material's flash point and toxicity. A packaging group designation may signal that the material poses a greater hazard than similar materials in a hazard class. There are three packaging group designations:

- Packaging group I: *High danger*
- Packaging group II: *Medium danger*
- Packaging group III: *Minor danger*

The shipping papers for railroad transportation are called a **waybill** (**FIGURE 13-20**). A list of the contents in every car on the train is called the **consist**, or **train list**. The conductor, engineer, or a designated member of the train crew will have a copy of the consist. If the incident happens in a railyard, then you may find a waybill as well.

On a marine vessel, shipping papers are called the **dangerous cargo manifest** (**FIGURE 13-21**). The manifest is generally kept in a tube-like container in the wheelhouse, in the custody of the captain or master.

For air transport, the **air bill** serves as the shipping papers (**FIGURE 13-22**). It is kept in the cockpit and is the pilot's responsibility.

Pipelines

Of all the various methods used to transport hazardous materials, the high-volume pipeline is the one that is most rarely involved in emergencies. A pipeline is "a length of pipe including pumps, valves, flanges, control devices, strainers, and/or similar equipment for conveying fluids" and gases over potentially long distances (NFPA 470). In many areas, large-diameter pipelines transport natural gas, gasoline, diesel fuel, and other products from delivery terminals to distribution facilities. Pipelines are often buried underground but may be aboveground in remote areas.

STRAIGHT BILL OF LADING
ORIGINAL - NOT NEGOTIABLE

BOL/Reference No.
RSI82715

CARRIER: NORFOLK SOUTHERN

Date: 12/23/2008

Shipper: RSI LOGISTICS, INC (OKEMOS, MI US)

The property described below, in apparent good order, except as noted (contents and condition of packages unknown), marked, consigned, and destined as indicated below, which said carrier (the word carrier being understood throughout this contract as meaning any person or corporation in possession of the property under the contract) agrees to carry to its usual place of delivery at said destination, if on its route, otherwise to deliver to another carrier on the route to said destination. It is mutually agreed, as to each carrier of all or any said property, that every service to be performed hereunder shall be subject to all the terms and conditions of the Uniform Domestic Straight Bill of Lading set forth (1) in Official, Southern, Western and Illinois Freight Classification in effect on the date hereof, if this is a rail or a rail-water shipment, or (2) in the applicable motor carrier classification or tariff if this is a motor carrier shipment

Shipper hereby certifies that he is familiar with all the terms and conditions or the said bill of lading, including those on the back thereof, set forth in the classification or tariff which governs the transportation of this shipment, and the said terms and conditions are hereby agreed to by the shipper and accepted for himself and his assigns.

Consignee Information: CONSIGNEE DEER PARK, TX Address: City: DEER PARK, TX US	
Route: NS-ESTL-BNSF	
Origin Switch Route:	
Destination Switch Route: HUSTN-PTRA	Rail Car No: GATX290861

For assistance in any transportation emergency involving chemicals, phone CHEMTREC, day or night, Toll Free 1-800-424-9300

DESCRIPTION		*WEIGHT
ONE TANK CAR	Contains: Methyl Esters STCC#2899415 BIODIESEL-15, Biodiesel Sales Order Contract No: RSI82715 Sales Order Contract No: AAT122308-4 Purchase Order Contract No: AAT122308-4	(Sub. To Correction) 204400 Lbs.

SEAL NUMBERS:

Gross

Tare

Net

Weighed By: ______________________

If charges are to be prepaid, write or stamp here, "To be Prepaid"
Prepaid

Subject to Section 7 of the conditions of applicable bill of lading, if this shipment is to be delivered to the consignee without recourse on the consignor, the consignor shall sign the following statement:: *The carrier shall not make delivery of this shipment without payment of freight and all other lawful charges.*

Not In Effect

* This is to certify that the above named materials are properly classified, described, packaged, marked, and labeled, and are in proper condition for transportation, according to the applicable regulations of the Department of Transportation.

FIGURE 13-19 A bill of lading, or freight bill.

Courtesy of RSI Logistics, Inc.

WAYBILL NO.

WAYBILL
NON-NEGOTIABLE

SHIP DATE:

SHIPPER NUMBER	SHIPPER REFERENCE NUMBER	CONSIGNEE REFERENCE NUMBER	P.O. NUMBER
SHIPPER		CONSIGNEE	
STREET ADDRESS		STREET ADDRESS	
STREET ADDRESS		STREET ADDRESS	
CITY, STATE AND ZIP CODE		CITY, STATE AND ZIP CODE	
CONTACT	PHONE NUMBER	CONTACT	PHONE NUMBER

3rd PARTY NUMBER	SHIPPING CO. LIABILITY IS LIMITED TO $50 PER SHIPMENT OR 50 CENTS PER POUND (U.S. DOLLARS), WHICHEVER IS HIGHER SUBJECT TO A MAXIMUM LIABILITY OF $25,000, UNLESS A HIGHER VALUE IS DECLARED AND APPLICABLE CHARGES (FOR DECLARING A VALUE ON WAYBILL) ARE PAID PRIOR TO SHIPPING (SUBJECT TO THE TERMS AND CONDITIONS ON REVERSE SIDE, AND THE SERVICE CONDITIONS FOUND IN THE SHIPPING CO. SERVICE CONDITIONS POLICY).
3rd PARTY NAME	
STREET ADDRESS	
CITY, STATE AND ZIP CODE	DECLARED VALUE
CONTACT NAME	$

				(SUBJECT TO CORRECTION)			
NO. OF PIECES	TYPE	HM	KIND OF PACKAGE, DESCRIPTION OF ARTICLES, SPECIAL MARKS & EXCEPTIONS	ACTUAL WEIGHT	LENGTH	WIDTH	HEIGHT

Special Instructions:

☐ SATURDAY DELIVERY ☐ SUNDAY DELIVERY ☐ APPOINTMENT DELIVERY ☐ INSIDE DELIVERY ☐ RESIDENTIAL DELIVERY

Services Requested:

☐ SAME DAY/ NEXT FLIGHT OUT ☐ NEXT DAY AM ☐ NEXT DAY PM ☐ 2ND DAY ☐ ECONOMY DEFERRED (3-5 DAYS)

FOR CHARTER AIR, TIME DEFINITE, OR GUARANTEED SERVICE, PLEASE CALL 1-800-000-XXXX FOR AVAILABILITY.

QUOTE NUMBER	DIM WEIGHT

SHIPPER CERTIFICATION: Shipper certifies by its signature, its agreement to all of the foregoing terms and conditions, and further certifies that the above named materials are properly classified, described, packaged, marked and labeled, and are in proper condition for transportation according to the applicable regulations of the DOT.

SHIPPER REPRESENTATIVE

SIGNATURE X ______________________ Print Name X ______________________ ____________ Date ____________

PICKED UP BY:	RECEIVED BY:	RECEIVED BY CONSIGNEE IN GOOD ORDER UNLESS NOTED BELOW:
DRIVER SIGNATURE ____________	CONSIGNEE SIGNATURE ____________	# S/W SKIDS DEL'D INTACT _____
PLEASE PRINT ____________	PLEASE PRINT ____________	# SKIDS DEL'D: _____
COMPANY ____________	COMPANY ____________	☐ GOOD ORDER ☐ SHORT ☐ OVER ☐ DAMAGED
DATE ____________	DATE ____________	DESCRIBE EXCEPTIONS:
TIME ____________	TIME ____________	

All rules as contained in Shipping Co. Services Conditions Policy will apply. Terms and conditions stated on any Bill of Lading used to transfer goods for carriage, other than an Shipping Co. Waybill, will be null and void. Quotes are based on the information provided and are only an estimate. Final charges are based on actual shipment pieces, weight, dimensions, and services performed as a requirement for delivery. Any changes in actual shipment details will affect the final charges.

Shipping Co. is a certified participant in compliance with the Transportation Security Administration Regulations, Part 109, a Federal Security program.

1 - SHIPPER'S COPY

FIGURE 13-20 A waybill.

Courtesy of a private source.

CARGO MANIFEST

AIR	AIRCRAFT DATA: CARRIER, A/C NO., A/C MODEL	DEST CODE	REF	DESTINATION	MISSION DATA: NO., SU, DATE	ALW WT	ALW CU
	MANIFEST ID: STA, FY, TY, NO.	PAGE NO.					
SUR-FACE	POE	DATE SAILED	VOYAGE DOCUMENT NO.	POD	REF	VESSEL NAME	STATUS, SUST, TRUCK NO., REMARKS, PAGE NO.

DOC ID	VEHICLE TRAILER OR CNTNR NUMBER	YR / MAKE; CNTNR NUMBER; COMMODTY DESCRIP	COM CODE	CAR-GO EXC	VOYAGE DOC NO.; A I R D M; TRNS PT	PORT OF DISCH	TYPE PACK	TRANSPORTATION CONTROL NUMBER	CONSIGNEE; S E R V; ACTVTY ADDRESS	P R I O R I T Y	NAME; AMMO LOT NO./NOMEN; DIMENSIONS; RDD, PROJ, STOW LOC, TRANS ACCT	IDENTIFICATION NO. OR REMARKS	PIECES	WEIGHT	CUBE

ITEMS HAVE BEEN LOADED:	ITEMS HAVE BEEN RECEIVED EXCEPT AS CIRCLED NOTED ON REVERSE SIDE	TOTALS	0	0.00	0.00
DATE / SIGNATURE OF LOADING AGENT	DATE / SIGNATURE OF UNLOADING AGENT	DATE / SIGNATURE OF RECEIVING AGENT			

DD FORM 1385, NOV 78 REPLACES EDITION OF 1 APR 66 WHICH MAY BE USED Adobe Professional 7.0

FIGURE 13-21 A dangerous cargo manifest.

Courtesy of the U.S. Department of Defense.

Pipeline incidents may present you with challenges and hazards not typically encountered at most hazardous materials incidents. Subject-matter experts from the company that owns the pipeline may be required to assist hazardous materials response teams from the local jurisdiction in such cases. These incidents, like rail incidents, could have far-reaching implications and present all types of incident response personnel with a challenging set of circumstances.

The **pipeline right-of-way** is an area, patch of land, or roadway that extends a certain number of feet on either side of the pipe itself. The company that owns the pipeline maintains this area. The company is also responsible for placing warning signs at regular intervals along the length of the pipeline. These pipeline warning signs must include a warning symbol, the pipeline owner's name, and an emergency contact phone number (**FIGURE 13-23**).

Information about the pipe's contents and owner is also often found at the vent pipes. These inverted J-shaped tubes provide pressure relief or natural venting during maintenance and repairs. **Vent pipes** are clearly marked and are located approximately 3 feet (1 meter [m]) above the ground.

Pipeline emergencies are complicated events that typically require specially trained responders who are trained well above the awareness level. If you suspect an incident involving a pipeline, contact the owner of the pipeline immediately. The company will dispatch a crew to assist with the incident.

U.S. DOT Marking System

In the United States, the DOT created a marking system consisting of placards and labels—that is, the **U.S. Department of Transportation (DOT) marking system**—that shippers must use to identify hazardous materials that are transported. (This system is also used in Canada by Transport Canada.) Placards are diamond-shaped indicators (10¾ inches [27 cm] on each side) that are placed on all four sides of highway transport vehicles, railroad tank cars, and other forms of transportation carrying hazardous materials (**FIGURE 13-24**). Labels are smaller versions (4-inch [10-cm] diamond-shaped indicators) of placards. Labels are placed on the four sides of individual boxes and smaller packages being transported (**FIGURE 13-25**).

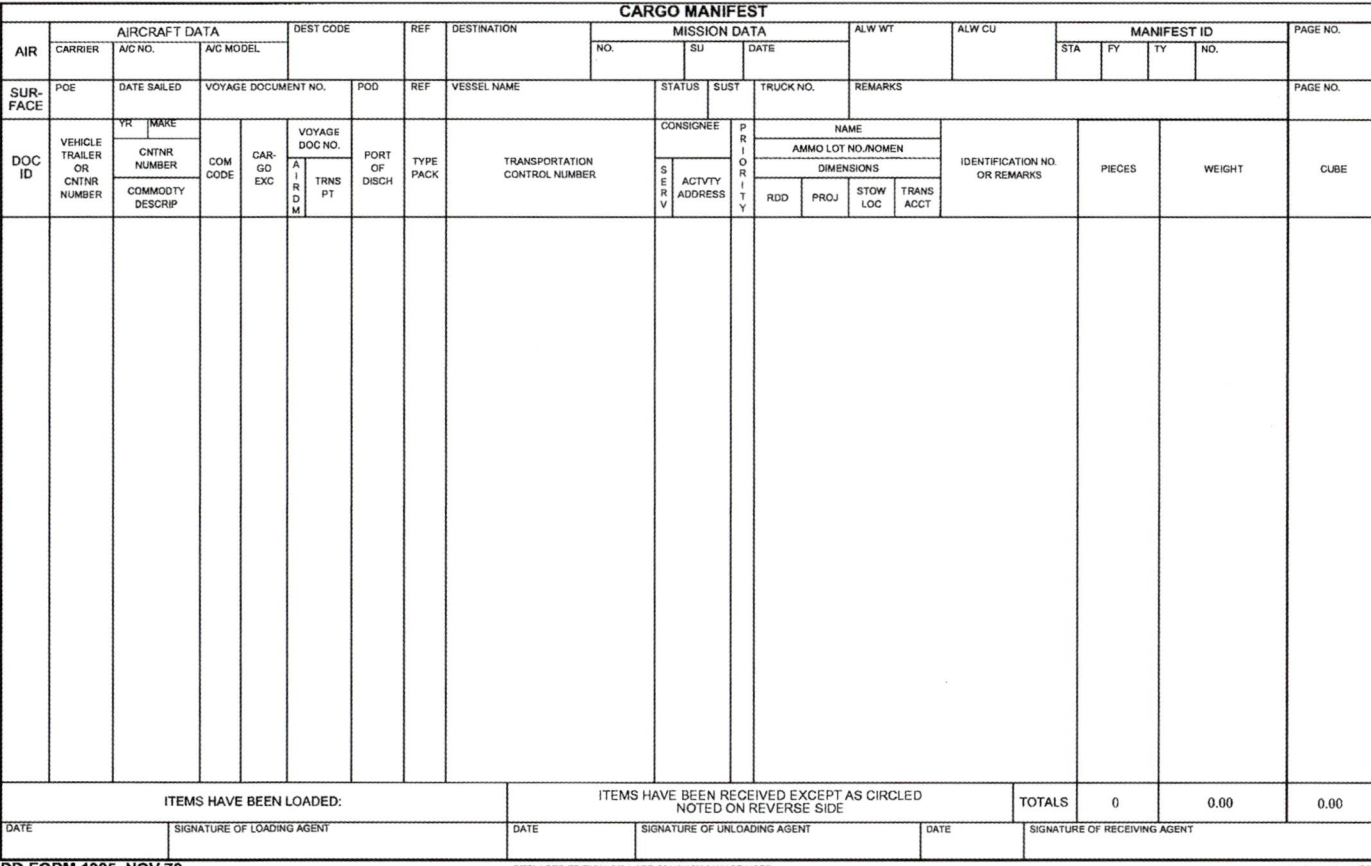

Set your tabulator stops here

STAPLE DOCUMENTS ABOVE PERFORATION

Line-up here

Shipper's Name and Address	Shipper's Account Number	Not Negotiable **Air Waybill** Issued by
		Copies 1, 2 and 3 of this Air Waybill are originals and have the same validity.
Consignee's Name and Address	Consignee's Account Number	It is agreed that the goods described herein are accepted in apparent good order and condition (except as noted) for carriage SUBJECT TO THE CONDITIONS OF CONTRACT ON THE REVERSE HEREOF. ALL GOODS MAY BE CARRIED BY ANY OTHER MEANS INCLUDING ROAD OR ANY OTHER CARRIER UNLESS SPECIFIC CONTRARY INSTRUCTIONS ARE GIVEN HEREON BY THE SHIPPER, AND SHIPPER AGREES THAT THE SHIPMENT MAY BE CARRIED VIA INTERMEDIATE STOPPING PLACES WHICH THE CARRIER DEEMS APPROPRIATE. THE SHIPPER'S ATTENTION IS DRAWN TO THE NOTICE CONCERNING CARRIER'S LIMITATION OF LIABILITY. Shipper may increase such limitation of liability by declaring a higher value for carriage and paying a supplemental charge if required.
Issuing Carrier's Agent Name and City		Accounting Information
Agent's IATA Code	Account No.	
Airport of Departure (Addr. of First Carrier) and Requested Routing		Reference Number — Optional Shipping Information

To	By First Carrier — Routing and Destination	to	by	to	by	Currency	CHGS Code	WT/VAL PPD	WT/VAL COLL	Other PPD	Other COLL	Declared Value for Carriage	Declared Value for Customs

Airport of Destination	Requested Flight/Date	Amount of Insurance	INSURANCE - If carrier offers insurance, and such insurance is requested in accordance with the conditions thereof, indicate amount to be insured in figures in box marked "Amount of Insurance".

Handling Information

SCI

These commodities, technology or software were exported from the United States in accordance with the Export Administration Regulations. Ultimate destination

Diversion contrary to U.S. law prohibited.

No. of Pieces RCP	Gross Weight	kg lb	Rate Class — Commodity Item No.	Chargeable Weight	Rate / Charge	Total	Nature and Quantity of Goods (incl. Dimensions or Volume)

Prepaid	Weight Charge	Collect	Other Charges
	Valuation Charge		
	Tax		
	Total Other Charges Due Agent		Shipper certifies that the particulars on the face hereof are correct and that **insofar as any part of the consignment contains dangerous goods, such part is properly described by name and is in proper condition for carriage by air according to the applicable Dangerous Goods Regulations.**
	Total Other Charges Due Carrier		Signature of Shipper or his Agent
Total Prepaid		Total Collect	
Currency Conversion Rates		CC Charges in Dest. Currency	Executed on (date) at (place) Signature of Issuing Carrier or its Agent
For Carriers Use only at Destination		Charges at Destination	Total Collect Charges

APPERSON K0419 (10/03) WHSE. #05640

FIGURE 13-22 An air bill.

Courtesy of Apperson Print Resources Inc.

FIGURE 13-23 A pipeline warning sign provides information about the pipe's contents and the owner's name and contact information.

© Photodisc/Getty Images.

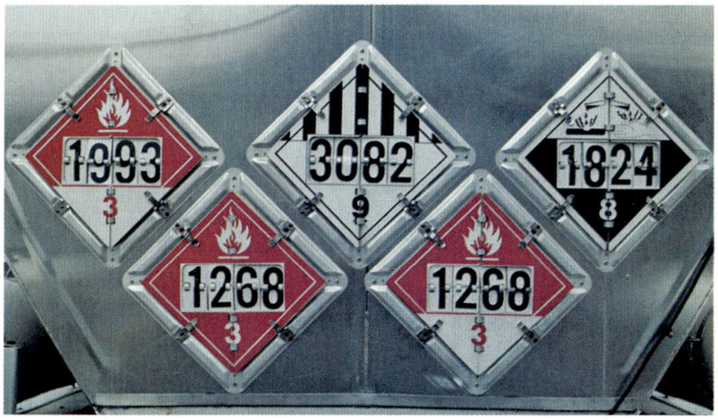

FIGURE 13-24 A placard is a large diamond-shaped indicator that is placed on all sides of transport vehicles that carry hazardous materials.

© Mark Winfrey/Shutterstock

Placards, labels, and markings are intended to give you a general idea of the hazard inside a container or cargo tank. A **placard** identifies the broad hazard class based on the chemical family (flammable, poison, corrosive) to which the material inside belongs.

The nine chemical families (hazard classes) and their respective divisions recognized in the *ERG* are outlined here (U.S. Department of Transportation, Transport Canada, and Secretariat of Communications and Transport of Mexico, 2020):

- DOT Class 1: Explosives
  - Division 1.1: Explosives that have a mass explosion hazard
  - Division 1.2: Explosives that have a projection hazard but not a mass explosion hazard
  - Division 1.3: Explosives that have a fire hazard and either a minor blast hazard or a minor projection hazard or both, but not a mass explosion hazard
  - Division 1.4: Explosives with no significant blast hazard
  - Division 1.5: Very insensitive explosives with a mass explosion hazard
  - Division 1.6: Extremely insensitive articles that do not have a mass explosion hazard
- DOT Class 2: Gases
  - Division 2.1: Flammable gases
  - Division 2.2: Nonflammable, nontoxic gases
  - Division 2.3: Toxic* gases
- DOT Class 3: Flammable liquids (and combustible liquids in the United States)
- DOT Class 4: Flammable solids, substances liable to spontaneous combustion, and substances that, on contact with water, emit flammable gases
  - Division 4.1: Flammable solids, self-reactive substances, and solid desensitized explosives
  - Division 4.2: Substances liable to spontaneous combustion
  - Division 4.3: Substances that, when in contact with water, will emit flammable gases
- DOT Class 5: Oxidizing substances and organic peroxides
  - Division 5.1: Oxidizing substances
  - Division 5.2: Organic peroxides
- DOT Class 6: Toxic substances and infectious substances
  - Division 6.1: Toxic substances
  - Division 6.2: Infectious substances
- DOT Class 7: Radioactive materials
- DOT Class 8: Corrosive substances
- DOT Class 9: Miscellaneous hazardous materials/products, substances, or organisms

*The words *poison* and *poisonous* are synonymous with the word *toxic*.

The *ERG* organizes chemicals into the nine basic hazard classes, or families, defined by the DOT. The members of each family exhibit similar properties. There is also a "Dangerous" placard, which indicates that the same load contains materials from more than one hazard class (**FIGURE 13-26**).

A label on a box inside a delivery truck, for example, relates only to the potential hazard or hazard class inside that package. In some cases, a four-digit United

Nations (UN) identification number may be required on some placards. This number identifies the specific material being shipped; a list of UN numbers is included in the *ERG*. You may see references to UN numbers that also include the letters *NA*, listed as UN/NA. *NA* stands for North America, but as a point of reference, UN and NA numbers are identical. These placards and labels can be viewed from a distance with binoculars, or shipping papers or other reference sources may provide the four-digit UN identification number or the name of the material, or you may see the colors on the placard. All of these methods can be used to determine the appropriate guide for managing a released material.

In most cases, the package or cargo tank must contain a certain amount of hazardous material before a placard is required. However, keep this point in mind: The absence of a placard does not mean no hazardous cargo is present. For example, the "1000-pound rule" applies to some explosives, flammable and nonflammable gases, flammable/combustible liquids, flammable solids, air-reactive solids, oxidizers and organic peroxides, poison solids, corrosives, and miscellaneous (class 9) materials. Placards are required for these materials only when the shipment weighs more than 1000 pounds. (More information on the 1000-pound rule can be found in the Code of Federal Regulations [CFR] Title 49, Subtitle B, Chapter 1, Subchapter C, Part 172, Subpart F.)

Conversely, some chemicals are so hazardous that shipping any amount of them requires the use of labels or placards. These materials include some explosives, poisonous gases, water-reactive solids, and high-level radioactive substances. Personnel at the scene should seek additional specifics about any material in question by consulting the appropriate response agency or using

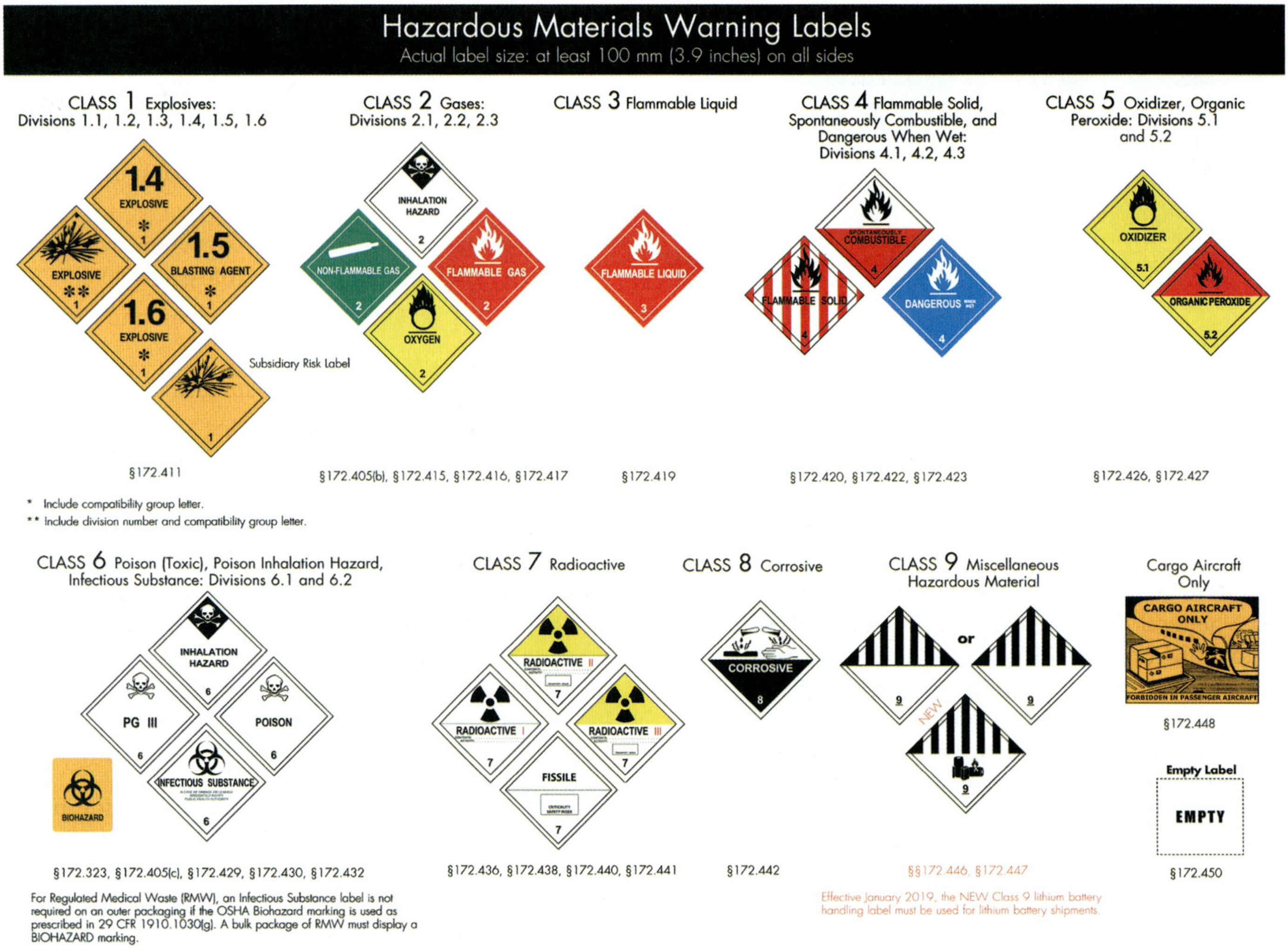

FIGURE 13-25 A label is a smaller version of the placard and is placed on boxes or smaller packages that contain hazardous materials.

Courtesy of the U.S. Department of Transportation.

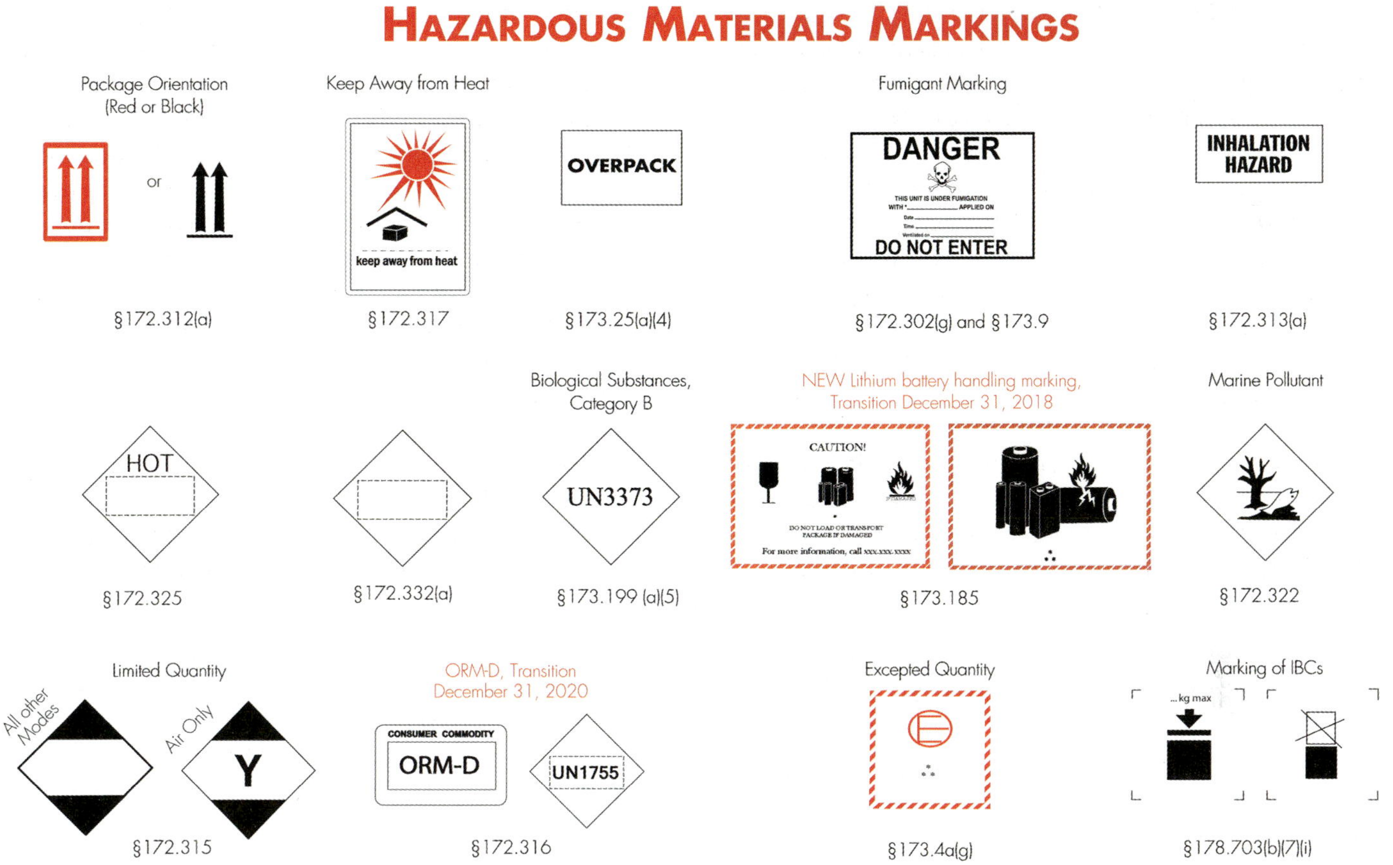

FIGURE 13-25 *(Continued)*

Courtesy of the U.S. Department of Transportation.

FIGURE 13-26 A "Dangerous" placard indicates that the load contains materials from more than one hazard.

© Jones & Bartlett Learning. Courtesy of Rob Schnepp.

the emergency response number on a shipping document, if applicable, to gather more information.

Thus, hazardous materials can be identified in several ways so that if a spill or release occurs, personnel and responders know what they are dealing with. Personnel should consult as many sources as possible (preferably at least three) to gather information about a released substance. Moreover, information sources should not be reserved just for incident response: It is important to review your common "go-to" sources of information on a regular basis to check for updates and refresh your familiarity with their layout and content. Think through a hypothetical situation you could encounter, and go through the information sources you think would be the most useful and appropriate.

The *Emergency Response Guidebook*

The U.S. DOT, the Secretariat of Communications and Transportation of Mexico, and Transport Canada jointly developed the *ERG* (**FIGURE 13-27**). The *ERG* (both the printed and online versions) offers a certain amount of guidance for anyone operating at a hazardous materials incident. This reference source merges the DOT labels and placards with the nine hazard classes and the UN identification system to provide

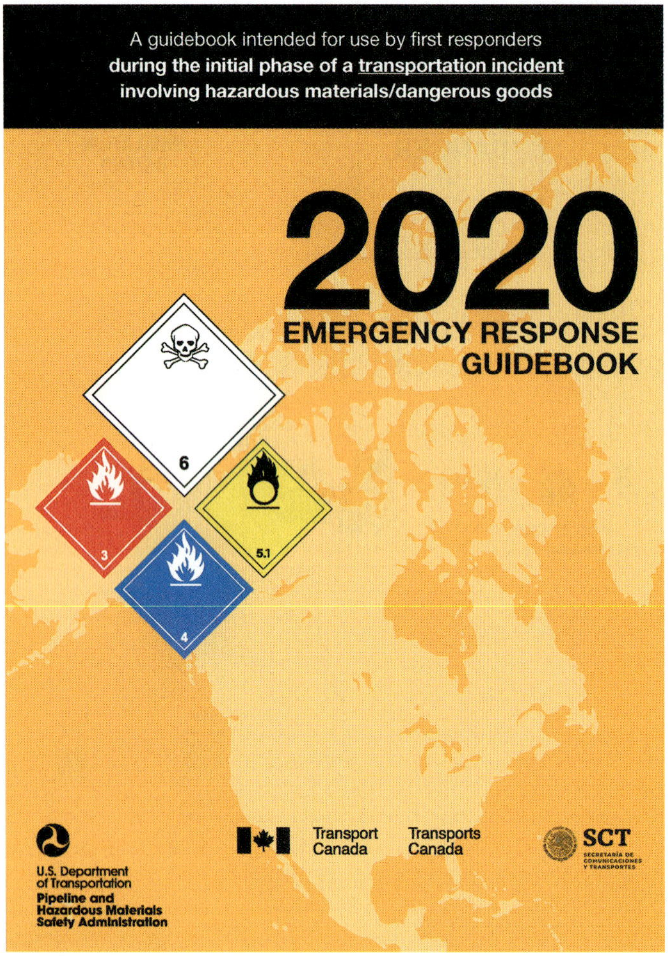

FIGURE 13-27 The *ERG* is a reference that can be used as the basis for your initial actions at a hazardous materials incident.

Reproduced from U.S. Department of Transportation, Transport Canada, and Secretariat of Communications and Transport of Mexico. 2020. *2020 Emergency Response Guidebook*. Pipeline and Hazardous Materials Safety Administration, U.S. Department of Transportation. www.phmsa.dot.gov/sites/phmsa.dot.gov/files/2021-01/ERG2020-WEB.pdf

guidance on initial response actions regarding issues such as the following:

- Isolation and protective action distances
- Recommended PPE for general hazard classes
- Fire control measures
- Information on the GHS
- Hazard identification numbers on some types of intermodal containers
- Pipeline information
- Many other good sources of on-scene guidance

Information on specific substances appears in the yellow-, blue-, and orange-colored sections of the *ERG* (more on this later in the section). Note that the *ERG* does not list all chemicals that could potentially be shipped by land, sea, air, or rail. The guide, which is revised and issued every 4 years and provides information on more than 4000 chemicals, is intended to help you decide which preliminary actions to take.

It is important for all users of the ERG *to know and understand how to use the guide and to know which new information has been updated in each revision cycle.* This text does not go into the details of each revision cycle. Rather, it focuses on the fundamentals of the guide and how it is used by awareness-level personnel and other responders.

Using the *ERG*

At a hazardous materials/WMD incident, if it is possible to identify the UN identification number, the chemical name, marking, placard, label, or perhaps the container shape in the *ERG*, users of the *ERG* may be able to determine the initial emergency actions to be taken as well as make some initial choices about PPE (**FIGURE 13-28**). For example, based on the recommendations in the *ERG*, would a normal work uniform provide adequate protection? Is chemical-protective clothing and equipment indicated? Is a self-contained breathing apparatus (SCBA) needed? Is structural firefighting–protective clothing adequate, or is **high-temperature–protective clothing**—"protective clothing designed to protect the wearer for short-term high-temperature exposures" (NFPA 470)—indicated? Answering these types of questions also assists awareness-level personnel in determining whether the nonintervention or defensive mode of operation is indicated.

As mentioned earlier, the *ERG* is divided into four colored sections: yellow, blue, orange, and green.

- **Yellow section:** More than 4000 chemicals are found in this section, listed numerically by their four-digit UN number (ID No.). Use the yellow section when the UN number is known or can be identified (**FIGURE 13-29**). Each entry also includes the chemical name. Entry number 1005, for example, identifies "ammonia, anhydrous." This section is useful because it links the UN number to the chemical name of the substance and also directs the user toward the orange section, emergency action guide number (Guide No.). For example:

ID No.	Guide No.	Name of Material
1005	125	Ammonia, anhydrous

  Some substances are highlighted in green in this section. Those highlighted substances are either a toxic inhalation hazard, a chemical warfare agent, or a water-reactive material and should be further referenced in the green section.

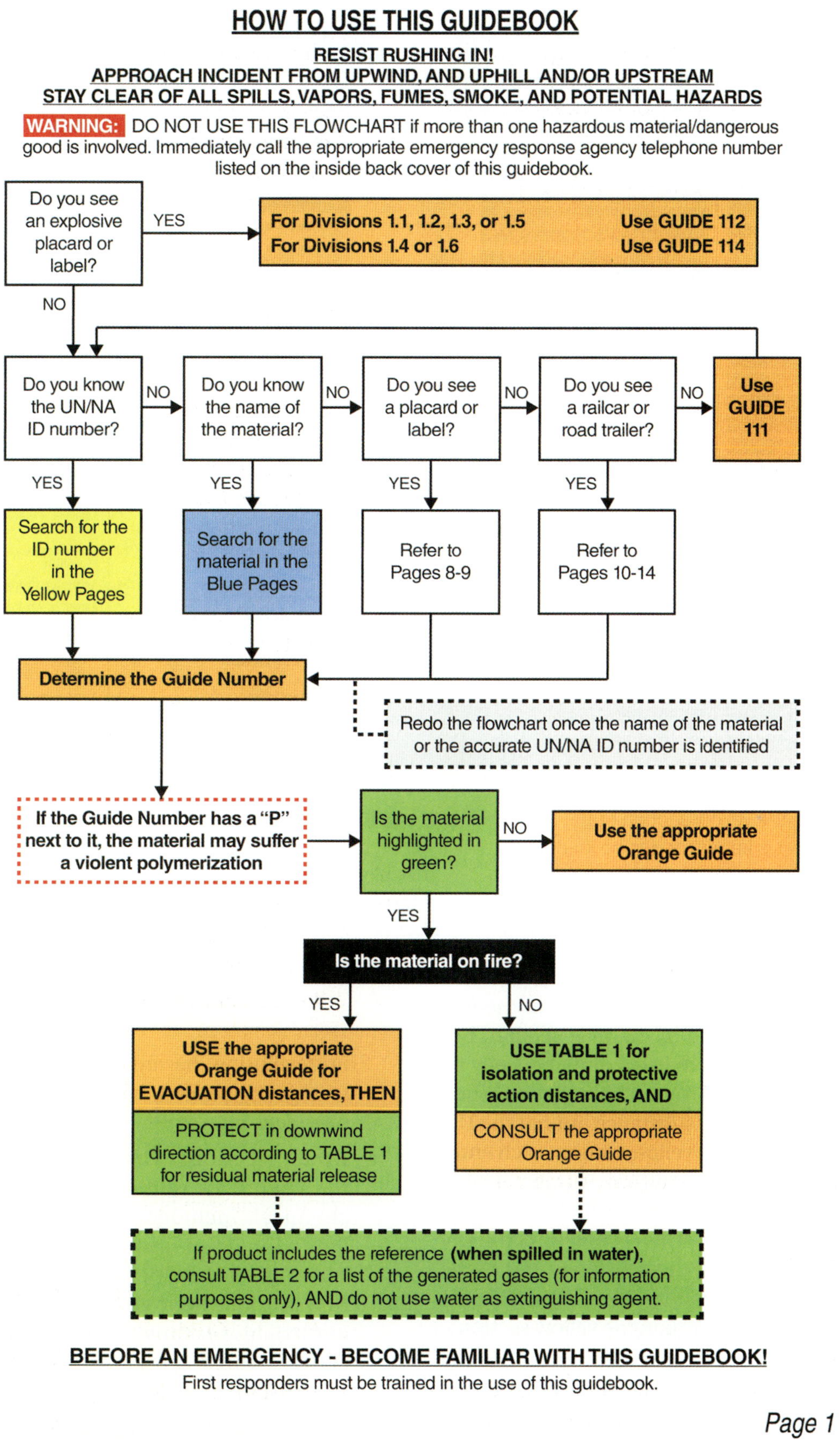

FIGURE 13-28 Flowchart illustrating how to use the *ERG*.

Reproduced from U.S. Department of Transportation, Transport Canada, and Secretariat of Communications and Transport of Mexico. 2020. *2020 Emergency Response Guidebook*. Pipeline and Hazardous Materials Safety Administration, U.S. Department of Transportation. www.phmsa.dot.gov/sites/phmsa.dot.gov/files/2021-01/ERG2020-WEB.pdf

ID No.	Guide No.	Name of Material
—	117	AC
—	154	Adamsite
—	112	Ammonium nitrate-fuel oil mixtures
—	158	Biological agents
—	112	Blasting agent, n.o.s.
—	153	Buzz
—	153	BZ
—	159	CA
—	125	CG
—	125	CK
—	153	CN
—	153	CS
—	154	CX
—	151	DA
—	153	DC
—	154	DM
—	125	DP
—	151	ED
—	112	Explosives, division 1.1, 1.2, 1.3 or 1.5
—	114	Explosives, division 1.4 or 1.6
—	153	GA
—	153	GB
—	153	GD
—	153	GF
—	153	H
—	153	HD
—	153	HL
—	153	HN-1
—	153	HN-2
—	153	HN-3
—	153	L (Lewisite)
—	153	Lewisite
—	152	MD
—	153	Mustard
—	153	Mustard Lewisite
—	152	PD
—	119	SA
—	153	Sarin
—	153	Soman
—	153	Tabun
—	153	Thickened GD
—	153	Toxins
—	153	VX
1001	116	Acetylene, dissolved
1002	122	Air, compressed
1003	122	Air, refrigerated liquid (cryogenic liquid)
1005	125	Ammonia, anhydrous
1005	125	Anhydrous ammonia
1006	120	Argon
1006	120	Argon, compressed
1008	125	Boron trifluoride
1008	125	Boron trifluoride, compressed
1009	126	Bromotrifluoromethane
1009	126	Refrigerant gas R-13B1
1010	116P	Butadienes, stabilized
1010	116P	Butadienes and hydrocarbon mixture, stabilized
1010	116P	Hydrocarbon and butadienes mixture, stabilized
1011	115	Butane
1012	115	Butylene

Page 29

FIGURE 13-29 Use the yellow section of the *ERG* when the UN number (ID No.) is known or can be identified.

Reproduced from U.S. Department of Transportation, Transport Canada, and Secretariat of Communications and Transport of Mexico. 2020. *2020 Emergency Response Guidebook*. Pipeline and Hazardous Materials Safety Administration, U.S. Department of Transportation. www.phmsa.dot.gov/sites/phmsa.dot.gov/files/2021-01/ERG2020-WEB.pdf

Name of Material	Guide No.	ID No.
Aluminum borohydride	135	2870
Aluminum borohydride in devices	135	2870
Aluminum bromide, anhydrous	137	1725
Aluminum bromide, solution	154	2580
Aluminum carbide	138	1394
Aluminum chloride, anhydrous	137	1726
Aluminum chloride, solution	154	2581
Aluminum dross	138	3170
Aluminum ferrosilicon powder	139	1395
Aluminum hydride	138	2463
Aluminum nitrate	140	1438
Aluminum phosphide	139	1397
Aluminum phosphide pesticide	157	3048
Aluminum powder, coated	170	1309
Aluminum powder, pyrophoric	135	1383
Aluminum powder, uncoated	138	1396
Aluminum remelting by-products	138	3170
Aluminum resinate	133	2715
Aluminum silicon powder, uncoated	138	1398
Aluminum smelting by-products	138	3170
Amines, flammable, corrosive, n.o.s.	132	2733
Amines, liquid, corrosive, flammable, n.o.s.	132	2734
Amines, liquid, corrosive, n.o.s.	153	2735
Amines, solid, corrosive, n.o.s.	154	3259
2-Amino-4-chlorophenol	151	2673
2-Amino-5-diethylaminopentane	153	2946
2-Amino-4,6-dinitrophenol, wetted with not less than 20% water	113	3317
2-(2-Aminoethoxy)ethanol	154	3055
N-Aminoethylpiperazine	153	2815
Aminophenols	152	2512
Aminopyridines	153	2671
Ammonia, anhydrous	125	1005
Ammonia, solution, with more than 10% but not more than 35% Ammonia	154	2672
Ammonia, solution, with more than 35% but not more than 50% Ammonia	125	2073
Ammonia solution, with more than 50% Ammonia	125	3318
Ammonium arsenate	151	1546
Ammonium bifluoride, solid	154	1727
Ammonium bifluoride, solution	154	2817
Ammonium dichromate	141	1439
Ammonium dinitro-o-cresolate, solid	141	1843
Ammonium dinitro-o-cresolate, solution	141	3424
Ammonium fluoride	154	2505
Ammonium fluorosilicate	151	2854
Ammonium hydrogendifluoride, solid	154	1727
Ammonium hydrogendifluoride, solution	154	2817
Ammonium hydrogen sulfate	154	2506
Ammonium hydrogen sulphate	154	2506
Ammonium hydroxide	154	2672
Ammonium hydroxide, with more than 10% but not more than 35% Ammonia	154	2672
Ammonium metavanadate	154	2859
Ammonium nitrate, liquid (hot concentrated solution)	140	2426

Page 97

FIGURE 13-30 The same chemicals listed in the yellow section of the *ERG* are found in the blue section, listed alphabetically by name.

Reproduced from U.S. Department of Transportation, Transport Canada, and Secretariat of Communications and Transport of Mexico. 2020. *2020 Emergency Response Guidebook*. Pipeline and Hazardous Materials Safety Administration, U.S. Department of Transportation. www.phmsa.dot.gov/sites/phmsa.dot.gov/files/2021-01/ERG2020-WEB.pdf

- **Blue section:** The same chemicals listed in the yellow section are also found in the blue section, but in the blue section, they are listed alphabetically by name (**FIGURE 13-30**). Use the blue section when you know the name of the substance but not the UN number (ID No.). As in the yellow section, each entry includes the emergency action guide number (Guide No.), found in the orange section, and the UN identification number (ID No.). This section links the UN number with the name of the substance and to the associated emergency action guide number (Guide No.), just in a different way than the yellow section.
- **Orange section:** The orange section describes the hazards associated with the chemicals listed in the yellow and blue sections. In this section, the chemicals are organized by emergency action guide number (**FIGURE 13-31**). The general hazard class, fire/explosion hazards, health hazards, and basic emergency actions, based on the hazard class, are provided.
- **Green section:** The green section organizes the chemicals numerically by UN number. The chemicals included in this section are those highlighted in green in the yellow and blue sections. These chemicals include water-reactive materials that produce toxic gases (e.g., calcium phosphide and trichlorosilane); **toxic inhalation hazards (TIHs)**, which are gases or volatile liquids that are extremely toxic to humans; chemical warfare agents (CWAs); and dangerous water-reactive materials (WRMs; e.g., anhydrous ammonia, sarin, and sodium cyanide). These gases or volatile

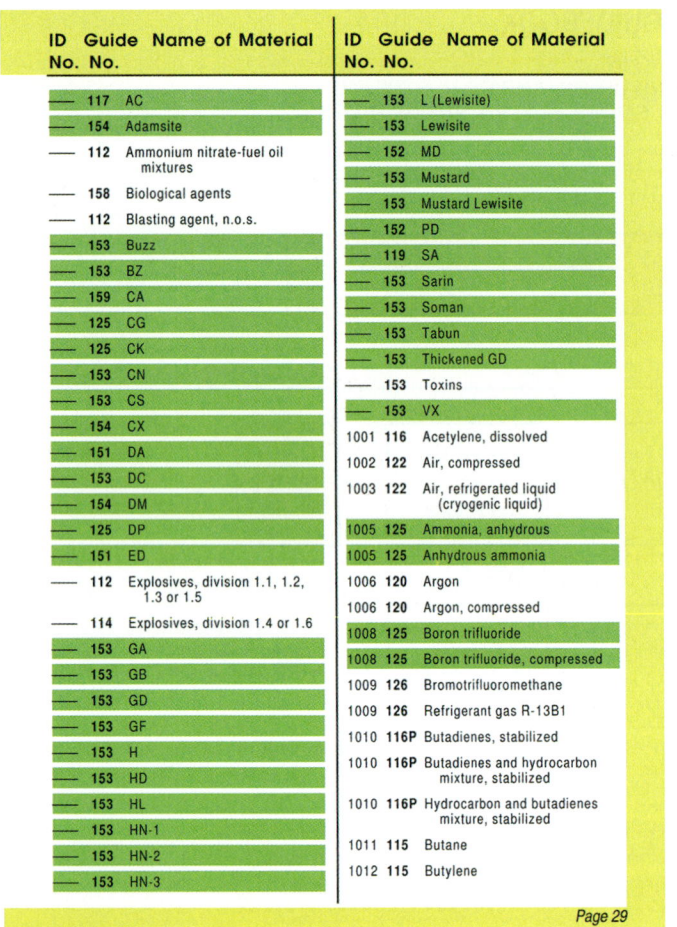

GUIDE 125 GASES - TOXIC AND/OR CORROSIVE

POTENTIAL HAZARDS

HEALTH
- **TOXIC; may be fatal if inhaled, ingested or absorbed through skin.**
- Vapors are extremely irritating and corrosive.
- Contact with gas or liquefied gas may cause burns, severe injury and/or frostbite.
- Fire will produce irritating, corrosive and/or toxic gases.
- Runoff from fire control or dilution water may cause environmental contamination.

FIRE OR EXPLOSION
- Some may burn but none ignite readily.
- Vapors from liquefied gas are initially heavier than air and spread along ground.
- Some of these materials may react violently with water.
- Cylinders exposed to fire may vent and release toxic and/or corrosive gas through pressure relief devices.
- Containers may explode when heated.
- Ruptured cylinders may rocket.
- For UN1005: Anhydrous ammonia, at high concentrations in confined spaces, presents a flammability risk if a source of ignition is introduced.

PUBLIC SAFETY
- **CALL 911. Then call emergency response telephone number on shipping paper.** If shipping paper not available or no answer, refer to appropriate telephone number listed on the inside back cover.
- Keep unauthorized personnel away.
- Stay upwind, uphill and/or upstream.
- Many gases are heavier than air and will spread along the ground and collect in low or confined areas (sewers, basements, tanks, etc.).
- Ventilate closed spaces before entering, but only if properly trained and equipped.

PROTECTIVE CLOTHING
- Wear positive pressure self-contained breathing apparatus (SCBA).
- Wear chemical protective clothing that is specifically recommended by the manufacturer **when there is NO RISK OF FIRE.**
- Structural firefighters' protective clothing provides thermal protection **but only limited chemical protection.**

EVACUATION

Immediate precautionary measure
- Isolate spill or leak area for at least 100 meters (330 feet) in all directions.

Spill
- For highlighted materials: see Table 1 - Initial Isolation and Protective Action Distances.
- For non-highlighted materials: increase the immediate precautionary measure distance, in the downwind direction, as necessary.

Fire
- If tank, rail car or tank truck is involved in a fire, ISOLATE for 1600 meters (1 mile) in all directions; also, consider initial evacuation for 1600 meters (1 mile) in all directions.

In Canada, an Emergency Response Assistance Plan (ERAP) may be required for this product. Please consult the shipping paper and/or the ERAP Program Section (page 390).

Page 186 **ERG 2020**

GASES - TOXIC AND/OR CORROSIVE **GUIDE 125**

EMERGENCY RESPONSE

FIRE

Small Fire
- Dry chemical or CO_2.

Large Fire
- Water spray, fog or regular foam.
- If it can be done safely, move undamaged containers away from the area around the fire.
- Do not get water inside containers.
- Damaged cylinders should be handled only by specialists.

Fire Involving Tanks
- Fight fire from maximum distance or use unmanned master stream devices or monitor nozzles.
- Cool containers with flooding quantities of water until well after fire is out.
- Do not direct water at source of leak or safety devices; icing may occur.
- Withdraw immediately in case of rising sound from venting safety devices or discoloration of tank.
- ALWAYS stay away from tanks engulfed in fire.

SPILL OR LEAK
- Do not touch or walk through spilled material.
- Stop leak if you can do it without risk.
- If possible, turn leaking containers so that gas escapes rather than liquid.
- Prevent entry into waterways, sewers, basements or confined areas.
- Do not direct water at spill or source of leak.
- Use water spray to reduce vapors or divert vapor cloud drift. Avoid allowing water runoff to contact spilled material.
- Isolate area until gas has dispersed.

FIRST AID
- Call 911 or emergency medical service.
- Ensure that medical personnel are aware of the material(s) involved and take precautions to protect themselves.
- Move victim to fresh air if it can be done safely.
- Give artificial respiration if victim is not breathing.
- **Do not perform mouth-to-mouth resuscitation if victim ingested or inhaled the substance; wash face and mouth before giving artificial respiration. Use a pocket mask equipped with a one-way valve or other proper respiratory medical device.**
- Administer oxygen if breathing is difficult.
- Remove and isolate contaminated clothing and shoes.
- In case of contact with liquefied gas, thaw frosted parts with lukewarm water.
- In case of contact with substance, immediately flush skin or eyes with running water for at least 20 minutes.
- **In case of skin contact with hydrogen fluoride, anhydrous (UN1052)**, if calcium gluconate gel is available, rinse 5 minutes, then apply gel. Otherwise, continue rinsing until medical treatment is available.
- Keep victim calm and warm.
- Keep victim under observation.
- Effects of contact or inhalation may be delayed.

ERG 2020 Page 187

FIGURE 13-31 The orange section of the *ERG* is organized by emergency action guide number (Guide No.) and describes the hazards associated with the chemicals in the guide.

Reproduced from U.S. Department of Transportation, Transport Canada, and Secretariat of Communications and Transport of Mexico. 2020. *2020 Emergency Response Guidebook*. Pipeline and Hazardous Materials Safety Administration, U.S. Department of Transportation. www.phmsa.dot.gov/sites/phmsa.dot.gov/files/2021-01/ERG2020-WEB.pdf

liquids are extremely toxic to humans and pose a hazard to health during transport; in fact, any material listed in the green section is extremely hazardous.

The green section of the *ERG* provides the initial isolation distances for certain materials, including both small and large spills (**FIGURE 13-32**). Additionally, it provides guidance on how to gauge the size of the release. For example, a small spill is a leak from one small package; a small leak in a large container (up to 208 liters or 55 gallons); a small cylinder leak; or any small leak, even one in a large package. A large spill is a large leak or spill (greater than 208 liters or 55 gallons) from a larger container or package; a spill from a number of small packages; or anything from a 1-ton cylinder, tank truck, or railcar.

The green section also offers recommendations on isolation distances and the size and shape of protective action zones. This section is useful when it is necessary to protect people from TIH vapors resulting from a release. The initial isolation zones may be used to define an area surrounding an incident where persons may be exposed to potentially dangerous or life-threatening concentrations of the vapor both upwind and downwind from the release. The orange-bordered guides differ from the green section in that the orange-bordered guides are relevant to evacuation distances required to protect against fragmentation hazards from a large container if it should fail due to explosion. The rationale is that if a certain material becomes involved with fire, the TIH hazard may be less than the fire or explosion hazard.

To use the *ERG*, follow the steps in **SKILL DRILL 13-1**.

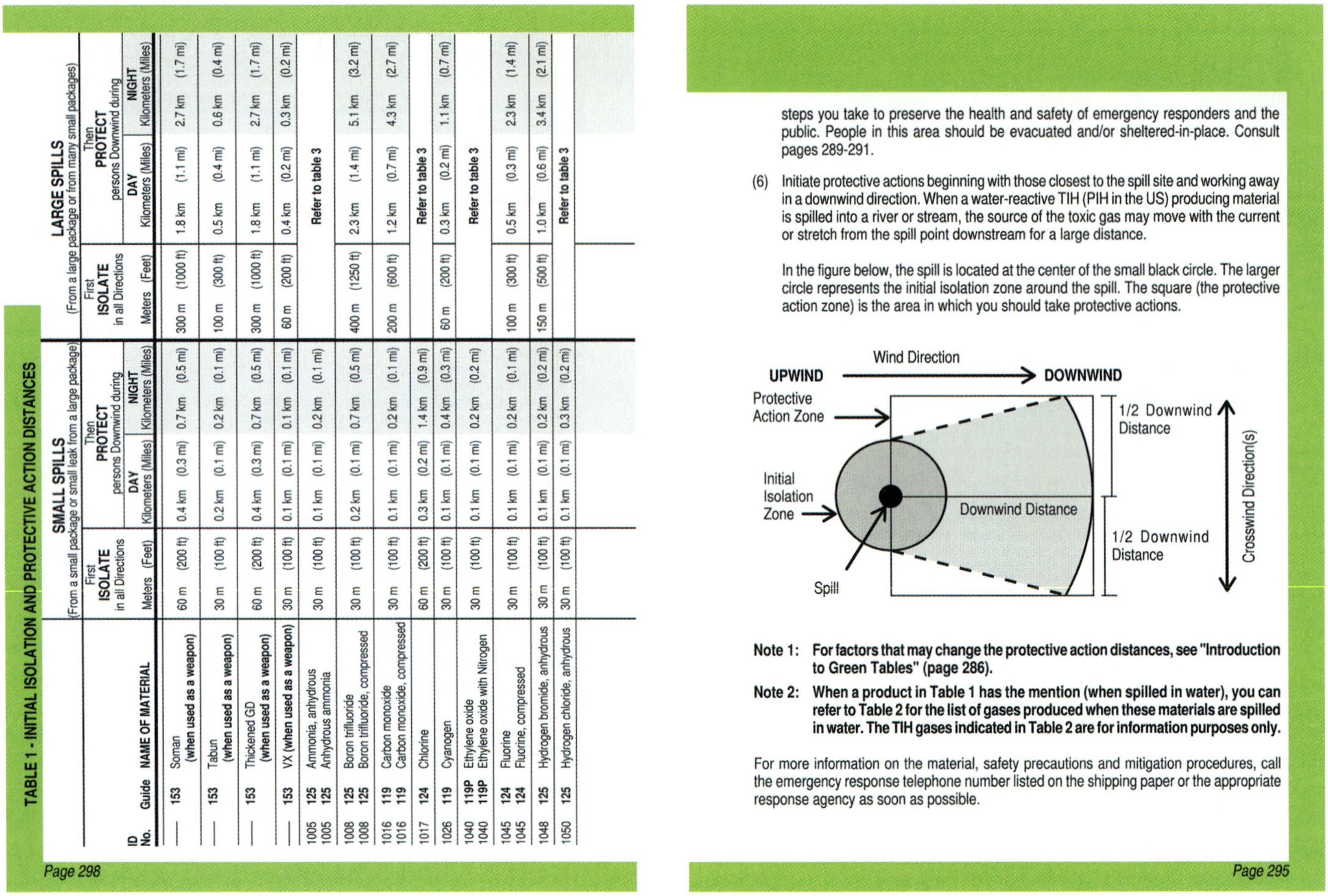

TABLE 1 - INITIAL ISOLATION AND PROTECTIVE ACTION DISTANCES

ID No.	Guide	NAME OF MATERIAL	SMALL SPILLS (From a small package or small leak from a large package) First ISOLATE in all Directions Meters (Feet)	Then PROTECT persons Downwind during DAY Kilometers (Miles)	NIGHT Kilometers (Miles)	LARGE SPILLS (From a large package or from many small packages) First ISOLATE in all Directions Meters (Feet)	Then PROTECT persons Downwind during DAY Kilometers (Miles)	NIGHT Kilometers (Miles)
—	153	Soman (when used as a weapon)	60 m (200 ft)	0.4 km (0.3 mi)	0.7 km (0.5 mi)	300 m (1000 ft)	1.8 km (1.1 mi)	2.7 km (1.7 mi)
—	153	Tabun (when used as a weapon)	30 m (100 ft)	0.2 km (0.1 mi)	0.2 km (0.1 mi)	100 m (300 ft)	0.5 km (0.4 mi)	0.6 km (0.4 mi)
—	153	Thickened GD (when used as a weapon)	60 m (200 ft)	0.4 km (0.3 mi)	0.7 km (0.5 mi)	300 m (1000 ft)	1.8 km (1.1 mi)	2.7 km (1.7 mi)
—	153	VX (when used as a weapon)	30 m (100 ft)	0.1 km (0.1 mi)	0.1 km (0.1 mi)	60 m (200 ft)	0.4 km (0.2 mi)	0.3 km (0.2 mi)
1005 1005	125 125	Ammonia, anhydrous Anhydrous ammonia	30 m (100 ft)	0.1 km (0.1 mi)	0.2 km (0.1 mi)	Refer to table 3		
1008 1008	125 125	Boron trifluoride Boron trifluoride, compressed	30 m (100 ft)	0.2 km (0.1 mi)	0.7 km (0.5 mi)	400 m (1250 ft)	2.3 km (1.4 mi)	5.1 km (3.2 mi)
1016 1016	119 119	Carbon monoxide Carbon monoxide, compressed	30 m (100 ft)	0.1 km (0.1 mi)	0.2 km (0.1 mi)	200 m (600 ft)	1.2 km (0.7 mi)	4.3 km (2.7 mi)
1017	124	Chlorine	60 m (200 ft)	0.3 km (0.2 mi)	1.4 km (0.9 mi)	Refer to table 3		
1026	119	Cyanogen	30 m (100 ft)	0.1 km (0.1 mi)	0.4 km (0.3 mi)	60 m (200 ft)	0.3 km (0.2 mi)	1.1 km (0.7 mi)
1040 1040	119P 119P	Ethylene oxide Ethylene oxide with Nitrogen	30 m (100 ft)	0.1 km (0.1 mi)	0.2 km (0.2 mi)	Refer to table 3		
1045 1045	124 124	Fluorine Fluorine, compressed	30 m (100 ft)	0.1 km (0.1 mi)	0.2 km (0.1 mi)	100 m (300 ft)	0.5 km (0.3 mi)	2.3 km (1.4 mi)
1048	125	Hydrogen bromide, anhydrous	30 m (100 ft)	0.1 km (0.1 mi)	0.2 km (0.2 mi)	150 m (500 ft)	1.0 km (0.6 mi)	3.4 km (2.1 mi)
1050	125	Hydrogen chloride, anhydrous	30 m (100 ft)	0.1 km (0.1 mi)	0.3 km (0.2 mi)	Refer to table 3		

Page 298

steps you take to preserve the health and safety of emergency responders and the public. People in this area should be evacuated and/or sheltered-in-place. Consult pages 289-291.

(6) Initiate protective actions beginning with those closest to the spill site and working away in a downwind direction. When a water-reactive TIH (PIH in the US) producing material is spilled into a river or stream, the source of the toxic gas may move with the current or stretch from the spill point downstream for a large distance.

In the figure below, the spill is located at the center of the small black circle. The larger circle represents the initial isolation zone around the spill. The square (the protective action zone) is the area in which you should take protective actions.

Note 1: For factors that may change the protective action distances, see "Introduction to Green Tables" (page 286).

Note 2: When a product in Table 1 has the mention (when spilled in water), you can refer to Table 2 for the list of gases produced when these materials are spilled in water. The TIH gases indicated in Table 2 are for information purposes only.

For more information on the material, safety precautions and mitigation procedures, call the emergency response telephone number listed on the shipping paper or the appropriate response agency as soon as possible.

Page 295

FIGURE 13-32 The green section of the *ERG* is organized numerically by UN number and provides the initial isolation distances, protective actions, and guidelines for gauging the size of a release for certain materials.

Reproduced from U.S. Department of Transportation, Transport Canada, and Secretariat of Communications and Transport of Mexico. 2020. *2020 Emergency Response Guidebook*. Pipeline and Hazardous Materials Safety Administration, U.S. Department of Transportation. www.phmsa.dot.gov/sites/phmsa.dot.gov/files/2021-01/ERG2020-WEB.pdf

SKILL DRILL 13-1

Using the *Emergency Response Guidebook* NFPA 470: 5.2.1, 5.3.1

1. Identify the chemical name and/or the chemical ID number for the placard.

© Technicsorn Stocker/Shutterstock

SKILL DRILL 13-1 CONTINUED

Using the *Emergency Response Guidebook* NFPA 470: 5.2.1, 5.3.1

ID No.	Guide No.	Name of Material
—	117	AC
—	154	Adamsite
—	112	Ammonium nitrate-fuel oil mixtures
—	158	Biological agents
—	112	Blasting agent, n.o.s.
—	153	Buzz
—	153	BZ
—	159	CA
—	125	CG
—	125	CK
—	153	CN
—	153	CS
—	154	CX
—	151	DA
—	153	DC
—	154	DM
—	125	DP
—	151	ED
—	112	Explosives, division 1.1, 1.2, 1.3 or 1.5
—	114	Explosives, division 1.4 or 1.6
—	153	GA
—	153	GB
—	153	GD
—	153	GF
—	153	H
—	153	HD
—	153	HL
—	153	HN-1
—	153	HN-2
—	153	HN-3

ID No.	Guide No.	Name of Material
—	153	L (Lewisite)
—	153	Lewisite
—	152	MD
—	153	Mustard
—	153	Mustard Lewisite
—	152	PD
—	119	SA
—	153	Sarin
—	153	Soman
—	153	Tabun
—	153	Thickened GD
—	153	Toxins
—	153	VX
1001	116	Acetylene, dissolved
1002	122	Air, compressed
1003	122	Air, refrigerated liquid (cryogenic liquid)
1005	125	Ammonia, anhydrous
1005	125	Anhydrous ammonia
1006	120	Argon
1006	120	Argon, compressed
1008	125	Boron trifluoride
1008	125	Boron trifluoride, compressed
1009	126	Bromotrifluoromethane
1009	126	Refrigerant gas R-13B1
1010	116P	Butadienes, stabilized
1010	116P	Butadienes and hydrocarbon mixture, stabilized
1010	116P	Hydrocarbon and butadienes mixture, stabilized
1011	115	Butane
1012	115	Butylene

Page 29

Reproduced from U.S. Department of Transportation, Transport Canada, and Secretariat of Communications and Transport of Mexico. 2020. *2020 Emergency Response Guidebook*. Pipeline and Hazardous Materials Safety Administration, U.S. Department of Transportation. www.phmsa.dot.gov/sites/phmsa.dot.gov/files/2021-01/ERG2020-WEB.pdf

2. Look up the material name in the appropriate section of the *ERG*. Use the yellow section to obtain information based on the UN identification number. Use the alphabetized blue section to obtain information based on the chemical name. *Note any green highlights, which indicate the substance also has an entry and recommendations in the green section of the guide.*

Continues.

SKILL DRILL 13-1 CONTINUED

Using the *Emergency Response Guidebook* NFPA 470: 5.2.1, 5.3.1

GUIDE 125 GASES - TOXIC AND/OR CORROSIVE

POTENTIAL HAZARDS

HEALTH
- **TOXIC; may be fatal if inhaled, ingested or absorbed through skin.**
- Vapors are extremely irritating and corrosive.
- Contact with gas or liquefied gas may cause burns, severe injury and/or frostbite.
- Fire will produce irritating, corrosive and/or toxic gases.
- Runoff from fire control or dilution water may cause environmental contamination.

FIRE OR EXPLOSION
- Some may burn but none ignite readily.
- Vapors from liquefied gas are initially heavier than air and spread along ground.
- Some of these materials may react violently with water.
- Cylinders exposed to fire may vent and release toxic and/or corrosive gas through pressure relief devices.
- Containers may explode when heated.
- Ruptured cylinders may rocket.
- For UN1005: Anhydrous ammonia, at high concentrations in confined spaces, presents a flammability risk if a source of ignition is introduced.

PUBLIC SAFETY
- **CALL 911. Then call emergency response telephone number on shipping paper.** If shipping paper not available or no answer, refer to appropriate telephone number listed on the inside back cover.
- Keep unauthorized personnel away.
- Stay upwind, uphill and/or upstream.
- Many gases are heavier than air and will spread along the ground and collect in low or confined areas (sewers, basements, tanks, etc.).
- Ventilate closed spaces before entering, but only if properly trained and equipped.

PROTECTIVE CLOTHING
- Wear positive pressure self-contained breathing apparatus (SCBA).
- Wear chemical protective clothing that is specifically recommended by the manufacturer **when there is NO RISK OF FIRE.**
- Structural firefighters' protective clothing provides thermal protection **but only limited chemical protection.**

EVACUATION

Immediate precautionary measure
- Isolate spill or leak area for at least 100 meters (330 feet) in all directions.

Spill
- For highlighted materials: see Table 1 - Initial Isolation and Protective Action Distances.
- For non-highlighted materials: increase the immediate precautionary measure distance, in the downwind direction, as necessary.

Fire
- If tank, rail car or tank truck is involved in a fire, ISOLATE for 1600 meters (1 mile) in all directions; also, consider initial evacuation for 1600 meters (1 mile) in all directions.

In Canada, an Emergency Response Assistance Plan (ERAP) may be required for this product. Please consult the shipping paper and/or the ERAP Program Section (page 390).

Page 186 **ERG 2020**

GASES - TOXIC AND/OR CORROSIVE **GUIDE 125**

EMERGENCY RESPONSE

FIRE

Small Fire
- Dry chemical or CO_2.

Large Fire
- Water spray, fog or regular foam.
- If it can be done safely, move undamaged containers away from the area around the fire.
- Do not get water inside containers.
- Damaged cylinders should be handled only by specialists.

Fire Involving Tanks
- Fight fire from maximum distance or use unmanned master stream devices or monitor nozzles.
- Cool containers with flooding quantities of water until well after fire is out.
- Do not direct water at source of leak or safety devices; icing may occur.
- Withdraw immediately in case of rising sound from venting safety devices or discoloration of tank.
- ALWAYS stay away from tanks engulfed in fire.

SPILL OR LEAK
- Do not touch or walk through spilled material.
- Stop leak if you can do it without risk.
- If possible, turn leaking containers so that gas escapes rather than liquid.
- Prevent entry into waterways, sewers, basements or confined areas.
- Do not direct water at spill or source of leak.
- Use water spray to reduce vapors or divert vapor cloud drift. Avoid allowing water runoff to contact spilled material.
- Isolate area until gas has dispersed.

FIRST AID
- Call 911 or emergency medical service.
- Ensure that medical personnel are aware of the material(s) involved and take precautions to protect themselves.
- Move victim to fresh air if it can be done safely.
- Give artificial respiration if victim is not breathing.
- **Do not perform mouth-to-mouth resuscitation if victim ingested or inhaled the substance; wash face and mouth before giving artificial respiration. Use a pocket mask equipped with a one-way valve or other proper respiratory medical device.**
- Administer oxygen if breathing is difficult.
- Remove and isolate contaminated clothing and shoes.
- In case of contact with liquefied gas, thaw frosted parts with lukewarm water.
- In case of contact with substance, immediately flush skin or eyes with running water for at least 20 minutes.
- **In case of skin contact with hydrogen fluoride, anhydrous (UN1052)**, if calcium gluconate gel is available, rinse 5 minutes, then apply gel. Otherwise, continue rinsing until medical treatment is available.
- Keep victim calm and warm.
- Keep victim under observation.
- Effects of contact or inhalation may be delayed.

ERG 2020 Page 187

3. Determine the correct emergency action guide to use for the chemical identified.

Reproduced from U.S. Department of Transportation, Transport Canada, and Secretariat of Communications and Transport of Mexico. 2020. *2020 Emergency Response Guidebook*. Pipeline and Hazardous Materials Safety Administration, U.S. Department of Transportation. www.phmsa.dot.gov/sites/phmsa.dot.gov/files/2021-01/ERG2020-WEB.pdf

SKILL DRILL 13-1 CONTINUED

Using the *Emergency Response Guidebook* NFPA 470: 5.2.1, 5.3.1

TABLE 1 - INITIAL ISOLATION AND PROTECTIVE ACTION DISTANCES

ID No.	Guide	NAME OF MATERIAL	SMALL SPILLS (From a small package or small leak from a large package): First ISOLATE in all Directions, Meters (Feet)	SMALL SPILLS: Then PROTECT persons Downwind during DAY, Kilometers (Miles)	SMALL SPILLS: Then PROTECT persons Downwind during NIGHT, Kilometers (Miles)	LARGE SPILLS (From a large package or from many small packages): First ISOLATE in all Directions, Meters (Feet)	LARGE SPILLS: Then PROTECT persons Downwind during DAY, Kilometers (Miles)	LARGE SPILLS: Then PROTECT persons Downwind during NIGHT, Kilometers (Miles)
——	153	Soman (when used as a weapon)	60 m (200 ft)	0.4 km (0.3 mi)	0.7 km (0.5 mi)	300 m (1000 ft)	1.8 km (1.1 mi)	2.7 km (1.7 mi)
——	153	Tabun (when used as a weapon)	30 m (100 ft)	0.2 km (0.1 mi)	0.2 km (0.1 mi)	100 m (300 ft)	0.5 km (0.4 mi)	0.6 km (0.4 mi)
——	153	Thickened GD (when used as a weapon)	60 m (200 ft)	0.4 km (0.3 mi)	0.7 km (0.5 mi)	300 m (1000 ft)	1.8 km (1.1 mi)	2.7 km (1.7 mi)
——	153	VX (when used as a weapon)	30 m (100 ft)	0.1 km (0.1 mi)	0.1 km (0.1 mi)	60 m (200 ft)	0.4 km (0.2 mi)	0.3 km (0.2 mi)
1005 1005	125 125	Ammonia, anhydrous Anhydrous ammonia	30 m (100 ft)	0.1 km (0.1 mi)	0.2 km (0.1 mi)	Refer to table 3		
1008 1008	125 125	Boron trifluoride Boron trifluoride, compressed	30 m (100 ft)	0.2 km (0.1 mi)	0.7 km (0.5 mi)	400 m (1250 ft)	2.3 km (1.4 mi)	5.1 km (3.2 mi)
1016 1016	119 119	Carbon monoxide Carbon monoxide, compressed	30 m (100 ft)	0.1 km (0.1 mi)	0.2 km (0.1 mi)	200 m (600 ft)	1.2 km (0.7 mi)	4.3 km (2.7 mi)
1017	124	Chlorine	60 m (200 ft)	0.3 km (0.2 mi)	1.4 km (0.9 mi)	Refer to table 3		
1026	119	Cyanogen	30 m (100 ft)	0.1 km (0.1 mi)	0.4 km (0.3 mi)	60 m (200 ft)	0.3 km (0.2 mi)	1.1 km (0.7 mi)
1040 1040	119P 119P	Ethylene oxide Ethylene oxide with Nitrogen	30 m (100 ft)	0.1 km (0.1 mi)	0.2 km (0.2 mi)	Refer to table 3		
1045 1045	124 124	Fluorine Fluorine, compressed	30 m (100 ft)	0.1 km (0.1 mi)	0.2 km (0.1 mi)	100 m (300 ft)	0.5 km (0.3 mi)	2.3 km (1.4 mi)
1048	125	Hydrogen bromide, anhydrous	30 m (100 ft)	0.1 km (0.1 mi)	0.2 km (0.2 mi)	150 m (500 ft)	1.0 km (0.6 mi)	3.4 km (2.1 mi)
1050	125	Hydrogen chloride, anhydrous	30 m (100 ft)	0.1 km (0.1 mi)	0.3 km (0.2 mi)	Refer to table 3		

Page 298

TABLE 3 - INITIAL ISOLATION AND PROTECTIVE ACTION DISTANCES FOR LARGE SPILLS FOR DIFFERENT QUANTITIES OF SIX COMMON TIH (PIH in the US) GASES

TRANSPORT CONTAINER	First ISOLATE in all Directions: Meters	(Feet)	Then PROTECT persons Downwind during DAY: Low wind (< 6 mph = < 10 km/h) km	(Miles)	DAY: Moderate wind (6-12 mph = 10 - 20 km/h) km	(Miles)	DAY: High wind (> 12 mph = > 20 km/h) km	(Miles)	NIGHT: Low wind (< 6 mph = < 10 km/h) km	(Miles)	NIGHT: Moderate wind (6-12 mph = 10 - 20 km/h) km	(Miles)	NIGHT: High wind (> 12 mph = > 20 km/h) km	(Miles)
UN1005 Ammonia, anhydrous: Large Spills														
Rail tank car	300	(1000)	1.9	(1.2)	1.5	(0.9)	1.1	(0.6)	4.5	(2.8)	2.5	(1.5)	1.4	(0.9)
Highway tank truck or trailer	150	(500)	0.9	(0.6)	0.5	(0.3)	0.4	(0.3)	2.0	(1.3)	0.8	(0.5)	0.6	(0.4)
Agricultural nurse tank	60	(200)	0.5	(0.3)	0.3	(0.2)	0.3	(0.2)	1.4	(0.9)	0.3	(0.2)	0.3	(0.2)
Multiple small cylinders	30	(100)	0.3	(0.2)	0.2	(0.1)	0.1	(0.1)	0.7	(0.5)	0.3	(0.2)	0.2	(0.1)
UN1017 Chlorine: Large Spills														
Rail tank car	1000	(3000)	10.1	(6.3)	6.8	(4.2)	5.3	(3.3)	11+	(7+)	9.2	(5.7)	6.9	(4.3)
Highway tank truck or trailer	600	(2000)	5.8	(3.6)	3.4	(2.1)	2.9	(1.8)	6.7	(4.3)	5.0	(3.1)	4.1	(2.5)
Multiple ton cylinders	300	(1000)	2.1	(1.3)	1.3	(0.8)	1.0	(0.6)	4.0	(2.5)	2.4	(1.5)	1.3	(0.8)
Multiple small cylinders or single ton cylinder	150	(500)	1.5	(0.9)	0.8	(0.5)	0.5	(0.3)	2.9	(1.8)	1.3	(0.8)	0.6	(0.4)

TABLE 3

"+" means distance can be larger in certain atmospheric conditions

Page 351

4. Identify the primary hazard (e.g., health hazards), potential fire and explosion hazards, recommended protective clothing, and evacuation recommendations. (For the substance used in this exercise, you should find a Table 1 recommendation of initial isolation and protective action distances as well as isolation distances if the substance is involved in fire. Also, take note of the firefighting recommendation, guidelines for handling spills or leaks, and first-aid measures.)

5. If necessary, identify the isolation distance and the protective actions required for the chemical substance in the green section.

Reproduced from U.S. Department of Transportation, Transport Canada, and Secretariat of Communications and Transport of Mexico. 2020. *2020 Emergency Response Guidebook*. Pipeline and Hazardous Materials Safety Administration, U.S. Department of Transportation. www.phmsa.dot.gov/sites/phmsa.dot.gov/files/2021-01/ERG2020-WEB.pdf

Harmful Substances' Routes of Entry into the Human Body

Even though awareness-level personnel are not considered to be incident responders, it is important for them to understand the pathways by which chemical substances can enter the human body and potentially cause harm. For any chemical or other harmful substance to injure a person or animal, it must first get into or onto the body. Chemical substances can enter the human body in four ways (**FIGURE 13-33**):

- Inhalation: Through the lungs
- Absorption: By permeating the skin
- Ingestion: Via the gastrointestinal tract
- Injection: Through cuts or other breaches in the skin

The following sections discuss potential routes of entry into the human body of chemical or other harmful substances and methods used to protect against these agents.

Inhalation Exposure

An **inhalation exposure** occurs when harmful substances enter the body through the respiratory system. The lungs are a direct point of access to the bloodstream, so they can quickly transfer an airborne substance into the circulatory system and onward to the rest of the body. In addition, the lungs cannot be decontaminated, so any exposure will result in some type of harm that cannot be addressed in the same way as an exposure to skin.

The respiratory system is vulnerable to attack from a wide range of substances, from corrosive materials such as chlorine and ammonia to solvent vapors such as gasoline and acetone, superheated air from a fire, or any other material finding its way into the air. In addition to gases and vapors, small particles of dust, fiberglass insulation, asbestos, or soot from a fire can become lodged in sensitive lung tissue, causing substantial irritation.

Given these facts, it is imperative that personnel, if properly trained and equipped to do so, wear appropriate respiratory protection when operating in the presence of airborne contamination. Fortunately, most firefighters have ready access to an excellent form of respiratory protection—namely, the positive-pressure, open-circuit SCBA. This equipment is by far the single most important piece of PPE that firefighters have at their disposal. Sometimes firefighters may need to use other forms of respiratory protection depending on the specific respiratory hazard they are facing while performing a given task. For example, full-face and half-face air-purifying respirators (APRs) offer specific degrees of protection if the chemical hazard present is known and the appropriate filter canister is used (**FIGURE 13-34**).

APRs do not provide oxygen, however. Thus, if the oxygen content of the atmosphere in the work area is known or suspected to be abnormal, full- or half-face respirators are not a viable option. Per OSHA, any work

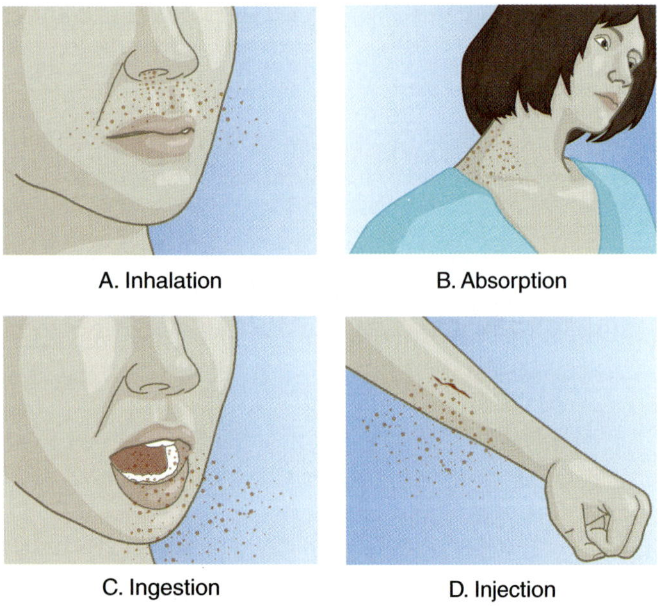

FIGURE 13-33 The four ways a chemical substance can enter the body.

© Jones & Bartlett Learning

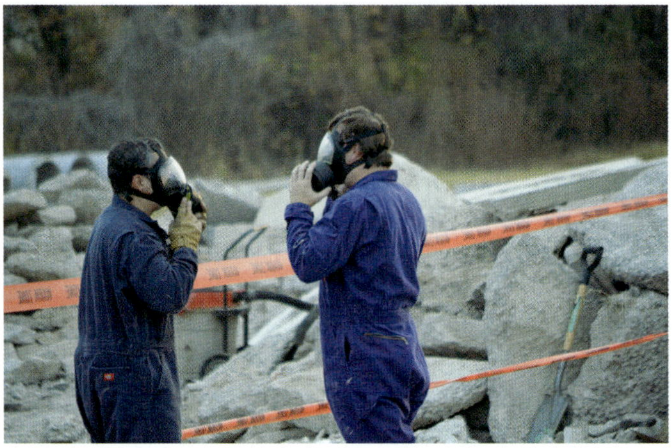

FIGURE 13-34 APRs offer protection against known and characterized airborne chemical hazards.

© Jones & Bartlett Learning. Courtesy of Rob Schnepp.

environment containing less than 19.5 percent oxygen is considered to be oxygen deficient and requires the use of SCBA or a supplied-air respirator.

Respirators are lighter than SCBA, are more comfortable to wear, and usually allow personnel to work for longer periods because they are not dependent on a limited source of breathing air. SCBAs certainly offer a higher level of respiratory protection, but in the right circumstances, APRs may represent a viable form of respiratory protection.

When considering protection against airborne contamination, it is important to understand the origin, concentration, and potential impact of the contamination relative to the oxygen levels in the area. In short, the appropriate type of respiratory protection is determined by looking at the overall situation, including the nature of the contaminant. In some cases, the anticipated particle size of the contamination may dictate the level of respiratory protection employed (**TABLE 13-2**).

Anthrax spores offer an excellent example to illustrate this point. Weaponized anthrax spores typically vary in size from 0.5 micron to 1 micron. Based on that size range, a typical full-face APR with a nuisance dust filter, or a surgical-type mask, would not offer sufficient protection against this hazard. Anyone operating in an area contaminated with anthrax should wear SCBA or, at a minimum, a full-face APR with P100 filtration (which filters out greater than 99 percent of 0.3-micron or larger particles).

Anthrax exposures also illustrate another important pair of terms and definitions that personnel should understand: **infectious** and **contagious**. Anthrax is a pathogenic microorganism capable of causing an illness (infectious). A person with an illness caused by an anthrax exposure, however, cannot pass the illness along to another person (contagious). In other words, anthrax is not contagious. Conversely, smallpox is both infectious and contagious, which is why this pathogen poses such a high risk in the event of an outbreak.

Consider this scenario: A container of gaseous helium is leaking inside a poorly ventilated storage room. Helium is nonflammable; the main threat it poses is the possibility of oxygen deficiency. In this case, SCBA is the appropriate respiratory protection based on the anticipated hazard. In many cases, when hazardous materials technicians respond to incidents, they use air-monitoring devices to characterize the work area prior to entry. When this step is taken, the choice of respiratory protection is based on more definitive information.

Particle size also determines where the inspired contamination will eventually end up. The larger particles that make up visible mists will be captured in the nose and upper airway, for example, whereas smaller particles may work their way deeper into the lungs (**TABLE 13-3**).

Respiratory protection is one of the most important PPE components. In all cases where airborne contamination is encountered, take the time to understand the nature of the threat, evaluate the respiratory protection available, and decide whether it provides adequate protection. When protecting the lungs, "good enough" is not an option.

Absorption Exposure

The skin is the largest organ in the body and is susceptible to the damage inflicted by many substances. In addition to serving as the body's protective shield against heat, light, and infection, the skin helps regulate body temperature, stores water and fat, and serves as a

TABLE 13-2 Particle Sizes of Common Types of Respiratory Hazards

Type of Particle	Size
Fume	< 1 micron
Smoke	≤ 1 micron
Dust	≥ 1 micron
Fog	> 40 microns
Mist	< 40 microns

© Jones & Bartlett Learning

TABLE 13-3 Location of Respiratory Trapping by Particle Size

Particle Size	Respiratory Area Affected
< 7 microns	Nose
5–7 microns	Larynx
3–5 microns	Trachea and bronchi
2–3 microns	Bronchi
1–2.5 microns	Respiratory bronchioles
0.5–1 micron	Alveoli

© Jones & Bartlett Learning

sensory center for painful and pleasant stimulation. Without this important organ, human beings would not be able to survive.

When discussing chemical exposures, however, absorption is not limited only to the skin. **Absorption exposure** occurs when substances travel through body tissues until they reach the bloodstream. The eyes, nose, mouth, and, to a certain degree, the intestinal tract are also part of the equation. The eyes, for example, can absorb a large amount of any liquids and vapors that encounter these sensitive tissues. This absorption is particularly problematic because the eyes connect directly to the optic nerve, which allows the chemical to follow a direct route to the brain and the central nervous system.

Although the skin functions as a shield for the body, that shield can be pierced by many chemicals. An aggressive solvent such as methylene chloride (found in paint stripper), for example, is readily absorbed through the skin. A secondary hazard associated with this chemical occurs when the body attempts to metabolize the substance after it is absorbed. A by-product of that metabolism reaction is carbon monoxide, a cellular asphyxiant. Asphyxiants are substances that prevent the body from using oxygen at the cellular level, thereby causing suffocation. With solvents such as methylene chloride, the initial chemical is broken down to form another substance that is potentially a greater health hazard than the original chemical because it can cause asphyxiation. Methylene chloride is also suspected to be a **carcinogen**—a human cancer-causing agent.

Absorption hazards are not just limited to solvents. Hydrofluoric acid, for example, poses a significant threat to life when it is absorbed through the skin. This unique corrosive can bind with certain substances in the body (predominantly calcium). Secondary health effects occurring after exposure can include muscular pain and potentially lethal cardiac arrhythmias.

In the field, personnel must constantly evaluate the possibility of chemical contact with their skin and eyes. Structural firefighting turnout gear provides little or no protection against liquid chemicals. Consult the *ERG* or an SDS for response guidance when deciding whether turnout gear is appropriate for the hazard you are facing. In the event the turnout gear does not offer adequate protection, you may be operating beyond the scope of awareness-level training.

Ingestion Exposure

In addition to absorption through the skin, chemicals can be brought into the body through **ingestion exposure**—that is, through the gastrointestinal tract. The water, nutrients, and vitamins the body requires are predominantly absorbed in this manner. For example, at a structure fire, firefighters generally have an opportunity to rotate out of the building for rest and refreshment. When they do so, they may not always take the time to wash up prior to eating or drinking. This leads to a high probability of spreading contamination from the hands to the food and subsequently to the intestinal tract. If you do not think about every situation where you might become exposed, you may put yourself in harm's way.

Injection Exposure

Chemicals brought into the body through open cuts and abrasions qualify as an **injection exposure**. To protect yourself from this route of exposure, begin by realizing when you are in a compromised state. Any cuts or open wounds should be addressed before reporting for duty. If they are significant, you may be excluded from operating in contaminated environments. Open wounds act as a direct portal to the bloodstream and subsequently to muscles, organs, and other body systems. If a chemical substance encounters this open portal, the health effects could be immediate and pronounced. Remember—intact skin is a good protective shield. Do not go into battle if your shield is not up to the task.

Managing Exposure to Harmful Substances

The damage that a hazardous material/WMD will inflict on human beings, animals, or the environment is a function of the physical and chemical properties of the released substance, as well as the conditions under which it was released and the duration of the exposure. Among other factors, characteristics such as the concentration of the material, the temperature of the material at the time of its release, and the pressure under which the substance was released affect both the release parameters and the potential health effects on those exposed to the material. Additionally, the age, sex, genetics, and underlying medical conditions of the exposed person will have some bearing on their outcome. Chemical exposures are complicated events because of the number of variables that may be present.

When it comes to rendering medical care to persons exposed to harmful substances, guidance can be found in NFPA 470, chapter 46, "Competencies for Hazardous Materials/WMD Basic Life Support (BLS) Responders," and chapter 47, "Competencies for Hazardous Materials/WMD Advanced Life Support (ALS) Responders." This standard outlines the basic set of hazardous materials response skills that all EMS responders, regardless of their scope of practice, need to work safely at a hazardous materials/WMD scene and deliver effective patient care. In most cases, that care will be performed in a safe area (cold zone) away from the hazard, after decontamination. There are very few circumstances where definitive advanced life support must be rendered in the hot zone. EMS responders, however, should understand the nature of the incident and look at the scene with a critical eye to pick up the clues that might assist with defining the nature of the exposure.

CASE STUDY

You Are the Support Person CONCLUSION

During an odor investigation in a light-industrial area of your district, you are directed by a citizen to an abandoned open-head 55-gallon steel drum just off the city street. The person points you to a local shopkeeper, who made the 911 call. The shopkeeper reports that the drum must have been illegally dropped off overnight and that he smelled an unusual odor when he walked by it. You notice a green label on the side of the drum reading "nonhazardous waste." You and the other members of the engine company are trained to the hazardous materials awareness level (based on NFPA 470). Even though you are not tasked to respond to a hazardous materials incident, you now find yourself on the scene of a potential hazardous materials incident.

1. **What initial actions would you take based on your level of training?**

   Answer: As awareness-level personnel, the on-scene engine company should maintain a safe distance from the drum, isolate the immediate area, and prevent civilians and other responders from entering the secured perimeter. Also, the company officer should request from the dispatch center that a higher level of hazardous materials responders be dispatched to the scene. These responders may be within the AHJ or requested from a neighboring agency.

2. **What does the label on the drum signify?**

   Answer: Although the label may identify the contents as a nonhazardous material, it is not a 100 percent accurate indicator. It is not uncommon for illegally dumped hazardous materials to be in improperly stored or labeled containers. Hazardous materials technicians or specialists should be called in to sample and identify the contents.

3. **How would you go about obtaining more information on the potential contents and potential hazards of the contents of the drum? What notifications would you make?**

   Answer: There is no additional information to be gained from the label, but the responders should take note of the fact the material is in a steel drum. This may indicate that the material inside is not corrosive, provided there are no visible signs of degradation of the steel or obvious corrosion. Responders could also make an effort to question other residents or business owners to determine if there are any additional clues as to the origin of the drum. Responders could also investigate the surrounding streets to see if public or private security cameras captured any information on suspicious vehicles or persons in the area prior to the discovery of the drum by the shopkeeper.

WRAP-UP

SUMMARY

- A careful scene size-up is especially important during hazardous materials incidents. Personnel and responders must interpret the available clues and weave them together to identify hazardous materials, make informed decisions, and operate safely.
- Containers of hazardous materials include cylinders (containers with a circular cross-section designed to store liquids or gases under high pressure), steel or polyethylene drums (barrel-like vessels used to store a variety of substances, including corrosives, flammable liquids, and grease), bags, railroad tank cars, plastic buckets, aboveground and underground storage tanks, cargo tanks, and pipelines.
- The presence of markings on buildings, packages, boxes, and containers can help personnel identify a released chemical. When used correctly, marking systems and information sheets indicate the presence of a hazardous material and provide clues about the substance.
- The four routes of entry that harmful substances can take in the human body are inhalation (through the respiratory system), absorption (through the skin), ingestion (through the gastrointestinal tract), and injection (through open cuts and abrasions).

KEY TERMS

absorption exposure Exposure to a hazardous material in which substances travel through body tissues until they reach the bloodstream.

air bill The shipping papers on an airplane.

asphyxiant A material that prevents the body from using oxygen at the cellular level, causing the victim to suffocate.

bill of lading The shipping papers used for transport of chemicals over roads and highways; also called *freight bill.*

BLEVE (boiling liquid/expanding vapor explosion) An explosion that occurs when pressurized liquefied materials, such as propane or butane, inside a closed vessel are exposed to a source of high heat.

bung An opening on top of a closed-head drum that is typically sealed with a threaded cap.

carboy A glass, plastic, or steel storage container, ranging in volume from 5 to 15 gallons.

carcinogen A human cancer-causing agent.

Chemical Abstracts Service (CAS) A division of the American Chemical Society that provides hazardous materials personnel and responders with access to the CAS Registry, an enormous collection of chemical substance information.

closed-head drum A drum with a lid that is permanently attached to the drum; the lid has one or more small openings called *bungs.*

consist A list of the contents of every car on a train; also called *train list.*

contagious Capable of transmitting a disease.

container A receptacle, piping, or pipeline used for storing or transporting material of any kind. (NFPA 470)

cryogen See *cryogenic liquid.*

cryogenic liquid A fluid, such as liquid helium, liquid nitrogen, or liquid argon, that has a boiling point lower than –130°F (–90°C) at an absolute pressure of 14.7 pounds per square inch (psi; 101.3 kilopascals [kPa]); also called *cryogen.*

cylinder A container that has a circular cross-section and is designed to store liquids or gases under pressure higher than 40 pounds per square inch (psi; 276 kilopascals [kPa]); this definition does not include portable tanks, multiunit tanks, car tanks, cargo tanks, or tank cars. (NFPA 1)

dangerous cargo manifest The shipping papers on a marine vessel.

Dewar container A cylinder container designed to preserve the temperature of cryogenic liquids.

drum A barrel-like storage vessel constructed of low-carbon steel, polyethylene, cardboard, stainless steel, nickel, or other materials that is used to store a wide variety of substances, including food-grade materials, corrosives, flammable liquids, and grease.

Emergency Response Guidebook (ERG) The reference book, written in plain language, to guide emergency responders in their initial actions at the incident scene, specifically the *Emergency Response Guidebook* from the U.S. Department of Transportation, Transport Canada, and the Secretariat of Transport and Communications of Mexico used to guide personnel and responders in their initial actions at the incident scene. (NFPA 470)

freight bill See *bill of lading.*

hazard control zones The areas at hazardous materials/WMD incidents within an established perimeter that are designated based on safety and the degree of hazard.

Hazardous Materials Information System (HMIS) A color-coded marking system that identifies the hazard level of chemicals in a container so that personnel can work safely around chemicals.

high-temperature–protective clothing Protective clothing designed to protect the wearer from short-term high-temperature exposures, such as protective clothing ensembles used by aircraft firefighters to fight fires in flammable liquids. (NFPA 470)

incident action plan (IAP) An oral or written plan approved by the incident commander (IC) containing incident objectives reflecting the overall strategy for managing an incident for a specific time frame and target location. (NFPA 470)

infectious Capable of causing an illness by entry of a pathogenic microorganism.

ingestion exposure Exposure to a hazardous material from swallowing the substance.

inhalation exposure Exposure to a hazardous material from breathing the substance into the lungs.

injection exposure Exposure to a hazardous material from the substance entering the body through cuts or other breaches in the skin.

label A smaller version of a placard (4 inches [10 centimeters] on each side) that is required to be placed on the four sides of individual boxes and smaller packages being transported.

NFPA 704 hazard identification system A hazardous materials marking system designed for fixed-facility use. It uses a diamond-shaped symbol of any size, which is itself broken into four smaller diamonds, each representing a particular property or characteristic of the material.

open-head drum A drum with a lid that is secured by a bolted clasp-type ring, which circles the entire head of the drum.

packaging group designation A label that describes a substance according to the degree of hazards it presents.

pipeline A length of pipe including pumps, valves, flanges, control devices, strainers, and/or similar equipment for conveying fluids. (NFPA 470)

pipeline right-of-way An area, patch, or roadway that extends a certain number of feet on either side of a pipeline and may contain warning and informational signs about hazardous materials carried in the pipeline.

placard A diamond-shaped indicator (10¾ inches [27 centimeters] on each side) that is required to be placed on all four sides of highway transport vehicles, railroad tank cars, and other forms of transportation carrying hazardous materials to identify the substance being transported.

risk-based response process A systematic process based on the facts, science, and circumstances of the incident, by which responders analyze a problem involving hazardous materials/weapons of mass destruction (WMD) to assess the hazards and consequences, develop an incident action plan (IAP), and evaluate the effectiveness of the plan. (NFPA 470)

safety data sheet (SDS) Formatted information, provided by chemical manufacturers and distributors of hazardous products, about a chemical's composition, physical and chemical properties, health and safety hazards, emergency response, and waste disposal of the material. (NFPA 470)

shipping papers A shipping order, bill of lading, manifest, or other shipping document serving a similar purpose and containing the information required by regulations of the U.S. Department of Transportation (DOT). (NFPA 498)

toxic inhalation hazards (TIHs) Gases or volatile liquids that are extremely toxic to humans.

train list See *consist.*

U.S. Department of Transportation (DOT) marking system A system of labels and placards that is used when materials are being transported from one location to another in the United States and Canada by Transport Canada.

vent pipes Inverted J-shaped tubes that allow for pressure relief or natural venting of a pipeline for maintenance and repairs.

waybill Shipping papers for railroad transport.

REVIEW QUESTIONS

1. What is one of the first actions awareness-level personnel should take at a hazardous materials incident?
2. Define *drum*, *carboy*, and *cylinder*.
3. Describe SDSs, the NFPA 704 marking system, HMIS, the military hazardous materials/WMD marking system, pipeline warning signs, and the U.S. DOT marking system.
4. How might a responder use the U.S. DOT marking system?
5. What is the *ERG*?
6. What are the four ways a chemical substance can enter the human body?
7. At a hazardous materials incident, where is patient care usually provided?

DISCUSSION QUESTIONS

1. How can awareness-level personnel and operations-level responders use the various marking systems?
2. What are the three parts of the SIN mnemonic, and how are they used?
3. Why is using respiratory protection so important?

REFERENCES

National Fire Protection Association. 2017. *NFPA 498: Standard for Safe Havens and Interchange Lots for Vehicles Transporting Explosives*. 2018 Edition. Quincy, MA: National Fire Protection Association.

National Fire Protection Association. 2023. *NFPA 1: Fire Code*. 2024 Edition. Quincy, MA: National Fire Protection Association.

National Fire Protection Association. 2020. *NFPA 704: Standard System for the Identification of the Hazards of Materials for Emergency Response*. 2022 Edition. Quincy, MA: National Fire Protection Association.

National Fire Protection Association. 2021. *NFPA 470: Hazardous Materials/Weapons of Mass Destruction (WMD) Standard for Responders*. 2022 Edition. Quincy, MA: National Fire Protection Association.

U.S. Department of Transportation, Transport Canada, and Secretariat of Communications and Transport of Mexico. 2020. *2020 Emergency Response Guidebook*. Washington, DC: Pipeline and Hazardous Materials Safety Administration, U.S. Department of Transportation. www.phmsa.dot.gov/sites/phmsa.dot.gov/files/2021-01/ERG2020-WEB.pdf.

Chapter Opener: © Eric Scruggs

Appendix A

NFPA 1010 and NFPA 470 Correlation Grids for Support Person

NFPA 1010: *Standard on Professional Qualifications for Firefighters, 2024 Edition*		
Job Performance Requirement	**Chapter(s)**	**Knowledge and Skills Objective(s)**
Chapter 5: Support Person (NFPA 1001)		
5.1: General		
5.1.1	1, 2, 5	▪ Describe the organization of the fire department. (p. 10–11) ▪ Explain the basic structure of the chain of command within the fire department. (p. 13) ▪ Outline the roles and responsibilities of the support person in the organization. (p. 16) ▪ Describe the mission of the fire service. (p. 2) ▪ Explain the concept of governance, and describe how regulations, standards, policies, and standard operating procedures affect it. (p. 8) ▪ Explain how fires are spread by conduction, convection, and radiation. (p. 141–143) ▪ Understand the basic principles of fire dynamics. (p. 147–150) ▪ Describe the purpose of the fire department's employee assistance program. (p. 37) ▪ Identify the signs and symptoms associated with behavioral and emotional distress. (p. 34–36) ▪ Explain the importance of physical fitness and a healthy lifestyle to the performance of the duties of a support person. (p. 14–15, 30–31)
5.1.2	3	▪ Demonstrate the ability to don and doff a protective ensemble. (p. 67–70) ▪ Perform field reduction of contaminants from a protective ensemble. (p. 72–73) ▪ Prepare the protective ensemble and equipment for reuse. (p. 70) ▪ Locate information in departmental documents, standards, and code materials. (p. 8)
5.2: Communications		
5.2.1	4	▪ Describe the procedures for reporting an emergency. (p. 119) ▪ Describe procedures for handling emergency calls. (p. 119) ▪ Identify other modes of fire service communication. (p. 114–116) ▪ Identify radio departmental procedures and codes for using fire department radios. (p. 125) ▪ Explain methods of receiving emergency and nonemergency fire department communications. (p. 113–116) ▪ Explain procedures for transmitting the emergency information to a dispatch center. (p. 119) ▪ Explain the importance of following departmental standard operating procedures for receiving and processing communications. (p. 119)

Job Performance Requirement	Chapter(s)	Knowledge and Skills Objective(s)
		▪ Identify the information to be obtained when taking a report of an emergency to enable necessary assistance to be dispatched. (p. 119) ▪ Receive a phone call and obtain, route, and document information according to department procedures. (p. 120) ▪ Demonstrate the ability to operate fire department communications equipment and technology. (p. 126) ▪ Demonstrate the relay of information to the dispatch center. (p. 120) ▪ Demonstrate the ability to record information according to department standard operating procedures. (p. 120)
5.2.2	4	▪ Describe the basic procedures and etiquette of effective radio communications. (p. 125) ▪ Recognize routine traffic, emergency traffic, and emergency evacuation signals. (p. 128) ▪ Outline the information provided in size-up and progress reports. (p. 127) ▪ Describe when to use plain language and how ten-codes are implemented in communications. (p. 125–126) ▪ Demonstrate how to send and receive messages over the fire department radio. (p. 126) ▪ Determine if a radio communication is routine or emergency traffic. (p. 126)
5.3: Incident Support Operations		
5.3.1	2, 3	▪ List the respiratory hazards posed by smoke and fire. (p. 74–77) ▪ List the conditions that require respiratory protection. (p. 73–74) ▪ Describe the types of breathing apparatus. (p. 78–79) ▪ Describe the differences between open- and closed-circuit breathing apparatus. (p. 80) ▪ Describe the limitations of self-contained breathing apparatus (SCBA). (p. 80–81) ▪ Describe the physical and psychological limitations of an SCBA user. (p. 81) ▪ Demonstrate the ability to identify potentially hazardous atmospheres and avoid them. (p. 48)
5.3.2	2	▪ Describe how to safely mount an apparatus. (p. 38–40) ▪ Describe how to safely ride on a fire apparatus. (p. 38–40) ▪ Describe how to safely dismount an apparatus. (p. 38–40) ▪ Describe hazards and safety measures associated with riding apparatus. (p. 38–40) ▪ List prohibited practices when riding in a fire apparatus. (p. 38–40) ▪ Identify types of department-issued personal protective equipment (PPE) and their usage. (p. 40–41) ▪ Demonstrate the ability to use each piece of provided safety equipment. (p. 48)
5.3.3	2, 3	▪ Describe how to manage traffic safely at an emergency scene. (p. 40–41) ▪ Explain considerations for hazard and scene control. (p. 47–48) ▪ List the common hazards at an emergency scene. (p. 47–48) ▪ Describe measures firefighters follow to ensure electrical safety at an emergency scene. (p. 49–50) ▪ Describe measures firefighters follow to ensure safe operation at an emergency scene. (p. 45–46) ▪ List protective equipment available for use when operating at an emergency scene. (p. 60–66)

Job Performance Requirement	Chapter(s)	Knowledge and Skills Objective(s)
		▪ Describe how to safely dismount an apparatus. (p. 40) ▪ Demonstrate how to properly wear a protective ensemble. (p. 67–68) ▪ Demonstrate how to safely deploy proper scene control devices to establish a protected work area. (p. 48) ▪ Operate within a protected work area as directed. (p. 48) ▪ Demonstrate how to mount an apparatus safely. (p. 39) ▪ Demonstrate how to dismount from an apparatus safely. (p. 40)
5.3.4	9, 10, 11	▪ Describe the equipment and procedures that are used to access static sources of water. (p. 291–295) ▪ Describe the two types of suction hose. (p. 315) ▪ Describe the advantages of a portable tank system. (p. 294–295) ▪ Describe the characteristics of a mobile water supply apparatus. (p. 295) ▪ List examples of suitable static water supply sources. (p. 291) ▪ Describe types of supply hose. (p. 314–315) ▪ Describe the various sizes of fire hose and how they are used. (p. 304–305) ▪ Describe the common techniques used to carry and advance supply hose. (p. 349–351) ▪ Describe the procedures used to connect supply lines to a fire hydrant. (p. 334) ▪ Describe the characteristics of dry-barrel hydrants. (p. 282–283) ▪ Describe the characteristics of wet-barrel hydrants. (p. 283–284) ▪ Properly demonstrate a forward hose lay. (p. 338–340) ▪ Properly demonstrate a reverse hose lay. (p. 341–342) ▪ Properly demonstrate a split hose lay. (p. 343) ▪ Attach a soft sleeve hose to a fire hydrant. (p. 336) ▪ Set up a portable tank and equipment necessary to use it. (p. 296) ▪ Demonstrate the safe operation of a fire hydrant. (p. 286–287, 289) ▪ Demonstrate how to safely shut down a fire hydrant. (p. 288–289) ▪ Assist the pump driver/operator with drafting. (p. 293–294) ▪ Perform the connection and proper placement of an intake hose for drafting operations. (p. 293–294)
5.3.5	5, 6	▪ Define Class A fires and the risks associated with them. (p. 144) ▪ Define Class B fires and the risks associated with them. (p. 144) ▪ Define Class C fires and the risks associated with them. (p. 144) ▪ Define Class D fires and the risks associated with them. (p. 145) ▪ Define Class K fires and the risks associated with them. (p. 145) ▪ Describe the four methods of extinguishing fires. (p. 143–144, 166) ▪ Describe the basic steps of fire extinguisher operation. (p. 181–184) ▪ Describe the three risk classifications for area hazards. (p. 168–170) ▪ Describe the types of agents and operating systems used in fire extinguishers. (p. 171–181) ▪ Explain the classification and rating system for fire extinguishers. (p. 166–168) ▪ Explain the labeling system for fire extinguishers. (p. 167–168) ▪ Explain the considerations used when selecting the proper class of fire extinguisher. (p. 165–166)

Job Performance Requirement	Chapter(s)	Knowledge and Skills Objective(s)
		■ State the primary purposes of fire extinguishers. (p. 162) ■ Demonstrate the safe extinguishment of a Class A fire with a multipurpose dry-chemical fire extinguisher. (p. 186) ■ Demonstrate the safe extinguishment of a Class A fire with a stored-pressure water-type fire extinguisher. (p. 186) ■ Demonstrate the safe extinguishment of a Class B flammable liquid fire with a dry-chemical fire extinguisher. (p. 187) ■ Select an appropriate extinguisher based on the size and type of fire, and transport the fire extinguisher to the location of the fire. (p. 183–184) ■ Operate a carbon dioxide fire extinguisher. (p. 188)
5.3.6	7	■ Describe how to operate lighting equipment to light exterior and interior scenes. (p. 211) ■ Describe the equipment used to illuminate an emergency scene. (p. 210–212) ■ Describe the safety precautions to take when working with lighting equipment. (p. 212–213) ■ Describe the types of lights used to illuminate exterior and interior scenes. (p. 211–212) ■ Demonstrate proper operation of power and lighting equipment, cords, connectors, and ground-fault interrupter (GFI) devices. (p. 213) ■ Demonstrate where to properly locate lights to illuminate an emergency scene. (p. 213)
5.3.7	2	■ Describe the hazards and safety concerns associated with building utilities. (p. 48–51) ■ List safety equipment used when working around building utilities. (p. 49–51) ■ Describe the methods used to turn off building utilities. (p. 49–51) ■ Describe when gas service should be shut off. (p. 50–51) ■ Describe when the electrical service should be shut off. (p. 49–50) ■ Describe when water service should be shut off. (p. 51) ■ Demonstrate how to locate and safely shut off electric utilities. (p. 50) ■ Demonstrate how to locate and safely shut off gas utilities. (p. 52)
5.3.8	8	■ List the common types of knots that are used in the fire service. (p. 244–245) ■ List the terminology used to describe the bends in rope that are formed when a knot is tied. (p. 244–245) ■ List the terminology used to describe the parts of a rope when tying knots. (p. 244–245) ■ Describe the four primary types of fire service rope. (p. 233–235) ■ Describe the characteristics of utility ropes. (p. 235, 237) ■ List the two types of life safety rope. (p. 233) ■ Describe the characteristics of webbing. (p. 235–236) ■ Describe how to preserve rope strength and integrity. (p. 239) ■ List the four components of the rope maintenance formula. (p. 239) ■ Describe the importance of properly maintaining tools and equipment. (p. 239) ■ Describe the reasons for placing rope out of service. (p. 241–242) ■ Describe how to store rope properly. (p. 242–244) ■ Describe how to clean rope. (p. 239) ■ Describe how to inspect rope. (p. 241–242) ■ Describe how to keep an accurate rope record. (p. 241–242)

Job Performance Requirement	Chapter(s)	Knowledge and Skills Objective(s)
		▪ Describe the methods used to hoist a tool. (p. 261–268) ▪ Describe the characteristics of a clove hitch. (p. 249) ▪ Describe the characteristics of a safety knot. (p. 245) ▪ Describe the characteristics of a figure eight knot. (p. 246) ▪ Describe the characteristics of a bowline knot. (p. 254) ▪ Describe the characteristics of a half hitch. (p. 248) ▪ Describe the characteristics of a sheet bend. (p. 257) ▪ Describe the characteristics of a water knot. (p. 259) ▪ Demonstrate how to properly tie a bowline. (p. 256) ▪ Demonstrate how to properly tie a clove hitch around an object. (p. 251–252) ▪ Demonstrate how to properly tie a clove hitch in the open. (p. 250) ▪ Demonstrate how to properly tie a figure eight bend. (p. 259–260) ▪ Demonstrate how to properly tie a figure eight follow-through. (p. 254) ▪ Demonstrate how to properly tie a figure eight knot. (p. 247) ▪ Demonstrate how to properly tie a figure eight on a bight. (p. 253) ▪ Demonstrate how to properly tie a half hitch. (p. 248–249) ▪ Demonstrate how to properly tie a safety knot. (p. 246) ▪ Demonstrate how to properly tie a sheet or Becket bend. (p. 257–258) ▪ Demonstrate how to properly tie a water knot. (p. 260–261) ▪ Demonstrate how to hoist a charged hose line. (p. 267–268) ▪ Demonstrate how to hoist a ladder. (p. 265–266) ▪ Demonstrate how to hoist a pike pole. (p. 264) ▪ Demonstrate how to hoist an axe. (p. 262–263) ▪ Demonstrate how to hoist an exhaust fan or power tool. (p. 270–271) ▪ Demonstrate how to hoist an uncharged hose line. (p. 269–270)
5.4: Rescue Operations		Not applicable for support person
5.5: Preparedness and Maintenance		
5.5.1	3	▪ Explain the procedures for refilling SCBA air cylinders. (p. 95–97) ▪ Explain the procedures for returning an SCBA air cylinder to service. (p. 87–90) ▪ List and describe the major components of SCBA cylinders. (p. 82–83) ▪ Refill an SCBA air cylinder from a compressor or a cascade system. (p. 98) ▪ Replace an SCBA air cylinder on another firefighter. (p. 95–96) ▪ Properly exchange/replace an SCBA air cylinder from a rack or apparatus. (p. 93–94)
5.5.2	3, 7, 8, 10	▪ Describe how to clean rope. (p. 239) ▪ Describe the supplies needed to clean and inspect tools and equipment. (p. 220–221) ▪ Describe how to clean and maintain tools and equipment. (p. 220–221) ▪ Identify procedures for documenting maintenance performed on tools and equipment. (p. 219–220)

Job Performance Requirement	Chapter(s)	Knowledge and Skills Objective(s)
		▪ Identify procedures, including reporting requirements, for removing a damaged tool from service. (p. 219–220) ▪ Describe why it is important for you to know where tools are stored. (p. 197) ▪ Explain the importance of replacing tools in their assigned locations. (p. 219–220) ▪ Demonstrate how to clean and inspect an SCBA. (p. 88–89, 90–92, 99–100) ▪ Demonstrate how to clean and inspect hand tools. (p. 221) ▪ Demonstrate how to clean and inspect fire department ropes. (p. 240, 242) ▪ Demonstrate how to clean and inspect ladders. (p. 224–225) ▪ Demonstrate how to clean and maintain hose. (p. 325–326) ▪ Demonstrate how to place a life safety rope in a rope bag. (p. 243) ▪ Properly document equipment maintenance following established guidelines. (p. 225, 242, 326)
5.5.3	10, 11	▪ List the common types of hose damage. (p. 319–323) ▪ Describe the importance of a hose inspection. (p. 325) ▪ Describe the procedure for documenting a defective hose and removing it from service. (p. 325) ▪ List the common methods for cleaning and drying hose. (p. 323–325) ▪ List the common types of hose rolls used to organize supply hose. (p. 316–318) ▪ List the common types of hose rolls used with attack hose. (p. 316–318) ▪ Describe the common techniques used to load supply hose. (p. 342–347) ▪ Describe the types of loads used to organize attack hose. (p. 352–356) ▪ Properly mark a defective section of hose and remove it from service. (p. 326) ▪ Properly demonstrate a flat hose load. (p. 345) ▪ Properly demonstrate a horseshoe hose load. (p. 346–347) ▪ Properly demonstrate a minuteman hose load. (p. 355–356) ▪ Properly demonstrate a preconnected flat hose load. (p. 353–354) ▪ Properly demonstrate a triple-layer hose load. (p. 357–358) ▪ Properly demonstrate an accordion hose load. (p. 348–349) ▪ Properly demonstrate a self-locking twin-doughnut hose roll. (p. 321–322) ▪ Properly demonstrate a single-doughnut hose roll. (p. 318–319) ▪ Properly demonstrate a straight hose roll. (p. 317) ▪ Properly demonstrate a twin-doughnut hose roll. (p. 320) ▪ Demonstrate the proper techniques for cleaning and drying different types of hose. (p. 325–326) ▪ Demonstrate replacing the swivel gasket on a fire hose. (p. 308–309) ▪ Drain a fire hose. (p. 359)

NFPA 470: ***Hazardous Materials/Weapons of Mass Destruction (WMD) Standard for Responders, 2022 Edition***		
Job Performance Requirement	**Chapter(s)**	**Knowledge and Skills Objective(s)**
Chapter 5: Awareness Level (NFPA 1072)		
5.1: General		
5.1.1	12	▪ Describe the different levels of hazardous materials training: awareness, operations, technician, specialist, and incident commander. (p. 369–374)
5.1.2	12	▪ Describe the different levels of hazardous materials training: awareness, operations, technician, specialist, and incident commander. (p. 369–374)
5.1.3: General Skill Requirements	12, 13	▪ Identify the location of both the emergency response plan and/or standard operating procedures. (p. 367, 380) ▪ Describe the different levels of hazardous materials training: awareness, operations, technician, specialist, and incident commander. (p. 369–374)
5.1.4 General Skills Requirements. (Reserved)		
5.2 *Recognition and Identification.		
5.2.1	12, 13	▪ Define the terms *hazardous materials* (or *dangerous goods*, in Canada) and *weapons of mass destruction*. (p. 367) ▪ Describe how to approach a scene size-up when potentially hazardous materials are involved. (p. 380–384) ▪ Identify and describe the types of containers that are often used to contain hazardous materials. (p. 384–387) ▪ Describe the purpose and types of various transportation and facility markings for hazardous materials. (p. 387) ▪ Identify and describe the four routes of entry harmful substances take in the human body. (p. 408–410) ▪ Use the *Emergency Response Guidebook (ERG)*. (p. 400–403)
5.2.2		
5.3 *Initiate Protective Actions.		
5.3.1	13	▪ Describe how to approach a scene size-up when potentially hazardous materials are involved. (p. 380–384) ▪ Describe the purpose and types of various transportation and facility markings for hazardous materials. (p. 387) ▪ Identify and describe the four routes of entry harmful substances take in the human body. (p. 408–410) ▪ Use the *Emergency Response Guidebook (ERG)*. (p. 400–403)
5.4 Notification.		
5.4.1	13	▪ Describe how to approach a scene size-up when potentially hazardous materials are involved. (p. 380–384)

Appendix B

Emergency Medical Care

KNOWLEDGE OBJECTIVES

After studying this appendix, you will be able to:

- Identify the different levels of medical providers.
- Describe how standard precautions reduce the risk of infectious diseases.
- Explain the importance of being able to identify a victim who is not responsive and not breathing adequately.
- Describe how to perform hands-only cardiopulmonary resuscitation (CPR) on a victim.
- Explain the reason for using an automated external defibrillator (AED).
- Describe how to position a victim in the recovery position to maintain an open airway.
- Identify severe bleeding and select an appropriate method to manage bleeding.
- Describe the basic management of shock.

SKILLS OBJECTIVES

After studying this appendix, you will be able to perform the following skills:

- Demonstrate how to prevent infection.
- Demonstrate how to perform hands-only CPR.
- Demonstrate how to use an AED.
- Demonstrate placing a victim in the recovery position.
- Demonstrate how to control bleeding.

CASE STUDY

You Are the Support Person

Your crew is dispatched to a motor vehicle accident involving one car that crashed into a telephone pole. You arrive to find one car with heavy front-end damage still up against the telephone pole. Your lieutenant informs you that the police have secured the roadway, and the pole appears stable. Your lieutenant directs you to check on the victims in the front of the car. You find the driver slumped over toward the passenger. The driver does not respond to you when you approach. The passenger is sobbing and has an arm **laceration** that is bleeding heavily. You can see blood inside the car on the passenger's side.

1. How will you protect yourself from contracting an infectious disease?
2. What is your highest priority for the driver?
3. Which concerns do you have for the passenger?

Introduction

This appendix covers the basic emergency medical care skills included in NFPA 1010, *Standard for Firefighter Professional Qualifications, 2024 Edition.* Because first responders encounter ill or injured victims on the job, a basic knowledge of these skills is essential. This chapter covers infection control, hands-only **cardiopulmonary resuscitation (CPR)**, controlling bleeding, and the management of shock. It emphasizes the importance of ensuring your safety and the safety of other rescuers, victims, and bystanders at incidents where emergency medical services (EMS) are required. These incidents may include calls where the fire department is activated to perform or assist with EMS duties, as well as other emergency calls where firefighters, other responders, or members of the community may become sick or injured.

As a support person, it is important for you to follow any state regulations and local ordinances that require that certain EMS tools and techniques be used only by those with specific certifications or licenses. CPR certification programs are available for professionals, such as first responders and healthcare providers, and laypeople. These courses, along with first-aid certification programs, allow fire service personnel to learn and practice using tools and techniques beyond what is presented in this appendix.

In most states, people who work in emergency medical care are categorized into four licensure levels, each with a different role and focus as well as different training requirements. The levels are emergency medical responder (EMR), emergency medical technician (EMT), advanced EMT (AEMT), and paramedic. An EMR has basic emergency care and operations training to manage the emergency scene and initiate immediate life-saving care before the ambulance arrives. EMRs may also perform roles under the direction of providers with more advanced training. An EMT has additional depth and breadth of training in basic emergency care and transportation of sick and injured victims. Although not always the first to arrive, EMTs most commonly focus on initial scene stabilization and fundamental emergency care. EMTs are the primary link between the emergency scene and the healthcare system. An AEMT has training in specific aspects of **advanced life support (ALS)**, such as intravenous (IV) therapy, advanced airway management, and the administration of certain emergency medications. AEMTs' primary focus is on more advanced assessment techniques and emergency interventions, as well as on helping to link more fundamental levels of care with more advanced levels. Paramedics have the greatest breadth and depth of training among emergency care providers, focusing on ALS assessment and diagnostic and treatment tools and techniques, such as interpreting heart rhythms, placing breathing tubes, administering emergency medication, and more.

Infectious Diseases and Standard Precautions

The COVID-19 pandemic; the **acquired immunodeficiency syndrome (AIDS)** epidemic; and growing concerns about hepatitis, influenza, tuberculosis (TB), and methicillin-resistant *Staphylococcus aureus* (MRSA) have increased awareness of infectious (communicable) diseases. As a support person, it is essential for you to have a basic understanding of the most common infectious diseases. This knowledge protects you from unnecessary exposure to these diseases. It also ensures that you do not become unduly alarmed when encountering persons with these diseases. Infectious diseases can be contracted in several ways, such as by eating contaminated food or coming into contact with infected blood or contaminated airborne droplets. Exposure can take place through a small cut, via direct contact with a mucous membrane, or through unprotected sex.

The three most common routes for transmission of infectious diseases that you will be exposed to during your work as a support person are contact with infected blood or other body fluids that may transmit infection or carry infected blood, referred to as **other potentially infectious materials (OPIM)**; breathing infectious airborne droplets; and direct contact with infectious agents. Disease-causing agents spread through contact with infected blood are called bloodborne **pathogens**. **Human immunodeficiency virus (HIV)**, the virus that causes AIDS, and the viruses that cause hepatitis B (**hepatitis B virus [HBV]**) and hepatitis C (**hepatitis C virus [HCV]**) are bloodborne pathogens. Other infectious diseases are spread through inhalation of airborne pathogens—COVID-19, influenza, TB, whooping cough, and **severe acute respiratory syndrome (SARS)** belong to this group. A third group of infectious diseases is spread by direct contact. One example is MRSA, an infection spread by direct contact with the pathogen on a victim's skin or with contaminated clothing or towels.

Bloodborne Pathogens

HIV is transmitted by contact with infected blood, semen, or vaginal secretions. There is no scientific documentation that the virus can be transmitted by contact with sweat, saliva, tears, sputum, urine, feces, vomitus, or nasal secretions unless these fluids contain visible signs of blood. Exposure can take place in the following ways:

- The victim's blood is splashed or sprayed into your eyes, nose, or mouth or an open sore or cut.
- You have blood from the infected victim on your hands and then touch your own eyes, nose, mouth, or an open sore or cut.
- A needle that was used to inject the victim breaks your skin.
- Broken glass from a motor vehicle collision or other incident that is covered with blood from an infected victim penetrates your glove and skin.

Some people who are infected with HIV do not know they are infected. Others who are infected do not show any symptoms. This is why healthcare workers are required to wear certain types of personal protective equipment (PPE) whenever they are likely to come into contact with any victim. In addition to wearing appropriate PPE, whenever you are on the job, cover any open wounds you have to reduce the chance of infection.

Like HIV, hepatitis B is spread by direct contact with infected blood, although it is far more contagious than HIV. Follow the standard precautions described in the following section to reduce your chance of contracting hepatitis B. Check with your medical director about receiving injections of hepatitis vaccine to protect you against this infection. This vaccine should be made available to you for free. Meningitis and syphilis are two other diseases that can be spread by contact with contaminated blood.

SAFETY TIP

The Centers for Disease Control and Prevention (CDC) recommends that all healthcare workers adhere to the following universal precautions:

- Always wear approved nonlatex gloves when handling victims, and change gloves after contact with each victim. Wash your hands immediately after removing gloves. Leather gloves are not considered safe because leather is porous and traps fluids, so it cannot be adequately decontaminated.
- Always wear protective eyewear and a face mask or a face shield when you anticipate that blood or other body fluids might splatter. Wear a gown or apron if you anticipate splashes of blood or other body fluids, such as those that occur with childbirth and major trauma.
- Wash your hands and other skin surfaces immediately and thoroughly if they become contaminated with blood or other body fluids. Change soiled clothes and wash exposed skin thoroughly.
- Place used needles directly in a puncture-resistant container designed for sharps. Do not recap, cut, or bend a used needle.
- Use a face shield, pocket mask, or other airway adjunct if the victim needs resuscitation.

Airborne Pathogens

COVID-19, seasonal influenza, and TB are contagious diseases spread by droplets from the respiratory system. These viruses spread through the air when an infected person coughs or sneezes. COVID-19 is a potentially life-threatening infectious disease caused by the severe acute respiratory syndrome coronavirus 2 (SARS-CoV-2) virus. COVID-19 began spreading in late 2019, quickly becoming a worldwide pandemic. The most common symptoms of COVID-19 include respiratory symptoms, such as those caused by a severe cold or the flu. However, a wide range of symptoms have been reported, including some that can lead to severe illness and death. COVID-19 can be spread by people who do not show symptoms.

Influenza is caused by viruses that change over time. When certain conditions are right, a new strain of the influenza virus may cause many people in a community to become sick. The H1N1 strain of influenza (swine flu) caused concern because few people have immunity to this strain of virus. When a new strain of an influenza virus develops, your department must follow the latest CDC recommendations for your protection.

TB, a bacterial disease typically affecting the lungs, is often difficult to distinguish from other conditions. However, those victims who pose the highest risk almost invariably have a cough. This disease is dangerous because drug-resistant strains of TB have evolved. You should have a skin test for TB every year.

To minimize your exposure when encountering a victim with an airborne disease, wear a face mask or an N95 high-efficiency particulate air (HEPA) respirator if possible (**FIGURE B-1**). These masks were designed to reduce the transmission of airborne diseases.

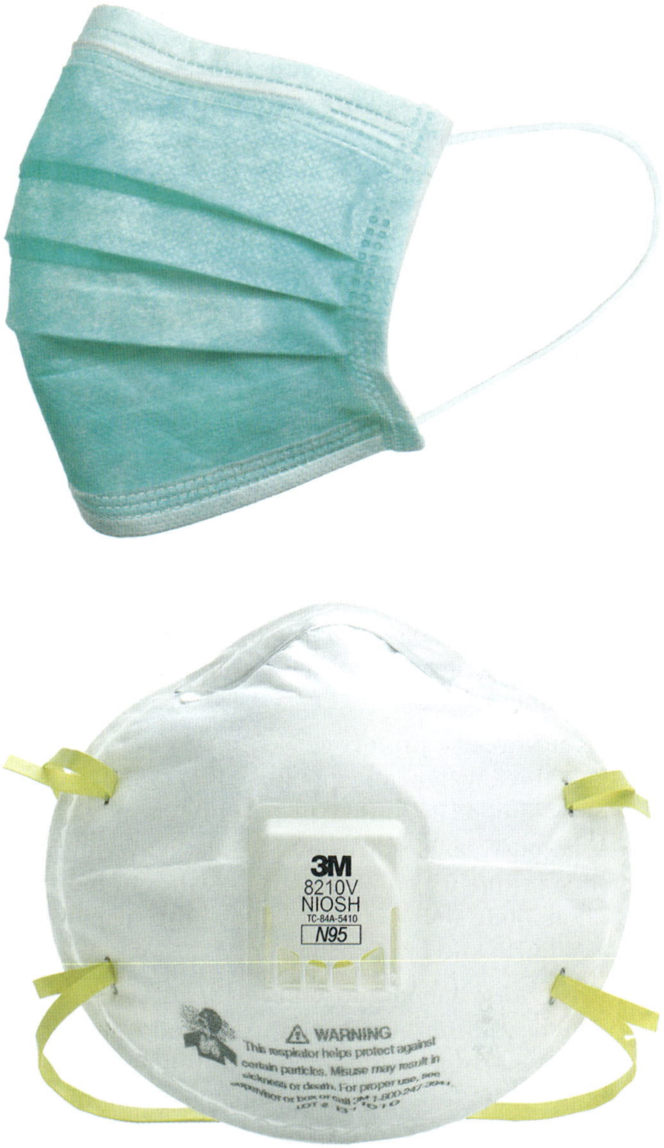

FIGURE B-1 A surgical mask or an N95 respirator will reduce the transmission of airborne diseases.

Top: © SanitFuangnakhon/Shutterstock. **Bottom:** © ACZ.Image/Shutterstock

SAFETY TIP

Simple, portable safety equipment can help prevent many injuries and illnesses. Medical gloves, masks, and eye protection prevent the spread of infectious diseases. Some situations may require additional safety equipment. Do not hesitate to call for additional equipment as needed. For further information, see NFPA 1581, *Standard on Fire Department Infection Control Program*.

Direct Contact

MRSA infection is caused by the bacterium *S. aureus*—often simply called "staph." MRSA is a strain of staph that is resistant to the broad-spectrum antibiotics commonly used for treatment. Most MRSA infections occur in healthcare settings such as hospitals, dialysis centers, and nursing homes, and they most commonly arise in people with weakened immune systems, in whom this disease can be fatal. Nevertheless, these infections can also occur in otherwise healthy people. In healthy people, MRSA may show up as a skin sore. As a support person, you need to follow standard precautions to avoid potentially contracting MRSA. In addition, avoid sharing your towels, razors, and other personal care items. To minimize your risk of infection, wash your towels in hot water and dry them thoroughly.

Standard Precautions

Federal regulations require all healthcare workers, including members of the fire service, to assume that all victims are potentially infected with bloodborne pathogens. These regulations require that all personnel use protective equipment to prevent possible exposure to blood and certain body fluids. Using protective equipment can also provide you with some protection against airborne pathogens. The CDC recommends following a set of **standard precautions**:

1. Always wear approved, protective nonlatex gloves when handling victims, and change gloves after contact with each victim. Wash your hands with soap and warm water immediately after removing gloves because it is still possible for your hands to have been contaminated before you put on the gloves or for some materials to penetrate the gloves. If necessary, hand sanitizer can be used as a temporary cleansing agent until soap and warm water are available.
2. Always wear a protective N95 mask, eyewear, or a face shield when you anticipate that blood or other body fluids, including respiratory droplets, may splatter. Wear a gown/apron, head covering, and shoe covers if you anticipate splashes of blood or other body fluids, such as those that occur with childbirth and major trauma.
3. Wash your hands and other skin surfaces immediately and thoroughly with soap and warm water if they become contaminated with blood or other body fluids (**FIGURE B-2**). Change contaminated clothes and wash exposed skin thoroughly.
4. Do not recap, cut, or bend used needles. Place them directly in a puncture-resistant container designed for "sharps."
5. Use a face shield, pocket mask, or other airway adjunct if the victim needs resuscitation.

FIGURE B-2 Wash your hands thoroughly with soap and warm water if you are contaminated with blood or other body fluids.

© Diy13/Shutterstock

Proper removal of gloves is important to minimize the spread of pathogens, as demonstrated in **SKILL DRILL B-1**.

Federal agencies, such as the CDC and the Occupational Safety and Health Administration (OSHA), and state agencies, such as state public health departments, develop regulations dealing with standard precautions. Because these regulations are constantly changing, it is important for your department to stay updated on these regulations and provide continuing education to keep you current with the latest changes related to infectious disease precautions.

Immunizations

Certain immunizations are recommended for emergency medical care providers, including influenza vaccine, tetanus prophylaxis, and hepatitis B vaccine. You also should check the status of your varicella (chickenpox) vaccine and your measles, mumps, and rubella (German measles) vaccines. Tuberculin testing may also be recommended. Your medical director can determine which immunizations and tests are needed for members of your department.

Cleaning and Decontamination for Bloodborne Pathogens

All spills of blood or OPIM must be cleaned up promptly by trained personnel using proper PPE and a disinfectant. Surfaces, tools, equipment, and any other contaminated items must be cleaned and sterilized to avoid potential infection. Decontamination can be accomplished with two types of disinfectant. The most common is household bleach diluted as 1 part bleach to 10 parts water. The other type of disinfectant is a tuberculocidal disinfectant registered with the U.S. Environmental Protection Agency (EPA). Check the labels of any disinfectant and follow the manufacturer's instructions, warnings, and recommendations.

Spray the disinfectant on the contaminated area and let it stand for a minimum of 10 minutes. If possible, cover a heavily contaminated area with paper towels or rags and then soak them with the disinfectant. If the contaminated area contains broken glass or other sharp materials, use equipment such as a dustpan and broom or tongs to pick it up. Do not attempt to pick up contaminated sharp objects with your hands, even

SKILL DRILL B-1

Proper Removal of Medical Gloves

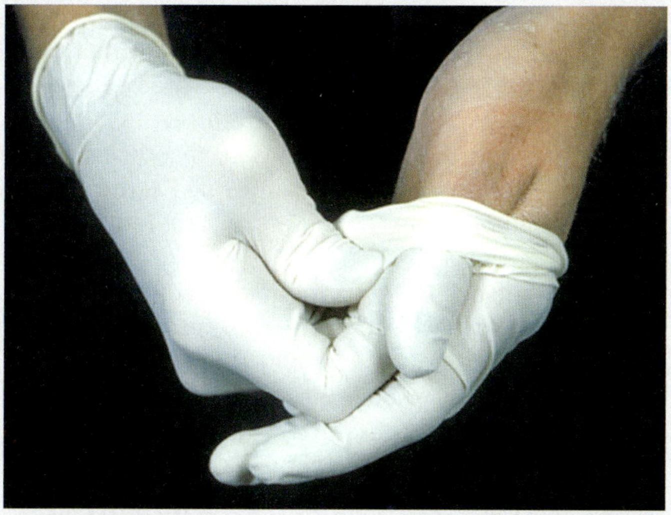

1. Begin by partially removing one glove. With the other gloved hand, pinch the first glove at the wrist, being careful to touch only the outside of the glove, and start to roll it back off the hand, inside out.

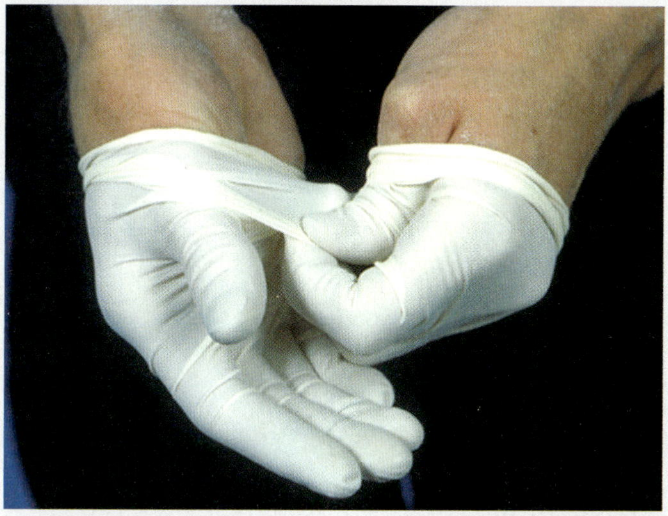

2. Remove the second glove by pinching the exterior with the partially gloved hand.

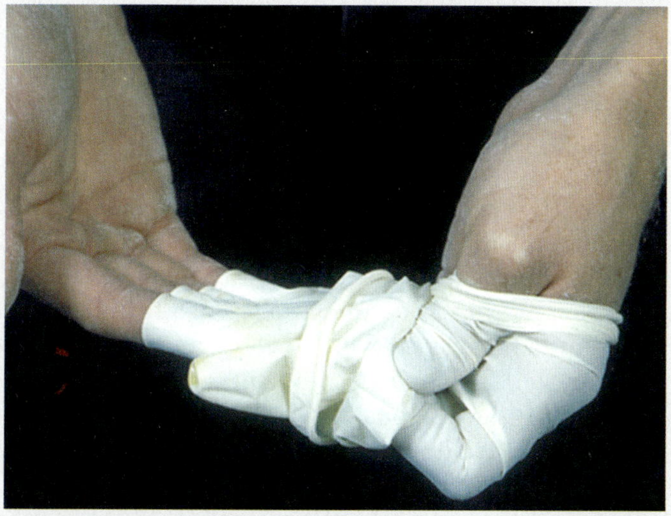

3. Pull the second glove inside out toward the fingertips.

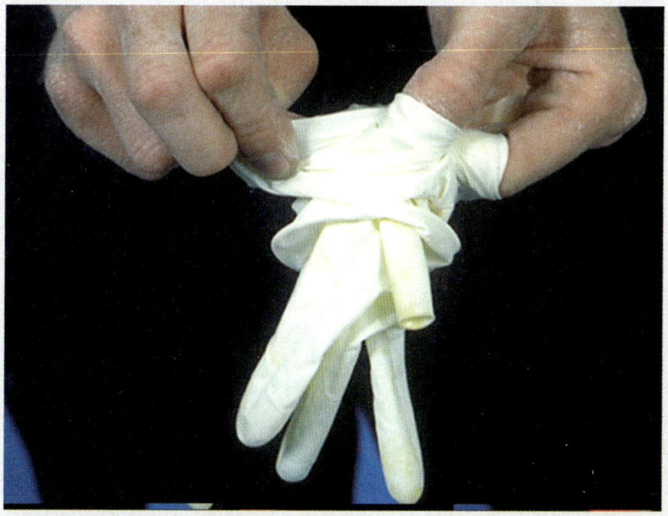

4. Grasp both gloves with your free hand, touching only the clean interior surfaces, and gently remove the gloves.

© Jones & Bartlett Learning. Courtesy of MIEMSS

FIGURE B-3 Dispose of soft items with potentially infected fluids in a biohazard bag.

© TheBlueHydrangea/Shutterstock

while wearing gloves. After cleanup is completed, any tools used must also be decontaminated or disposed of. Contaminated clothing must be appropriately disposed of or laundered by an approved service. Follow your department's approved policies and procedures.

Disposal of Biohazard Contaminated Waste

Federal regulations guide the proper disposal of gloves, bandages, clothing, glass, vehicle parts, and other items a member of the fire service might encounter that could be contaminated with blood or OPIM. It is essential for the waste to be handled properly and to be disposed of in the correct type of container.

Soft items such as soaked bandages that may drip blood or OPIM must go in a red bag marked with a biohazard symbol (**FIGURE B-3**). Soft items such as protective gloves that do not have excessive blood or OPIM may be disposed of in regular trash. Hard or sharp items such as needles, broken glass, hard plastic, or metal must be disposed of in a biohazard-marked sharps container. Once something is placed into a biohazard-marked bag or container, it must not be removed. For disposal of biohazard-marked bags and containers, follow your department's approved policies and procedures.

Immediate Life Threats

The first step in providing any level of medical care is to find out what kind of care the victim needs. This step is referred to as the *initial assessment*. Different levels of EMS providers will have different assessment and care capabilities. However, the first step is always to identify immediate threats to the victim's life so that you can begin to correct them.

Initial Assessment

If the first responder has already determined that it is safe to proceed, the initial assessment consists of checking if the victim is responsive and then determining if they are breathing normally. If the area is not safe for you to enter, call for appropriate resources. If the area is safe for you to enter but not safe for the victim, you will need to move the victim to a safe location to assess and treat them.

If you suspect that someone needs immediate medical care because they appear unresponsive or severely injured, follow your department's policies and procedures to notify someone, such as your officer or dispatcher, and don the appropriate PPE. If the EMS system has not already been activated, call 911 ("phone first"), especially if you are the only rescuer.

Checking Responsiveness

As you approach the victim, identify yourself as a member of the fire service and ask if they need assistance. If they acknowledge you by replying to you, making eye contact, or looking toward you, ask what help they need and provide what assistance you can. If the victim is awake but appears to be severely bleeding, this is an immediate life threat. The steps to follow in that type of situation are explained in the section on controlling bleeding later in this appendix.

Call for additional resources if the victim needs assistance that is beyond your ability to provide. Note that this guideline is the same for EMS situations as it is for special rescue, hazardous materials, and other incidents that require specialized training, tools, and techniques.

If the victim does not seem to be awake, determine if the victim is unresponsive by tapping or gently shaking the victim and loudly asking, "Are you OK?" This is sometimes referred to as "shake and shout." If the victim responds to you, ask what help they need and provide what assistance you can (**FIGURE B-4**). If the victim does not respond to your "shake and shout," they are unresponsive.

Unresponsive Victims

The following steps will help you provide care for an unresponsive victim. Procedures differ slightly depending on whether the victim is an adult, a teen, a young **child**, or an **infant**. This section assumes you are caring for an adult or a teenage victim.

First, move the unresponsive victim to a relatively flat, firm, and level surface and place them on their back if they are not already in that position. Obtain assistance if you need it. If the victim appears to have injured their neck or back, try to support their head and neck and minimize movement of their spine.

Check if the unresponsive victim is breathing normally. If they are unresponsive and not breathing normally (gasping or breathing very slowly) or are not breathing at all, you will need to perform hands-only CPR. If an automated external defibrillator (AED) is available in the area, ask someone to bring it to you. Other CPR techniques, including performing CPR on a child or infant, can be learned in a formal CPR program for professional rescuers or healthcare providers.

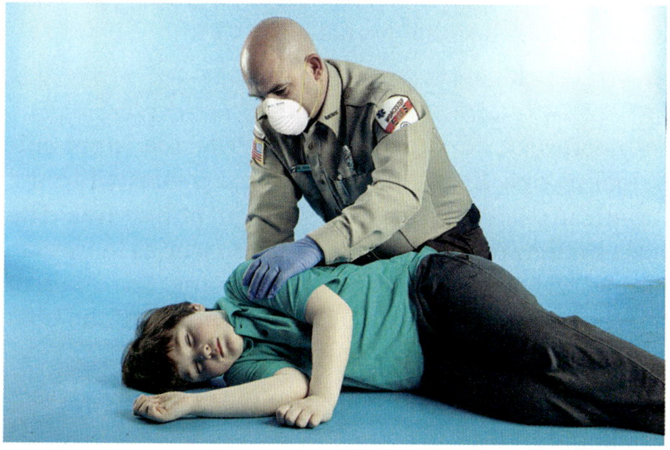

FIGURE B-4 Check to see if the victim is responsive or unresponsive.

© Jones & Bartlett Learning. Courtesy of MIEMSS.

Hands-Only Cardiopulmonary Resuscitation (CPR)

If a victim is not responding and is not breathing normally, their heart may have stopped pumping blood. This is called **cardiac arrest**, which is the leading cause of death in adults. Without a supply of blood, the cells of the body will die because they cannot get oxygen and nutrients and cannot eliminate waste products. Brain damage can begin within 4 to 6 minutes after a victim has gone into cardiac arrest. Within 8 to 10 minutes, the damage to the brain may become irreversible.

Cardiac arrest may have many different causes, including heart attack; untreated breathing problems; and medical emergencies such as allergic reactions, drowning, suffocation, poisoning, and traumatic injury. Regardless of the cause of cardiac arrest, the initial treatment is the same: provide CPR.

To perform hands-only CPR on an adult victim who is not responsive and not breathing normally, follow the steps in **SKILL DRILL B-2**.

SKILL DRILL B-2

Performing Hands-Only CPR

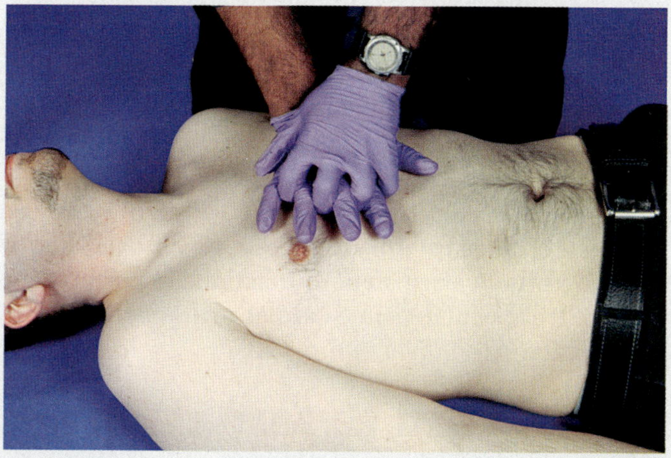

1. Place one hand on top of the other in the center of the victim's chest.

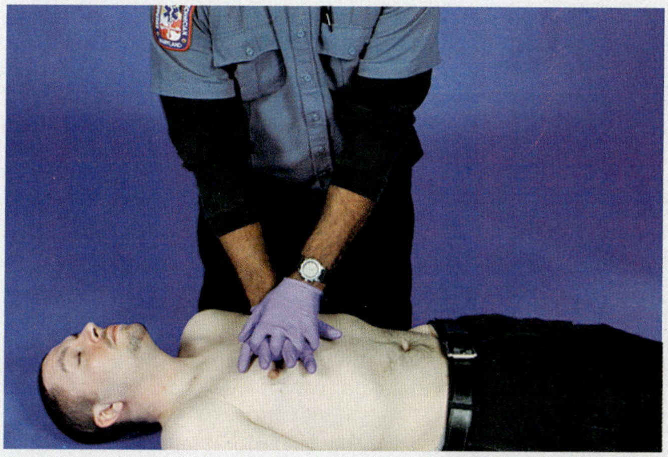

2. Compress the chest approximately 2 inches (in.; 5 centimeters [cm]) at a rate of approximately 100 compressions per minute—about as fast as counting "1 and 2 and 3 and 4."

© Jones & Bartlett Learning. Courtesy of MIEMSS.

When to Stop CPR

You should discontinue CPR only in the following circumstances:

1. The victim begins breathing or moving.
2. Someone else trained in CPR arrives to take over.
3. A physician orders you to stop.
4. You are too exhausted to continue.
5. Continuing CPR would place your life or the lives of others at risk.

Complications of CPR

A discussion of CPR is not complete without mentioning its complications. However, these complications can be minimized when proper care and technique are used. If your hands slip to the side of the sternum during chest compressions or if your fingers rest on the ribs, you may break the victim's ribs while delivering a compression. To prevent this complication, use proper hand positioning and do not let your fingers come in contact with the ribs. If you hear a cracking sound while performing CPR, check and correct your hand position but continue CPR. Sometimes you may break bones or cartilage even when using proper CPR technique.

Another potential complication can arise if you encounter a situation where you believe you should begin CPR, but a family member, caregiver, or bystander tells you that you should not because that would go against the wishes of the victim. This can be a difficult situation for anyone providing medical care. However, an individual's right to make decisions about their medical care does not end when they go into cardiac arrest.

People often discuss decisions about their care with their family, physician, and lawyer and draw up legal documents to help ensure their wishes are carried out, even in situations where they cannot communicate. Legal documents that specify a person's wishes regarding medical care include advance directives, living wills, medical orders for life-sustaining treatment (MOLST), do not resuscitate (DNR) orders, and durable powers of attorney for health care. These documents are all generally referred to as **advance directives**. Unfortunately, advance directives vary from state to state and are not always immediately available to rescuers. Even if they are, the instructions may not be obvious or easy to understand.

If you encounter a victim who appears to be in cardiac arrest and you are told about or presented with advance directives, you should comply with the law and the victim's wishes. However, if you are unclear about the advance directive or unsure whether you should begin CPR, you should generally start CPR. Follow your department's protocols regarding advance directives.

Another legal concept is abandonment—the discontinuation of CPR without the order of a licensed physician or without turning the victim over to someone who is at least as qualified as you are. Following the guidelines outlined in the "When to Stop CPR" section in this appendix can help you avoid this potential complication.

If you avoid these pitfalls, you need not be overly concerned about the legal implications of performing CPR. Your most important protection against a possible legal suit is to become proficient in CPR.

The Cardiac Chain of Survival

In most cases of cardiac arrest, CPR alone is not sufficient to save lives, but it is the first treatment in the American Heart Association's Chain of Survival. The links in this chain vary depending on who is in cardiac arrest and where the cardiac arrest occurs. The Chain of Survival for adults in cardiac arrest outside of the hospital includes six links:

1. Early recognition and prevention of cardiac arrest
2. Rapid activation of the 911 system
3. Immediate CPR with high-quality chest compressions
4. Rapid defibrillation
5. Integrated ALS and postarrest care
6. Recovery

We have already discussed the first three links in the Chain of Survival. As a support person, you can help the victim by identifying a life-threatening situation, ensuring the EMS system has been activated, and providing early CPR. You can also assist by operating an **automated external defibrillator (AED)**, as discussed in the next section. Keeping these links of the Chain of Survival strong will help keep the victim alive until paramedics and hospital personnel can administer early advanced care. Just as an actual chain is only as strong as its weakest link, the Chain of Survival is only as good as its weakest step. Your actions in performing early CPR are vital to giving cardiac arrest victims their best chance for survival

Automated External Defibrillator (AED)

More than 70 percent of all out-of-hospital cardiac arrest victims have an irregular heart electrical rhythm called **ventricular fibrillation**. This condition, often referred to as *V-fib*, is the rapid, disorganized, and ineffective quivering (fibrillation) of the heart. An AED will accurately identify ventricular fibrillation and advise you to deliver a shock if needed. An electric shock applied to the heart will defibrillate the heart to help it begin beating in an organized and effective manner. A victim in cardiac arrest stands the greatest chance for survival when early defibrillation is available.

As a member of the fire service, you will often be the first emergency care provider to reach a victim who has collapsed in cardiac arrest. When you perform effective CPR, you are helping to keep the victim's brain and heart supplied with oxygen until a defibrillator and ALS can be brought to the scene.

It is important to understand that AEDs do not defibrillate every victim in cardiac arrest. When an AED is first turned on, it will begin giving voice prompts (instructions) for what to do next. The AED must first analyze the victim's heart rhythm to check if defibrillation will help. If the victim's heart rhythm should be defibrillated, the AED should automatically begin charging to prepare to defibrillate, along with a voice prompt such as, "Charging! Continue CPR!" Once the AED is ready to deliver the shock, it will give a voice prompt such as, "Deliver shock now! Push the flashing button to deliver shock!"

Before delivering the shock, the first responder should ensure that CPR has stopped. They should then check the victim from head to toe to confirm that no one is touching the victim or any equipment directly attached to the victim. After confirming that, they should push the button on the AED as indicated to deliver the electrical shock. Once the shock is delivered, the first responder operating the AED should direct someone to resume CPR immediately.

If the victim's heart rhythm is not one that will be helped by defibrillation, the AED will give a voice prompt such as, "No shock advised! Continue CPR!" If the AED advises not to shock, continue chest compressions. The AED will prompt you to recheck the heart rhythm later.

Increasing numbers of fire departments are equipping personnel with AEDs to get defibrillators to cardiac arrest victims more quickly (**FIGURE B-5**). AEDs allow first responders to combine effective CPR with early defibrillation to restore an organized heartbeat in a victim.

Performing Automated External Defibrillation

The steps for using an AED are listed in **SKILL DRILL B-3**.

AEDs vary in size, shape, configuration, and operation. If AEDs are available in your department, learn how to use your specific AEDs.

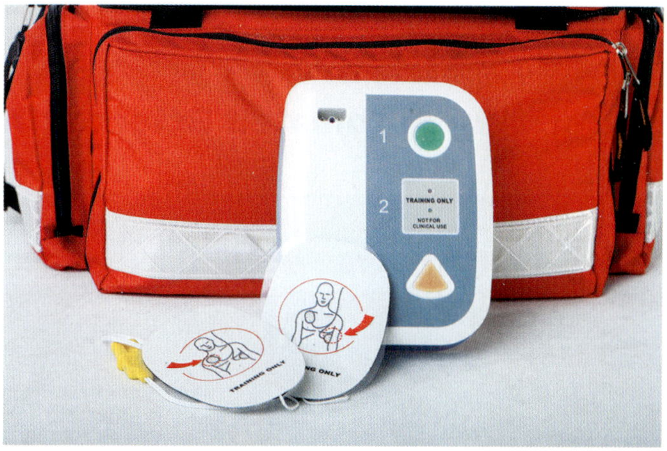

FIGURE B-5 An AED.

© Photographee.eu/Shutterstock

CPR and AED Training

As a support person, you should complete a CPR course through a recognized agency, such as the Emergency Care and Safety Institute (ECSI) or another national training organization. These courses teach and provide the opportunity to practice additional skills not covered here, including providing care to young children and infants and assisting victims who have a heartbeat but are not breathing and victims who are choking, among others. You cannot achieve proficiency in CPR unless you have adequate practice on adult, child, and infant manikins. These courses will help you practice, refine, and test these hands-on life-saving skills.

Recovery Position

If an unresponsive victim is breathing and has not suffered trauma, place the victim in the side-lying recovery position. This position helps keep the victim's airway open by allowing secretions to drain out of the mouth instead of into the trachea. It also uses gravity to prevent the victim's tongue and lower jaw from blocking the airway.

To place a victim in the recovery position, follow the steps in **SKILL DRILL B-4**.

SAFETY TIP

Many injuries involve some type of bleeding. When you approach a victim who is bleeding, you need to maintain standard precautions.

SKILL DRILL B-3

Performing Automated External Defibrillation

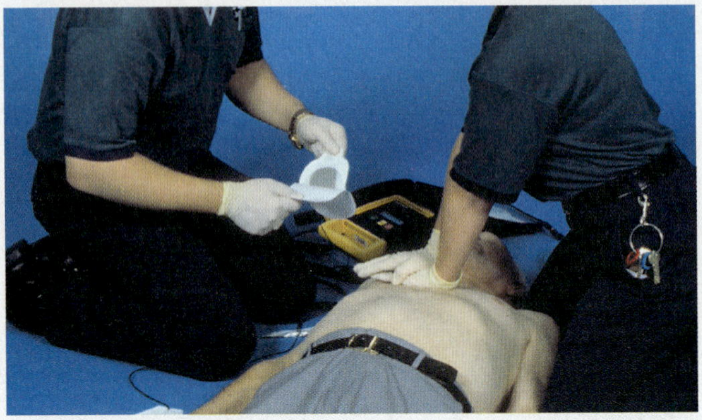

1. Only apply an AED to a victim in cardiac arrest. If CPR is in progress when you arrive, it is crucial to limit the amount of time compressions are interrupted. If the victim is unresponsive and not breathing normally, begin providing chest compressions. Continue chest compressions until an AED arrives and is ready for use. It is important to start chest compressions and use the AED as soon as possible. Compressions provide vital blood flow to the heart and brain, improving the victim's chance of survival.

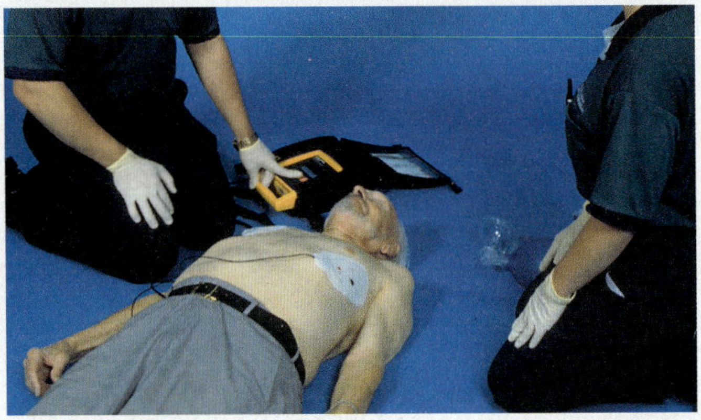

2. Turn on the AED. This will start the AED voice prompts. Remove clothing from the victim's chest area. Apply the pads to the chest: one to the right of the breastbone (sternum) just below the collarbone (clavicle), and the other on the left lower chest area with the top of the pad 2 to 3 in. (5 to 8 cm) below the armpit. Plug the pads' connector into the AED if not already attached.

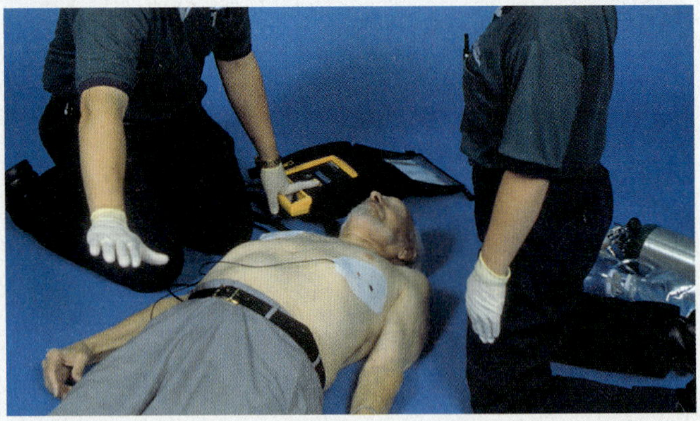

3. Stop CPR. State aloud, "Clear the victim," and ensure that no one is touching the victim. Push the analyze button, if there is one, and wait for the AED to determine if a shockable rhythm is present. Most AEDs will indicate this with a voice prompt such as, "Analyzing! Do not touch the patient!" The AED will be able to check if a defibrillation will help.

If a shock is not advised, resume chest compressions. If a shock is advised, the machine will charge and emit a voice prompt such as, "Charging! Continue CPR!" Continue chest compressions while the machine charges. Once the device has charged, it will emit a voice prompt such as, "Clear patient! Push red button to shock!" At that point, stop chest compressions, reconfirm that no one is touching the victim, and push the shock button.

SKILL DRILL B-3 CONTINUED

Performing Automated External Defibrillation

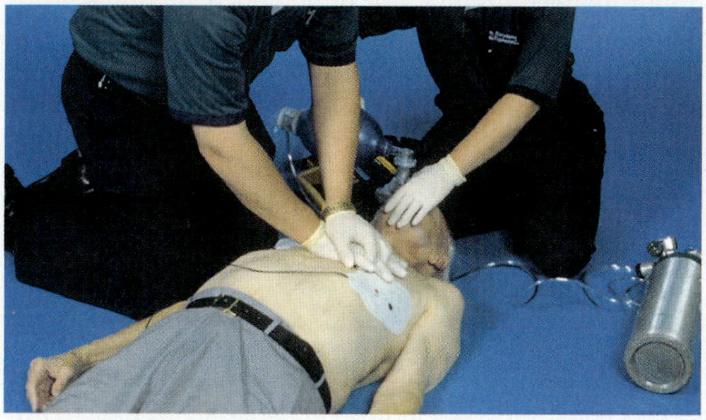

4. After the shock is delivered, immediately resume chest compressions. After about 2 minutes, the AED will prompt you to reanalyze the victim's cardiac rhythm.

© Jones & Bartlett Learning. Courtesy of MIEMSS.

SKILL DRILL B-4

Placing a Victim in the Recovery Position

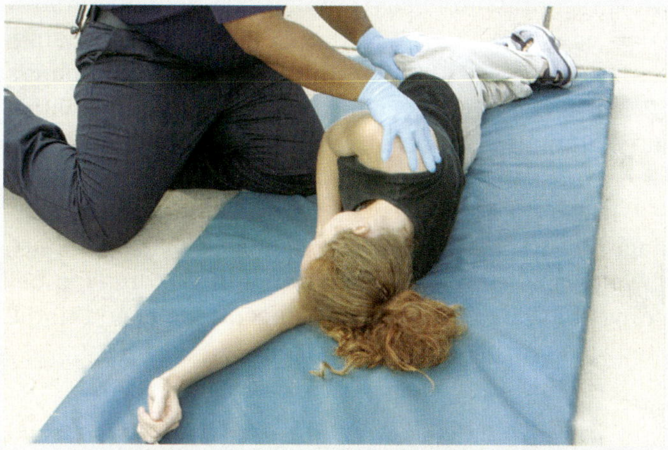

1. Carefully roll the victim onto one side as you support the victim's head. Roll the victim as a unit without twisting the body.

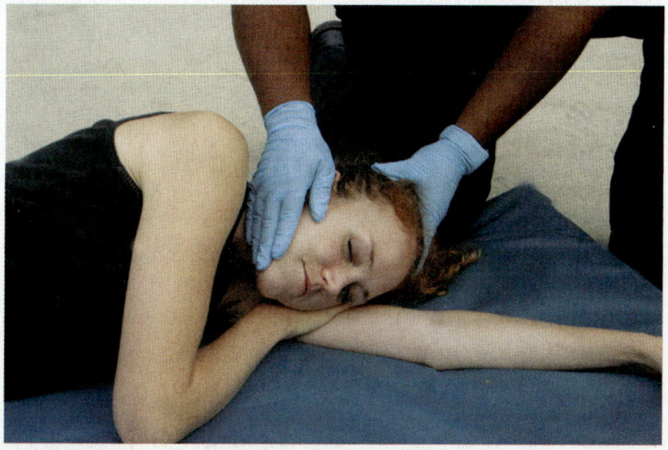

2. Place the victim's head on its side and extend the chin while pointing the mouth downward so that any secretions drain out of the mouth.

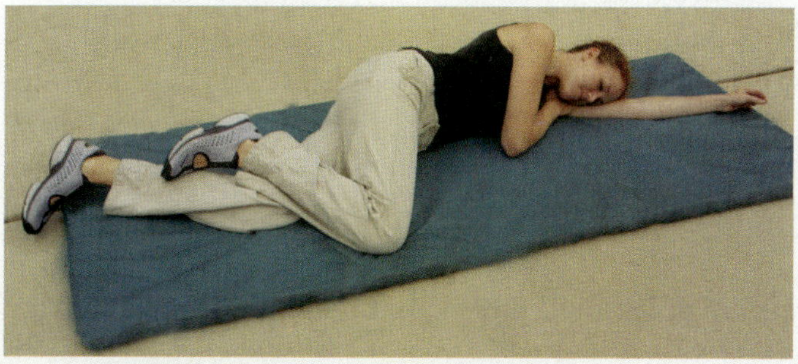

3. Place the victim's top hand under their cheek to help hold their head in a good position. Bend the victim's knee to help maintain the victim in the recovery position. Continue to monitor the victim's airway.

© Jones & Bartlett Learning. Photographed by Glen E. Ellman.

Bleeding Control

The steps for initial assessment apply to any encounter with a victim. These steps include making sure the area is safe to enter, removing the victim if necessary, following your department's policies and procedures to notify an officer or dispatcher, donning appropriate PPE, and calling 911 if it has not already been activated. Fire service personnel must quickly recognize life-threatening bleeding, also referred to as **hemorrhage**, and be prepared to control it rapidly. Severe, life-threatening bleeding can occur from a variety of injuries.

Identifying Life-Threatening Blood Loss

Whereas external life-threatening bleeding is visible, life-threatening internal bleeding, which can lead to shock, can be easy to miss. If a victim has obvious bleeding or injuries that may result in bleeding, search for the source of bleeding on the victim. Remember that there may be more than one source of bleeding, and some bleeding may be hidden.

Check the victim from their head to their toes, searching for bleeding on their arms, legs, neck, armpits, and groin—down the front and back of their body. Try to identify life-threatening bleeding by looking and feeling for a continuous flow of blood, blood pumping from the body like a sprinkler, or blood pooling in clothes or on the ground.

Controlling External Blood Loss

External blood loss can come from three types of blood vessels: **capillaries**, **veins**, and **arteries** (**FIGURE B-6**).

The most common type of external blood loss is **capillary bleeding**, in which the blood oozes out (such as from a paper cut or scrape). Capillary bleeding is virtually never, by itself, life-threatening. The next most common type of bleeding is **venous bleeding**. This type of bleeding has a steady flow. Bleeding from a large vein may be profuse and life-threatening. The third type of bleeding is **arterial bleeding**. Arterial blood is under higher pressure and often spurts or surges from the wound with each heartbeat. Arterial bleeding is often life-threatening because of the large amounts of blood that can quickly be lost.

Regardless of the type of bleeding, the steps to control it remain the same:

1. Apply direct, continuous pressure.
2. Pack the wound.
3. Apply a tourniquet.

These three life-saving procedures are regularly taught to first responders, including firefighters and police officers, as well as non–first responder adults and children.

Direct Pressure

Most external bleeding can be controlled by applying direct pressure to the wound. First, identify the sources of the external bleeding. If necessary, cut the victim's clothes or otherwise uncover them. Wipe away blood or other materials to find exactly where you need to apply direct pressure. Place enough dry, clean material—such as gauze—to cover the source of the bleeding, and press on it with one or both of your gloved hands (**FIGURE B-7**). However, do not cover the wound with a large amount of dry, clean material. That will make the direct pressure you are applying ineffective.

If you do not have a sterile dressing or gauze bandage, use the cleanest cloth available, such as a T-shirt, towel, socks, or other soft material. If the dressing becomes blood-soaked, you may need to take additional steps to control the bleeding.

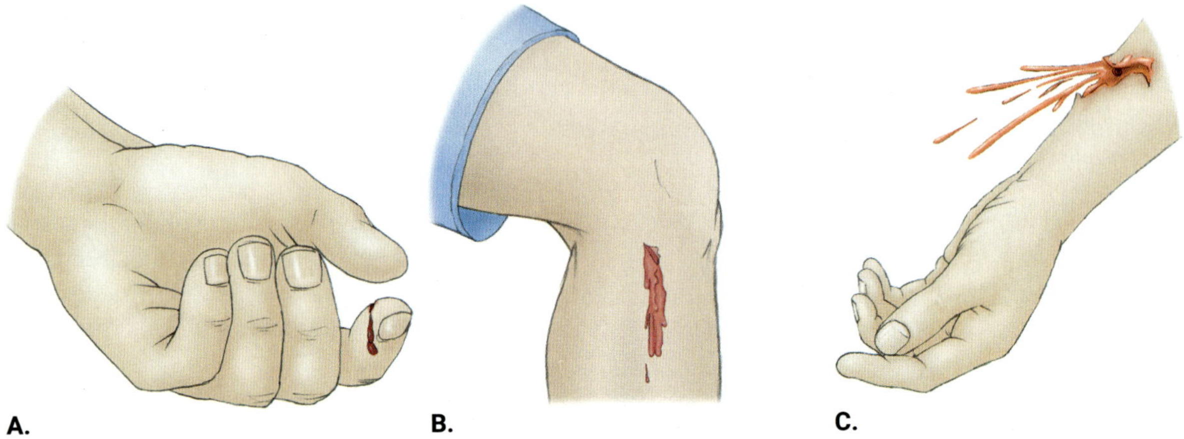

FIGURE B-6 Recognizing the three types of external bleeding. **A.** Capillary. **B.** Venous. **C.** Arterial.

© Jones & Bartlett Learning

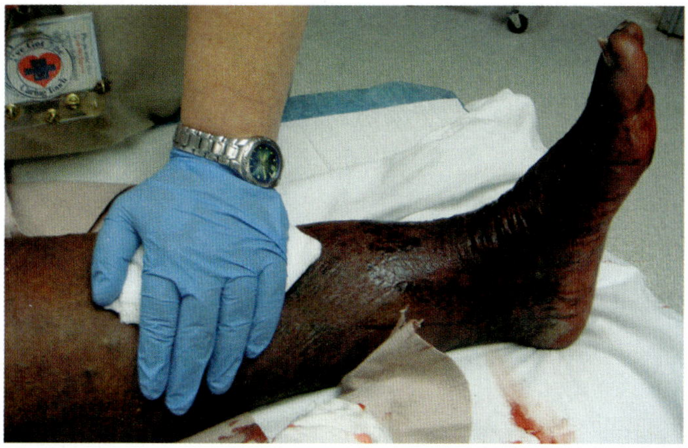

FIGURE B-7 Applying direct pressure to a wound.

Courtesy of Rhonda Hunt.

Wound Packing

If direct pressure is not enough to stop the bleeding because the wound is too large or deep, you may need to pack the wound with clean, dry material such as sterile gauze. Wound packing can be applied to wounds on the arms, legs, neck, shoulder, and groin areas. Do not pack wounds in the chest or abdomen. To pack a wound, follow the steps in **SKILL DRILL B-5**.

Tourniquets

If the source of life-threatening bleeding is a wound on an arm or leg and the bleeding cannot be controlled by direct pressure, a **tourniquet** should be used if available. "Tourniquet" is the generic term for any device

SKILL DRILL B-5

Wound Packing

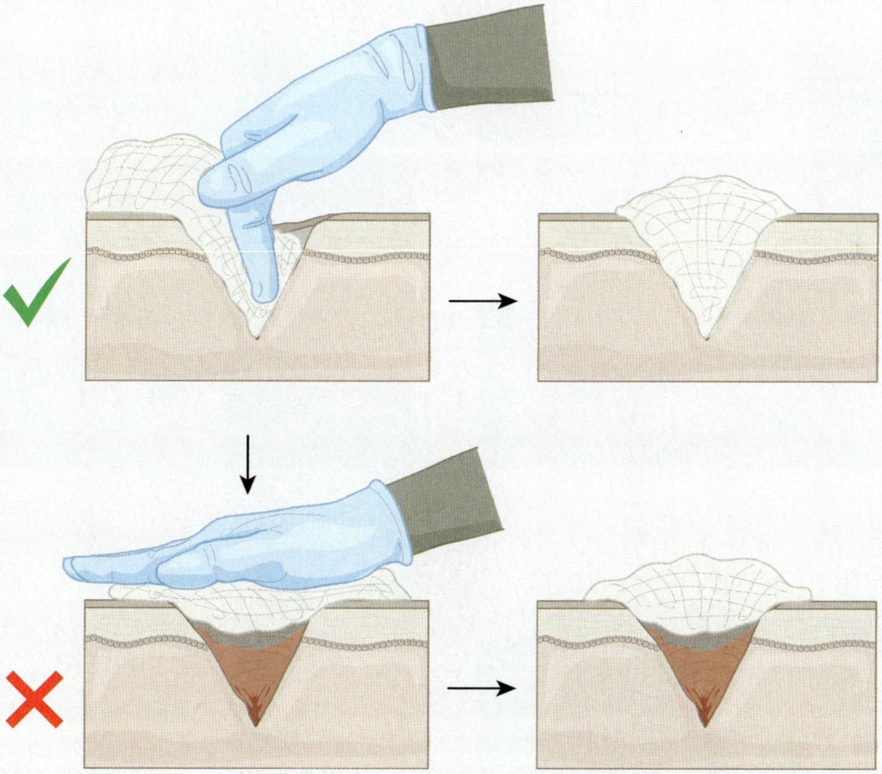

1. With one gloved finger, do your best to maintain direct pressure inside the wound, roughly gauging the depth you will need to pack.
2. With the fingers of your other hand, pack the wound with clean, dry material, such as sterile gauze, until the wound is filled.
3. Once the wound is fully packed, continue to apply direct pressure on the packing to control the bleeding. This will help apply pressure on the source of bleeding deep inside the wound.

© Jones & Bartlett Learning

designed to constrict part of an arm or leg so tightly that blood cannot get through. Tourniquets from different manufacturers can have drastically different designs, from classic stick-and-belt–style tourniquets to ones that use dials and ratchets and others that are simply wide elastic bands. It is crucial to be familiar and practiced with the types of tourniquets available in your department.

To control bleeding with a tourniquet, follow the steps listed in **SKILL DRILL B-6**.

Shock

The three parts of the **circulatory system** are the pump (heart), the pipes (blood vessels), and the fluid (blood). **Shock** is the failure of the circulatory system. Circulatory failure has three primary causes: pump failure, pipe failure, and fluid loss.

Pump Failure

Cardiogenic shock occurs when the heart cannot pump enough blood to supply the body's needs. Pump failure can result if a heart attack has weakened the heart.

Pipe Failure

Pipe failure is caused by the expansion (dilation) of the blood vessels to as much as three or four times their normal size. This condition causes blood to pool in the vessels (pipes) instead of circulating throughout the system. When blood pools in the vessels, the rest of the body, including the heart and other vital organs,

SKILL DRILL B-6

Controlling Bleeding with a Tourniquet

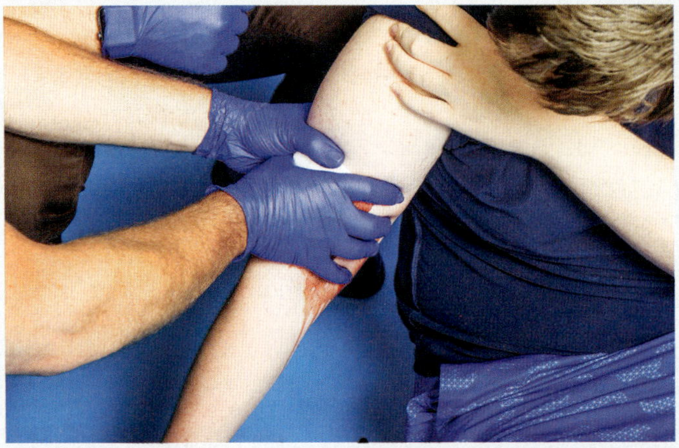

1. Apply the tourniquet above the level of bleeding (closer to the body than the wound).

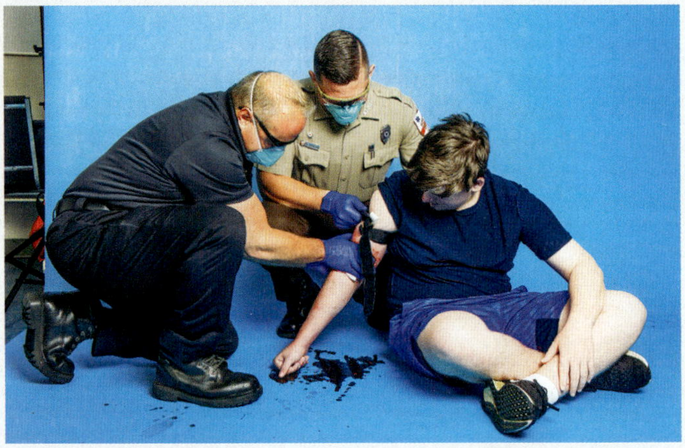

2. Snug the tourniquet band as tightly as possible around the injured limb. It should be tight enough that you cannot slide your finger under the band.

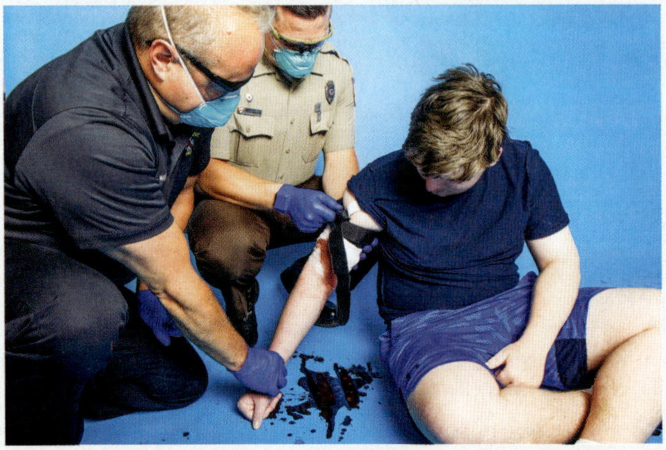

3. Tighten the tourniquet until bleeding from the extremity stops.

Check for bleeding. If bleeding continues or a distal pulse is felt in the injured limb, consider tightening the tourniquet further or applying a second tourniquet 1 to 2 in. (2.5 to 5 cm) above the first.

© Jones & Bartlett Learning. Photographed by Darren Stahlman.

is deprived of blood and the oxygen and nutrients it provides to sustain life. **Blood pressure** falls, and shock results. Blood pressure is the pressure of the circulating blood against the walls of the arteries. Blood pressure must be sufficiently high for blood to flow through the circulatory system.

Fluid Loss

The third general type of shock is caused by fluid loss. Fluid loss caused by excessive bleeding is the most common cause of shock. In this condition, blood escapes from the normally closed circulatory system through an internal or external wound. The system's total fluid level (blood volume) drops until the pump cannot operate efficiently. To compensate for fluid loss, the heart pumps faster to maintain pressure in the pipes. However, as the fluid continues to drain out, the pump eventually stops pumping altogether, resulting in cardiac arrest.

External bleeding is not difficult to detect because you can see blood escaping from the circulatory system to the outside of the victim's body. With internal bleeding, however, blood escapes from the system, but the bleeding usually cannot be seen. Sometimes, you may see signs of internal blood loss, such as bruising, swelling, and rigidity in the affected area. If the victim is conscious, they may report severe pain in the immediate area. Even though the escaped blood remains inside the body, it is now outside the circulatory system and is not available to be pumped by the heart. Whether the bleeding is external or internal, if it remains unchecked, the result will be shock, eventual pump failure, and death.

Signs and Symptoms of Shock

Shock deprives the body of sufficient blood to function normally. As this condition progresses, the body alters some of its functions to maintain sufficient blood supply to its vital parts. A victim who is in shock may exhibit some or all the following signs and symptoms:

- Confusion, restlessness, or anxiety
- Cold, clammy, sweaty, pale skin
- Rapid breathing
- Rapid, weak pulse
- Increased capillary refill time
- Nausea and vomiting
- Weakness or fainting
- Thirst

Initially, the victim's breathing may be rapid and deep, but as shock progresses, breathing becomes slow and shallow.

Changes in mental status may be the first sign of shock; therefore, monitoring the overall mental status of a victim can help you detect shock. The change in mental status may be significant. In severe cases of shock, the victim loses consciousness. If a trauma victim who has been quiet suddenly becomes agitated, restless, and vocal, you should suspect shock. If a trauma victim who has been loud, verbal, and belligerent becomes quiet, you should also suspect shock and begin treatment.

You may be unable to use changes in skin color to help you detect shock quickly. Therefore, you must be alert for other signs of shock. The capillary refill test and skin condition (cool and clammy) will also help you recognize shock in victims. To perform the capillary refill test, squeeze the victim's nail bed and see how fast the blood returns. This rate can give you an indication of the status of the victim's circulatory system. If the blood is slowly returning, the victim may be in shock.

General Treatment for Shock

You can combat shock from any cause and keep it from worsening by taking several simple but important steps. Remember that the protocols for use of these skills may vary. Always follow the protocols as approved by your medical director.

Position the Victim. If the victim has no head injury, extreme discomfort, or difficulty breathing, place the victim flat on their back on a horizontal surface. Place a blanket under the victim, if available. If the victim is having chest pain or difficulty breathing, place the victim in a sitting or semi-reclining position. If the victim is unconscious, place them in the recovery position. Check on the victim at least every 5 minutes. Treat any life-threatening injuries.

Maintain the Victim's Body Temperature. Attempt to keep the victim comfortably warm. A victim with cold, clammy skin should be covered. It is as essential to place blankets under the victim to keep body heat from escaping into the ground as it is to cover the victim with blankets.

Do Not Allow the Victim to Eat or Drink. Even though a victim in shock is often very thirsty, do not give liquids by mouth. There are two reasons for this rule:

1. A victim in shock may be nauseated, and eating or drinking may cause vomiting.
2. A victim in shock may need emergency surgery. Victims should not have anything in their stomachs before surgery.

CASE STUDY

You Are the Support Person CONCLUSION

Your crew is dispatched to a motor vehicle accident involving one car that crashed into a telephone pole. You arrive to find one car with heavy front-end damage still up against the telephone pole. Your lieutenant informs you that the police have secured the roadway, and the pole appears stable. Your lieutenant directs you to check on the victims in the front of the car. You find the driver slumped over toward the passenger. The driver does not respond to you when you approach. The passenger is sobbing and has an arm laceration that is bleeding heavily. You can see blood inside the car on the passenger's side.

1. **How will you protect yourself from contracting an infectious disease?**

   Answer: If possible, approach the scene and accomplish your duties as a support person without making contact with blood or OPIM. If your assignment includes caring for the bleeding passenger or working in an area where there is blood or OPIM, you must use standard precautions and wear appropriate PPE, including medical gloves, eye protection, and possibly more.

2. **What is your highest priority for the driver?**

   Answer: The driver must be checked to identify any immediate life threats so that you can begin to correct them. If the driver continues to be unresponsive and they are not breathing normally, you must remove them from the vehicle as quickly as possible and begin CPR.

3. **Which concerns do you have for the passenger?**

   Answer: The primary concern for the passenger is the bleeding. If direct pressure cannot stop the bleeding, you should consider applying a tourniquet and/or packing the wound.

WRAP-UP

SUMMARY

- In most states, people who work in emergency medical care are categorized into one of the four following licensure levels, each with a different role and focus as well as different training requirements: EMR, EMT, AEMT, and paramedic.
- To protect yourself from exposure, it is essential to have a basic understanding of the most common infectious diseases. The three most common routes for transmission of infectious diseases that you will be exposed to during your work as a support person are contact with infected blood or other body fluids, breathing infectious airborne droplets, and direct contact with infectious agents.
- It is crucial that all first responders follow all CDC recommendations when coming in contact with victims.
- All emergency responders must be certified and fully capable of performing CPR and using emergency medical equipment such as an AED when required.

KEY TERMS

acquired immunodeficiency syndrome (AIDS) An immune disorder caused by infection with HIV, resulting in an increased vulnerability to infections and certain rare cancers.

advance directives Healthcare instructions intended to help ensure a person's healthcare wishes are carried out even if they cannot actively make or communicate their own decisions about medical care.

advanced life support (ALS) Life-saving procedures performed by advanced emergency medical technicians (AEMTs) and paramedics, including administration of medications and other advanced procedures, under the direction of a physician.

arterial bleeding Serious bleeding from an artery in which blood frequently pulses or spurts from an open wound.

arteries Blood vessels that carry blood away from the heart.

automated external defibrillator (AED) A medical device used to diagnose and treat cardiac arrest. An AED identifies ventricular fibrillation and advises the user to deliver a shock if needed.

blood pressure The pressure exerted by the circulating blood against the walls of the arteries.

capillaries The smallest blood vessels that connect small arteries and small veins. Capillary walls serve as the membranes through which oxygen and carbon dioxide exchange occurs.

capillary bleeding Bleeding in which blood oozes from the open wound.

cardiac arrest A sudden ceasing of heart function.

cardiopulmonary resuscitation (CPR) The artificial circulation of the blood and movement of air into and out of the lungs in a pulseless, nonbreathing victim.

child Anyone between 1 year of age and the onset of puberty (12 to 14 years of age).

circulatory system The heart and blood vessels, which together are responsible for the continuous flow of blood throughout the body.

hemorrhage Excessive bleeding.

hepatitis B virus (HBV) A viral infection that causes inflammation and loss of liver function. It is transmitted through blood.

hepatitis C virus (HCV) A viral infection that causes inflammation and loss of liver function. It is transmitted through blood.

human immunodeficiency virus (HIV) The virus that causes acquired immune deficiency syndrome (AIDS).

infant A child younger than 1 year.

laceration An irregular cut or tear through the skin.

other potentially infectious materials (OPIM) Body fluids, other than blood, that may transmit infection or carry blood that can carry infection, such as semen, vaginal secretions, cerebral spinal fluid, saliva, and others.

pathogens Microorganisms capable of causing disease.

severe acute respiratory syndrome (SARS) A viral respiratory illness caused by a coronavirus.

shock A state of collapse of the cardiovascular system; the state of inadequate delivery of blood to the organs of the body.

standard precautions An infection control concept that treats all body fluids as potentially infectious.

tourniquet The generic term used for any device designed to constrict part of an arm or leg so tightly that blood cannot get through.

veins Blood vessels that carry blood toward the heart.

venous bleeding External bleeding from a vein, characterized by steady flow. Venous bleeding may be profuse and life-threatening.

ventricular fibrillation An uncoordinated muscular quivering of the heart; the most common abnormal rhythm causing cardiac arrest.

REFERENCES

American Heart Association. 2023. "2020 American Heart Association Guidelines for CPR and ECC." Accessed October 31, 2023. https://cpr.heart.org/en/resuscitation-science/cpr-and-ecc-guidelines

Berry, Cherisse, John M. Gallagher, Jeffrey M. Goodloe, Warren C. Dorlac, Jimm Dodd, and Peter E. Fischer. 2023. "Prehospital Hemorrhage Control and Treatment by Clinicians: A Joint Position Statement." *Prehospital Emergency Care* 27, no. 5, 544–551. https://doi.org/10.1080/10903127.2023.2195487.

Bobrow, Bentley J., Tyler F. Vadeboncoeur, Daniel W. Spaite, Jerald Potts, Kurt Denninghoff, VatsalChikani, Paula R. Brazil, Bob Ramsey, and Benjamin S. Abella. 2011. "The Effectiveness of Ultrabrief and Brief Educational Videos for Training Lay Responders in Hands-Only Cardiopulmonary Resuscitation: Implications for the Future of Citizen Cardiopulmonary Resuscitation Training." *Circulation. Cardiovascular Quality and Outcomes* 4, no. 2, 220–226. https://doi.org/10.1161/CIRCOUTCOMES.110.959353

REFERENCES CONTINUED

Dragset, Erik, Sigurd Blix, Jørgen Melau, Thomas Wilson, and Inger Lund-Kordahl. 2023. "Assessing Firefighters' Tourniquet Skill Attainment and Retention: A Controlled Simulation-Based Experiment." *Disaster Medicine and Public Health Preparedness* 17, e409. https://doi.org/10.1017/dmp.2023.68

Fisher, Andrew D., and Brandon M. Carius. 2019. "Stopping the Bleed: Hemorrhage Control and Fluid Resuscitation." *Physician Assistant Clinics* 4, no. 4, 781–793. https://doi.org/10/gf9rrx.

Jamal, Leila, Aman Saini, Keith Quencer, Izzet Altun, Hassan Albadawi, Aditya Khurana, Sailendra Naidu, Indravadan Patel, Sadeer Alzubaidi, and Rahmi Oklu. 2021. "Emerging Approaches to Pre-Hospital Hemorrhage Control: A Narrative Review." *Annals of Translational Medicine* 9, no. 14, 1192. https://doi.org/10.21037/atm-20-545

Lei, Roy, Michael D. Swartz, John A. Harvin, Bryan A. Cotton, John B. Holcomb, Charles E. Wade, and Sasha D. Adams. 2019. "Stop the Bleed Training Empowers Learners to Act to Prevent Unnecessary Hemorrhagic Death." *American Journal of Surgery* 217, no. 2, 368–372. https://doi.org/10.1016/j.amjsurg.2018.09.025.

Ludhwani, Dipesh, Amandeep Goyal, and Mandar Jagtap. 2023. "Ventricular Fibrillation." *StatPearls*. Accessed October 31, 2023. www.ncbi.nlm.nih.gov/books/NBK537120/.

National Institutes of Health, National Heart, Lung, and Blood Institute. 2022. "Cardiac Arrest: What Is Cardiac Arrest?" Accessed October 31, 2023. www.nhlbi.nih.gov/health/cardiac-arrest.

Glossary

A

absorption exposure Exposure to a hazardous material in which substances travel through body tissues until they reach the bloodstream.

accordion hose load A method of loading hose on a vehicle that results in a hose appearance that resembles accordion sections. This is achieved by standing the hose on its edge and laying it side to side in the hose bed.

acquired immunodeficiency syndrome (AIDS) An immune disorder caused by infection with HIV, resulting in an increased vulnerability to infections and certain rare cancers.

activity logging system A device that keeps a detailed record of every incident and activity that occurs.

adapter Any device that allows fire hose couplings to be safely interconnected with couplings of different sizes, threads, or mating surfaces, or that allows fire hose couplings to be safely connected to other appliances. (NFPA 1960)

adjustable wrench An open-ended wrench whose opening can be adjusted to accommodate bolts of different sizes.

advanced emergency medical technician (AEMT) Emergency medical services (EMS) personnel who can do everything an emergency medical technician (EMT) can do and who have advanced training in specific areas of advanced life support, including intravenous (IV) therapy, interpretation of cardiac rhythms, and advanced airway management.

advance directives Healthcare instructions intended to help ensure a person's healthcare wishes are carried out even if they cannot actively make or communicate their own decisions about medical care.

advanced life support (ALS) Life-saving procedures performed by advanced emergency medical technicians (AEMTs) and paramedics, including administration of medications and other advanced procedures, under the direction of a physician.

adze The curved or straight wedge part of a Halligan tool.

aerial apparatus See *aerial fire apparatus.*

aerial fire apparatus A vehicle equipped with an aerial ladder, elevating platform, or water tower that is designed and equipped to support firefighters and rescue operations by positioning personnel, handling materials, providing continuous egress, or discharging water at positions elevated from the ground. (NFPA 1900)

aerial See *aerial fire apparatus.*

aerosol An intimate mixture of a liquid or a solid in a gas; the liquid or solid, called the *dispersed phase,* is uniformly distributed in a finely divided state throughout the gas, which is the continuous phase or dispersing medium. (NFPA 99)

air-aspirating nozzle A nozzle that draws air into the water stream, creating an aerated or foamy spray that increases the surface area of water droplets, allowing for better heat absorption and faster cooling of a fire, and when used with firefighting foam solutions, aerates the foam mixture.

air bill The shipping papers on an airplane.

air-cylinder pressure gauge The device on an SCBA that measures and displays pressure readings to indicate the quantity of breathing air available.

air cylinder See *breathing air cylinder.*

air-line respirator See *supplied-air respirator.*

airport firefighter The Firefighter II who has demonstrated the skills and knowledge necessary to function as an integral member of an aircraft rescue and firefighting (ARFF) team. (NFPA 1010)

air-purifying respirator (APR) A respirator that removes specific air contaminants by passing ambient air through one or more air purification components. (NFPA 1984)

ambulance A vehicle used for out-of-hospital medical care and patient transport that provides a driver's compartment; a patient compartment to accommodate an emergency medical services provider (EMSP) and at least one patient located on the primary cot positioned so that the primary patient can be given emergency care during transit; equipment and supplies at the scene as well as during transport; safety, comfort, and avoidance of aggravation of the patient's injury or illness; two-way radio communication; and audible and visual warning devices. (NFPA 1900)

ammonium phosphate See *monoammonium phosphate.*

apparatus See *fire apparatus.*

aqueous film-forming foam (AFFF) A concentrate based on fluorinated surfactants plus foam stabilizers to produce a fluid aqueous film for suppressing hydrocarbon fuel vapors and usually diluted with water to a 1 percent, 3 percent, or 6 percent solution. (NFPA 11)

area of origin The room or general area where a fire started. Also called *fire compartment, fire seat,* or *seat of the fire.*

arterial bleeding Serious bleeding from an artery in which blood frequently pulses or spurts from an open wound.

arteries Blood vessels that carry blood away from the heart.

asphyxiant A material that prevents the body from using oxygen at the cellular level, causing the victim to suffocate.

assistant chief A midlevel chief who often has a functional area of responsibility, such as training, or who is responsible for a group of battalions or districts and who answers directly to the fire chief. Also called *deputy chief* or *division chief.*

atmosphere-supplying respirator (ASR) A respirator that supplies the respirator user with breathing air from a source independent of the ambient atmosphere and includes self-contained breathing apparatus (SCBA) and supplied-air respirators (SARs). (NFPA 1970)

atom The smallest particle of an element that retains the properties of that element.

attack engine An engine used to pump water through attack lines at the fireground.

attack hose Hose designed to be used by trained firefighters and fire brigade members to combat fires beyond the incipient stage. Also called *attack line.* (NFPA 1962)

attack line evolutions The delivery of water from an attack engine to a handline, which discharges the water onto the fire. Also called *attack line operations.*

attack line operations See *attack line evolutions.*

attack line See *attack hose.*

autoignition Initiation of combustion by heat but without a spark or flame. (NFPA 921)

automated external defibrillator (AED) A medical device used to diagnose and treat cardiac arrest. An AED identifies ventricular fibrillation and advises the user to deliver a shock if needed.

automatic location identification (ALI) A series of data elements that informs the recipient of the location of the alarm. (NFPA 1225)

automatic number identification (ANI) A series of alphanumeric characters that informs the recipient of the source of the alarm. (NFPA 1225)

awareness-level personnel Personnel who, during the course of their normal duties, could encounter an emergency involving hazardous materials/weapons of mass destruction (WMD) and who are expected to recognize the presence of the hazardous materials/WMD, protect themselves, call for trained personnel, and secure the scene. (NFPA 470)

axes Cutting tools that have a wide cutting blade that can be used to chop into a wall, roof, or door.

B

backdraft A deflagration (explosion) resulting from the sudden introduction of air into a confined space containing oxygen-deficient products of incomplete combustion. (NFPA 1403)

backpack fire extinguisher A portable fire extinguisher usually consisting of a 5-gal (19-L) water tank that is worn on the user's back and features a hand-powered piston pump for discharging the water and that is primarily used to fight brush and grass fires.

band saw An electrically powered saw that has a toothed metal blade stretched over two pulleys in a loop.

Bangor ladder An extension ladder with a staypole. Also called a tormentor ladder.

barrel The upright steel casing that is the main part of a fire hydrant.

base station A stationary radio transceiver with an integral alternating-current (AC) power supply. (NFPA 1225)

battalion chief Usually the first level of chief, the person in charge of running calls and supervising multiple stations or districts within a city.

battering ram A tool made of hardened steel with handles on the sides used to force doors and breach walls. Larger versions may be used by as many as four people; smaller versions are made for one or two people.

bend knot A knot that joins two ropes or webbing pieces together. (NFPA 2500)

bib The lower part of the protective hood that is part of the structural firefighting ensemble.

bight The open loop in a rope or piece of webbing formed when it is doubled back on itself. (NFPA 1006)

bill of lading The shipping papers used for transport of chemicals over roads and highways; also called *freight bill.*

BLEVE (boiling liquid/expanding vapor explosion) An explosion that occurs when pressurized liquefied materials, such as propane or butane, inside a closed vessel are exposed to a source of high heat.

block creel construction Rope constructed without knots or splices in the yarns, ply yarns, strands or braids, or rope. (NFPA 2500)

blood pressure The pressure exerted by the circulating blood against the walls of the arteries.

boiling liquid/expanding vapor explosion (BLEVE) An explosion that occurs when pressurized liquefied materials (e.g., propane or butane) in a closed container are exposed to a source of high heat, releasing the fuel, which instantly vaporizes and ignites.

boiling point The temperature at which the vapor pressure of a liquid equals the surrounding atmospheric pressure. (NFPA 1)

bolt cutter A cutting tool used to cut through thick metal objects such as bolts, locks, and wire fences.

bonnet The top of a hydrant.

booster hose A non-collapsible hose used under positive pressure having an elastomeric or thermoplastic tube, a braided or spiraled braided reinforcement, and an outer protective cover. Also called *booster line.* (NFPA 1962)

booster line See *booster hose.*

box-end wrench A hand tool used to tighten or loosen bolts. The end is enclosed, as opposed to an open-end wrench. Each wrench is a specific size, and most have ratchets for easier use.

braided reinforcement A hose reinforcement consisting of one or more layers of interlaced spiraled strands of yarn or wire, with a layer of rubber between each braid. (NFPA 1962)

braided rope Rope constructed by intertwining strands in the same way that hair is braided.

branch line See *secondary feeder.*

breathing air cylinder The pressure vessel or vessels that are an integral part of the self-contained breathing apparatus (SCBA) and that contain the breathing gas supply; can be configured as a single cylinder or other pressure vessel or as multiple cylinders or pressure vessels. (NFPA 1970)

British thermal unit (BTU) The amount of heat energy required to raise 1 pound of water at sea level by 1 degree Fahrenheit.

brush company See *wildland company.*

bung An opening on top of a closed-head drum that is typically sealed with a threaded cap.

bunker coat See *structural firefighting protective coat.*

bunker gear See *structural firefighting protective clothing.*

bunker pants See *structural firefighting protective trousers.*

bunny tool See *rabbet tool.*

C

call box A system of telephones connected by phone lines, radio equipment, or cellular technology to a communications center or fire department.

call classification and prioritization The process of assigning a response category based on the nature of the reported problem. (NFPA 1225)

call receipt The process of receiving a call for service and obtaining the information necessary to initiate a response.

calorie The amount of heat energy required to raise 1 gram of water (at sea level) by 1 degree Celsius.

candidate A person who applies to become a firefighter.

capillaries The smallest blood vessels that connect small arteries and small veins. Capillary walls serve as the membranes through which oxygen and carbon dioxide exchange occurs.

capillary bleeding Bleeding in which blood oozes from the open wound.

captain The second rank of promotion in the fire service, between the lieutenant and the battalion chief. Captains are responsible for managing a fire company and for coordinating the activities of that company among the other shifts.

carbon dioxide (CO_2) A nontoxic gas produced when sufficient oxygen is available for complete combustion that can displace oxygen in the atmosphere. Also, a colorless, odorless, electrically nonconductive inert gas that is a suitable medium for extinguishing Class B and Class C fires. (NFPA 10)

carbon dioxide (CO_2) fire extinguisher A fire extinguisher that uses carbon dioxide gas as the extinguishing agent. It is rated for use on Class B and C fires.

carbon monoxide (CO) A toxic gas produced through incomplete combustion.

carboy A glass, plastic, or steel storage container, ranging in volume from 5 to 15 gallons.

carcinogen A human cancer-causing agent.

cardiac arrest A sudden ceasing of heart function.

cardiopulmonary resuscitation (CPR) The artificial circulation of the blood and movement of air into and out of the lungs in a pulseless, nonbreathing victim.

carpenter's handsaw A saw designed for cutting wood.

cartridge-/cylinder-operated fire extinguisher A fire extinguisher in which the expellant gas is in a separate container from the agent storage container. (NFPA 10)

cascade system A method of piping air tanks together to allow air to be supplied to the self-contained breathing apparatus (SCBA) fill station using a progressive selection of tanks, each with a higher pressure level. (NFPA 1900)

ceiling hook A tool with a long wooden or fiberglass pole that has a metal point with a spur at right angles at one end. It can be used to probe ceilings and pull down plaster lath material.

ceiling jet A strong, turbulent convection current that rose to the ceiling and traveled along it.

chafing block A sturdy rubber, plastic, or wooden block placed under a fire hose where it lays on the ground or rests against hard surfaces to raise the hose off the ground and provide a smooth contact surface and protect the hose from abrasion, friction, and wear against rough surfaces like asphalt, concrete, or debris.

chain of command A rank-based, hierarchical structure that creates an orderly line of authority.

chainsaw A power saw that uses the rotating movement of a chain equipped with sharpened cutting edges. It is typically used to cut through wood.

charged hose line A hose line filled with water and under pressure from the pump.

charge To fill with water under pressure.

Chemical Abstracts Service (CAS) A division of the American Chemical Society that provides hazardous materials personnel and responders with access to the CAS Registry, an enormous collection of chemical substance information.

chemical energy Potential energy in molecular bonds and the kinetic energy created by a chemical reaction.

chief's bugle See *chief's trumpet*.

chief's trumpet An obsolete amplification device that was a precursor to a bullhorn and that enabled a chief officer to give orders to firefighters during an emergency. Also called *chief's bugle*.

child Anyone between 1 year of age and the onset of puberty (12 to 14 years of age).

chisel A metal tool with one sharpened end that is used to break apart material in conjunction with a hammer, mallet, or sledgehammer.

circulatory system The heart and blood vessels, which together are responsible for the continuous flow of blood throughout the body.

Class A fire A fire in ordinary combustible materials, such as wood, cloth, paper, rubber, and many plastics. (NFPA 1)

Class A foam concentrate A concentrate that, when combined with water, reduces the surface tension of the water and creates a foam. Also called *wet water*.

Class A foam fire extinguisher A fire extinguisher that contains a solution of water and Class A foam concentrate.

Class B fire A fire in flammable liquids, combustible liquids, petroleum greases, tars, oils, oil-based paints, solvents, lacquers, alcohols, and flammable gases. (NFPA 1)

Class C fire A fire that involves energized electrical equipment. (NFPA 1)

Class D fire A fire in combustible metals, such as magnesium, titanium, zirconium, sodium, lithium, and potassium. (NFPA 1)

Class K fire A fire in a cooking appliance that involves combustible cooking media (vegetable or animal oils and fats). (NFPA 1)

claw bar A tool with a pointed claw hook on one end and a forked- or flat-chisel pry on the other end. It is often used for forcible entry.

clean agent Electrically nonconducting, volatile, or gaseous fire extinguishant that does not leave a residue upon evaporation. (NFPA 10)

clean-agent fire extinguisher A fire extinguisher that uses a halogenated extinguishing agent. Also called *halogenated-agent fire extinguisher*.

Clemens hook A multipurpose tool that can be used for several forcible entry and ventilation applications because of its unique head design.

closed-circuit self-contained breathing apparatus (closed-circuit SCBA) A recirculation-type self-contained breathing apparatus (SCBA) in which the exhaled gas is rebreathed by the wearer after the carbon dioxide has been removed from the exhalation gas and the oxygen content within the system has been restored from sources, such as compressed breathing air, chemical oxygen, liquid oxygen, or compressed gaseous oxygen. Also called *rebreather*. (NFPA 1970)

closed-head drum A drum with a lid that is permanently attached to the drum; the lid has one or more small openings called *bungs*.

closet hook A type of pike pole intended for use in tight spaces, commonly 2 to 4 ft (0.6 to 1.2 m) in length.

code A standard that is an extensive compilation of provisions covering broad subject matter or that is suitable for adoption into law independent of other codes and standards. (NFPA 1)

Code of Federal Regulations (CFR) A collection of permanent rules published in the *Federal Register* by the executive departments and agencies of the U.S. federal government. Its 50 titles represent broad areas of interest that are governed by federal regulation. Each volume of the CFR is updated annually.

collapsible fire hose Fire hose typically made from synthetic materials that make the hose flexible and foldable.

combination hose load A hose-loading method used when one long hose line is needed; the end of the last length of a hose in one bed of a split hose bed is coupled to the beginning of the first length of hose in the opposite bed.

combination wrench A hand tool with an open-end wrench on one end and a box-end wrench on the other.

combustible Capable of undergoing combustion. (NFPA 1700)

combustion A chemical process of oxidation that occurs at a rate fast enough to produce heat and usually light in the form of either a glow or a flame. (NFPA 1)

come along A hand-operated tool used for dragging or lifting heavy objects that uses pulleys and cables or chains to multiply a pulling or lifting force.

communications center See *public safety communications center.*

company officer The individual responsible for command of a company, a designation not specific to any particular fire department rank (can be a firefighter, lieutenant, captain, or chief officer, if responsible for command of a single company). (NFPA 1026)

company The basic firefighting organizational unit staffed by various grades of firefighters under the supervision of an officer and assigned to one or more specific pieces of apparatus. (NFPA 1410)

compassion fatigue A disorder characterized by a gradual lessening of compassion over time

compressor A device used for increasing the pressure and density of a gas. (NFPA 853)

computer-aided dispatch (CAD) A combination of hardware and software that provides data entry; makes resource recommendations; and notifies and tracks those resources before, during, and after fire service alarms, preserving records of those alarms and status changes for later analysis. (NFPA 1225)

conduction Heat transfer to another body or within a body by direct contact. (NFPA 921)

consist A list of the contents of every car on a train; also called *train list.*

contagious Capable of transmitting a disease.

container A receptacle, piping, or pipeline used for storing or transporting material of any kind. (NFPA 470)

convection Heat transfer by circulation within a medium such as a gas or a liquid. (NFPA 921)

coping saw A saw designed to cut curves in wood.

coupler A hose appliance that allows two hoses to be connected together and flow into a single hose.

coupling A connection device that connects (couples) individual lengths of fire hose together or connects a hose to a fire hydrant or a pump or to nozzles and hose appliances.

coupling One set or pair of connection devices attached to a fire hose that allow the hose to be interconnected to additional lengths of hose or adapters and other firefighting appliances. (NFPA 1960)

crew A collective term used casually to refer to a group of firefighters in a department with similar duties or responsibilities. See also *company* and *unit.*

critical incident stress debriefing (CISD) A postincident meeting designed to assist rescue personnel in dealing with psychological trauma resulting from an emergency. (NFPA 1006)

critical incident stress management (CISM) A program designed to reduce acute and chronic effects of stress related to job functions. (NFPA 450)

cross-band repeater A repeater that boosts a weak signal and then retransmits it over a different radio band.

crowbar A straight bar made of steel or iron with a forked chisel on the working end that is suitable for performing forcible entry.

cryogenic liquid A fluid, such as liquid helium, liquid nitrogen, or liquid argon, that has a boiling point lower than –130°F (–90°C) at an absolute pressure of 14.7 pounds per square inch (psi; 101.3 kilopascals [kPa]); also called *cryogen.*

cryogen See *cryogenic liquid.*

cutting torch A torch that produces a high-temperature flame capable of heating metal to its melting point, thereby cutting through an object. Because of the high temperatures (5700°F [3148°C]) that these torches produce, the operator must be specially trained before using this tool.

cylinder A container that has a circular cross-section and is designed to store liquids or gases under pressure higher than 40 pounds per square inch (psi; 276 kilopascals [kPa]); this definition does not include portable tanks, multiunit tanks, car tanks, cargo tanks, or tank cars. (NFPA 1)

D

dangerous cargo manifest The shipping papers on a marine vessel.

dead-end water main A water main that is supplied from only one direction.

decay stage The stage of fire development within a structure characterized by either a decrease in the fuel load or available oxygen to support combustion, resulting in lower temperatures and lower pressure in the fire area. (NFPA 1410)

deck gun A device that is permanently mounted on and operated from a vehicle and equipped with a piping system that delivers water to the gun.

defensive mode The mode of operation in which direct contact with the material or container is avoided and responders' efforts focus on controlling or limiting the effects of the release.

Department of Transportation (DOT) The U.S. government agency that publicizes and enforces rules and regulations that relate to the transportation of many hazardous materials.

deploy hose To remove hose from the hose bed or other storage location. Also called *pull hose.*

deputy chief See *assistant chief.*

Dewar container A cylinder container designed to preserve the temperature of cryogenic liquids.

diagonal cutter See *wire cutter.*

digital radio A radio that transmits information via radio waves using digital data or analog (voice) signals that have been converted to a digital signal and compressed.

direct-line phone A phone that connects two predetermined points and does not require the user on either end to dial to cause the phone at the other end to ring. Also called *ring-down phone.*

discharge cap See *hydrant cap.*

dispatcher See *telecommunicator.*

dispatch To send out emergency response resources promptly to an address or incident location for a specific purpose. (NFPA 450)

distributor pipe The smallest-diameter underground water main pipes in a water distribution system that deliver water to local users within a neighborhood. Also called *distributor.*

distributor See *distributor pipe.*

division chief See *assistant chief.*

doff The process of properly removing a member's personal protective equipment (PPE) and respiratory protection to limit additional contamination and exposure. (NFPA 1700)

don The process of properly dressing in full personal protective equipment (PPE), ensuring all exposed skin and the airway are protected. (NFPA 1700)

door banger See *walk-in.*

double-female adapter A hose adapter that is used to join two male hose couplings.

double-jacket hose A hose constructed with two layers of woven fibers.

double-male adapter A hose adapter that is used to join two female hose couplings.

drafting hydrant See *dry hydrant.*

draft The use of suction to move a liquid (such as water) from a vessel or source that is below the intake of a pump. (NFPA 1910)

drag rescue device (DRD) A fabric handle integrated just below the collar at the back of the protective coat that a rescuer can grab to drag an incapacitated firefighter to safety.

dress To tighten and remove twists, kinks, and slack from the rope after tying a knot.

driver/operator A person qualified to operate a fire apparatus. Also called *engineer* or *technician.* (NFPA 1910)

drum A barrel-like storage vessel constructed of low-carbon steel, polyethylene, cardboard, stainless steel, nickel, or other materials that is used to store a wide variety of substances, including food-grade materials, corrosives, flammable liquids, and grease.

dry-barrel hydrant A type of hydrant with the main control valve below the frost line between the footpiece and the barrel. Also called *frost-proof hydrant.* (NFPA 24)

dry chemical A powder composed of very small particles, usually sodium bicarbonate, potassium bicarbonate, or ammonium phosphate based, with added particulate material supplemented by special treatment to provide resistance to packing, resistance to moisture absorption (caking), and the proper flow capabilities. (NFPA 10)

dry-chemical fire extinguisher A fire extinguisher that uses a dry-chemical extinguishing agent and is usually rated for use on Class B and C fires and sometimes on Class A fires.

dry hydrant An arrangement of pipe permanently connected to a water source other than a piped, pressurized water supply system that provides a ready means of water supply for firefighting purposes and that utilizes the drafting (suction) capability of a fire department pump. Also called *drafting hydrant.* (NFPA 1142)

dry-powder fire extinguisher A fire extinguisher that uses solid materials in powder or granular form to extinguish Class D combustible metal fires by crusting, smothering, or heat-transferring means.

dry powder Solid materials in powder or granular form designed to extinguish Class D combustible metal fires by crusting, smothering, or heat-transferring means. (NFPA 10)

drywall hook A specialized version of a pike pole that can remove drywall more effectively because of its hook design.

dual hose lines Two parallel hose lines laid at the same time.

dual-path pressure reducer A feature that automatically provides a backup method for air to be supplied to the regulator of a self-contained breathing apparatus (SCBA) if the primary passage malfunctions.

dump valve A large opening from the water tank of a mobile water supply apparatus for unloading purposes. (NFPA 1900)

dynamic rope A rope typically used for climbing that is designed to be elastic and stretch when loaded.

E

electrical energy Energy is produced by an electrical charge.

element A substance that cannot be chemically broken down into a simpler substance.

elevated water storage tower An aboveground water storage tank that is designed to maintain pressure on a water distribution system.

emergency medical dispatcher (EMD) Personnel specifically trained and certified in interviewing techniques, prearrival instructions, and call prioritization. (NFPA 450)

emergency medical responder (EMR) Emergency medical services (EMS) personnel who have basic training for providing initial medical assistance, have training in bleeding control and cardiopulmonary resuscitation (CPR), and often perform in an assistant role within the ambulance.

emergency medical services (EMS) company A company that may include medical units and first-response vehicles and that responds to and assists in the transport of medical and trauma victims to medical facilities for further treatment. Also called *emergency medical services (EMS) squad* or *squad.*

emergency medical services (EMS) personnel Personnel responsible for administering care to people who are sick and injured.

emergency medical services (EMS) squad See *emergency medical services (EMS) company.*

emergency medical technician (EMT) Emergency medical services (EMS) personnel who can do everything an emergency medical responder (EMR) can do and who have training in basic emergency care skills, including oxygen therapy, bleeding control, cardiopulmonary resuscitation (CPR), automated external defibrillation, use of basic airway devices, and assisting patients with certain medications.

Emergency Planning and Community Right-to-Know Act (EPCRA) Legislation that requires a business that handles chemicals to report on those chemicals' type, quantity, and storage methods to the fire department and the local emergency planning committee.

***Emergency Response Guidebook* (ERG)** The reference book, written in plain language, to guide emergency responders in their initial actions at the incident scene, specifically the *Emergency Response Guidebook* from the U.S. Department of Transportation, Transport Canada, and the Secretariat of Transport and Communications of Mexico used to guide personnel and responders in their initial actions at the incident scene. (NFPA 470)

emergency scene The area encompassed by the incident and the surrounding area needed by the emergency forces to stage apparatus and mitigate the incident. Also called *on scene*. See also *fire ground* and *fire scene*. (NFPA 901)

emergency traffic An urgent message, such as a call for help or evacuation, transmitted over a radio that takes precedence over all normal radio traffic.

emergency vehicle technician (EVT) The individual who repairs and performs service on emergency vehicles.

employee assistance programs (EAPs) An employer-sponsored service designed for personal or family problems, including mental health, substance abuse, various addictions, marital problems, parenting problems, emotional problems, or financial or legal concerns. (NFPA 450)

end-of-service-time-indicator (EOSTI) A warning device on a self-contained breathing apparatus (SCBA) that alerts the user that the reserved air supply is being utilized. (NFPA 1970)

endothermic A chemical reaction that absorbs heat.

energy The ability to do work.

engine company A piece of fire apparatus staffed with firefighters that has the primary responsibility to deliver a fire stream or streams to extinguish the fire in coordination with ventilation (truck company) and rescue operations. (NFPA 1700)

engineer See *driver/operator*.

engine See *pumper*.

entrain To encircle, draw along, and transport.

Environmental Protection Agency (EPA) The U.S. federal agency that ensures safe manufacturing, use, transportation, and disposal of hazardous substances.

escape rope Rope dedicated solely for the purpose of supporting people during emergency self-escape (self-rescue); not intended for use in a hazardous environment involving fire or fire products; not classified as a life safety rope. (NFPA 2500)

evacuation signal A distinctive signal intended to be recognized by the occupants as requiring evacuation of the building. (NFPA 1900)

evolution A set of prescribed actions that result in an effective fireground activity. (NFPA 1410)

exothermic A chemical reaction that produces heat.

explosion A violent and pressurized release of energy.

explosive limits See *flammable range*.

extinguishing agent A material used to stop the combustion process. Extinguishing agents may include liquids, gases, dry-chemical compounds, and dry-powder compounds.

extra hazard area An occupancy where the total amount of Class A combustibles and Class B flammables is greater than expected in occupancies classed as ordinary (moderate) hazards and the combustibility and heat release rate of the materials are high. Also called *high hazard area*.

F

face piece The part of the SCBA consisting of the face mask and exhalation valve that delivers breathing air to the firefighter and protects the face from high temperatures and smoke.

Federal Communications Commission (FCC) The federal regulatory authority that oversees radio communications in the United States.

film-forming fluoroprotein (FFFP) foam A protein-foam solution that uses fluorinated surfactants to produce a fluid aqueous film for suppressing liquid fuel vapors. (NFPA 10)

fingers Individual, adjustable vanes or protrusions on the face of a fog-stream nozzle that are responsible for shaping the water stream into a specific pattern.

fire alarm box A device connected via an underground cable to a municipal fire alarm system.

fire and life safety educator (FLSE) An individual who has demonstrated the ability to coordinate, create, administer, prepare, deliver, and evaluate educational programs and information.

fire apparatus A vehicle designed to be used under emergency conditions to transport personnel and equipment or to support the suppression of fires and mitigation of other hazardous situations. Also called *apparatus*. (NFPA 1010)

fireball A burst of flames that rapidly ignites available flammable vapors but is not under pressure.

firebreak A swath where the fuel (trees and brush) is removed.

fire chief The highest-ranking officer in charge of a fire department. (NFPA 1550)

fire code The code that specifies practices and procedures to prevent fires, prevent fires that start from spreading by suppressing them and blocking them, and protect lives in the event of a fire by specifying how occupants will be evacuated.

fire compartment The area of origin when the fire is in a structure.

fire department An organization providing rescue, fire suppression, and related activities, including any public, governmental, private, industrial, or military organization engaging in this type of activity. (NFPA 1010)

fire department connection (FDC) A connection through which the fire department can pump supplemental water into the sprinkler system, standpipe, or other system furnishing water for fire extinguishment to supplement existing water supplies. (NFPA 13)

fire escape rope Rope dedicated solely for the purpose of supporting people during emergency self-escape (self-rescue) from an immediately hazardous environment involving fire or fire products; not classified as a life safety rope. (NFPA 2500)

firefighter A member of a fire department who is assigned to do routine cleaning and maintenance, place hose line to extinguish fires, and assist with a public fire-prevention program.

Firefighter I A person at the first level of progression, as defined in Chapter 6 of NFPA 1010, who has demonstrated the knowledge and skills to function as an integral member of a firefighting team under direct supervision in hazardous conditions. (NFPA 1010)

Firefighter II A person at the second level of progression, as defined in Chapter 7 of NFPA 1010, who has demonstrated the skills and depth of knowledge to function under general supervision. (NFPA 1010)

fire ground Another name for *emergency scene* when the incident is a fire. Also called *fire scene.*

fire hook A tool used to pull down burning structures.

fire hose A flexible conduit used to convey water or other extinguishing agents (NFPA 1960).

fire hose appliance A piece of hardware (excluding nozzles) generally intended for connection to fire hose to control or convey water. Also called *hose appliance.* (NFPA 1962)

fire hose tool A device that assists firefighters with handling, manipulating, connecting, and using a fire hose.

firehouse See *fire station.*

fire inspector An individual who conducts fire code inspections and applies codes and standards. (NFPA 1030)

fire investigator An individual who has demonstrated the skills and knowledge necessary to conduct, coordinate, and complete an investigation. (NFPA 1030)

fire load The total energy content of combustible materials in a building, space, or area, including furnishings and contents and combustible building elements; expressed in megajoules (MJ). (NFPA 557)

fire mark Historically, a plaque displayed on a building with the name or logo of a fire insurance company informing firefighters that the building was insured by that insurance company, which meant that insurance company would pay the firefighters for extinguishing the fire.

fire marshal A person designated to provide delivery, management, and/or administration of fire protection- and life safety-related codes and standards, investigations, education, and/or prevention services for local, county, state, provincial, federal, tribal, or private sector jurisdictions as adopted or determined by that entity. (NFPA 1030)

fireplug Historically speaking, a plug installed to control water accessed from wooden pipes but today is slang for *fire hydrant.*

fire point The lowest temperature at which a liquid will ignite and achieve sustained burning when exposed to a test flame in accordance with ASTM 92, S*tandard Test Method for Flash and Fire Points by Cleveland Open Cup Tester.* Also called *flame point.* (NFPA 1)

fire police officer An individual officially deployed who provides scene security, directs traffic, and conducts other duties. (NFPA 1091)

fire protection engineer A member of the fire department or an employee of an architectural firm who is responsible for reviewing building plans and working with building owners to ensure that the design of and systems for fire detection and suppression meet applicable codes and function as needed.

fire scene Another name for *emergency scene* when the incident is a fire. Also called *fireground.*

fire seat See *area of origin.*

fire station A building that houses fire apparatus and equipment for a geographic area within a fire department. Also called *fire house.*

fire stream A stream of water or extinguishing agents.

fire suppression The activities involved in controlling and extinguishing fires. (NFPA 1500)

fire tetrahedron A geometric shape used to depict the four components—fuel, oxygen, heat, and chemical chain reactions—required for a fire to occur.

fire The visible result of combustion.

fire triangle The three components—fuel, oxygen, and heat—required for combustion.

fire warden An individual charged with enforcing fire regulations in colonial America.

flake out See *lay.*

flame impingement Flames in direct contact with the surface of a material transferring radiant heat.

flame inhibitor A chemical extinguishing agent that reacts with the fuel to chemically disrupt the combustion process.

flameover See *rollover.*

flame point See *fire point.*

flammable range The range in concentration between the lower and upper flammable limits. Also called *explosive limits.* (NFPA 67)

flashover A transition phase in the development of a compartment fire in which surfaces exposed to thermal radiation reach ignition temperature more or less simultaneously, and fire spreads rapidly throughout the space, resulting in full-room involvement or total involvement of the compartment or enclosed space. (NFPA 921)

flash point The minimum temperature at which a liquid or a solid emits vapor sufficient to form an ignitable mixture with air near the surface of the liquid or the solid. (NFPA 115)

flat bar A specialized type of prying tool made of flat steel with prying ends suitable for performing forcible entry.

flat-head axe A tool that has a head with an axe on one side and a flat head on the opposite side.

flat hose load A hose-loading method in which the hose is laid flat and stacked on top of the previous section.

flow path The movement of heat and smoke from the higher pressure within the fire area toward the lower-pressure areas accessible via doors, window openings, and roof structures. (NFPA 1410)

flow rate The quantity of water flowing, usually measured in gallons (or liters) per minute.

FLSE See *fire and life safety educator.*

forcible entry Techniques used by fire personnel to gain entry into buildings, vehicles, aircraft, or other areas of confinement when normal means of entry are locked or blocked. (NFPA 440)

forestry fire hose A hose designed to meet specialized requirements for fighting wildland fires. (NFPA 1960)

forward hose lay A method of laying a supply line where the line starts at the water source and ends at the attack engine. Also called *straight hose lay.*

freelancing The dangerous practice of acting independently of command instructions.

freight bill See *bill of lading.*

Fresno ladder A narrow, two-section extension ladder specifically designed to provide attic access and to be used in any tight space.

frost-proof hydrant See *dry-barrel hydrant.*

fuel A material that will maintain combustion under specified environmental conditions. (NFPA 53)

fuel-limited fire A fire in which the heat release rate and fire growth are controlled by the characteristics of the fuel because there is adequate oxygen available for combustion. (NFPA 1410)

fully developed stage The stage of fire development where the heat release rate has reached its peak within a compartment. (NFPA 1410)

G

gas The physical state of a substance that has no shape or volume of its own and will expand to take the shape and volume of the container or enclosure it occupies. (NFPA 921)

gate valve A valve firefighters attach to an outlet on a dry hydrant that can turn off the flow of water at that outlet.

general use life safety rope A life safety rope with a diameter that is at least 7/8 in. (11 mm) but not larger than 5/8 in. (16 mm), with a minimum breaking strength of 8992 lbf (40 kN).

geographic information systems (GIS) A system of computer software, hardware, data, and personnel to describe information tied to a spatial location. (NFPA 450)

global positioning system (GPS) A satellite-based radio navigation system composed of three segments: space, control, and user. (NFPA 1900)

governance The framework and procedures for managing and operating an organization.

gravity-feed system A water distribution system that depends on gravity to provide the required pressure. The system storage is usually located at a higher elevation than the end users of the water.

gripping pliers A hand tool with a pincer-like working end that can be used to bend wire or hold smaller objects.

growth stage The stage of fire development where the heat release rate from an incipient fire has increased to the point where heat transferred from the fire and the combustion products are pyrolyzing adjacent fuel sources, and the fire begins to spread across the ceiling of the fire compartment (rollover). (NFPA 1410)

H

hacksaw A cutting tool designed for use on metal. Different blades can be used for cutting different types of metal.

Halligan bar See *Halligan tool.*

Halligan See *Halligan tool.*

Halligan tool A prying tool that incorporates a sharp tapered pick, a blade (either an adze or wedge), and a fork; it is specifically designed for use in the fire service. Also called a *Halligan* and a *Halligan bar.*

halocarbon See *halogenated hydrocarbon.*

halogen A family of elements (chemicals listed in the periodic table), including fluorine (F), bromine (Br), iodine (I), and chlorine (Cl), that are chemically related.

halogenated-agent fire extinguisher A fire extinguisher that uses a halogenated extinguishing agent. Also called *clean-agent fire extinguisher.*

halogenated extinguishing agent A liquefied gas extinguishing agent that extinguishes fire by chemically interrupting the combustion reaction between fuel and oxygen. Halogenated agents leave no residue. (NFPA 402)

halogenated hydrocarbon A hydrocarbon in which at least one hydrogen atom of the hydrocarbon is replaced by a halogen. Also called *halocarbon.*

hammer A striking tool.

handle The grip used for holding and carrying a portable fire extinguisher.

hand light A small, portable light carried by firefighters to improve visibility at emergency scenes. It is often powered by rechargeable batteries.

handline A hose and nozzle that can be held and directed by hand. (NFPA 11)

handsaw A manually powered saw designed to cut different types of materials. Examples include hacksaws, carpenter's handsaws, keyhole saws, and coping saws.

hard suction hose A short section of supply hose that is used to draft water from a static source such as a river, lake, or portable drafting basin to the suction side of the fire pump on a fire department engine or into a portable pump.

hazard control zones The areas at hazardous materials/WMD incidents within an established perimeter that are designated based on safety and the degree of hazard.

hazardous material Matter (solid, liquid, or gas) or energy that, when released, is capable of creating harm to people, the environment, and property, including weapons of mass destruction (WMD) as defined in 18 U.S. Code, Section 2332a, as well as any other criminal use of hazardous materials, such as illicit labs, environmental crimes, or industrial sabotage. (NFPA 470)

hazardous materials company A company that responds to and controls scenes where hazardous materials have spilled or leaked and whose members wear special suits and are trained to deal with most chemicals.

Hazardous Materials Information System (HMIS) A color-coded marking system that identifies the hazard level of chemicals in a container so that personnel can work safely around chemicals.

hazardous materials specialist Defined in the OSHA HAZWOPER regulation only, a hazardous materials specialist responds with and provides support to hazardous materials technicians and acts as the incident-site liaison with federal, state, local, and other government authorities regarding site activities.

hazardous materials technician A person who responds to hazardous materials/weapons of mass destruction (WMD) incidents using a risk-based response process to analyze a problem involving hazardous materials/WMD, plan a response to the problem, implement the planned response, evaluate the progress of the planned response and adjust accordingly, and assist in terminating the incident. (NFPA 470)

hazardous waste A substance that remains after a process or manufacturing plant has used some of the material and the substance is no longer pure.

HAZWOPER (HAZardous Waste OPerations and Emergency Response) The federal OSHA regulation that governs hazardous materials waste site and response training (specifics to emergency response can be found in 29 CFR 1910.120[q]).

heads-up display (HUD) Visual display of information and system conditions status that is visible to the wearer. (NFPA 1970)

heat energy The potential energy of a combustible material and the kinetic energy released when heat is applied to the material. Also called *thermal energy.*

heat flux The measure of the rate of heat transfer to a surface, typically expressed in kilowatts per meter squared (kW/m^2) or British thermal units per square feet (BTU/ft^2). (NFPA 268)

heat-induced tear (HIT) A tear in a nonpressurized container that occurs when the container is exposed to direct or indirect flame impingement for a prolonged period, causing the structural integrity of the container to fail.

heat release rate (HRR) The rate at which heat energy is generated by burning. (NFPA 921)

heat stratification See *thermal layering.*

heat transfer The movement of heat energy from a hotter medium to a cooler medium by conduction, convection, or radiation.

hemorrhage Excessive bleeding.

hepatitis B virus (HBV) A viral infection that causes inflammation and loss of liver function. It is transmitted through blood.

hepatitis C virus (HCV) A viral infection that causes inflammation and loss of liver function. It is transmitted through blood.

Higbee indicator An indicator on both the male and female threaded couplings that indicates where the threads start. These indicators should be aligned before firefighters start to thread the couplings together.

high hazard area See *extra hazard area.*

highline system A system of using rope or cable suspended between two points for movement of persons or equipment over an area that is a barrier to the rescue operation, including systems capable of movement between points of equal or unequal height. (NFPA 1006)

high-temperature–protective clothing Protective clothing designed to protect the wearer from short-term high-temperature exposures, such as protective clothing ensembles used by aircraft firefighters to fight fires in flammable liquids. (NFPA 470)

hitch A knot that attaches to or wraps around an object so that when the object is removed, the knot will fall apart. (NFPA 2500)

horn The tapered discharge nozzle of a carbon dioxide fire extinguisher.

horseshoe hose load A hose-loading method in which the hose is laid on its edge around the perimeter of the hose bed so that it resembles a horseshoe.

hose appliance See *fire hose appliance.*

hose bed The main storage area on an apparatus for carrying hose.

hose bridge A device that protects a hose when it is necessary for a vehicle to drive over a hose. Also called *hose ramp.*

hose clamp A device used to compress a fire hose to stop water flow.

hose hoist See *hose roller.*

hose jacket A device used to stop a leak in a fire hose or to join hose lines that have damaged couplings.

hose liner The inside portion of a hose that is in contact with the flowing water; also called *hose inner jacket.*

hose ramp See *hose bridge.*

hose record A written history of each individual length of fire hose.

hose roller A device that is placed on the edge of a roof and is used to protect hose as it is hoisted up and over the roof edge. Also called *hose hoist.*

hose size An expression of the internal diameter of the hose. (NFPA 1962)

hose tool See *fire hose tool.*

human immunodeficiency virus (HIV) The virus that causes acquired immune deficiency syndrome (AIDS).

hux bar A multipurpose tool that can be used for several forcible entry and ventilation applications because of its unique design. It also may be used as a hydrant wrench.

hydrant cap The cover on a fire hydrant outlet that is in place when the hydrant is not in use. Also called *discharge cap.*

hydrant wrench A hand tool that is used to operate the valves on a hydrant; it also may be used as a spanner wrench. Some models are plain wrenches, whereas others have a ratchet feature.

hydra-ram A one-piece integrated hydraulic forcible entry tool.

hydraulic cutters See *hydraulic shears.*

hydraulic shears A lightweight, hand-operated tool that can produce up to 10,000 lb (4500 kg) of cutting force. Also called *hydraulic cutters.*

hydraulic spreader A lightweight, hand-operated tool that can produce up to 10,000 lb (4500 kg) of prying and spreading force.

hydraulic tool A power tool that uses pressurized fluid to exert force.

hydrofluorocarbon (HFC) A common type of halocarbon used in fire suppression.

hydrogen cyanide (HCN) An extremely toxic gas produced by the incomplete combustion of many common plastic-based materials.

hydrophobic The quality of repelling or being unable to mix with water.

hydrostatic test A test performed by filling pressure-containing components completely with water or other incompressible fluid while expelling all contained air, closing or capping all open ports of the pressure-containing components, and then raising and maintaining the contained pressure to pressurize the pressure-containing components to a prescribed value through an externally supplied pressure-generating device. (NFPA 1900)

hydrostatic testing Pressure testing of a fire extinguisher to verify its strength against unwanted rupture. (NFPA 10)

I

ignition temperature Minimum temperature a substance should attain in order to ignite under specific test conditions. (NFPA 402)

immediately dangerous to life and health (IDLH) Any condition that would pose an immediate or delayed threat to life, cause irreversible adverse health effects, or interfere with an individual's ability to escape unaided from a hazardous environment. (NFPA 1700)

incident action plan (IAP) An oral or written plan approved by the incident commander (IC) containing incident objectives reflecting the overall strategy for managing an incident for a specific time frame and target location. (NFPA 470)

incident commander (IC) The individual responsible for all incident activities, including the development of strategies and tactics and the ordering and release of resources. (NFPA 470)

incident safety officer (ISO) A member of the command staff responsible for monitoring and assessing safety hazards and unsafe situations and for developing measures for ensuring personnel safety. (NFPA 1700)

incipient stage The early stage of fire development where the fire's progression is limited to a fuel source and the thermal hazard is localized to the area of the burning material. (NFPA 1410)

incomplete combustion A combustion process during which the fuel is not completely consumed, usually because of a limited supply of oxygen.

infant A child younger than 1 year.

infectious Capable of causing an illness by entry of a pathogenic microorganism.
ingestion exposure Exposure to a hazardous material from swallowing the substance.
inhalation exposure Exposure to a hazardous material from breathing the substance into the lungs.
initial attack apparatus Fire apparatus with a fire pump of at least 250-gpm (946-L/min) capacity, water tank, and hose body, whose primary purpose is to initiate a fire-suppression attack on structural, vehicular, or vegetation fires and to support associated fire department operations. Also called *quick attack apparatus.* (NFPA 1900)
injection exposure Exposure to a hazardous material from the substance entering the body through cuts or other breaches in the skin.
International Fire Service Accreditation Congress (IFSAC) A national organization that accredits or recognizes emergency service certification systems.
interoperability The ability to communicate across different radio bands and between different agencies.

J

joule A measure of heat energy equal to 0.4 calorie.
junction box An electrical enclosure that houses one or more wiring connections. The box protects the connections from environmental conditions and accidental contact.

K

Kelly tool A steel bar with two main features: a large pick and a large chisel or fork.
kern In a kernmantle rope, the center or core of the rope.
kernmantle rope Rope made of two parts, the kern and the mantle.
keyhole saw A saw designed to cut keyhole circles in wood and drywall.
kinetic energy The energy possessed by an object as a result of its motion.
knot A fastening made by tying rope or webbing in a prescribed way. (NFPA 2500)
K tool A tool that is used to remove lock cylinders from structural doors so that the locking mechanism can be unlocked.

L

label A smaller version of a placard (4 inches [10 centimeters] on each side) that is required to be placed on the four sides of individual boxes and smaller packages being transported.
laceration An irregular cut or tear through the skin.
ladder company See *truck company.*
ladder pipe A monitor that attaches to the rungs of a vehicle-mounted aerial ladder. (NFPA 1960)
ladder truck See *aerial fire apparatus.*
laid rope See *twisted rope.*
laminar smoke flow The smooth or streamlined movement of smoke.
large-diameter hose (LDH) A hose 3.5 in. (89 mm) or larger that is designed to move large volumes of water to supply master stream appliances, portable hydrants, manifolds, standpipe and sprinkler systems, and fire department pumpers from hydrants and in relay. (NFPA 1410)
lay To arrange hose on the ground. Also called *flake out.*
lieutenant The first level of officer and the person who is usually responsible for a single fire company on a shift.
life safety rope Rope dedicated solely for the purpose of supporting people during rescue, firefighting, other emergency operations, or training evolutions. (NFPA 2500)
light-emitting diodes (LEDs) Electronic semiconductors that emit a single-color light when activated. LEDs are used for operational displays in self-contained breathing apparatus (SCBA).
light energy Energy produced by electromagnetic waves packaged in discrete bundles called *photons.*
light hazard area An occupancy where the quantity, combustibility, and heat release of the materials are low and the majority of materials are arranged so that a fire is not likely to spread. Also called *low hazard area.*
line-of-sight system See *simplex communication system.*
liquid Matter that has a specific volume but does not have a specific size or shape.
loaded-stream fire extinguisher A stored-pressure water-type fire extinguisher that uses an alkali metal salt as a freezing-point depressant.
local emergency planning committee (LEPC) A committee comprising members of industry, transportation, the public at large, media, and fire and law enforcement agencies that gathers and disseminates information on hazardous materials stored in the community and ensures that there are adequate local resources to respond to a chemical event in the community.
locking mechanism A device that locks a fire extinguisher's trigger to prevent its accidental discharge.
loop A piece of rope formed into a circle by crossing the rope.
loop knot A knot that forms a secure loop in the end of a rope.
lower explosive limit (LEL) The minimum concentration of a combustible vapor or combustible gas in a mixture of the vapor or gas and gaseous oxidant, above which propagation of flame will occur on contact with an ignition source. (NFPA 115)
lower flammable limit (LFL) See *lower explosive limit.*
low hazard area See *light hazard area.*
lug A protrusion or indentation on a hose coupling that aids in securing and tightening the connection between two couplings.

M

mallet A short-handled hammer.
mantle In a kernmantle rope, the braided covering that protects the kern from dirt and abrasion. Also called *sheath.*
master stream appliance A device that discharges high-volume water streams, usually between 350 gpm (1325 L/min) and 1500 gpm (5678 L/min), though much larger capacities are available. Also called *master stream device.*
master stream device See *master stream appliance.*
matter Anything that has mass and volume.
maul A specialized striking tool, weighing 6 lb (3 kg) or more, with an axe on one side of the head and a sledgehammer on the other side.
mayday A verbal declaration indicating that a firefighter is lost, missing, or trapped and requires immediate assistance.
mechanical energy The potential energy stored because of the position of an object or the kinetic energy of an object in motion.

medium-diameter hose (MDH) A hose 2½-in. (64-mm) or 3-in. (76-mm) in diameter most often used as attack hose, but can be used as supply hose.

minimum breaking strength (MBS) The result of subtracting 3 standard deviations from the mean result of the lot being tested using the formula in 8.2.5.2. (NFPA 2500)

minuteman hose load A hose-loading method that allows a single firefighter to deploy and advance the necessary amount of hose from the shoulder; avoids having to maneuver the hose lines around obstacles and helps to prevent sharp kinks.

mobile data terminal (MDT) Technology that allows firefighters to receive data while in the fire apparatus or at the station.

mobile radio A two-way radio that is permanently mounted in a fire apparatus.

mobile repeater A repeater used at an incident scene to boost signals at the scene by capturing the weak signals from portable radios and rebroadcasting them.

mobile water supply apparatus A vehicle designed primarily for transporting (pickup, transporting, and delivering) water to fire emergency scenes to be applied by other vehicles or pumping equipment. Also called *tanker, tender,* or *water tender.* (NFPA 1900)

moderate-hazard area See *ordinary-hazard area.*

molecular bond The connection between atoms in a molecule.

molecule Atoms chemically bonded together.

monoammonium phosphate A finely ground substance that looks like yellow talcum powder that is used as an extinguishing agent in dry-chemical fire extinguishers that are rated for Class A, B, and C fires. Also called *ammonium phosphate.*

multiple-jacket A construction consisting of a combination of two separately woven reinforcements (double jacket) or two or more reinforcements interwoven. (NFPA 1962)

multiplex channel Simultaneous transmission of multiple data streams, most often voice signals, in either or both directions over the same frequency.

multipurpose dry-chemical fire extinguisher A fire extinguisher that uses a monoammonium phosphate–based extinguishing agent that is effective on fires involving ordinary combustibles, such as wood or paper, and fires involving flammable liquids and that is rated to fight Class A, B, and C fires.

multipurpose hook A long pole with a wooden or fiberglass handle and a metal hook on one end used for pulling.

multi-tool A compact, pocket-size, multiple-function tool that combines several different tools, such as a knife, scissors, wire cutters, pliers, and screwdriver.

municipal fire alarm system A network of fire alarm boxes and emergency telephones on street corners or in public places.

municipal water system A system having water pipes servicing fire hydrants and designed to furnish, over and above domestic consumption, a minimum of 250 gpm (946 L/min) at 20 psi (138 kPa) residual pressure for a 2-hour duration. (NFPA 1140)

N

National Board on Fire Service Professional Qualifications A national organization that accredits or recognizes emergency service certification systems. Also called *Pro Board.*

National Fire Protection Association (NFPA) A nonprofit association that develops and maintains nationally recognized minimum consensus standards and fire codes for fire safety and handling of hazardous materials.

neutral plane The interface at a vent, such as a doorway or a window opening, between the hot gas flowing out of a fire compartment and the cool air flowing into the compartment where the pressure difference between the interior and exterior is equal.

New York hook See *roofman's hook.*

Next Generation 911 (NG911) A system designed to replicate traditional Enhanced 911 (E911) features and functions and provide additional capabilities to provide access to emergency services from all connected communications sources and provide multimedia data capabilities for public safety answering points (PSAPs) and other emergency service organizations. (NFPA 1225)

NFPA 704 hazard identification system A hazardous materials marking system designed for fixed-facility use. It uses a diamond-shaped symbol of any size, which is itself broken into four smaller diamonds, each representing a particular property or characteristic of the material.

non-collapsible fire hose Fire hose typically made from durable materials such as PVC or synthetic rubber so that it maintains its shape and structure. Also called *rigid fire hose.*

nonintervention mode The operating mode in which responders do not operate near the hazardous materials container and focus their efforts on public protection actions only, allowing the container or product to take its natural course.

nose cups An insert inside the face piece of a self-contained breathing apparatus (SCBA) that fits over the user's mouth and nose.

nozzle A constricting appliance attached to the end of a fire hose or monitor to increase the water velocity and form a stream. (NFPA 1960)

nozzle A device for use in applications requiring special water discharge patterns, directional spray, or other unusual discharge characteristics. (NFPA 13)

nuclear energy Potential energy stored in the nucleus of an atom or the kinetic energy released by splitting the nucleus of an atom into two smaller nuclei (fission) or by combining two small nuclei into one large nucleus (fusion).

O

Occupational Safety and Health Administration (OSHA) Part of the Department of Labor, the U.S. federal agency that regulates worker safety and, in some cases, responder safety.

offensive mode The mode of operation in which responders have direct contact with the material or container and take aggressive action to control the release.

off-gas To emit harmful chemicals in the form of a gas.

on scene See *emergency scene.*

open-circuit self-contained breathing apparatus (open-circuit SCBA) An SCBA in which the exhaled air is released into the atmosphere and is not reused.

open-end wrench A hand tool that is used to tighten or loosen bolts. The end is open, as opposed to a box-end wrench. Each wrench is a specific size.

open-head drum A drum with a lid that is secured by a bolted clasp-type ring, which circles the entire head of the drum.

operating stem The steel rod extending from the top of a dry-barrel hydrant to the hydrant valve that firefighters can turn using the stem nut at the top to open and close the main valve.

operations-level responder A person who responds to hazardous materials/weapons of mass destruction (WMD) incidents for the purpose of implementing or supporting actions to protect nearby persons, the environment, or property from the effects of the release. (NFPA 470)

ordinary dry-chemical fire extinguisher A dry-chemical fire extinguisher rated for only Class B and Class C fires.

ordinary hazard area An area that contains more Class A and Class B materials than a light hazard area and where the combustibility and heat release rate of the materials are moderate. Also called *moderate hazard area.*

other potentially infectious materials (OPIM) Body fluids, other than blood, that may transmit infection or carry blood that can carry infection, such as semen, vaginal secretions, cerebral spinal fluid, saliva, and others.

outlet An opening on a fire hydrant through which water is discharged.

overhaul A firefighting term involving the process of final extinguishment after the main body of the fire has been knocked down. All traces of fire must be extinguished at this time. (NFPA 1700)

oxidation Reaction with oxygen, either in the form of the element or in the form of one of its compounds. (NFPA 53)

P

packaging group designation A label that describes a substance according to the degree of hazards it presents.

paramedic Emergency medical services (EMS) personnel who can do everything an advanced emergency medical technician (AEMT) can do and who have extensive training in advanced life support, including administering drugs, cardiac monitoring, inserting advanced airways, manual defibrillation, and other advanced assessment and treatment skills.

PASS Acronym for the steps involved in operating a portable fire extinguisher: Pull pin, Aim nozzle, Squeeze trigger, Sweep across burning fuel.

pathogens Microorganisms capable of causing disease.

personal alert safety system (PASS) A device that continually monitors for lack of movement of the wearer and automatically activates an alarm signal, indicating the wearer is in need of assistance; can also be manually activated to trigger the alarm signal. (NFPA 1970)

personal protective equipment (PPE) The full complement of garments firefighters are required to wear while on an emergency scene, including turnout coat, protective trousers, firefighting boots, firefighting gloves, a protective hood, self-contained breathing apparatus (SCBA), a personal alert safety system (PASS) device, and a helmet with eye protection. (NFPA 1010)

personnel accountability report (PAR) Periodic reports verifying the status of responders assigned to an incident or planned event. (NFPA 1026)

personnel accountability system A system that readily identifies both the location and function of all members operating at an incident scene. (NFPA 1550)

phosgene A gas formed from incomplete combustion of many common household products.

pick-head axe A tool that has a head with an axe on one side and a pointed end ("pick") on the opposite side.

pike pole A pole with a sharp point ("pike") on one end coupled with a hook. It is used to make openings in ceilings and walls. Pike poles are manufactured in different lengths for use in rooms of different heights.

pin lug A lug that looks like a small cylinder that extends outward from a hose coupling.

pipeline A length of pipe including pumps, valves, flanges, control devices, strainers, and/or similar equipment for conveying fluids. (NFPA 470)

pipeline right-of-way An area, patch, or roadway that extends a certain number of feet on either side of a pipeline and may contain warning and informational signs about hazardous materials carried in the pipeline.

pipe wrench A wrench having one fixed grip and one movable grip that can be adjusted to fit securely around pipes and other tubular objects.

placard A diamond-shaped indicator (10¾ inches [27 centimeters] on each side) that is required to be placed on all four sides of highway transport vehicles, railroad tank cars, and other forms of transportation carrying hazardous materials to identify the substance being transported.

plaster hook A long pole with a pointed head and two retractable cutting blades on the side.

plume The column of hot gases, flames, and smoke rising above a fire. Also called *convection column, thermal updraft,* or *thermal column.* (NFPA 921)

polar solvent A water-soluble, flammable liquid, such as alcohol, acetone, ester, and ketone.

policies Formal statements that outline expectations for performance and procedures in different circumstances but usually require personnel to make judgments to determine the best course of action within the stated study.

portable monitor A monitor that can be lifted from a vehicle-mounted bracket and moved to an operating position on the ground by not more than two people. (NFPA 1960)

portable radio A battery-operated, hand-held transceiver. (NFPA 1225)

portable tanks Folding or collapsible tanks that are used at the fire scene to hold water for drafting.

posttraumatic stress disorder (PTSD) A behavioral disorder that develops after a person has experienced a critical incident; characterized by reexperiencing the event and overresponding to stimuli that recall the event; symptoms include depression, startle reactions, flashback phenomena, and dissociative episodes (e.g., amnesia of the event).

potential energy Energy stored by an object as a result of its position or condition.

powered air-purifying respirator (PAPR) An air-purifying respirator that uses a powered blower to force the ambient air through one or more air-purifying components to the respiratory inlet covering. (NFPA 1984)

power saw A saw that is usually powered by an electric motor or a gasoline engine. The three primary types of mechanical saws are chainsaws, rotary saws, and reciprocating saws.

preconnected attack line Attack hose that travels on the engine already equipped with a nozzle and connected to a pump discharge outlet so it is ready for immediate use. Also called *preconnect.*

preconnected flat hose load A hose-loading method in which the hose is laid flat and stacked on top of the previous section, similar to the flat hose load but with one or two loops sticking out of the load to make it easier to deploy.

preconnect See *preconnected attack line.*

preincident plan A document developed by gathering general and detailed data that is used by responding personnel in effectively managing emergencies for the protection of occupants, responding personnel, property, and the environment. (NFPA 1660)

pressure indicator A gauge on a pressurized portable fire extinguisher that indicates the internal pressure of the expellant.

primary feeder The largest-diameter water main pipe in a water distribution system that carries the greatest amounts of water. Also called *trunk line.*

private water system A privately owned water system that operates separately from the municipal water system.

Pro Board See *National Board on Fire Service Professional Qualifications.*

pry bar A specialized prying tool made of a hardened steel rod with a tapered end that can be inserted into a small area.

public information officer (PIO) A member of the command staff responsible for interfacing with the public and media or with other agencies with incident-related information requirements. (NFPA 1550)

public safety answering point (PSAP) A facility equipped and staffed to receive emergency and nonemergency calls requesting public safety services via telephone and other communication devices. (NFPA 1225)

public safety communications center A building or portion of a building that is specifically configured for the primary purpose of providing emergency communications services or public safety answering point (PSAP) services to one or more public safety agencies under the authority or authorities having jurisdiction. Also called *communications center.* (NFPA 1225)

pull hose See *deploy hose.*

pumper Fire apparatus with a permanently mounted fire pump of at least 750-gpm (1300-L/min) capacity, water tank, and a hose body whose primary purpose is to control structural and associated fires. (NFPA 1900)

pump tank fire extinguisher A nonpressurized, manually operated water-type fire extinguisher that is rated for use on Class A fires. Discharge pressure is provided by a hand-operated, double-acting piston pump.

pyrolysis A process in which material is decomposed, or broken down, into simpler molecular compounds by the effects of heat alone; pyrolysis often precedes combustion. (NFPA 921)

Q

quick attack apparatus See *initial attack apparatus.*

quint apparatus Fire apparatus with a permanently mounted fire pump, a water tank, a hose storage area, an aerial device or elevating platform with a permanently mounted waterway, and a complement of ground ladders. Also called *quint.* (NFPA 1710)

quint See *quint apparatus.*

R

rabbet tool A hydraulic spreading tool designed to pry open doors that swing inward. Also called a *bunny tool.*

rapid intervention crew/company (RIC) A dedicated crew of at least one officer and three members, positioned outside the immediately dangerous to life and health (IDLH) area, trained and equipped as specified in NFPA 1407, who are assigned for rapid deployment to rescue lost or trapped members. Also called *rapid intervention team (RIT).* (NFPA 1550)

rapid intervention crew/company universal air connection (RIC UAC) A system that allows emergency replenishment of breathing air to the self-contained breathing apparatus (SCBA) of disabled or entrapped fire or emergency services personnel. (NFPA 1970)

rapid intervention team (RIT) See *rapid intervention crew/company (RIC).*

rebreather See *closed-circuit self-contained breathing apparatus (closed-circuit SCBA).*

recessed lug A lug that is circular indentation and requires a specially designed spanner wrench called a booster hose wrench to engage.

reciprocating saw A saw that is powered by an electric motor or a battery motor and whose blade moves back and forth.

recruit A candidate whose application to become a firefighter is accepted.

reducer A fitting used to connect a small hose line or pipe to a larger hose line or pipe. (NFPA 1142)

regulation A mandate issued and enforced by governmental bodies such as the U.S. Occupational Safety and Health Administration (OSHA) and the U.S. Environmental Protection Agency (EPA).

regulator purge/bypass valve A device or devices designed to bypass a regulator.

remote pressure gauge The device on a self-contained breathing apparatus (SCBA) that measures and displays pressure readings to indicate the quantity of breathing air available and that is located on the shoulder strap or in another location where it can be seen by the user while the SCBA is in use.

repeater A combination of a radio receiver and transmitter that receives radio signals, usually over multiple channels, and retransmits—repeats—them to a wider geographic location.

repeater channel A radio channel that transmits to a repeater.

rescue apparatus Apparatus that carry an extensive array of regular and specialized tools and equipment that are used to rescue victims.

rescue company A piece of fire apparatus staffed with firefighters that is generally used for search and rescue at fire incidents. (NFPA 1700)

rescue knife A spring-assisted folding knife that can be used with one hand; often includes a seat belt cutter and a window breaker.

rescue technician See *technical rescuer.*

residual pressure The pressure that exists in the distribution system, measured at the residual hydrant at the time the flow readings are taken at the flow hydrants. (NFPA 24)

response Immediate and ongoing activities, tasks, programs, and systems to manage the effects of an incident that threaten life, property, operations, or the environment. (NFPA 1600)

reverse hose lay A method of laying a supply line where the supply line starts at the attack engine and ends at the water source.

rigging The process of building a system to move or stabilize a load. (NFPA 1006)

rigid fire hose See *non-collapsible fire hose.*

ring-down phone See *direct-line phone.*

risk-based response process A systematic process based on the facts, science, and circumstances of the incident, by which responders analyze a problem involving hazardous materials/weapons of mass destruction (WMD) to assess the hazards and consequences, develop an incident action plan (IAP), and evaluate the effectiveness of the plan. (NFPA 470)

rocker lug A rectangular-shaped, bevel-edged lug that extends from a coupling. Also called *rocker pin.*

rocker pin See *rocker lug.*

rollover The condition in which unburned fuel (pyrolysate) from the originating fire has accumulated in the ceiling layer to a sufficient concentration (i.e., at or above the lower flammable limit) that it ignites and burns. This can occur without ignition of, or prior to the ignition of, other fuels separate from the origin. Also called *flameover.* (NFPA 921)

roofman's hook A long pole with a solid metal hook used for pulling. Also called a *New York hook.*

rope A compact but flexible, torsionally balanced, continuous structure of fibers produced from strands that are twisted, plaited, or braided together and that serve primarily to support a load or transmit a force from the point of origin to the point of application. (See also 3.3.153.2, Life Safety Rope.) (NFPA 1006)

rope bag A bag used to protect and store rope so that the rope can be easily and rapidly deployed without kinking.

rope record A record for each piece of rope that includes a history of when the rope was placed in service, when it was inspected, when and how it was used, and which types of loads were placed on it.

rotary saw A saw that is powered by an electric motor or a gasoline engine and that uses a large rotating blade to cut through material. The blades can be changed depending on the material being cut.

round turn A piece of rope looped to form a complete circle with the two ends parallel.

rubber-covered hose A hose whose outside covering is made of rubber, which is said to be more resistant to damage.

run cards Information prepared in advance and stored that describes a predetermined response to an emergency.

running end The part of a rope that is not used to form a knot.

run The act of a unit traveling to an incident.

S

safety data sheet (SDS) Formatted information, provided by chemical manufacturers and distributors of hazardous products, about a chemical's composition, physical and chemical properties, health and safety hazards, emergency response, and waste disposal of the material. (NFPA 470)

safety knot A knot used to back up another knot that has a tendency to become loose when not continuously loaded or that can slip when loaded by securing the leftover working end of the rope.

San Francisco hook A multipurpose tool that can be used for several forcible entry and ventilation applications because of its unique design, which includes a built-in gas shut-off and directional slot.

saponification The process of converting the fatty acids in cooking oils or fats to soap or foam; the action caused by a Class K fire extinguisher.

SCBA harness The backpack or frame for mounting the working parts of the self-contained breathing apparatus (SCBA) and the straps and fasteners used to attach the SCBA to the firefighter.

SCBA regulator The part of the self-contained breathing apparatus (SCBA) that reduces the high pressure in the cylinder to a usable lower pressure and controls the flow of air to the user.

screwdriver A tool used for turning screws.

seat-belt cutter A specialized cutting device that cuts through seat belts.

seat of the fire See *area of origin.*

secondary feeder The smaller-diameter water main pipe in a water distribution system that connects a primary feeder to a distributor. Also called *branch line.*

self-contained breathing apparatus (SCBA) An atmosphere-supplying respirator that supplies a respirable air atmosphere to the user from a breathing-air source that is independent of the ambient environment and designed to be carried by the user. (NFPA 1970)

self-expelling agent An agent that has sufficient vapor pressure at normal operating temperatures to expel itself from a fire extinguisher.

severe acute respiratory syndrome (SARS) A viral respiratory illness caused by a coronavirus.

sheath See *mantle.*

shipping papers A shipping order, bill of lading, manifest, or other shipping document serving a similar purpose and containing the information required by regulations of the U.S. Department of Transportation (DOT). (NFPA 498)

shock A state of collapse of the cardiovascular system; the state of inadequate delivery of blood to the organs of the body.

shock load An instantaneous load that places a rope under extreme tension, such as when a falling load is suddenly stopped as the rope becomes taut.

shove knife A forcible entry tool used to trip the latch on outward-swinging doors.

shut-off valve A valve whose primary function is to operate in either a fully shut-off or a fully open condition. (NFPA 1960)

simplex communication system A radio system that allows communication to flow in only one direction at a time. Also called *line-of-sight system* or *talk-around channel.*

single-doughnut hose roll A hose roll that has both female couplings on the outside of the roll and the male coupling is protected by the hose rolled on top of it and that is used when the hose will be put into use directly from its rolled state.

single-jacket hose A construction consisting of one woven jacket. (NFPA 1962)

size-up The process of gathering and analyzing information to help fire officers make decisions regarding the deployment of resources and the implementation of tactics. (NFPA 1410)

sledgehammer A hammer that can be one of a variety of weights and sizes.

small-diameter hose (SDH) A hose 1½-in. (38-mm) or 1¾-in. (44-mm) hose that is most often used as the primary attack hose for most fires.

smoke explosion A violent release of energy that occurs when smoke travels away from its source to a void area or other area separate from the fire compartment and comes in contact with a source of ignition without any change to the ventilation profile.

smoke particles The unburned, partially burned, and completely burned substances found in smoke.

smoke The airborne solid and liquid particulates and gases evolved when a material undergoes pyrolysis or combustion, together with the quantity of air that is entrained or otherwise mixed into the mass. (NFPA 1700)

socket wrench A wrench that fits over a nut or bolt and uses the ratchet action of an attached handle to tighten or loosen the nut or bolt.

soft sleeve hose A short section of large-diameter supply hose that is used to provide water from the large steamer outlet (the large-diameter port) on a fire hydrant or other pressurized water source to the suction side of the fire pump. Also called *soft suction hose.*

soft suction hose See *soft sleeve hose.*

solid Matter that has a specific size and shape; one of the three states of matter.

soot Black particles of carbon produced in a flame. (NFPA 1700)

spanner wrench A type of tool used to couple or uncouple hose by turning the rocker lugs or pin lugs on the connections.

specific gravity The density of a liquid compared to water (which is 1.0).

speed socket wrench A long, curved handle with a socket wrench that fits over a nut or bolt at the end. To tighten or loosen the nut or bolt, the handle is rotated.

split hose bed A hose bed arranged to enable the engine to lay out either a single supply line or two supply lines simultaneously.

split hose lay A scenario in which the attack engine forward lays a supply line from an intersection to the fire, and the supply engine reverse lays a supply line from the hose left by the attack engine to the water source.

spring-loaded center punch A spring-loaded punch used to break automobile glass.

squad See *emergency medical services (EMS) company.*

stand-alone communications center A public safety communications center that serves and dispatches a single emergency response agency.

standard Documents, the main text of which contains only mandatory provisions using the word "shall" to indicate requirements, and are in a form generally suitable for mandatory reference by another standard or code or for adoption into law. Nonmandatory provisions are not to be considered a part of the requirements of a standard and shall be located in an appendix or annex, footnote, informational note, or other means as permitted in the NFPA Manuals of Style. (NFPA 1)

standard operating guidelines (SOGs) Written organizational directives that establish or prescribe specific operational or administrative methods to be followed routinely, which can be varied because of operational needs in the performance of designated operations or actions. (NFPA 1550)

standard operating procedures (SOPs) Written organizational directives that establish or prescribe specific operational or administrative methods to be followed routinely for the performance of designated operations or actions. (NFPA 1550)

standard precautions An infection control concept that treats all body fluids as potentially infectious.

standing part The part of a rope between the working end and the running end.

standpipe A pipe and attached hose valves and hose (if provided) used for conveying water to various parts of a building for fire-fighting purposes. (NFPA 1140)

State Emergency Response Commission (SERC) A group that acts as a liaison between the local and state levels by collecting and disseminating information relating to hazardous materials emergencies. SERC includes representatives from agencies such as the fire service, law enforcement services, and elected officials.

state of matter The physical state of a material (solid, liquid, or gas).

static rope A rope that stretches very little under load.

static water source A water source that is not under pressure, such as a pond, a lake, a stream, or a swimming pool.

steamer port The large-diameter port on a fire hydrant.

stem nut The large nut at the top of the operating stem in a dry-barrel hydrant that is turned to open the hydrant valve.

storage hose roll See *straight hose roll.*

stored-pressure fire extinguisher A fire extinguisher in which both the extinguishing agent and expellant gas are kept in a single container and that includes a pressure indicator or gauge. (NFPA 10)

stored-pressure water-type fire extinguisher A fire extinguisher in which water or a water-based extinguishing agent is stored under pressure.

Storz hose coupling A hose coupling that has the property of being both the male and the female coupling. It is connected by engaging the lugs and turning the coupling a one-third turn.

straight hose lay See *forward hose lay.*

straight hose roll A hose roll that has the male coupling at the center of the roll and the female coupling on the outside of the roll and that is used for general handling and transportation of hose as well as to prepare hose to be stored on a rack. Also called *storage hose roll.*

strategy The general course of action, direction, or plan to accomplish incident objectives. (NFPA 470)

structural firefighting protective clothing All of the clothing elements of the structural firefighting protective ensemble.

structural firefighting protective coat The element of the protective ensemble that provides protection to the upper torso and arms, excluding the hands and head. Also called *bunker coat* or *turnout coat.* (NFPA 1970)

structural firefighting protective ensemble Multiple elements of compliant protective clothing and equipment that, when worn together, provide protection from some risks, but not all risks, of emergency incident operations. (NFPA 1970)

structural firefighting protective footwear The element of the protective ensemble that provides protection to the foot, ankle, and lower leg. (NFPA 1970)

structural firefighting protective glove The element of the protective ensemble that provides protection to the hand and wrist. (NFPA 1970)

structural firefighting protective helmet The element of the protective ensemble that provides protection to the head. (NFPA 1970)

structural firefighting protective hood The interface element of the protective ensemble that provides limited protection to the coat/helmet/self-contained breathing apparatus (SCBA) face-piece interface area. (NFPA 1970)

structural firefighting protective trousers The element of the protective ensemble that provides protection to the lower torso and legs, excluding the ankles and feet. Also called *bunker pants* or *turnout pants.* (NFPA 1970)

structural firefighting The activities of rescue, fire suppression, and property conservation in buildings or other structures, vehicles, railcars, marine vessels, aircraft, or like properties. (NFPA 1010)

suction hose A hose that is designed to prevent collapse under vacuum conditions so that it can be used for drafting water from below the pump (lakes, river, wells, etc.) (NFPA 1960)

Superfund Amendments and Reauthorization Act of 1986 (SARA) One of the first U.S. laws to affect how fire departments respond in a hazardous materials emergency.

supplied-air respirator (SAR) An atmosphere-supplying respirator for which the source of breathing air is not designed to be carried by the user. Also called *air-line respirator.* (NFPA 1852)

supply engine An engine used to pump water to an attack engine using supply lines that may be connected to a pressurized water source such as a fire hydrant or to a static (unpressurized) water source.

supply hose Hose designed for the purpose of moving water between a pressurized water source and a pump that is supplying attack lines. Also called *supply line.* (NFPA 1960)

supply line evolutions The process of laying supply hose to deliver water from a water supply source to an attack engine. Also called *supply line operations.*

supply line operations See *supply line evolutions.*

supply line See *supply hose.*

support person A fire department member who is not a firefighter but who assists members of a fire department by performing duties in environments that are not hazardous.

surfactant A compound that lowers the surface tension (or interfacial tension) between two liquids, between a gas and a liquid, or between a liquid and a solid and that can act as a detergent, wetting agent, emulsifier, foaming agent, and dispersant. (NFPA 1700)

swivel A ring around female couplings that you turn to secure the female coupling around the male coupling without twisting the hose.

swivel gasket An O-shaped piece of rubber inside the swivel section of a female hose coupling that forms a seal that stops water from leaking when a male hose coupling is tightened against it.

T

tactics Specific actions or tasks taken to achieve strategies involving the deployment and directing of resources at an incident. (NFPA 470)

tag line A rope that personnel on the ground can use to guide an object that is being hoisted or lowered.

talk-around channel See *simplex communication system.*

tamper seal A retaining device that breaks when the locking mechanism is released.

tanker See *mobile water supply apparatus.*

target hazard Any occupancy type or facility that presents a high potential for loss of life or serious impact to the community resulting from a fire, explosion, or chemical release.

technical rescuer A person who is trained to perform or direct a technical rescue. Also called *rescue technician.* (NFPA 1006)

technical use life safety rope A life safety rope with a diameter that is at least 3/8 in. (9.5 mm) but not larger than 1/2 in. (12.5 mm), with a minimum breaking strength of 4496 lbf (20 kN) and that is used by members of highly trained rescue teams who deploy to technical environments such as mountainous or wilderness terrain.

technician See *driver/operator.*

telecommunicator An individual whose primary responsibility is to receive, process, or disseminate information of a public safety nature via telecommunication devices. Also called *dispatcher.* (NFPA 1225)

telephone interrogation The phase in a 911 call during which the telecommunicator asks questions to obtain vital information, such as the location of the emergency.

temperature The measurement of the movement of molecules used to describe how hot or cold something is.

ten-codes A system of predetermined coded messages, such as "What is your 10-20?" used by responders over the radio.

tender See *mobile water supply apparatus.*

thermal column See *plume.*

thermal conductivity The ability of a material to conduct heat.

thermal energy See *heat energy.*

thermal layering The phenomenon of gases forming into layers according to their temperatures. Also called *heat stratification.*

thermal radiation The means by which heat is transferred to other objects.

threaded hose coupling A type of coupling that requires a male fitting and a female fitting to be screwed together.

throw bag A water rescue system that includes 50 ft to 75 ft (15.24 m to 22.86 m) of water rescue rope, an appropriately sized bag, and a closed-cell foam float. (NFPA 1006)

throwline A floating rope that is intended to be thrown to a person during water rescues or as a tether for rescuers entering the water. (NFPA 2500)

time marks Status updates provided to the communications center every 10 to 20 minutes. Such an update should include the type of operation, the progress of the incident, the anticipated actions, and the need for additional resources.

tourniquet The generic term used for any device designed to constrict part of an arm or leg so tightly that blood cannot get through.

tower ladder company See *truck company.*

tower ladder truck See *aerial fire apparatus.*

toxic inhalation hazards (TIHs) Gases or volatile liquids that are extremely toxic to humans.

trailer A label that travels the entire length of a life safety rope under the outer sheath that identifies the rope as a life safety rope.

training officer The person designated by the fire chief with authority for overall management and control of the organization's training program. (NFPA 1401)

train list See *consist.*

trigger The button or lever used to discharge the agent from a portable fire extinguisher.

triple-layer hose load A hose-loading method in which the hose is folded back onto itself to reduce the overall length to one-third before loading the hose into the hose bed. This load method reduces deployment distances.

truck company A company of firefighters who are equipped with one or more pieces of aerial fire apparatus. Also called *ladder company* or *tower ladder company.* (NFPA 1700)

truck See *aerial fire apparatus.*

trunked radio system A repeater system that uses a computerized shared bank of frequencies to make the most efficient use of radio resources.

trunk line See *primary feeder.*

TTY/TDD systems User devices that allow speech- and/or hearing-impaired citizens to communicate over a telephone system. TTY stands for teletype, and TDD stands for telecommunications device for the deaf; the displayed text is the equivalent of a verbal conversation between two hearing persons.

turbulent smoke flow The agitated, boiling, and angry movement of smoke caused by rapid molecular expansion of the gases within the smoke and the restrictions of the box containing the smoke.

turnout coat See *structural firefighting protective coat.*

turnout gear See *structural firefighting protective clothing.*

turnout pants See *structural firefighting protective trousers.*

twisted rope Rope constructed of fibers twisted into strands, which are then twisted together. Also called *laid rope.*

two-way radios Portable communication devices used by firefighters. Every firefighting team should carry at least one radio to communicate distress, progress, changes in fire conditions, and other pertinent information.

U

ultrahigh-frequency (UHF) band Radio frequencies between 300 and 3000 MHz.

uncharged hose line A hose line that is not filled with water and not under pressure from a pump.

unified command A team effort that allows all agencies with jurisdictional responsibility for an incident or planned event, either geographical or functional, to manage the incident or planned event by establishing a common set of incident objectives and strategies. (NFPA 1026)

unit A collective term used casually to refer to a group of firefighters in a department with similar duties or responsibilities. See also *company* and *crew.*

unit selection The process of determining exactly which unit or units should be dispatched after a call to 911 is received by a communications center, based on the location and classification of the incident.

unity of command The concept that each person reports to only one direct supervisor.

universal emergency breathing safety system (EBSS) A device on a self-contained breathing apparatus (SCBA) that allows users to share their available air supply in an emergency situation.

upper explosive limit (UEL) The maximum amount of gaseous fuel that can be present in the air if the air–fuel mixture is flammable or explosive.

upper flammable limit (LFL) See *upper explosive limit.*

U.S. Department of Transportation (DOT) marking system A system of labels and placards that is used when materials are being transported from one location to another in the United States and Canada by Transport Canada.

utility rope Rope used for securing objects, hoisting equipment, or securing a scene to prevent bystanders from being injured; utility rope must never be used in life safety operations.

V

vapor density The weight of a gas compared to an equal volume of dry air.

vapor See *gas.*

veins Blood vessels that carry blood toward the heart.

venous bleeding External bleeding from a vein, characterized by steady flow. Venous bleeding may be profuse and life-threatening.

ventilation-limited fire A fire in which the heat release rate and fire growth are regulated by the available oxygen within the space. (NFPA 1410)

ventilation saws Cutting tools designed for roof ventilation that have a different cutting chain than chainsaws used to cut wood; also may have a depth gauge on the bar.

ventilation The controlled and coordinated removal of heat and smoke from a structure, replacing the escaping gases with fresh air. (NFPA 1410)

vent pipes Inverted J-shaped tubes that allow for pressure relief or natural venting of a pipeline for maintenance and repairs.

ventricular fibrillation An uncoordinated muscular quivering of the heart; the most common abnormal rhythm causing cardiac arrest.

very high-frequency (VHF) band Radio frequencies between 30 and 300 MHz; the VHF spectrum is further divided into high and low bands.

Voice over Internet Protocol (VoIP) Technology that converts a person's voice into a digital signal that can be sent via the Internet to another device.

voice recording system A system that records communications over the phones and the radio.

voting receiver A device in a repeater system that receives signals from radio transmissions and then sends the strongest signal to the repeater.

W

walk-in A person who comes to a fire station seeking assistance rather than calling 911. Also called *door banger.*

water flow The volume of water moving through a pipe, hose, or nozzle over a period of time, usually expressed in gallons (liters) per minute (gpm or L/min).

water hammer The surge of pressure that occurs when a high-velocity flow of water is abruptly shut off. The pressure exerted by the flowing water against the closed system can be seven or more times that of the static pressure. (NFPA 1962)

water main A generic term for any underground water pipe.

water mist fire extinguisher A fire extinguisher containing distilled or de-ionized water and employing a nozzle that discharges the agent in a fine spray. (NFPA 10)

water pressure The application of force by one object against another. When water is forced through the distribution system, it creates water pressure.

water shuttle operations A method of transporting water from a source to a fire scene using a number of mobile water supply apparatus.

water supply A source of water for firefighting activities. (NFPA 1140)

water tender See *mobile water supply apparatus.*

waybill Shipping papers for railroad transport.

weapon of mass destruction (WMD) Any weapon or material that is designed to cause death or serious injury or damage to buildings, structures, or the environment, such as an explosive or incendiary bomb, rocket, or grenade containing or delivering a toxic or dangerous chemical, biological agent, toxin, or vectors or a weapon designed to release dangerous levels of radiation. (NFPA 470)

webbing Woven material of flat or tubular weave in the form of a long strip. (NFPA 2500)

wet-barrel hydrant A type of hydrant that is intended for use where there is no danger of freezing weather and where each outlet is provided with a valve and an outlet. (NFPA 24)

wet-chemical extinguishing agent Normally, an aqueous solution of organic or inorganic salts or a combination thereof that forms an extinguishing agent. (NFPA 10)

wet-chemical fire extinguisher A fire extinguisher containing a wet-chemical extinguishing agent for use on Class K fires.

wetting-agent fire extinguisher A fire extinguisher that expels water combined with a concentrate to reduce the surface tension and increase its ability to penetrate and spread.

wet water See *Class A foam concentrate.*

wheeled fire extinguisher A portable fire extinguisher equipped with a carriage and wheels intended to be transported to the fire by one person. (NFPA 10)

wildland apparatus A four-wheel-drive vehicle used to transport firefighters closer to wildfires over rough, uneven terrain and that carries a tank of water and a pump that enables the firefighters to pump water while the truck is moving and special firefighting equipment, such as portable pumps, rakes, shovels, and chainsaws.

wildland company A company of firefighters who fight vegetation fires where larger pumpers cannot gain access and who are equipped with four-wheel-drive vehicles and special firefighting equipment. Also called *brush company.*

wire cutter A hand tool used to cut wire and small-diameter cable. Also called a *diagonal cutter.*

working end The part of the rope used for forming a knot.

wrench A hand tool that comes in several sizes and is used to tighten or loosen bolts.

Index

Note: Page numbers followed by *f*, *t* denote figures and tables, respectively.

D

E

F

S

T